Late 1800s	Early 1900s	Middle 1900s	
Discovery of electron, toxins and causes for tuberculosis, cholera, tetanus, and diphtheria, and twenty-one additional elements, including helium, neon, and radium; Development of aspirin, X-ray diagnosis, and cure for beri-beri with vitamins; Emergence of experimental embryology, and the periodic table of the elements.	Discovery of the neutron, positron (neutrino postulated), insulin, cause of yellow fever, penicillin, and five new elements; Formulation of theory of atomic structure, theory of mutation, theory of continental drift, and Einstein's theory of relativity; Understanding of conditioned reflexes and chemical physiology of muscle contractions.	Discovery of DNA, quasars, the nutrino, sulfa, cortisone, polio and measles vaccines, thirteen new elements, including plutonium; Formulation of theory of atomic fission, heart-lung machine, first radio map of Milky Way, world's largest reflector telescope, cure for meningitis; First open-heart surgery and pacemaker.	About two million U.S. scientists and engineers; First human heart transplant; Human hormones, enzymes, and genes synthesized; First "test tube baby"; Smallpox eliminated; Pulsars detected; First human walks on the moon; Satellites photograph Jupiter, Mars, Venus, Mercury; Human bone fragments 2.5 million years old found.
Development of the telephone, microphone, phonograph, motion picture machine, wireless telegraphy, automobile, trolley car, gasoline motor, electric fan and stove, color film, electric welding, and first recording adding machine.	Development of airplane, caterpillar tractor, removable auto tires, roller bearings, stainless steel, assembly line production of Ford Model T car, electric refrigerator, air conditioning, gas mask, helicopter, sonar, radar, submarine detector, and depth bomb.	Development of automatic dishwasher, self-cleaning oven, LP records, stereophonic sound, commercial TV, first all-electronic digital computer, transistor, electron microscope, laser, solar battery, civilian nuclear power station, jet air travel, nylon and dacron; First satellite launched and "space walks"; World's first atomic and hydrogen bombs explode.	Development of gene-splitting techniques, first "jumbo jet" carrying nearly 400 people; Computer industry explosion; Air and water pollution accelerate; Alaskan pipeline built; Fiber optics used for telephone transmission and medicine; Solar and nuclear energy explored; Three-Mile Island nuclear energy plant accident.
Recovery from Civil War; Economic depression; Wild West was tamed as new state governments were established; New railroads and highways tied country together; Unionism movement began; Millions of European immigrants began to flow into country; Transition from agrarian to industrialized society firmly in place; *Plessy v Ferguson* "separate but equal" court decision.	Wright brothers' first flight; World War I; Panama Canal built; North Pole reached; Wall Street crashed; Major economic depression; San Francisco earthquake; Women gain right to vote; Child labor laws enacted; Prohibition Act passed and repealed; Selective Service for military draft; Amelia Earhart flies Atlantic solo; Progressive Education movement became popular; League of Nations founded; U.S. National Park Service created.	World War II; United Nations formed; Economic stability and growth; Suburban shopping centers affected fabric of central city; Cold War and Korean War; Charge card economy became popular; Labor moved into technological age; Hawaii became 50th state; Civil Rights movement and desegregation of schools aimed at equity of economic and educational opportunity; John F. Kennedy assassinated.	Treaty for peaceful use of outer space adopted; Electronic music and games revolution; Terrorist activity; U.S. military in Vietnam; Assassination of Robert F. Kennedy, Martin Luther King, and other leaders; Abortion legalized; Equal educational opportunity for handicapped persons; Endangered species protected; Many women employed outside the home; Richard Nixon resigns; U.S. population center moves westward.
Emphasis on utilitarian aspects of science; Special interest groups (e.g., temperance) exert influence on science instruction; Student workbooks and teacher editions of textbooks appear; Eminent scientists give public lectures and advocate science education for children.	"New modern science" developed with significant ideas of science to be applied to help solve social and economic problems associated with the industrial revolution; Graded reading texts were organized to help students learn main ideas and generalizations of science.	Federal government invested billions of dollars in educational improvements ($650 million to improve science education programs for K–12 grades); Emphasis on "hands-on" science experiences; increase in number of students who chose science courses in high school and college.	Reduction in federal support of education and science education; Emphasis on accountability and "back-to-basics" in education; Computers flood schools.

	Late 1700s	Early 1800s	Middle 1800s
Science	Few scientists in United States, most were educated abroad. Discovery of photosynthesis, vitamin cure for scurvy and first fourteen elements, including oxygen, hydrogen, nitrogen, and uranium; Vaccinations introduced; Scientific explanation of combustion.	Discovery of nature of infrared rays, polarization of light, the first asteroid, and twenty-two new elements, including sodium, potassium, and iodine; Formulation of atomic theory, hypothesis on composition of gases, laws of electromagnetism, principles of induction and electrical circuits.	Discovery of the nature of fermentation and pasteurization, antiseptic surgery, chemical valence, and eight new elements; Formulation of Mendel's laws of heredity, germ theory, and cell theory; Publication of Darwin's *Origin of Species by Natural Selection*.
Technology	Development of the steam engine, boat, carriage, cotton gin, power loom, drill plowing, food canning, gas lighting, gas engine, lithography, and interchangeable musket parts.	Development of sewing machine, mowing machine, magnetic telegraph, stethoscope, photography, submarine, waterproof fabric, and phosphorus match.	Development of McCormick reaper, ammonia refrigerator, passenger elevator with brakes, and electric motor; Building of first transcontinental railroad and metal ship; Development of revolver-type pistol, nitroglycerin, dynamite, and underwater torpedo.
Society	Declaration of Independence; American War of Independence; Formation of the U.S. democratic system of government; Settlement of the eastern coastal states; Agrarian society; U.S. Bill of Rights ratified.	Louisiana Purchase; Lewis and Clark expedition; Canal transportation system began; War of 1812; Concern for equal opportunities for all citizens; Expansion to settle eastern half of the country; Indian removals began; Monroe Doctrine and policy of noninterference.	Industrialization gained momentum; California gold rush; Suez Canal completed; Great westward movement began; J.S. Mill's *On Liberty*; Slavery issues erupted into the Civil War; Lincoln assassinated; Ku Klux Klan organized; Reconstruction began; United States established gold standard.
Science Education	Schooling was primarily to prepare for clergy and medicine among wealth; Study of nature reinforced moral and ethical values.	Taxation provided schooling for rich and poor; Group instruction replaced individualized instruction; Emphasis on practical science knowledge for daily life; Science became talking about nature rather than experiencing it.	Schooling became compulsory; Land-grant colleges began to flourish; Emphasis in science teaching about objects of nature; Teachers questioned students about objects whether they had seen them or not; Curriculum materials were carefully organized in sequence from simplest to most complex ideas. Reasoning skills were introduced.

Science and Society

Science and Society

A Source Book for Elementary and Junior High School Teachers

RITA PETERSON
University of California, Irvine

JANE BOWYER
Mills College

DAVID BUTTS
University of Georgia

RODGER BYBEE
Carleton College

Charles E. Merrill Publishing Company
A Bell & Howell Company
Columbus Toronto London Sydney

Published by
Charles E. Merrill Publishing Co.
A Bell & Howell Company
Columbus, Ohio 43216

This book was set in Italia
Cover design: Tony Faiola
Cover photo by Jean Greenwald, courtesy of COSI
(Center of Science and Industry), Columbus
Text design: Cynthia Brunk
Production coordination: Linda Hillis Bayma

Library of Congress Catalog Card Number:
83-63504
International Standard Book Number:
0-675-20022-9
1 2 3 4 5 6 7—88 87 86 85 84
Printed in the United States of America

Cartoon on p. 21 by Don Robison
Creative art: Danmark & Michaels
Line art: Alice Thiede/Ayres Associates, Inc.
Part and chapter opening photos: p. 308, Donald
Birdd; pp. 54, 174, Paul Conklin; p. 153, Larry
Hamill; p. 526, © Joanne Meldrum; pp. 4, 438,
NASA; p. 231, Harvey R. Phillips/Phillips Photo
Illustrators; pp. 118, 244, © C. Quinlan; p. 354,
Rich Smolan; pp. 2, 200, 411, 494, 562, Strix Pix;
p. 27, David Tull; p. 247, © Zoological Society of
San Diego, 1977

Preface

With each decade science and society become increasingly dependent upon each other. We live in a world greatly influenced by the findings of science. The quality of life and the welfare of the world's population of 4.6 billion individuals are closely tied to science. Thus, as society increases its dependence on science to solve the problems related to survival, science becomes more dependent on society for wise decisions about the use of the scientific information it produces.

Future decisions will demand that the public and its leaders have an adequate understanding of science because their views will influence policy affecting the survival of society and the environment. Teachers and others concerned about the quality of life must play an important part in providing an understanding of science to future students. Toward this end, *Science and Society* has been written.

Future teachers share an additional responsibility for teaching about science: to prepare students for earning a living in a world where much of the goods and services are dependent upon knowledge of science and advanced technology. With this responsibility in mind, teachers need to present science in a manner that is easily understood and made relevant to commonplace experiences so that all students will see a collection between science, technology, and their own personal lives.

But why should another science methods textbook be written at a time when more than twenty methods texts are available to help approximately sixty thousand readers learn how to teach science to children and early adolescents? *Science and Society* differs in significant ways from other science methods textbooks.

First and most importantly, readers must meet their responsibility as teachers to enable children and early adolescents to understand enough about science and make informed decisions about their own welfare and about the environment in which they live. While we hope to prepare tomorrow's citizens for their social and environmental responsibilities, this book is directly concerned with the science-related issues that children and adolescents can influence today. At a time when many school classrooms have de-emphasized science, *Science and Society* presents a serious directive to teach science. Further, readers are introduced to skills designed specifically to help them teach students how to make decisions about issues that affect their lives.

One of the greatest needs that elementary, intermediate, and middle or junior high school teachers have identified in surveys is the need to know more science themselves. Some need only a review, others need a bit of updating, but many prospective teachers need a comprehensive overview of modern science.

Because science content is so important to all these groups, we have provided up-to-date information in four science areas: Environmental Biology (chap. 9), Health Sciences and the Human Body (chap. 11), Earth and Planetary Sciences (chap. 13), and Physical Science (chap. 15).

Each science content chapter is followed by a chapter showing how the concepts of that area of science might be taught in kindergarten through eighth grade classrooms. A selection of activities is provided, illustrating for teachers how teaching objectives, methods, and content can be built into a real program for instruction. There are 13 activities for primary (K-2), 12 for intermediate (3-5), and 14 for middle school (6-8). Planning Charts, which appear at the beginning of each activities chapter, provide models to assist in planning a cohesive program in science.

Many science methods textbooks offer a list of activities from which readers select those that are most familiar, or those that re-quire the least material or preparation, or simply those that sound like the most fun. In our experience, lasting commitment to teaching science is best achieved when teachers plan cohesive programs, rather than select activities haphazardly. The activities chosen for inclusion in *Science and Society,* when used with the Planning Charts, will provide a foundation for a science program based on the integration of teaching objectives, science content, and methods of instruction.

A paperback activities book, *Teaching About Science and Society: Activities for Elementary and Junior High School,* is available for use as a supplement to this text or for independent use. It includes over 82 activities (a full year's curriculum) organized by grade level and specific instructional objectives. As in *Science and Society,* the activities are preceded briefly by their societal implications and appropriate science background.

Another way that this book differs from many other science methods textbooks is that it presents a variety of ways to teach science. This acknowledges our concern for the continuous changes that must take place in schools and society. During the sixties and seventies, the trend in science education was to advocate a particular method of teaching science. This method was usually based on one or more theories about learning and development such as those proposed by Piaget, Bruner, and others. Inquiry and discovery learning dominated the scene.

It is our view that books about teaching science need to introduce readers to a wide variety of approaches to teaching and communicating information about science rather than to advocate a single method or approach. In response to that need, we have provided a taxonomy of common teaching methods (chap. 4) and created lessons and activities which illustrate how to use a variety of these methods.

It is important for future teachers to do well in the classroom. Most of them will teach many other subjects in addition to science. This text provides a broad foundation of teaching methods to enhance every teacher's repertoire of teaching methods and to make it easier for teachers to integrate science into the total K-8 curriculum.

Some traditional features of science methods textbooks have been retained because they are especially valuable. These include a chapter about learning, development, and motivation and the implications of these theories for science teaching (chap. 3). The book also includes a synthesis of previous ideas about science concepts, processes, attitudes, and psychomotor skills. To this synthesis we have added a much-needed new component on decision-making processes for science education (chap. 2).

Finally, to make this book useful to students long after they are teaching in their own classrooms, we have included an annotated Resource Guide which describes over 100 films, games, activity guides, and professional books for teachers (see Appendix A). In addition, we have provided guidelines for introducing computers in the classroom to teach science and keeping up-to-date on the available computer software for teaching science (see Appendix B).

We have found writing this text an exciting challenge and hope a sense of our excitement is communicated to students. It has taken us four years to gather information and refine ideas that will meet the needs of prospective teachers for the eighties. We wish to express our sincere thanks to the following persons for their reviews of our manuscript: Glen Berkheimer, Michigan State University; George Dawson, Florida State University; Robert Shrigley, Pennsylvania State University; Robert Moore, Indiana University of Pennsylvania, and Timothy Gerne, William Paterson College. We are also indebted to our many fine students and colleagues for their helpful suggestions, and we invite your support in sharing these ideas with your students.

Contents

PART TWO
SCIENCE CONTENT AND ACTIVITIES

12 Teaching the Health Sciences 411

13 Earth and Planetary Sciences 438

14 Teaching the Earth Sciences 494

15 Physical Science   526

16 Teaching the Physical Sciences   562

Appendix A: Elementary and Junior High Science Resources: A Guide   610

Science and Society

ONE

Methods for Teaching Science

Part one provides a basic foundation in *how* to teach science. It is pedagogical in nature, focusing on the knowledge and skill required to be a successful classroom teacher. In contrast, part two focuses on the *content* of science itself. It is designed to update your knowledge of science and to provide examples of classroom science lesson plans.

Part one consists of eight chapters. In chapter 1 you will discover why it is urgent yet exciting to teach about science and society. Chapter 2 then provides a model to help you establish goals and objectives for your own science class. Familiar theories about learning, development, and motivation are examined in chapter 3 for their implications for teaching science. To help you develop a variety of teaching skills, a taxonomy of common teaching methods has been compiled in chapter 4; you can refer to it often when you wish to explore alternative methods of teaching to reach your objectives. A valuable step-by-step guide, or checklist, and helpful hints for planning science lessons, units, and year-round programs appear in chapter 5; chapter 6 offers suggestions for managing instruction in the classroom. Practical suggestions for evaluating the outcomes of your science lessons and program are found in chapter 7. Finally, chapter 8 closes part one with a thoughtful look at the challenges and rewards of teaching, in general, and teaching science, in particular.

1

Science and Society

The purpose of chapter 1 is to convey a sense of the excitement that comes with discoveries in science and a sense of responsibility for teaching science. You will take a brief trip back through history to capture glimpses of how science, society, and science teaching have changed during the past two hundred years in America. You will grapple with the important question of what science to teach for the year 2000. And, finally, you will be introduced to the theme, Science and Society.

KEY CONCEPTS AND IMPORTANT IDEAS

☐ the impact of science on society
☐ medical research ☐ population growth
☐ food shortages ☐ energy ☐ technology
☐ exploration of space ☐ science as a search for knowledge ☐ the social context of science ☐ the history of science education ☐ object teaching ☐ nature study movement ☐ elementary science movement ☐ interdependence of science and society ☐ identity and organization of matter ☐ change and interaction of matter ☐ limits of change ☐ conservation of matter

SCIENCE AND OUR BEAUTIFUL PLANET

We live on a beautiful planet where oceans, mountains, rivers, and valleys support an enormous variety of plants and animals that meet all of our physical needs. The weather and seasons change so little that we can be active all year round. This hospitable planet is situated in a well-ordered universe in which sun, moon, and stars behave so predictably that we can make fine distinctions in time, temperature, and distance.

It is an exciting place to be alive! Every day we increase our understanding of the natural forces and features of this planet and the larger universe, thanks to the curiosity of thousands of scientists. If we were to stop and think about the astonishing scientific discoveries that have occurred over the past two hundred years, we would see how profoundly science has changed the quality of our lives.

Consider your health for a moment. Today we are likely to live more than twenty years longer than our ancestors did at the turn of the century. Many diseases which were once dreaded are now controlled, largely through immunization. Improvements in the purity of food, water, and air have raised the standard of health throughout the world. The diets of most Americans have changed; carbohydrate consumption has been reduced by more than 20 percent and substituted with calories in the form of protein and fat, more than doubling the amount of beef consumed since 1930. The effect of this change in diet is now the subject of intense medical research.

Much of the credit for the improved state of health in the United States can be attributed to the pioneering and unrelenting efforts of scientists who have wanted to understand how the human body works and how it reacts to the world around it. Today the combination of new knowledge about the body's immune system on the one hand, and new technology such as heart-lung machines on the other hand, has made possible unprecedented surgical accomplishments like the implantation of artificial pacemakers and joints. During the coming decades immunologists will seek to bring new and improved vaccines into use for hepatitis, bacterial meningitis, gonorrhea and other sexually transmitted diseases, dental caries, and certain respiratory diseases such as infantile bronchiolitis. Medical researchers will continue to study links between medical disorders and exposure to specific chemicals: lung cancer and asbestos, liver sarcomas and vinyl chloride, male sterility and the pesticide DBCP, and neurological conditions and mercury and lead. It is a major task to identify chemicals that should be controlled when there are nearly 48,000 chemicals used commercially in the United States today, and as many as 1,000 new chemicals introduced each year.

Of course scientific discoveries alone cannot save lives. The number of physicians in the United States has increased dramatically during the past century. In 1880 there were only 82,000 doctors in this country; by 1970 that number had increased to nearly 350,000. Yet the number of physicians has just managed to keep up with population growth. In 1880 there were 163 physicians for every 100,000 people; in 1970 there were 166 for every 100,000 persons. In spite of medical breakthroughs and physicians' care, there are always new problems that science tries to understand and, if possible, to solve. For example, the importance of overcoming world hunger and malnutrition has become widely recognized.

Think for a moment about the world population. In 1950 there were about 2.5 billion people in the world. Today there are about 4.6 billion. By 1990, scientists predict the world population will reach 5.3 billion and in

Using the speed-of-light capabilities of body scanners, scientists are able to detect heart disease, manic depression, and many other illnesses. (UPI)

the year 2000 (move over!), we can expect 6.2 billion people. There is general consensus that the population growth rate is likely to slow down in all areas of the world except Africa between 1980 and 2000. And experts predict an ultimate world population of between 8 and 10 billion. When 125 of the nation's most knowledgeable authorities met in 1967 to discuss world food output in relation to world population growth, they estimated that food deficits would reach 84 million tons by 1990 and that the greatest deficits would be found in developing countries. Figure 1–1 indicates that the population increases will be the greatest in Africa, Asia, and Latin America. Grain consumption in those areas (see figure 1–2) shows the smallest increases, resulting in deficient diets and some starvation.

One of the most promising strategies for dealing with potential world food shortages is one that depends upon aerospace technology. Two experimental, remote sensing satellites, Landsat 2 and 3, are used increasingly in agricultural planning. Information collected from these satellites is used to monitor weather and crop acreage, inventory forests, map soils and fires, and determine land use. Such information is crucial for providing early warnings of winter kill, assessing wind and water damage, detecting insect and disease infestations, measuring soil moisture, and modeling crop growth and yield. Other promising solutions that will contribute to easing world food shortages come from agricultural science and technology. For example, the highly contagious foot-and-mouth disease that causes major losses of cattle, swine, and sheep in many foreign countries may soon be under control due to the recent discovery of a vaccine developed through genetic engineering. By coordinating efforts in satellite mapping, genetic engineering, new agricultural techniques, and methods of population growth control, scientists hope to ease food shortages and malnutrition on our planet. But our bodies need more than food.

Consider the energy used to heat, cool, and light your home, the energy used to travel, and the energy used to produce and deliver foods and furnishings to your home. It

Population

Latin America
United States
Western Europe
USSR and Eastern Europe
Asia and Oceania
Africa
Other industralized countries including Australia, Canada, Japan, and New Zealand

Boundary representation is not necessarily authoritative.

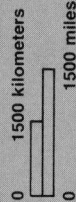

0   1500 kilometers
0   1500 miles

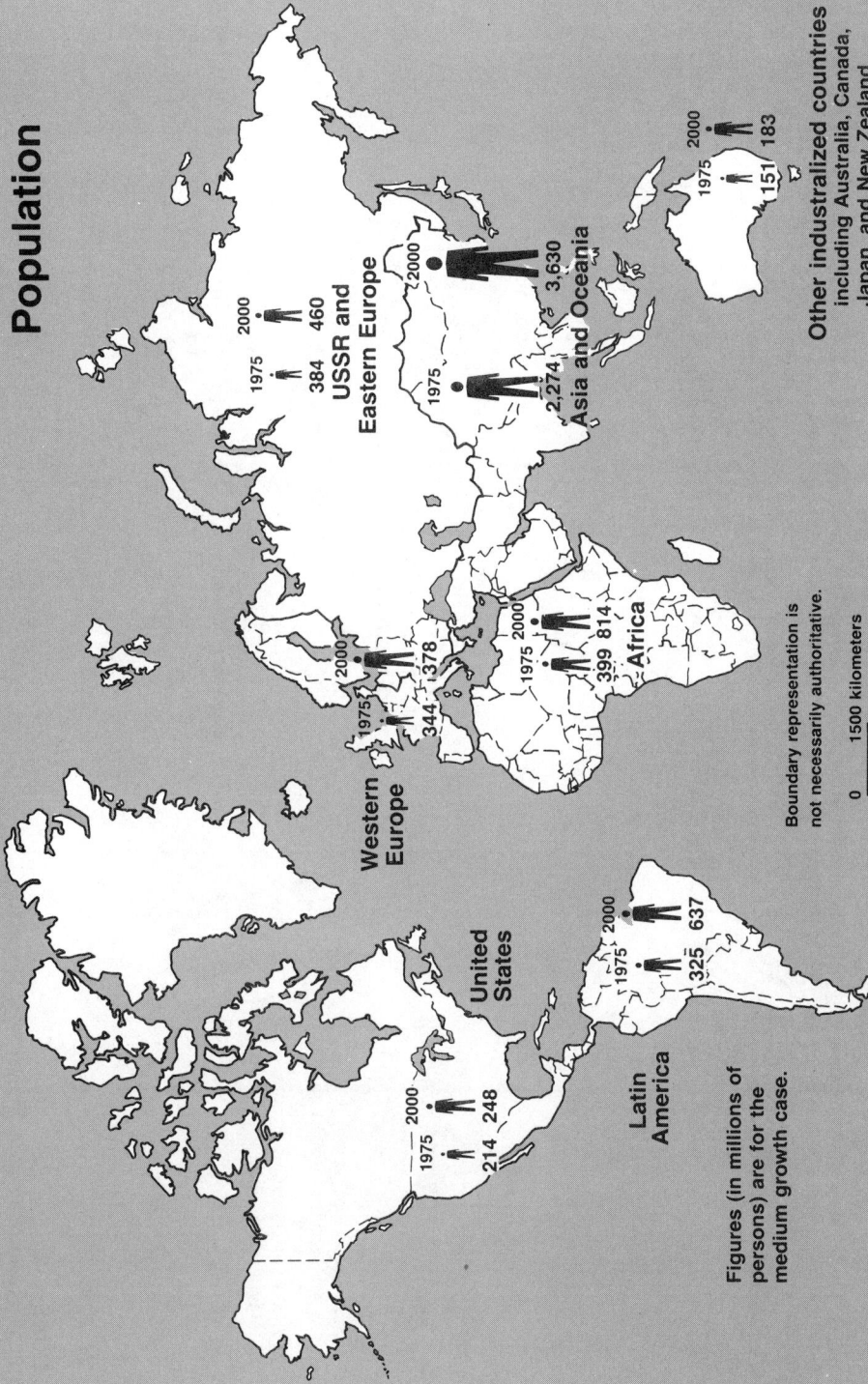

Figures (in millions of persons) are for the medium growth case.

FIGURE 1–1
Distribution of the world's population, 1975 and 2000.

Reprinted, by permission of the publisher, from G. O. Barney, *The Global 2000 Report to the President* (New York: Penguin Books, 1982).

7

Grain Consumption Per Capita

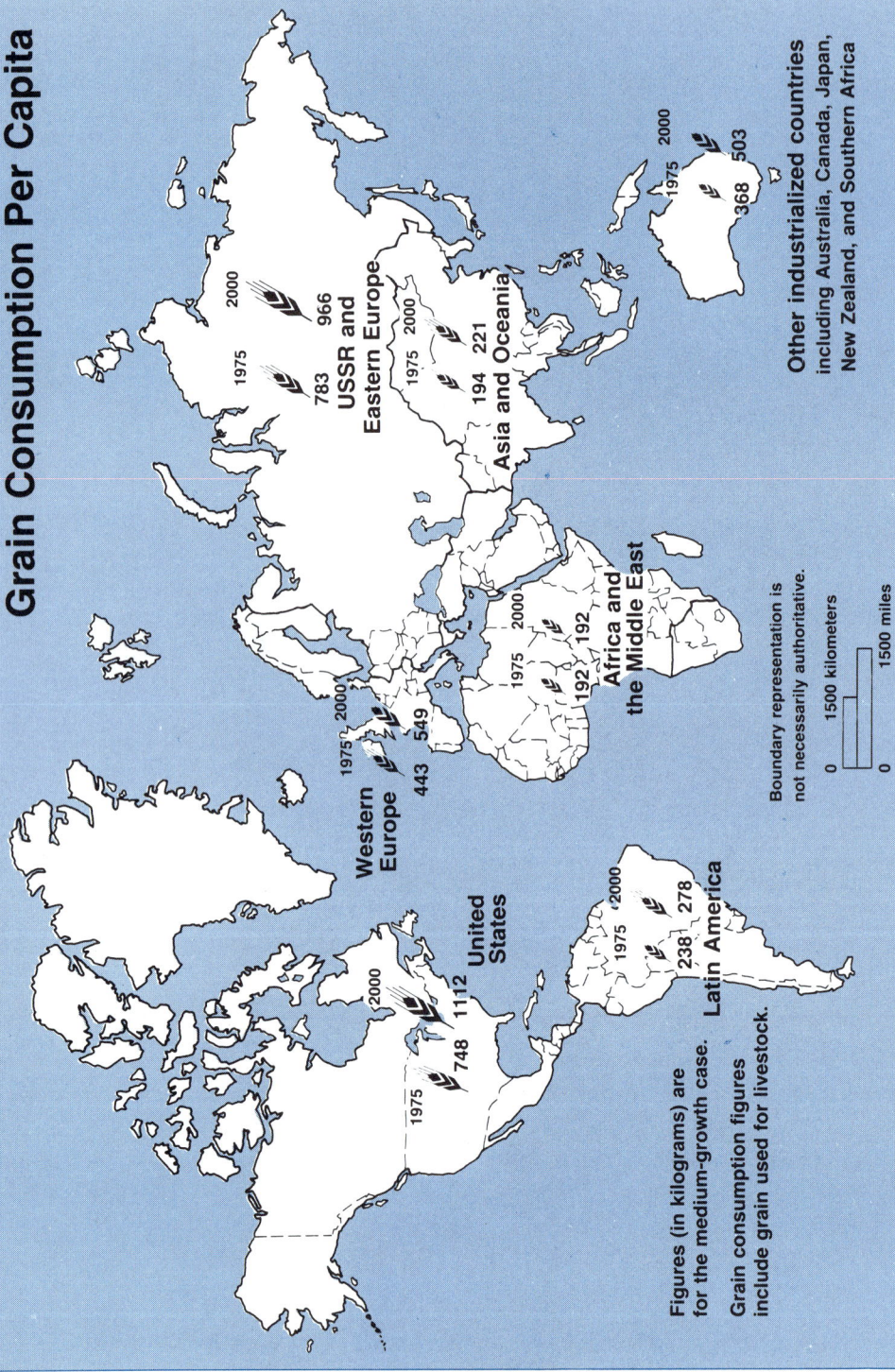

USSR and Eastern Europe
1975 — 783
2000 — 966

Asia and Oceania
1975 — 194
2000 — 221

Africa and the Middle East
1975 — 192
2000 — 192

Western Europe
1975 — 443
2000 — 549

United States
1975 — 748
2000 — 1112

Latin America
1975 — 238
2000 — 278

Other industrialized countries including Australia, Canada, Japan, New Zealand, and Southern Africa
1975 — 368
2000 — 503

Figures (in kilograms) are for the medium-growth case.

Grain consumption figures include grain used for livestock.

Boundary representation is not necessarily authoritative.

0   1500 kilometers

0   1500 miles

FIGURE 1–2

Per capita grain consumption, by regions, 1975 and 2000.

Reprinted, by permission of the publisher, from G. O. Barney, *The Global 2000 Report to the President* (New York: Penguin Books, 1982).

The costs of solar energy systems are very high when compared to current fossil fuel costs. Researchers are exploring solar flat plate collector designs which use low-cost, high-performance materials; eventually, the cost of solar energy may be comparable to other types of energy. (Brookhaven National Laboratory)

is easy to think of the personal energy savings we make in our homes; but we are unaware of the enormous amount of energy required to operate our technological industries, and often we forget how many millions of jobs are dependent upon energy.

If you had lived in the 1850s, wood would have been your primary source of energy. But large-scale wood gathering to supply energy needs led to the first energy crisis. Coal became the accepted substitute by the turn of the last century. Had you lived during the early 1900s, you and your family would have found that coal was easier to store and that it produced more heat. But by the middle of the current century, you would have turned to cleaner-burning natural gas and oil to supply most of your energy needs. Today, as we face the most recent energy crisis, nuclear and solar energy are being explored as substitutes for oil and gas.

The United States usually takes about six decades to switch from one energy source to

another; the world as a whole takes about one hundred years. The implication from these findings is that new energy sources must be actively explored long before current supplies near exhaustion. Scientists are now considering energy needs for the year 2030. In the past, most major studies of energy approached the problem on a national scale. However, since nearly 25 percent of current energy supplies come from one place on the globe, the Middle East, scientists now plan for global energy needs and sources.

Have you thought about all of the technology that improves the quality of your life? Scientists who work in industry contribute to the rich variety of products you enjoy, the manufacture of which is responsible for a great share of the nation's economy. In the middle of the last century, there were no industrial laboratories in the United States. Modern industrial research really began in the 1880s with Thomas Edison's laboratory in West Orange, New Jersey. His laboratory was the

Thomas Edison (1847–1931)

he! If balloons fly because they are filled with gas, so would his friend if he took a triple dose of Seidlitz powders.

Couple an inquiring mind of a boy with a mother who believed that learning was fun, and you have a winning combination. She made a game of teaching,—calling it exploring—the world of what was known. Edison became the greatest inventor in American history although he had had only three months of formal schooling.

Everything in the world around him interested Edison. He was constantly developing things that would work for ordinary people in ordinary conditions, things that would not easily malfunction and could conveniently be repaired. His inventions included the telephone, typewriter, generator, phonograph, and movies.

Edison once admitted that he had probably tried most everything while working on his eleven hundred inventions in sixty years. He usually was not concerned with the theoretical bases of their operations but rather made great intuitional leaps. Failure was not a stopping point. It is said that after more than ten thousand experiments to construct a battery had failed, Edison's response was not that he had failed—he had just discovered "ten thousand ways that won't work."

Imagine a child in your life whose attention focused on such questions as "How does a hen hatch chickens?" "What makes birds fly?" "Why does water put out a fire?" Think more about such a child who, when answers were not convincing, would attempt to find out on his own. If chickens could hatch eggs by sitting on them—so would

forerunner of what is now General Electric Laboratories. During the first few decades of the twentieth century, industrial research laboratories mushroomed throughout the country, introducing such wonders as the light bulb, the telephone, the phonograph, and the automobile. By 1913, there were 50 industrial laboratories, including Eastman Kodak, E. I. Du Pont, U.S. Steel, and Westinghouse. By 1930, 1600 industrial research laboratories had been established, and in 1979 more than 360,000 American scientists and engineers

Radio telescopes have revealed such spectacular events as the collision of two galaxies and the discovery of quasars. This 300-meter radio telescope at Arecibo, Puerto Rico, is a bowl carved in a natural depression where viewing is possible 24 hours a day. (Cornell University Photograph by Russ Hamilton)

were engaged in industrial research and development. Many times that number of nonscientists are engaged in work related to science and technology.

Until now we have been thinking about scientific discoveries that are earth-bound. Somewhere on an isolated plain about fifty miles west of Socorro, New Mexico, is the world's largest radio telescope. It consists of twenty-seven dish antennas that use the earth's rotation to sweep the skies. Called VLA for Very Large Array, this imaginative system is equivalent to a single radio telescope with a diameter of twenty-one miles! The radio pictures of celestial objects produced here help astronomers to answer puzzling questions about black holes, quasars, galaxies, and the nature of matter in the vast space between stars.

Magnificent telescopes that probe the distant edges of the universe represent one kind of space exploration; satellites and space ships represent another. America's space exploration of the moon and the planets in the solar system reads like a fairy tale. Beginning with Apollo, the first astronauts landed on the moon on July 20, 1969, and our television sets brought 600 million viewers in forty-nine countries so close that we felt as though we, too, could have been the first astronauts to set foot on the surface of the moon. Two years later, Mariner 9 orbited Mars, representing the first man-made object to circle another planet. By 1979, Pioneer 10 and 11 flew by Jupiter, sending back pictures of unbelievable beauty and causing us to revise previous notions of its nature. It is easy to forget that, within a single century, scientists and engineers designed the very first airplane and a space shuttle!

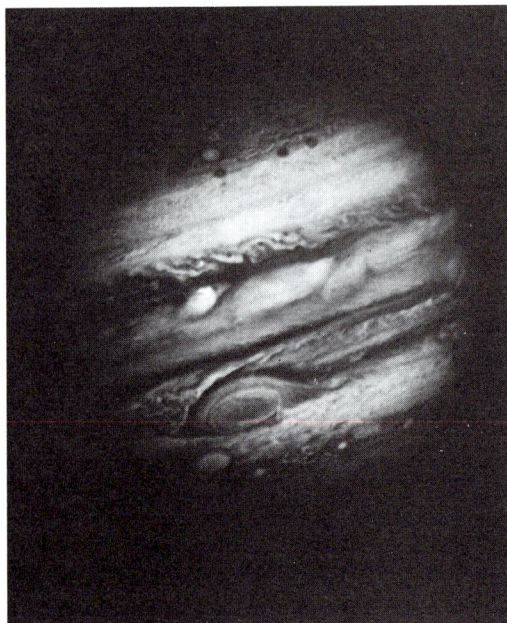

Within a single century scientists and engineers have designed the first airplane, the first satellite, and the first space shuttle. Views such as this one of Jupiter were undreamed of a century ago. (NASA)

SCIENCE: THE SEARCH FOR KNOWLEDGE

The world of the scientist is an interesting place. Today there are about 2 million U.S. scientists and engineers who spend their days in the pursuit, preservation, or application of scientific knowledge. Their searches range from the microscopic structure of matter to the expanse of the universe. In communicating their observations through writing, computers, tapes, pictures, television, books, and journals, individuals and teams of scientists have created the enormous, and intricate, structure of knowledge that we know as science. Each scientist who peers into a microscope, oscilloscope, or telescope; who digs into the surface of the earth or the moon; and

who watches two liquids form a crystal or two animals feed their young is captivated by objects and events that may never before have been seen by the human eye. And each one must mentally travel up, down, or across this great lattice-like structure of knowledge called science before he or she can give meaning to that which lies before the eye.

Years and sometimes lifetimes are spent trying to add a single bit of new knowledge to the structure of science; yet every day, we see new additions to the structure. From time to time parts of the structure have to be revised because scientists find a better way to describe what they observe, but the struggle to know continues. It is a struggle that is spurred on by human curiosity and made exciting by the possibility of new discovery.

During the first one hundred years of American history, there were very few scientists in the United States and most of them were educated abroad. That picture began to change by 1880 and since then, the number of scientists has increased. In 1900, there were about 100 doctorates awarded in the natural sciences; by 1970, nearly 11,000 students were awarded a degree in the natural sciences.

A century ago it was possible for scientists to read the world's scientific literature and to keep up to date with all active research in this country. Today many scientists try to keep the full picture in focus with the help of publications such as *Scientific American* and *Science,* but most are forced to specialize in a small area of the total structure of science.

If it is difficult for scientists to keep up with the rapid new developments in science, you might ask how you, as a teacher, can ever manage to do so. You will learn from this book how to teach science without feeling that you are always behind in what you should know. It is helpful to know that when scientists are asked a question outside their area of specialty, most of them feel no embarrassment at saying, "I'm sorry, I don't know. That's not really in my field of science." (See the endpaper for a brief history of science.)

THE SOCIAL CONTEXT OF SCIENCE

Science has always had an impact on society. In every age, the knowledge produced by scientists has created both benefits and problems for humanity. Contemporary society varies little from its predecessors in this regard. New scientific knowledge results in technological advances which simultaneously improve the quality of life in some respects and threaten it in others. Through medical technology, lives are saved, lengthened, or freed from pain and hunger. Medical science has improved the diagnosis and treatment of disease and has provided temporary solutions to world food shortages. For these reasons society values the scientific enterprise. But problems also grow from scientific discoveries. When issues related to nuclear energy, genetic engineering, and insecticides emerge, society

questions the value of science. One has only to read a history of science to realize that science—or the search for knowledge—brings with it possibilities for both improving and worsening the quality of life. The responsibility for deciding how new knowledge will be used is up to each of us.

If science has always had an impact on society, then what has changed? With each decade in this century, science and society have become increasingly interdependent. Scientific knowledge now accumulates at an unprecedented rate as a result of the electronic technology of computers, microscopes, and other monitoring or information processing systems. Scientists have compared the impact of the electronic revolution to that of the industrial revolution.

At the same time that more scientific knowledge is being produced, advances in electronic and print media have vastly increased the capacity to bring news of scientific discoveries and problems to world attention, often within minutes of their occurrences.

In addition to these changes, the world population has increased to approximately 4.6 billion individuals. World population

Rapid advances in computer technology have radically changed our ideas about information storage and processing. (Charles Gatewood/Art Resource, Inc.)

growth has caused an accelerated depletion of natural resources and a simultaneous growth in the development of synthetic products. Growing dependence upon synthetics has given rise to disposal problems of such products and their by-products.

For these reasons, society depends increasingly upon science to solve problems related to its survival. In turn, science depends increasingly upon society for support of the research it must conduct to solve these problems.

Are the problems faced by science and society more serious now? This question cannot be answered with a simple yes or no. Societies have always faced life-threatening situations; starvation due to overpopulation and contamination of the environment are serious problems that societies have encountered throughout history. But because the world population is greatest now, these same life-threatening conditions have the potential to affect greater numbers of people. The problem seems to be more serious today because the increase in population has created greater demands on the earth's limited resources; therefore, we are forced to rely more and more upon synthetic products. The waste produced by the manufacture of these products, in turn, poisons the environment and increases the probability that greater numbers of people will die of diseases such as cancer. Another dimension unique to today's problems is the time span required to solve problems such as cleaning up an environment once it is damaged seriously. Pesticides, hydrocarbons, and radioactive wastes are examples of substances that present a more serious threat to life due to both the duration and the scope of their contamination.

Many of the most serious problems facing society today require teams of scientists with diverse backgrounds and with abilities to integrate bodies of knowledge that were once

Many of the most serious problems facing society today require cooperation among science and society for solutions. (United Nations)

separately called biology, chemistry, physics, and so forth. The search for new sources of energy and the exploration of space are examples of challenges that require this multidisciplinary approach. The emergence of fields such as biochemistry, biophysics, geophysics, and astrophysics is evidence of the complexity of the discoveries, challenges, and problems that characterize the present and foretell the future in science.

The most serious problems facing society (and those for which scientists can be of some help) also have economic and political implications. For example, competition for the world's oil resources has not only affected the amount of fossil fuel consumed by the public, but it has also affected the costs of all goods and services dependent on fossil fuel for their production and delivery. Politically, world leaders compete for, and attempt to regulate, this dwindling resource.

The quality of life on this planet and the welfare of the world's population are closely tied to science, economics, and politics. Future decisions and problems demand that society and its leaders recognize the interdependence between science and society. As teachers, we play an important role in fostering this understanding. We have a special responsibility to be informed about science. Why? Because scientific knowledge has the potential to improve or deteriorate the quality of life for ourselves and the world of our students.

But what science shall we teach to prepare students to live in society by the year 2000? Before grappling with this important question, let's take a brief look at the past. As we take an imaginary trip back through our nation's history and capture glimpses of how science was taught during the past two hundred years in America, remember that teachers in every decade asked themselves the same question: What science ought to be taught? For them as

for you, their present world indeed looked different from its past. Science had changed their view of the world, too!

SCIENCE EDUCATION USA: A BRIEF HISTORY

Science teaching between 1750 and 1850 was reflected primarily in books for children. These "didactic writings" from England focused on the study of things—not the study of words about things. In the late 1700s the political purpose for schools was to serve as preparation for the clergy and medicine. Only the wealthy could afford to send their children to private schools.

Science teaching mirrored views about the study of nature and how nature reinforced moral and ethical values. By using many firsthand experiences, teachers and parents helped children to develop a natural theology. Emphases on thinking skills and a methodology that was satisfying to children illustrated teaching strategies that are found in science programs today. Firsthand experience with nature is also among current teaching strategies. Science was used as a vehicle to develop morals and ethical or theological understanding, and it was not unlike contemporary emphases on environmental ethics.

In the early 1820s, a distinct shift in schooling occurred. Informal schools and institutions developed with a primary focus on the teaching of reading, spelling, natural history, and geography. The political view of schooling was to furnish the minds of children with useful information. Lessons were designed to help select proper clothing for the season by experiments with melting snow under clothes of different colors, and to teach students how to behave in a thunderstorm. They acquired information about such practical matters as the souring of milk and the fermentation of beer.

Science teaching in American schools has always reflected society's views of nature and the values of science. (St. Louis Art Museum)

But with the proliferation of schools came larger groups of students and the advent of group instruction. Books were developed to meet the needs of this instructional change, reflecting a shift from the study *of* things (nature study) to the study *about* things. Instruction changed from the inductive, direct observation of objects and events to a deductive, indirect discussion about objects and events.

The political focus in the early 1800s illustrated a concern for greater equality of opportunity for people in a democracy. Money would no longer represent an essential difference to individual worth. Taxation provided means whereby those able to pay provided the education for those unable to pay. A parental and community force began to emerge and demand that schools serve everyone—not just the bright minds or the children from privileged homes. An initial source of profes-

sional pressure was seen in the demand for efficient management of group instruction. Since the infant and grammar schools served more people, teaching methods evolved to fit group instruction rather than individual exploration. Science became a process of talking about nature rather than one of experiencing nature—a trend still observable today.

In the mid-1800s, a new science curriculum seemed to emerge. Called **Object Teaching,*** it was a method whereby the teacher would question a child about an object. The child would respond with descriptive words about the object. Whether the child had seen the object or not seemed to be unimportant. Just saying the right words to describe it was thought to sharpen the child's observation skills and at the same time provide him or her

---

*Terms in boldface are defined in the glossary.

with a feeling that there was a harmonious purpose to the universe and a theologically correct view of the purpose of creation.

Object Teaching was used in schools that themselves were part of dramatic political and economic changes in the United States. A devastating Civil War, the emergence of factories and large industries, and an alarming inflation that undercut the value of the dollar were some of these pressures. Compulsory school attendance for everyone emerged with a debate as to whether it was legal to use tax monies for public schools. Parental pressures seemed consistent in that parents wanted schools to prepare their children for later employment. Attendance in school was defined by the law and courts as a parental responsibility. The teacher was to develop general student skills rather than scientific skills. Thus teachers had to cope with the demand to keep students in school (a custodial responsibility) while generally preparing them to be successful with any job. Cooperation, discipline, and obedience were essential skills or outcomes of schooling for all students.

The late 1800s were again a time of rapid change. New ideas and inventions revolutionized life everywhere. It was a time of exciting firsts:

☐ New technologies using heat engines and steam
☐ New developments in electricity, communication devices, transportation, and power
☐ New breakthroughs in optics, atomic theory, and genetics
☐ New ideas in geology and astronomy
☐ Key developments in technologies used in agriculture

These technological changes brought with them clear messages to the schools to incorporate a utilitarian dimension in science instruction. Schooling was available universally

and attendance was compulsory. Along with the new science technology, our country was experiencing a financial depression. Taxpayers demanded to know what their tax money purchased in schools. Widely diverse special interest groups demanded that their concerns be incorporated into science instruction.

☐ Agriculture wanted science to emphasize the beauty and goodness of nature (in contrast to the evil and ugliness of city living).
☐ Temperance groups wanted science to emphasize the evils of alcohol and its destruction of the body.
☐ Other health groups wanted the evils of stimulants and narcotics to be emphasized.
☐ Humane societies wanted the need for kindness to animals to be stressed.
☐ Parents wanted schooling to emphasize practical work skills toward a career or vocation.
☐ Teachers wanted a curriculum that they could teach.

From these widely diverse pressures, two conflicting approaches to science education emerged: the **Nature Study Movement** and the **Elementary Science Movement.** It was quite a challenge for a teacher in the late 1800s. That teacher was usually a woman between twenty and forty years old who had completed high school and maybe two years of normal school training. In normal school training, she would have been fortunate to have had a course in botany and, in a few cases, a course in zoology. To help teachers make science part of schooling, the first teacher editions of texts began to appear along with workbooks for students, inservice workshops, and service institutions. Eminent scientists gave popular lectures to large community audiences and called for science to be part of children's educational experience. Educators became interested in science as a vehicle for accomplishing the goals of schooling.

But here was a point of separation, and teachers were caught in the middle. If the goal was the emotional development of the child, then Nature Study was the choice. If the goal was the intellectual development of the child, then the generalization or core approach of the Elementary Science Movement was the direction to be taken. Thus teachers used materials which themselves represented quite diverse assumptions about the purpose of science in the classroom and *how* science was best learned or taught. Table 1–1 illustrates the basic differences between these two approaches.

In this turmoil of political, parental, and professional pressures, the "modern science" programs of the period 1910–1940 emerged. From a social and political view, there was stress to make science more practical. The world was changing rapidly with the advent of electricity and the industrial revolution. The transfer of the applied thinking of science to the social and economic problems that followed the industrial revolution was held to be a key function of science teaching. Established behavior patterns for the farm-based society seemed to function no longer. Modern science demanded that teachers use explicit instruction instead of the Nature Study approach which permitted a time in the week when formal instruction could be stopped and teachers and pupils could relax and talk informally about some topic of interest.

A significant focal point for the newest modern science was Craig's study in 1927. In a creative strategy, he generated a science course of study for the Horace Mann Elementary School. This course was based on the assumption that it should be a continuous unified program aimed at understanding the significant ideas of science. If you understand the significant ideas, you will be able to function as an effective problem solver in your daily life. To determine what were the significant ideas, he used groups of parents, teachers, and school managers to nominate science ideas. These were then ranked or placed in order of importance. The result was a description of ideas, topics, and generalizations about science as people experienced it in the 1920s. That structure has remained virtually unchanged in many science textbooks of the 1980s! The science ideas were then translated into graded readers for the students, instead of remaining exclusively in teachers' manuals as was seen in the Nature Study approach. The result was *reading about* the applications of science (electric switches, steam engines, telegraph keys, and many similar and exciting new technologies of the 1920s) with very little *doing* of science.

Then, in the 1950s, an event occurred that prompted a major reform in science education in the United States: Russia launched Sputnik, the world's first satellite. The next two decades proved to be a period of unprecedented federal support of science education. Between 1955 and 1975, the federal government invested over $650 million to improve the quality of science instruction. More than five hundred different projects were funded during this period in order to develop new science teaching materials and to upgrade teachers' skills and knowledge for teaching science, math, and social studies.

Many of the new science programs were unique in several ways:

☐ They were developed by teams which included scientists, educational psychologists, and teachers, rather than the single author approach characteristic of previous commercial science textbooks.

☐ The programs emphasized learning by doing, but used the most recent concepts in science rather than using nature trips as a starting point.

☐ The teaching methods used in the teachers' guides were based on the most up-to-date theories about how children and adolescents learn.

TABLE 1–1
Science Teaching Issues
in the Late 1800s

	Nature Study	Elementary Science
Content to be studied	Observe nature in a more or less random way—a naturalistic approach.	Work with a systematic body of knowledge—a scientific approach.
Goal	To develop in the child personal enjoyment of nature.	To further extend knowledge in science.
Approach	Primarily inductive with hands-on experience first. Use vocabulary only as needed.	Primarily deductive or didactic—learn the ideas or knowledge first. Use experience if time and resources permit.
Motivation for learning	Children will learn if permitted to do what is natural for them. This is essentially a romantic view of the unfolding goodness in the nature of the child. The best student is the best evidence.	Children will learn if exposed to a structured organized whole. This is essentially a behavioristic view. We learn to do what we do.
Curriculum choices	Nonsequential. Not structured or organized, and based on children's interest. High emotional and aesthetic impact, Emphasizing firsthand experience and not books. Closely related to student's daily life, e.g., topic selected based on season. Highly imaginative—author propaganda with animals talking and inanimate objects feeling. Mostly biological science content.	Highly sequential. Obvious structure based on scientific knowledge with explicit behaviors expected. Emphasis on understanding the conclusions of science. Closely related to scientific topics, e.g., topic based on chemistry or biology. No frills, "hard" science with equal emphasis in all disciplines.
Outcomes	Develop emotions and understanding.	Train the mind.

Variety and color were the hallmarks of the new science programs. Many came in cabinets, kits, boxes, shelves, and bags, and all contained materials that were inviting to touch, watch, and explore. The new programs came to be known by the acronyms of their project titles. The Intermediate Science Curriculum Study, for example, became ISCS. It was paralleled by Science Curriculum Improvement Study (SCIS), Science—A Process Approach (SAPA), Elementary Science Study (ESS), and so forth. Teachers often joked about the "alphabet soup" science programs but that did not hide their pride or the excitement that came from using these new programs which had been developed with the assistance of famous scientists and designed by teams of authors.

Because the science programs of the period 1950 to 1970s emphasized students *doing* science rather than *reading about* science, teachers had to become more skillful in guiding students' activities. To assist teachers, workshops were set up throughout the United States, usually offered in school districts or on university campuses, and sponsored by the National Science Foundation. In these workshops, teachers had opportunities to explore the materials themselves and learn new techniques for teaching science at the same time. Science received so much national attention that it was not surprising to find students and teachers rating science as their favorite subject. It is probably fair to say that this attention was spurred on by a spirit of U.S.-Russian competition after Russia launched their first spacecraft.

During the 1960s and early 1970s something else was occurring at the same time. Just as Sputnik had provided the visible symbol to justify federal support for the improvement of science education, the assassinations of John F. Kennedy, Martin Luther King, and Robert Kennedy provided the visible justification for allocating billions of dollars toward nationally supported economic and educational equity programs. Thus, without fanfare, the greatest nationally supported reformation in science education was gradually and completely overshadowed by the human rights

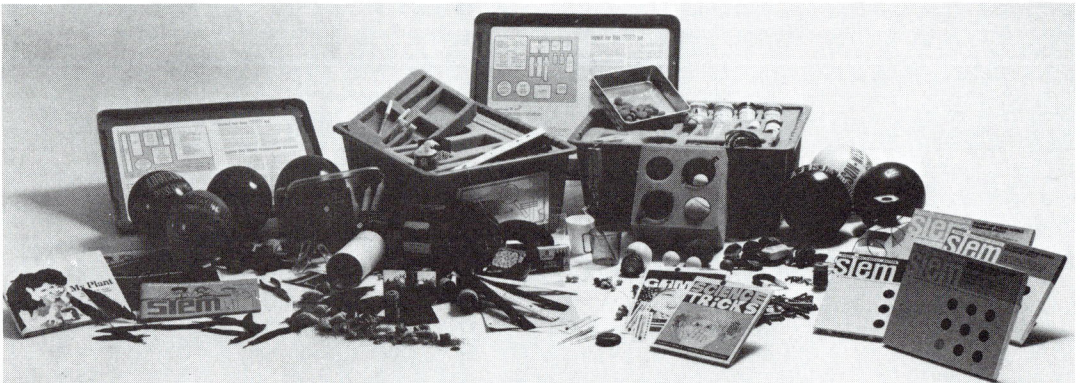

During the post-Sputnik era (1955–1965) dozens of new science programs were developed for elementary and secondary grades. They were published as books, in boxes, and as kits with many opportunities for students to explore materials and ideas firsthand. (© 1977, 1975, 1972 by Addison-Wesley Publishing Company, Inc. Reproduced by permission.)

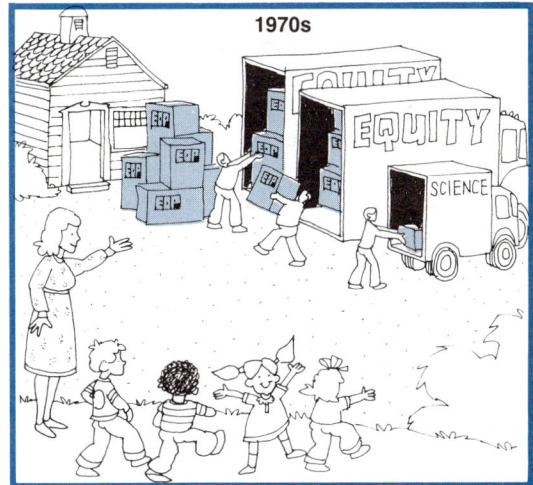

FIGURE 1–3
Federal support gradually shifted from science education to programs related to equity as
the human rights reform swept through American schools.

movement. As figure 1–3 illustrates, money,
concern, time, and energy were poured into a
new social reform in American schools.

By 1976, when the nation was two
hundred years old, many national leaders of
science education became concerned about
the status of science in the schools. The National Science Foundation commissioned
three studies to assess the status of science
education in the country. The results of the
studies were clear: although many fine examples of science teaching could be found,
science in general had all but disappeared
from the elementary school classroom! In its
place were competency-based programs in
reading, writing, and arithmetic, as well as
multicultural and bilingual programs—all with
high commitment to accountability for results
as measured through test scores. Gradually
progress was made in the war on educational
inequality. By the dawn of the new decade,
1980, science educators began to rekindle the
spirit needed to put science back into the curriculum of elementary schools and to rejuven-

ate science in the middle or junior high
schools of America. (Refer to the endpaper for
a brief history of science education.)

THE THEME: SCIENCE AND SOCIETY

We return from our brief trip through American history during which we looked at the
way science was taught in the past. Let us
now address the question raised earlier: What
science shall we teach for the year 2000?
What new knowledge and skills will be important in order to live and work in the world in
the future?

As the world population grows, it will be
important for all members of society to think
of themselves as living within and functioning
as a part of a global environmental system.
The earth as we experience it is a finite, complex, and balanced system that requires careful management if 4.6 billion people are to
make this planet a habitable place for 6.2 billion neighbors in the year 2000! The actions

and personal decisions of every individual now affect the quality of our lives and our environment, the earth.

Therefore, the first challenge to teachers in the 1980s—as well as to the students and all citizens—is to acquire knowledge of science and the way science and society depend upon each other. If students are to understand the limits of science, and the impact that people have upon each other and the environment, they will need to learn to think in new ways and try to understand the attitudes and values of others in society. Only then can students make informed decisions.

The second challenge for teachers in the 1980s and 1990s will be to educate students sufficiently in science and mathematics so that they can become productive citizens in a scientific and technological society. This will make it feasible to produce sufficient food and other essentials for life.

In the remainder of this chapter you will be introduced to six ideas or concepts that help clarify how science and society are related and interdependent. They provide structure for the theme, Science and Society.

Six Concepts in Science

When the subject of science is mentioned, a flood of science concepts often comes to the minds of most of us; perhaps terms like photosynthesis, meiosis, plate tectonics, and electromagnetism may be recalled and associated with particular science courses. Such concepts are part of the intricate structure of knowledge known as science noted earlier. Because these concepts are so numerous (remember, they represent centuries of work), it would be impossible to relate all or even the majority of them to the theme of this book. Therefore, we have identified six concepts that cut across all areas of science and, at the same time, have social significance for all citizens, now and in the future. (See figure 1–4.)

To introduce these six important ideas, or concepts, let us construct an imaginary scale. At one end of the scale is the individual, or you, as a person. At the other end of the scale is the universe; between, imagine the rest of the world, all other living and nonliving matter on and around the earth. Because of the psychological distance between these two points on the scale, it is very easy for individuals to be more concerned about their own identity and existence, and to be less concerned about the other things in the universe. The six concepts below characterize ways that scientists think about you and other things in the universe.

A personal attribute of value to all of us is **identity,** the personal meaning of one's existence as an individual. Because you are alive, you have an identity. When you cease to exist as an individual, your identity—as distinct from recollections of you or your image—ceases to exist. But identity not only applies to you and other human beings; it is the basis for learning more about all that can be observed in the universe.

The universe and all living and nonliving things in it are organized in various ways by scientists into species, or populations of plants and animals, and groups, classes, and systems of objects as well as events. Scientists study the structural **organization** of individual things in the universe to help them understand how nature works. Thus, in the study of science, identity and organization are two key concepts.

All individuals (referring to living matter) **interact** with the environment as they grow and **change.** They take from the environment what they need. The result of that process is that both the individual and the environment are changed. Each individual makes a relatively small change in the total environment during a lifetime, but the total world population has a substantial impact on the en-

ORGANIZATION

IDENTITY

INTERACTION

LIMITATIONS

CHANGE

CONSERVATION

FIGURE 1–4
The study of science in
schools can be thought to
revolve around six con-
cepts that are basic to all
fields of science: Identity,
Organization, Interaction,
Change, Conservation, and
Limitations.

vironment, especially when many are living in civilizations characterized by advanced technology. When individual plants and animals die, the environment is changed. A gradual decomposition takes place as materials are transformed into more elementary substances which are then recycled. Thus, interaction and change are two more important concepts to all areas of science.

The capacity of individuals to endure or adapt to changes in the environment is not well understood. Scientists have begun to find the **limits** of adaptability for some species of plants and animals. For example, the decline in birth rate among endangered species of birds has been linked to subtle changes in the environment. Scientists do not fully understand the capacity of the physical environment to absorb changes made by all the world's people. For example, how much can the lithosphere (land), hydrosphere (oceans, lakes, and rivers), or the atmosphere be changed and still support life as we know it? Least well understood or perceived by individuals in society is the cumulative impact that humanity can have on an environment when each individual is primarily concerned with his or her own existence or identity. There are few graphic estimates which illustrate the substantial nature of change to the total environment. We are a visually oriented society, and if we cannot see the impact we make in the environment, changes can go unnoticed for a very long time. Further, scientists have demonstrated that matter and energy in the universe can be exchanged but not destroyed. In other words, matter and energy are **conserved** when one can become the other. Therefore, there are no limits regarding the existence of matter and energy; they continue to exist but the identity of individuals may cease to exist in the universe. Thus, limitations and conservation are two final key concepts of significance to all areas of science.

The major theme for science presented here then is that individuals and groups do change and adapt to the environment; in the process they change the environment. More importantly for the future, there are limits to change. When environmental changes exceed the limits or the capacity of individuals, groups, or systems, survival is threatened. Identity is not conserved, even though matter and energy are ultimately conserved. Conversely, when individuals change the environment beyond its capacity to change, the iden-

tity of the environment on which they depend can no longer be conserved. Put most simply, people can destroy the environment that supports their lives if they are not careful. Education enables people to be careful and to make informed decisions.

We have chosen the theme, Science and Society, for this book because neither science nor society can exist without the other at this time in our world's history. Because we depend upon science to solve our survival problems, it is tempting for us to engage in a child-parent relationship in which society assumes the role of the dependent child while scientists, as parents, solve our problems. But society needs to understand that science alone cannot "bail us out" of future, major problems. The most serious problems will require intelligent decisions made by informed and responsible adults. Thus, our responsibility as adults and as teachers is to make ourselves and others aware of the responsibility to seek the knowledge needed to make informed, science-related decisions.

COMMENTARY

Many students in your classes will have jobs as adults that, in some way, will depend upon or contribute to science or the technology derived from science. The development of their appreciation for science is important to their preparation for earning a living. What science to teach, then, is the important question and it is one that we must ask of ourselves repeatedly because science and society change; accomplishments and responsibilities change. In the next chapter we will explore this question further and focus on how to establish goals and objectives for teaching science.

BIBLIOGRAPHY

Abelson, P. H. "Scientific Communication." *Science* 209 (1980):60–62.

Bartlett, M. *Common School Manual.* Utica, NY: Northway and Bennett, 1927.

Beauchamp, W.; Crampton, G.; and Gray, W. *Science Stories.* Chicago: Scott, Foresman & Co., 1933.

Boulding, K. "The Next 100 Years?" *Science* 209 (1980):19.

Cleaver, T. J. and Hickman, F.. "Energy and Society: Investigations in Decision-Making." *Biological Sciences Curriculum Study*. Chicago: Hubbard Scientific Company, 1977.

David, E., Jr. "Industrial Research in America: Challenge of a New Synthesis." *Science* 209 (1980):133–139.

Davis, B. D. "Frontiers of the Biological Sciences." *Science* 209 (1980):78–89.

Häfele, W. "A Global and Long-Range Picture of Energy Developments." *Science* 209 (1980):174–182.

Hopkins, L. *The Spirit of the New Education*. Boston: Lee and Shepard, 1892.

Ingelfinger, F. "Medicine." *Science* 209 (1980):123–126.

Joyce, R. "Frontiers in Chemistry." *Science* 209 (1980):89–95.

Kevles, D.; Sturchio, J.; and Carroli, P. "The Sciences in America, Circa 1880." *Science* 209 (1980):27–32.

Kohlstedt, S., "Science: The Struggle for Survival, 1880 to 1894." *Science* 209 (1980):33–42.

Mauldin, W. "Population Trends and Prospects." *Science* 209 (1980):148–157.

Mosteller, F. "The Next 100 Years of Science." *Science* 209 (1980):21.

National Science Board Commission on Precollege Education in Mathematics, Science and Technology, National Science Foundation. *Educating Americans for the 21st Century: A Report to the American People and the National Science Board*. Washington, D.C.: U.S. Government Printing Office, 1983.

National Science Foundation. *Annual Science and Technology Report to the Congress, 1980*. Washington, D.C.: U.S. Government Printing Office, 1981.

National Science Foundation. *Scientists, Engineers and Technicians in Private Industry: 1978–80*. Special Report No. NSF 80–320. Washington, D.C.: U.S. Government Printing Office, 1980.

Oettinger, A. "Information Resources: Knowledge and Power in the 21st Century." *Science* 209 (1980):191–198.

Ponzio, R.; Bowyer, J.; and Peterson, R. "Science Education: New Target Population." In *The Future of Education: Policy Issues and Challenges*, edited by K. Cirincione-Coles. Beverly Hills, CA: Sage Publications, 1981.

Revelle, R. "Energy Dilemmas in Asia: The Needs for Research and Development." *Science* 209 (1980):164–174.

Rubin, V. "Stars, Galaxies, Cosmos: The Past Decade, The Next Decade." *Science* 209 (1980):64–71.

Scott, C. *Nature Study and the Child*. Boston: D. C. Heath, 1900.

Simon, H. "The Behavioral and Social Sciences." *Science* 209 (1980):72–78.

Sokal, M. "Science and James McKeen Cattell, 1894 to 1945." *Science* 209 (1980):43–52.

Underhill, O. *The Origins and Development of Elementary School Science*. Chicago: Scott, Foresman & Co., 1941.

U.S. Department of Commerce, Bureau of the Census. *Historical Statistics of the United States: Colonial Times to 1970.* Bicentennial Edition. Washington, D.C.: U.S. Government Printing Office, 1976.

U.S. Office of Education. *Circular of Information,* No. 4, 1884.

Walsh, J. "Science in Transition, 1946 to 1962." *Science* 209 (1980):52–57.

Wetherill, G., and Drake, D. "The Earth and Planetary Sciences." *Science* 209 (1980):96–104.

Wolfle, D. "Science: A Memoir of the 1960's and 1970's." *Science* 209 (1980):57–60.

Wortman, S. "World Food and Nutrition: The Scientific and Technological Base." *Science* 209 (1980):157–164.

2

Setting Goals and Objectives

The purpose of chapter 2 is to provide guidance for setting goals and objectives to teach science. You will find a rationale or reason for setting goals and objectives, along with several suggestions for setting your own goals or priorities and writing objectives. Finally, you will find an illustrative set of goals and objectives for science education in the 1980s; they were developed and used to write the science activities in chapters 10, 12, 14, and 16 of this book.

KEY CONCEPTS AND IMPORTANT IDEAS

☐ rationale for goals and objectives ☐ how to write goals and objectives ☐ behavioral objectives ☐ conceptual knowledge ☐ identity ☐ organization ☐ interaction ☐ change ☐ conservation ☐ limitations ☐ learning processes ☐ psychomotor skills ☐ attitudes and values ☐ decision-making processes

WHY WRITE GOALS AND OBJECTIVES?

Most of us who have in our heads an ideal image about teaching which includes ourselves and a group of students vitally interested in what we are doing. In our image everyone is probably happy; their needs and ours are being met. Close your eyes for a moment and think about the image in your head. . . . Back to reality! There is very little chance of having things turn out the way we have them pictured unless we decide in advance what we want to achieve in a science lesson, activity, or program whether it is for the next hour, week, or year.

The importance of setting goals and objectives is to provide a target for our actions. Once we decide what we want to accomplish, it is easier to develop plans to reach those goals and objectives. Having a clear picture of our goals and objectives is like knowing ahead of time where we want to go before we look at a map.

A second reason for setting goals and objectives is that it makes evaluation easier.

In the mind's eye of every teacher is an image of self and a classroom full of students.

Without goals and objectives, it is difficult to measure or evaluate what students have learned or how successful we were as a teacher. Do you remember how angry you were the first time that you studied for a test, thought that you knew what was important to know about the subject, and then found that the test dealt with material you hadn't studied?

The connection between goals, objectives, and evaluation is important because it not only helps you know where you are going, but it helps students know what you expect of them; and it helps parents, your colleagues, and the administration know what you want to accomplish.

TIPS FOR BEGINNING TEACHERS

It's quite helpful to have in written form your goals for a science program and written objectives for science lessons. There are many uses:

☐ They are available when you are absent and someone must teach your class.

☐ They can be handed to a visitor, parent, or the principal who comes to observe your science class.

☐ They can be sent home in a letter to parents when you need support, assistance, or materials.

☐ They can be used when you are making a presentation at open house or at a parent-teachers meeting.

HOW DO YOU SET GOALS AND OBJECTIVES?

Setting goals and objectives for teaching science is like making plans for a vacation—there are many decisions to make. Some decisions are general and are thought of as **goals.** For example, you may wish to decide if you'd like to spend your vacation exploring a new place or if you'd rather return to a favorite spot. Your decision will probably be based on a general feeling about whether you want to meet new people, try new experiences, or relax in familiar surroundings, perhaps with old friends. Once you have chosen your destination, other decisions about your vacation will be as specific as planning what to do on a particular day. Do you want to go for a swim or walk, write postcards, or stretch out and take a nap? Decisions about such specific or short-term outcomes involve **objectives.**

As you begin to think about what you want to teach in science (or how you will teach science if the text or kit has been chosen by someone else), undoubtedly you will have general or long-range outcomes you would like to achieve. Think of them as *goals.* Goals are usually chosen and written at the time decisions are being made about science programs for the year. Goals answer the question, "What major outcomes do I/we want the students to accomplish this year in science?" Several examples of goals for science programs are listed in figure 2–1. You might think about your own goals for teaching science and list them in the spaces provided for your goals.

Many experienced teachers decide on their own science goals for the year; then they ask various groups and individuals in the school community what are their goals, priorities, and expectations for science during the coming year. Teachers may ask their students, parents, other teachers, the school principal, or all of these people. Experienced teachers do not expect to achieve everyone's goals, but they do evaluate the goals that are important to others. Within the context of their own goals, they usually attempt to include some of

GOAL 1 To have students gain new knowledge in the life and physical sciences

GOAL 2 To teach students to think in new ways

GOAL 3 To help students acquire new skills in science that require manual and perceptual dexterity and precision

GOAL 4 To enable students to consider attitudes and values in science other than their own

GOAL 5 To develop students' confidence in applying science to everyday situations and in making decisions about issues related to science

YOUR GOALS

GOAL 1 _____
_____
_____

GOAL 2 _____
_____
_____

GOAL 3 _____
_____
_____

FIGURE 2–1
Example goals for science programs.

those suggested by others, especially if the goals seem to be in the best interest of the students.

It is important to remember that there may be differences in the expectations of others or conflicts among the suggested goals. For example, one group of parents may say their goal is to have their children gain new knowledge in science, and they may want them to score well on the state science achievement test. For another group of parents, the most important outcome of the year may be to have their children learn to enjoy science; these parents believe that the way to accomplish this end is to have their children take many field trips. Conflicts between goals or confusion between goals and objectives (illustrated in the second example above) sometimes can be resolved by allocating a reasonable amount of class time toward each of the goals that are competing or in conflict.

Another way to resolve such conflicts is to allow different groups to pursue different science activities during the day. Parents who may not want their children exposed to controversial issues in science may prefer to have their children read a noncontroversial science book. Other parents might like the idea of having their children discuss controversial issues in an attempt to understand contradictory points of view. For this latter group of students you might assign a different book to read or you may guide a discussion of such issues when the former group is in the library.

There is a second point worth remembering about goals: people (including yourself, students, other teachers, parents, or the administration) frequently forget goals or change goals without realizing it. If you put your goals in writing, you can reevaluate them from time to time or you can use them to remind others what you are trying to achieve as you teach science.

TIPS FOR MORE ADVANCED TEACHERS

Ask your students, their parents, or other teachers and the school principal, what they think the goals for science ought to be and ask them to explain their choices.

Be ready to tell them what your goals for teaching science are, and be prepared to explain the reasons for your choices.

Make a list of these various goals and try to choose five or six that seem to have the greatest merit. Then attempt to rank them in terms of their importance.

After you have decided upon your goals for teaching science, or inherited them with the position, it is time to consider objectives that will help you reach your goals. Objectives are more specific than goals and generally they are more numerous. Objectives suggest what outcomes students will accomplish that will lead to specific goals. Examine the example which shows how three objectives might lead to the achievement of one goal.

GOAL: To help students acquire new skills in science that require manual and perceptual dexterity and precision.

Objective 1

Teach students to coordinate their senses by showing them how to make small adjustments on tools and equipment: thermometer, eyedropper, simple balance, and so forth.

Objective 2

Teach students to control and handle a variety of nontoxic substances and to discriminate between toxic and nontoxic substances.

Objective 3

Teach students to handle and care for plants and animals in the classroom with supervision.

There are three important points to remember about writing objectives. First, objectives are usually written for specific lessons or groups of lessons. A single objective may appear only once in a single lesson or it may be repeated in several lessons throughout the week, month, or year. The three objectives in the example above would probably reappear in various lessons in the science program throughout the year.

A second point to remember is that most lessons have several objectives. For example, if you were going to have students explore bulbs and batteries, your objectives might be:

☐ To have students enjoy the science activity
☐ To improve students' manual dexterity
☐ To have students gain new factual knowledge

A final point to remember is that objectives can be written to accommodate various levels of specificity. They can be written so that objectives are only somewhat specific (like those for the physical science activity above) or written so that the objective is very specific. The more specific an objective is, the more easily it can be used to evaluate student performance. The following example illustrates two levels of specificity for the same objective.

Objective (version A)
To have students learn how to classify living things.

Objective (version B)
Given a list of animals, students will be able to sort them into three groups—birds, mammals, and insects— with 80 percent accuracy.

Version A is less specific than version B. Although version A specifies what skill is expected from students, only version B specifies how the ability to classify is measured and what level of performance is expected from students at the end of the science activity. Version B is called a **behavioral** (or **performance**) **objective** because an outside observer can measure how well students have achieved the behavior or performance expected. Behavioral objectives not only describe what knowledge or skill is to be learned but they also specify how much knowledge or skill students are expected to acquire. Chart 2–1 illustrates how to write behavioral objectives.

How specific should your objectives be? The answer to this question generally depends upon the purpose of the objective. Version A objectives are somewhat easier to use in conversations with parents who are interested in the kinds of science activities their children are experiencing. Version B objectives are more easily understood by other teachers and administrators. Some teachers say they prefer to write their objectives without too much specificity while they are planning a science unit. Later, when evaluation becomes important, they convert their (version A) objectives into behavioral (version B) objectives. Other teachers say they "think in terms of specific behaviors" from the start and believe that they teach with greater precision when they write their behavioral objectives as they plan.

If you decide to use behavioral objectives to guide your teaching, find out what materials on the subject are already available. Many school districts have developed and published sets of behavioral objectives for every subject in the school curriculum. If your district does not have such documents, try the county curriculum library or commercially published science programs.

CHART 2–1
Components for Writing Behavioral Objectives

	Stimulus	Action	Object or Product	Performance Level
	Describe the activity or situation	Specify who exhibits what behavior	Describe the object acted upon	Specify the level of accuracy
Example 1	Following a field trip to the zoo . . .	children will be able to draw . . .	a picture of an animal they saw . . .	well enough that most peers can guess its identity.
Example 2	After reading about electricity in the textbook . . .	students will be able to answer in writing . . .	the questions at the end of the chapter . . .	with 80 percent accuracy in content only.

HOW DO YOU ORGANIZE GOALS AND OBJECTIVES INTO A SCIENCE PROGRAM?

Most science programs—whether they consist of a series of textbooks, workbooks, or materials in a kit—will have identified a set of goals for the program and specific objectives for each individual lesson or for groups of lessons. If you browse through one or two of these you will discover how their goals and objectives are organized. One of the best known models for establishing educational objectives is one developed by Benjamin Bloom, Thomas Hastings, and George Madaus; it is described in their book, *Handbook on Formative and Summative Evaluation of Student Learning,* which is listed in the Bibliography.

In the remainder of this chapter, we provide a model with many practical examples of goals and objectives for a science program. Our model contains a cohesive set of goals and objectives that were developed to illustrate the theme, Science and Society. We propose them as worthwhile goals and objectives

for science education in the 1980s. Before deciding which goals and objectives to use, we discussed our ideas about teaching science and what we thought was most important to teach elementary and middle or junior high school students. Out of that discussion came chapter 1 which represents a *rationale* for choosing the theme Science and Society. Our views may change as science, society, and education change during the next decade, but establishing a theme was the first step before deciding upon the goals and objectives for the science activities in part two of the book. (You may want to reread "The Theme: Science and Society," in chapter 1.)

From the theme, we developed five goals that represent desirable, long-range outcomes for students who live in our science-oriented society. The goals for *Science and Society* are summarized in figure 2–2.

For each of the five goals, several objectives were developed for each of the three grade levels: Primary (kindergarten through grade two), Intermediate (grades three through five), and Middle (grades six through eight). These objectives appear in five tables,

> GOAL 1 To enable students to gain new conceptual knowledge in the life and physical sciences.
>
> GOAL 2 To enable students to acquire or use a variety of learning processes in science.
>
> GOAL 3 To enable students to develop new psychomotor skills in science which require manual dexterity and perceptual precision.
>
> GOAL 4 To enable students to consider attitudes and values in science other than their own.
>
> GOAL 5 To enable students to gain skill and confidence in applying science to everyday situations and in making decisions related to science.

FIGURE 2–2
Goals for the 1980s with a theme: Science and Society

one for each goal. For each table, a brief rationale for the choice of objectives is provided.

Rationale for Selection of Conceptual Knowledge Objectives

Earlier you learned that the six concepts—*identity, organization, interaction, change, conservation,* and *limitations*—are important to all fields of science. But what is the educational significance of these concepts in science? If students are to look at themselves in a broader context—as individuals who are important parts of a total environmental system—then they need to have experiences that make that knowledge a reality to them. Table 2–1 outlines the objectives for conceptual knowledge; at the end of the program, students will understand these six concepts.

Identity is an important idea for the future because it focuses upon the existence and significance of living and nonliving things in the environment. Students recognize what constitutes their own identities. They also recognize the identities of their family members, friends, pets, and favorite possessions and they value them. Identity is a beginning point of reference to the real world; thus it is a beginning point of learning about the identity and value of less familiar things in the world. In a large and technological society, it is easy to lose sight of the worth of the individual. When decisions are made, they are often made on behalf of minors or for the benefit of the greatest number of individuals in society. When such decisions are contrary to individual preference, a part of one's self-worth is unexpressed. The concept of identity, therefore, reinforces the self-worth of the individual as the conceptual context broadens.

The concept of *limits* is an equally important idea to understand. It is at the opposite end of the imaginary scale from the identity of individuals or things. The identity of the individual is best understood and valued when its limits are understood. It makes little difference to people who learn that matter and energy are conserved in the environment or in the universe, unless that understanding includes them as individuals within the system. Their identity within that system is crucial to them.

Therefore, children, adolescents, and adults need to understand that objects and individuals, however organized in the environmental system, have limits to their identities, and that people or things can only be changed up to a point before they lose their identities. Developmental experiences need to be made a part of a science curriculum so that this concept is presented at early levels. And it needs to be expanded at advanced levels as learners gain more experience with environmental groups and systems.

TABLE 2–1
Conceptual Knowledge Objectives

Grade Level	Identity and Organization	Interaction and Change	Conservation and Limitations
Primary (grades K–2)	Individual things (living and nonliving matter) have identities based on the way they are put together (the structural and spatial organization)	All things (living and nonliving matter) change because they interact with other things (matter)	Things (living and nonliving matter) can change and still keep (conserve) their identities There are limits to how much things can change and still keep their identities When things are changed beyond their limits, the stuff they are made of (matter and energy) is still there (conserved) but their identities are lost
Intermediate (grades 3–5)	Groups (of living and nonliving matter) have identities based on the way they are put together (the structural or spacial organization of the members)	All groups change: members of groups interact with each other, with other groups, or with the environment (living and nonliving matter)	Groups can change and still conserve their identities There are limits to the capacity of groups to change and still conserve their identities When groups are forced to change beyond their limits, matter and energy are conserved but groups lose their identities
Middle (grades 6–8)	Environments and systems have identities based on the way they are put together (the structural and spatial organization of matter and energy)	All environments and systems change because living and nonliving matter interact	Environments and systems can change and still conserve their identities There are limits to the capacity of environments and systems to change and still conserve their identities When the capacity to change is exceeded, environments and systems lose their identities even though matter and energy are conserved

The concept of *identity* is the beginning point of all reference to the world as we know it: knowledge of self, familiar faces and objects, and ultimately the unknown. (Rohn Engh)

The concept of *limitation* grows out of our experience with self, others and the environment. As we discover the limits of our own existence or capabilities (physical and emotional) and the limits of those people and things around us, we recognize that other living and nonliving things in the environment have limits beyond which they cannot change, or they will cease to exist as we know them. (© Leonard Lee Rue III/Tom Stack & Associates)

Between the two extremes on the imaginary scale are the other supporting ideas or concepts: organization, interaction, change, and conservation. *Organization* is not a new concept in science education. The way scientists organize plants, animals, rocks, and so forth into groups has not only been a major theme in science education, but often has become an end in itself. Likewise, the internal organization or structure of living and nonliving matter has been a major science theme at the high school and university levels. As science has been taught in the past, both views of organization have focused students' attention on a specific specimen (like the frog) or group (like rocks and minerals) and have not related specific content to the broader picture. If students are to capture the essence of their relationship to the environment, they need ex-

periences that will help them develop the ability to change frames of reference from the parts to the whole—from self or individual things to the environment or the larger system.

Interaction and *change* are two concepts that also have been a part of science programs in the past. These concepts have helped elementary and junior high school students focus on and describe observed changes, and have helped them understand the importance of cause-and-effect events.

Linking the concepts interaction and change with *conservation* opens new possibil-

The concept of *organization* underlies how we view the world. Patterns in nature, the structure of living and nonliving matter, and the natural occurrence or groupings of things lead scientists and nonscientists to appreciate nature's organization. (Smithsonian Institution)

ities for learning. The permanence of objects can be understood in a new way. When students interact with and change things, making either reversible or irreversible changes, they experience limits of change; that is, they see the points at which things lose their identities. The power behind this chain of concepts can apply to them as individuals, as well as to the environment.

To summarize, six ideas or concepts presented here offer scientific and educational promise for the future. In science, these concepts span the scale from the identity of the individual or single object to the conservation of matter and energy in the universe. That is not to imply that we teach all the sciences between these extremes, but rather these partic-

ular concepts have the advantage of enlarging the students' focus from the self or single object to the total environment or system. Students become aware of the impact of their actions on others and the environment; they become aware of the limits to change. Moreover, from an educational standpoint these concepts can be meaningful to children since the concepts pertain to immediate experience with familiar objects and events in the environment. And this approach progresses from a concrete level to an abstract level that is meaningful to adolescents and adults as well.

The six concepts presented in table 2–1, Conceptual Knowledge Objectives, are introduced in a developmental sequence thought to approximate the students' capacity to understand the concept at each grade level.

Rationale for Selection of Learning Processes Objectives

The question of how conceptual knowledge in science is learned can be addressed by describing how the student *processes* the information he or she receives. Learning process approaches describe what we think goes on inside the student's head; they tend to vary in complexity. During the past two decades, researchers have attempted to find out what basic learning processes take place, when they take place, and how many people use or integrate these processes.

Ten learning processes are discussed; as you read through them, try to visualize how they occur and what people do when they receive and process information.

1. *Observing* usually includes using any one or a combination of the senses.
2. *Comparing* begins with real objects or things when children are young but gradually develops to include comparison of ideas or concepts as students mature.

The concept of *interaction* is the basis for understanding how the world works. Raindrops hitting a windowpane, wind blowing through a forest of fir trees, and a frog capturing an insect for lunch—all remind us that nature is composed of many parts which act upon one another. (A. Cosmos Blank for Photo Researchers)

3. *Measuring* refers to quantitative descriptions, made either by the learner or by a piece of equipment.

4. *Classifying* or forming groups of real objects for children becomes a more advanced process when it includes classifying ideas.

5. *Gathering* and *organizing information* may be in verbal (oral or written) or pictorial form.

6. *Constructing* and *interpreting graphs* generally includes numerical representation.

7. *Inferring* means suggesting more about a situation or condition than can be observed.

8. *Predicting* usually consists of making estimates or reasonable guesses based on observations.

9. *Forming hypotheses* means that one makes a statement of conditionality: if . . . , then. . . .

10. *Describing relationships among variables* occurs when one makes clear how a condition or event is similar to or different from other conditions or events.

By the time children have completed the primary grades (kindergarten through second grade) most are able to demonstrate the first four learning processes. This is a result of a wide range of experiences involving the description of things in their environment.

The next four learning processes can be developed with guidance during the intermediate grades as children gain further experience in describing environmental patterns. (Remember that a child can describe the environment or experience it depending upon whether he/she acts as a spectator or participant.) The last two learning processes are most associated with abstract thought or reasoning. With guidance these can begin to de-

The concept of *change* guides our expectations of nature and our search for more information. The way we feel about reversible (temporary) and irreversible (permanent) changes affects our responses to ourselves, those around us, and the environment. (Paul Conklin)

velop during the middle grades (grades six through eight), as children and adolescents try to explain environmental patterns.

What is the significance of these learning processes to education? You can see immediately that they have wide applicability to areas outside of science, and you may be interested in reading about these and other learning processes described in science education. Because these processes lead developmentally to learning, it becomes important for teachers and others to understand their re-

sponsibility to help students develop these processes. And what is the significance of learning processes to science? They are, of course, the main processes by which problems are solved and new knowledge is produced. Since these processes are generic, they lead to the development of still more advanced skills used by scientists.

An important new function of the learning processes presented here relates to decision-making processes. Because scientists are trained to use these learning processes, they

The concept of *conservation* helps us to understand the permanence of matter and energy in nature. Things around us—the sun, lakes, vegetation, animal life—may change in appearance or even seem to disappear; but matter and energy are not destroyed. They change from one form to another. (© C. Quinlan)

are able to apply them easily whenever they need them. In the section on decision-making processes you will see why it is important for

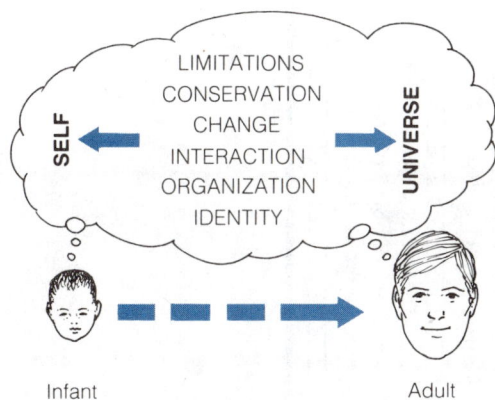

As students mature, their awareness of the universe expands, and concepts are understood at a broader, more abstract level.

others in society to acquire skill in the learning processes described in this section.

Table 2–2, Learning Processes Objectives, summarizes these ten learning processes that are also used for the activities in part two of this book. They are organized in a sequence that is thought to approximate students' capabilities through the grade levels.

Rationale for Selection of Psychomotor Skills Objectives

When people explore the environment by coordinating muscular activity with their intellect and senses in order to learn about the environment, the resulting process is called **psychomotor activity.** Reaching out to touch a dewdrop, walking closer to the edge of a cliff to see what lies below, leaning down to smell a rose—all of these are examples of psychomotor activity.

TABLE 2–2
Learning Processes Objectives

Grade Level	Observe	Compare	Measure	Classify	Gather/Organize Information	Construct/Interpret Graphs	Infer	Predict	Form Hypotheses	Separate Variables
Primary (grades K–2)	. . . Using real objects and events in the environment									
Intermediate (grades 3–5)	. . . Using real objects and verbal descriptions of objects and events in the environment									
Middle (grades 6–8)	. . . Using real objects and verbal descriptions of objects, events, and patterns in the environment									

What is the significance of this kind of activity to science? Consider the work of many scientists: they use tools and equipment; handle substances, plants, or animals; and construct models. Not all scientists work in laboratories, field stations, or observatories, of course, but many do. Their activities involve some of the basic skills that are essential to scientific research.

Is it necessary for students to have these skills since very few will actually become scientists? Psychomotor skills are an important way in which new knowledge is gained. It is not simply a question of having students be-

have as scientists in order to appreciate how new knowledge is produced. Rather, the conceptual knowledge students gain through the combination of psychomotor activity and learning processes such as observing, comparing, and measuring is qualitatively different from knowledge gained through secondhand sources such as reading and discussion. Most reseachers agree that children require some psychomotor activity before they can understand abstract concepts. Young children need to pour liquid from one container to another before they understand that the amount of liquid poured remains unchanged regardless of

The knowledge that students gain through the combination of psychomotor activity and learning processes such as observing and comparing is qualitatively different from knowledge gained through secondhand sources such as reading and discussion. (© C. Quinlan)

John Wells

Benjamin S. Bloom (b. 1913)

Expectations play a powerful role in classrooms. And, as it turns out, teachers commonly expect about one-third of their students to do very well, mastering everything taught; they expect about one-third to do moderately well, mastering some of the material; and finally, they expect one-third of the students to fail, mastering little material. These expectations, and the actual results in classrooms, are important flaws in our educational system.

According to Benjamin Bloom, the expectations just described represent a wasteful and destructive aspect of education. It reduces the aspirations of both teachers and students, it lowers the motivational level of students, it confines the aims of teachers, it falls short of students' potentials, and it slowly and systematically diminishes the self-concept of many students. Benjamin Bloom maintains that most students are capable of learning everything that is taught. If teachers adapted to the needs of students, provided more time for learning, and implemented some simple methodological changes, teachers would be more effective in the task of teaching. The basic solution to many educational problems lies not in more funds and new techniques, but in changing our views of students and their capacity to learn. This is

the size or shape of the container into which they pour it. The name for this abstract concept is **conservation of volume.** Providing children with a simple verbal definition would not ensure their understanding the concept, even if you were able to induce them to memorize the words and repeat the definition. For decades teachers have presumed that something has been learned when students are able to repeat key phrases or give definitions on cue. It is only when teachers discover that their students cannot seem to apply what they have learned to solve problems that teachers question how much their students have really learned. Students who are unable to explain a concept in their own words after reading about it, memorizing it, or hearing the class discuss it, may need to have "hands-on" or psychomotor experience.

One of the possible benefits of psychomotor activity in the classroom is its effect on students who have difficulty reading. Poor read-the simple, and radical, conclusion of Benjamin Bloom's educational work.

Benjamin Bloom was born in Lansford, Pennsylvania in 1913. His B.A. and M.S. degrees are from Pennsylvania State University. His Ph.D. was granted by the University of Chicago in 1942, where he has taught for over forty years. Since 1970, he has been the Charles H. Swift Distinguished Service Professor of Education. Recently, he also held the position of professor of education at Northwestern University.

Over his years of teaching Bloom has also investigated the ways students are educated. His work has included being a consultant on evaluation and curriculum to projects in the United States and many other nations around the world. He is one of the founders of the International Association for the Evaluation of Educational Achievement. He is also past president of the American Educational Research Association (AERA). Bloom's work in the 1960s included studies of the effects of home and school environment on learning achievement of children. His studies contributed directly to the development of Project Headstart.

Perhaps the most important concept Bloom has contributed to education is mastery learning. Mastery learning is based on the idea that failure is not a necessary outcome of education. Student learning is the central goal, and this goal is within the capabilities of students. With fairly simple modifications of instructional time and sequence, and more diagnostic use of evaluation, greater numbers of students can master the concepts taught. There is an added benefit: students have increased interests and better attitudes about themselves, others, and school.

Benjamin Bloom is one of our most notable educational scholars and writers. He has authored or coauthored numerous articles, book chapters, and several major books, including *Taxonomy of Educational Objectives: Cognitive Domain, Taxonomy of Educational Objectives: Affective Domain, Stability and Change in Human Characteristics, Human Characteristics and School Learning,* and *All Our Children Learning.*

ers are handicapped in classrooms where science is taught by textbooks; but when they are given the opportunity to engage in psychomotor activity in science, they develop greater feelings of satisfaction with their success.

At every grade level you will find a wide range of differences in students' needs and preferences for psychomotor activity. And although it is not yet clear from research how much and when students should have psychomotor activity to accompany the learning processes, it is clear that many students need the opportunity to learn abstract concepts in this way. Thus, the implication for teachers is that such opportunities need to be available. Part two of this book will help you to structure those opportunities.

The psychomotor skills summarized in table 2–3 are commonly associated with science activity. They are presented in a sequence that is thought to be developmentally appropriate; that is, progressing chronologically as students' sensory and muscular coordination become more refined and focused. To develop psychomotor skills, students use tools and equipment, handle substances, handle plants and animals, and construct projects and models. Looking across table 2–3 at any grade level, you will find variations in skills directed toward different things in the environment (tools, animals, and so on). Looking down the table, you will see how the refinements of sensory perception and muscular

coordination combine with the intellect to produce advanced psychomotor skills.

Rationale for Selection of Attitude and Value Objectives

When scientists are asked what should be taught in science, their replies are usually immediate and confident: teach the spirit of science; help people to understand why we ask the questions; capture the sense of excitement that comes from having the courage to

TABLE 2–3
Psychomotor Skills Objectives for Learners

Grade Level	Use Tools/ Equipment	Handle Substances	Handle Plants/ Animals	Construct Projects/ Models
Primary (grades K–2)	Use and care for simple tools/ equipment that require no adjustments: ruler, hammer, container, brush, etc.	Control and handle nontoxic substances: have them pour water, ladle sand, divide animal food, etc.	Touch and care for plants/animals while being supervised	Construct simple projects with guidance: have them make a coffee can house for animals, create a rock display, etc.
Intermediate (grades 3–5)	Use and care for tools/equipment that require small adjustments or coordination of senses: simple balance, eyedropper, thermometer, etc.	Control and handle variety of nontoxic substances: help them discriminate between toxic and nontoxic substances	Handle and care for plants/animals in the classroom without close supervision	Construct simple projects/models without close guidance: have them construct material for an experiment, construct shelter from raw material, etc.
Middle (grades 6–8)	Use and care for tools/equipment that require fine adjustment and discrimination: microscope, camera, etc.	Control and handle, toxic and nontoxic substances in the classroom: help them handle solvents, purifiers, sprays, etc.	Handle and care for plants and animals in new situations: on field trips, in other classrooms, etc.	Construct complex projects/models with some guidance: build cages, runways, connect electrical circuits, etc.

ask the question and find the answer! Carl Sagan put it this way:

> Science is a way of thinking much more than it is a body of knowledge. Its goal is to find out how the world works, to seek what regularities there may be, to penetrate to the connections of things—from subnuclear particles, which may be the constituents of all matter, to living organisms, the human community, and thence to the cosmos as a whole. . . . We are an intelligent species and the use of our intelligence quite properly gives us pleasure. In this respect the brain is like a muscle. When we think well, we feel good. Understanding is a kind of ecstasy (Carl Sagan, *Broca's Brain* [New York: Random House, 1974], pp. 13–14).

For over half a century, those concerned with teaching science and training science teachers have attempted to capture this spirit of science. But like any spirit, it is vibrant only when it comes from an individual. Bottled up, it loses its vitality and words like curiosity, creativity and objectivity can lose their meaning. These qualities are referred to as **attitudes.** Many teachers have found a way to communicate that spirit of science. For example, the science programs that were funded by the National Science Foundation during the 1960s and 1970s were able to capture and transmit these attitudes in two ways. Scientists worked with teachers in classrooms until the spark of inquiry was "caught" by the teacher and students. Colorful and exciting films captured students and teachers sharing this sense of wonder and discovery.

Although it is not clear just how these qualities, called attitudes, are engendered in people, the scientific significance of qualities like curiosity, creativity, and objectivity is clear. They are ways to solve the problems that will face science and society in the future, and they are ways of looking at the world that have led to our present civilization.

If curiosity is exploring our surroundings and asking questions about them, then you will find ways to encourage students to be curious. If you think of creativity as putting things and ideas together in new ways, then you will be able to encourage your students to be creative. And, if by objectivity you imagine people seeking, listening, and comparing alternative points of view, then you will find ways to help students become objective.

When *you* value these attitudes, you will find that your students value them, too. People generally adopt some of the attitudes of those they like and admire. Because of this fact, you will influence your students' attitudes whether you are conscious of it or not.

The feelings that people have are related to the **values** they hold, values that affect the way they react to information about science, music, art, or mathematics on television and radio or in newspapers. Values in science determine the degree to which people care about the environment, and the way people vote or fail to vote on science-related issues.

All of us are part of an environmental system in which each person's actions affect others and the environment. But just knowing that is not enough; concern is needed, and concern involves values.

Which values should be taught? That question cannot be answered in a textbook. It depends upon whether your values are similar to those held by the school community in which you teach, and whether you are supported for teaching a particular set of values. Thus, you need to recognize not only your own values related to science, but the values of those around you as well. Then you will have to decide for yourself what values to teach.

The number and diversity of values to consider as relevant to the classroom is enormous. Predominant American values that appear to have some relation to science are (1)

TABLE 2–4
Attitudes and Values Objectives

Grade Level	Attitudes			Values		
	Curiosity	Creativity	Objectivity	Responsibility for Self/Others	Responsibility for Living Things	Responsibility for Environment
Primary (grades K–2)	Learner actively explores environment and asks questions	Learner combines ideas, objects, and materials in new ways	Learner listens to alternative points of view	Learner accepts direction in caring for health and welfare of self and others in class	Learner accepts guidance in caring for plants and animals	Learner accepts assistance in learning how to use natural resources wisely
Intermediate (grades 3–5)	Learner actively explores environment and asks questions	Learner combines ideas, objects, and materials in new ways	Learner attempts to compare alternative points of view with evidence	Learner assumes responsibility for contributing to health and welfare of self and others	Learner assumes responsibility for providing basic needs to designated plants and animals at school	Learner assumes responsibility for wise use of natural resources in environment
Middle (grades 6–8)	Learner actively explores environment and asks questions	Learner combines ideas, objects, and materials in new ways	Learner seeks alternative points of view and multiple sources of evidence	Learner acts as model for others by protecting health and welfare of self and others at school	Learner acts as model for others by providing basic needs to plants and animals at school	Learner demonstrates to others the need and means to use natural resources wisely

concern for self and other people, (2) concern for other animals and plants, and (3) concern for the physical environment. This list could be expanded but it is a good place to start. Because people quite often think of themselves first and are socially reinforced for doing so, it becomes easy to distance concerns about the rest of the world, the life it supports, and the consequences of our actions. Recall that the conceptual knowledge proposed earlier was based upon the need to refocus attention on the consequences of actions and the values from which they originate. Thus, thoughtful consideration of one's own values and those of others involved in decision making and the sharing of resources will become more important in the future.

As you think about your own future and your students' futures, and you realize that the quality of life on this planet will be affected by the actions of billions of people, it suddenly becomes exceedingly important to become aware of values. In fact, that is what is different about the attitude and value objectives proposed here. Table 2–4, Attitudes and Values Objectives, summarizes the attitudes and values students should be able to demonstrate or accept in others. But the real intent here is to show the need to identify your own values and to help your students to become aware of their values and those of the people around them.

Rationale for Selection of Decision-Making Process Objectives

Once you and your students become familiar with knowledge in science that will affect your future, and you recognize that the public needs to develop a sense of responsibility and concern for the environment, then you will discover that the future demands more than these if individuals are to act on their knowledge and feelings of responsibility. Individuals of all ages will be faced with decisions about their own welfare, that of others, and the environment that supports their life. If these decisions are to be intelligent, then some guidance and practice in decision making seems essential. Guidance is not telling people what to decide; it is helping them to understand how to make decisions.

Decision making appears to be a complex process that is composed of several skills or processes. Table 2–5 summarizes the processes at three grade levels. The fact that most of us have said at one time or another, "I can't make up my mind," illustrates the widespread nature of decision-making difficulties. Difficulties in deciding may come from a lack of information, a conflict between facts and values, or from an inability to see alternatives. It may stem from being faced with two equally attractive or unattractive alternatives. Just knowing why the decision is difficult often helps us to make the decision. That is why understanding the thought processes associated with decision making can be helpful. Knowing the process will not make difficult decisions easy, however; and the responsibility for difficult decisions and their consequences remains. But understanding why a decision is difficult can help to clarify issues and responsibilities.

Each process described here clarifies decision making. Each process is discussed in relation to assumptions about the developmental capabilities of students. However, empirical evidence to support the discussion is not available at this time. Try to avoid thinking of the processes as steps. Otherwise they will become ritualized just as the scientific method became ritualized years ago, and the continuous, fluid movement back and forth between

TABLE 2–5
Decision-Making Process Objectives

Grade Level	Describe Decision to Be Made	Identify/ Gather Information	Convert Information into Alternatives	Clarify Consequences of Alternatives	Select Among Alternatives
Primary (grades K–2)	Recognize personal decisions that involve self	Become aware of information needed	Identify alternatives with help from teacher	Discuss immediate consequences of various alternatives	Select alternative and accept consequences
Intermediate (grades 3–5)	Describe decisions that involve self and others	Gather some information needed and become aware of differences between facts and attitudes or values	Form alternatives which reflect facts, and attitudes or values of self & others	Describe consequences of various alternatives to self and others	Rank alternatives, select one, and accept consequences
Middle (grades 6–8)	Describe situations that require decisions involving self, others, and environment	Collect and differentiate between facts, attitudes, and values, before making decisions that affect self, others, and environment	Form alternatives after considering facts, attitudes, and values of individuals affected by the decision	Describe immediate and long term consequences of various alternatives to self, others, and environment	Rank and select alternatives on basis of most positive effect for those concerned

processes—a movement essential for making wise decisions—would be lost.

Describing the Decision to Be Made.

This process requires that a person recognize who will be affected by the decision and what form of response is needed in the decision. A person must ask, "Will my decision affect only me, or other people, wildlife, and the environment as well?" "How am I supposed to respond: Do I need to vote, complete a questionnaire, write to a government official,

report information to someone, or just change my own behavior?"

Children aged five through seven years have difficulty understanding another person's point of view and therefore are probably unaware of the existence of alternative views. It is unlikely that children in the primary grades will be able to see that their decisions or actions relate to others or the environment. Older children and adolescents should be able to develop this skill, although not with the scope of an adult.

> Will my decisions affect only me, or other people, wildlife, the environment?

Understanding the thought processes that are associated with decision making can be helpful in making decisions.

Identifying and Gathering Information.

This process requires a person to recognize the need for different kinds of information and to be able to gather it. You may recognize this as one of the *learning processes* described earlier and see why it becomes so important in decision making. First, a person must seek factual information (about science) and learn to search for recognized sources of authority. But beyond factual knowledge, individuals have to consider the attitudes and values of others who will be affected by the decision.

Young children will have difficulty identifying and differentiating between sources of authority, will be unable to deal with conflicting information or authorities, and will be unable to appreciate the differences between fact and opinion or attitudes and values. If primary grade teachers conduct this process for children, and describe it as they do so, then children will probably be able to participate in the succeeding processes. Older children and adolescents will undoubtedly require varying amounts of guidance. In fact, many adults

prefer to have someone else, such as a news commentator, conduct this process for them. Teachers need to help students focus on the thought processes involved, with a sort of "thinking out loud" approach so that students are not simply waiting for the outcome.

Converting Information into Alternatives.

This process probably requires a person to ask, "Since (fact) is true, what should I do?" Among adults, the process of generating alternatives may occur simultaneously with the next process, *clarifying consequences*. However, since children and many adolescents will generate alternatives or choices without imagining consequences, these two processes have been separated here. A great deal of research will be needed before we understand how this thought process takes place.

Decision making lends itself well to discussions, especially when the teacher carefully structures the discussion so that every serious suggestion is treated with equal consideration. The process must be within the grasp of all students.

Clarifying Consequences.

This process requires a person to imagine the outcome of an action. A person must be able to say or think, "If I do (action), then (outcome) will happen." This process is called hypothetical-deductive reasoning. It corresponds to another learning process—*forming hypotheses*.

Research suggests that the process develops gradually and begins to appear about the ages of twelve to fifteen years, when students are conducting science experiments. It might be interesting for you to find out whether students can participate in *if-then* thinking at earlier ages, especially when the outcomes of their actions have a direct impact on themselves rather than on science objects in an experiment. Naturally, young children will need assistance with this process, but with an adult's guidance, they will be able to participate in the next, and last, process.

Selecting Among Alternatives. This process is frequently seen as the only component of decision making. It requires a person to compare two or more alternatives and make a choice between or among them. Actually, people simultaneously reject alternatives with negative consequences and select those with positive consequences. Your students will vary considerably, between and within age groups, in their abilities to select alternatives. Some individuals think impulsively while others think reflectively.

The major educational significance of the decision-making processes lies in the fact that these processes will have wide applicability to many other areas of the curriculum and to students' lives outside of school. Your students will be going through many changes during their formative years. They will be learning to make personal decisions about their appearance and behavior, their work and

Responsibility for decisions and their consequences can be difficult sometimes. Understanding why a decision is difficult can be comforting.

leisure activity, and their health. But students will need guidance and practice if they are to develop the capacity to make wise decisions that will affect their lives in significant ways. The development of this important capacity cannot be left simply to chance. The personal and environmental stakes of the future are too high.

Why is an understanding of the decision-making process significant to science? There are several reasons. First, science is no longer an isolated enterprise where scientists in white lab coats just "do" experiments and are expected to withhold their opinions about values that relate to the application of scientific discoveries. The scientific method itself led to the old view, shared by many scientists and the public, that the knowledge produced by the scientific method was value-free and that the scientist's job was to produce such knowledge.

Today scientists recognize that while the scientific method is still an effective approach for deductively solving problems and producing new knowledge the method finds its limits in solving global problems, especially those requiring decisions about human values.

Second, because future problems will be more global in nature, the decisions will require factual information from many areas of science and will have economic as well as political repercussions. Such decisions will affect the welfare of large portions of the world's population and the environment and will have longer lasting effects. Thus, the public's attitudes and values as well as their safety must be taken into account. Therefore, using the holistic approach required for many decisions in the future will mean that the deductive scientific method will become only one of many processes in decision making; it will be that process that provides factual, scientific information. Within this new frame of reference,

The central and significant idea presented in this book is that future science education programs should help people understand the impact of science upon society.

To meet this central purpose, people will need to acquire *conceptual knowledge* in science that allows them to understand their own and other's *identity* in the *organization* of all living and nonliving parts of the environment. Moreover, they need to understand how all living and nonliving things in the environment *interact* and *change* each other, and realize that there are *limits* to the capacity of all things to adapt to change. When these limits are exceeded, the identity of individuals, groups, and systems within and including the physical environment cannot be *conserved*. This is true even though matter and energy in the universe are ultimately conserved.

To acquire this conceptual knowledge, people will need experience with many *learning processes* used by scientists including *observation, comparison, measurement, classification, organization of information, construction* and *interpretation of graphs, inference, prediction, formulation of hypotheses*, and *description of relationships among variables*. These processes can be applied to verbal learning (as in textbooks and discussions), spectator learning, and participant learning. *Psychomotor skills* will be essential for learning abstract concepts but may not be required for all concepts or for all students in a given situation.

Attitudes and values will need to take on a new significance in the future. Not only do individuals need to develop concern and responsibility for living things and the physical environment, but they also need to develop an awareness of the attitudes and values of other people who are affected by decisions made in the future.

And, finally, knowledge of the *decision-making processes* will be essential if people are to act intelligently and responsibly in the future. Since scientific knowledge and the problems themselves will change, people will need to understand *how* to make decisions, not *what* to decide.

knowledge of the decision-making process is essential.

Third, it is also clear that neither scientists nor the public can afford to treat the future as a scientific experiment in which various alternatives will be tested. Many of the options for solving problems in the future will have irreversible outcomes and therefore do not lend themselves to testing. In other cases, viable alternatives will go undetected simply because they are not visible. Fostering creativity, the ability to imagine different alternatives or new combinations, will be important for your students. They will have to make decisions without having all the information they need.

A final reason why knowledge of the decision-making process is important to science is that the public, along with scientists, must take responsibility for making decisions that have serious consequences. The use of nuclear energy is just one example. The public cannot afford to assume a childlike faith that scientists will solve all of their future problems. Unlike scientists, the public is less prepared for making decisions related to science. From their traditional training in public schools, much of the public lacks experience in the more complex learning processes applied to science (i.e., inferring, predicting, forming hypotheses, and describing relationships among variables). Facts related to science and the scientific method, were emphasized and generally the social context (or the relationship of scientific knowledge to human values) was never discussed.

Therefore, to simply update public knowledge of science will not usually enable people to make intelligent, responsible, and confident decisions about future science-related issues. Most adults, adolescents, and children will need to understand the decision-making processes if they are to benefit humanity and living things in the environment.

Now that we have presented the model developed for goals and objectives for science education in the 1980s, we recognize the difficulty in keeping them all in mind at one time. The synopsis in figure 2–3 will help you see the entire set of goals and objectives in relation to the parts.

COMMENTARY

Setting goals and objectives in science can be a rewarding experience for teachers. It allows you to influence the direction of students' learning. Once you have established your own goals and objectives, or find those that you wish to adopt, you are ready to consider students' capabilities so that you can begin to plan activities to fit their capabilities. In the next chapter you will gain a better understanding of students when you see how theories about motivation, learning, and development can be applied to teaching science.

BIBLIOGRAPHY

Allison, G. T. *The Essence of Decision: Explaining the Cuban Missle Crisis.* Boston: Little, Brown, & Co., 1971.

Bloom, B. S.; Hastings J. T.; and Madaus, G. F. *Handbook on Formative and Summative Evaluation of Student Learning.* New York: McGraw-Hill, 1971.

Butts, David P. *Teaching Science in the Elementary School.* New York: The Free Press, Macmillian Publishing Co., 1973.

Esler, W. K. *Teaching Elementary Science.* Belmont, CA: Wadsworth, Inc., 1977.

Funk, J.; Okey, J.; Fiel, R., Jaus, H.; and Sprague, C. *Learning Processes in Science.* Dubuque, IA: Kendall/Hunt Publishing Co., 1979.

Gardner, P. L. "Attitudes to Science: A Review." *Studies in Science Education 2* (1975): 1–41.

Harrow, A. J. *A Taxonomy of the Psychomotor Domain.* New York: David McKay Co., 1972.

Inhelder, B., and Piaget, J. *Growth of Logical Thinking From Childhood to Adolescence.* New York: Basic Books, 1958.

Kagan, J. "Impulsive and Reflective Children." In *Learning and the Educational Process.* Edited by J. D. Krumbolz. Chicago: Rand McNally, 1964.

Karplus, R.; Karplus, E.; Formisano, M.; and Christian-Paulsen, A. "A Survey of Proportional Reasoning and Control of Variables in Seven Countries." *Journal of Research in Science Teaching* 14 (1977): 411–417.

Karplus, R., and Peterson, R. *Formal Reasoning Patterns,* San Francisco: Davidson Films, 1977.

Livermore, A. H. "The Process Approach of the AAAS Commission on Science Education." *Journal of Research in Science Teaching* 2:4 (1964): 271–282.

Lowery, L. F. *The Everyday Science Sourcebook.* Boston: Allyn & Bacon, 1978.

Murphy, J. T. "Musings on the Utility of Decision-Making Models." *Harvard Educational Review* (November 1977): 565–569.

Peterson, P. *School Politics Chicago Style.* Chicago: University of Chicago Press, 1976.

Peterson, R. W. "Changes in Curiosity Behavior from Childhood to Adolescence." *Journal of Research in Science Teaching* 16 (1979): 185–192.

Piaget, J. *Science Education and the Psychology of the Child.* New York: Viking Press, 1974.

Piaget, J. *The Grasp of Consciousness: Action and Concept in the Young Child.* Harvard, MA: Harvard University Press, 1976.

Sagan, C. *Broca's Brain.* New York: Random House, 1974.

Schwartz, T. *The Art of Logical Reasoning.* New York: Random House, 1980.

SCIS (Science Curriculum Improvement Study). *Don't Tell Me: Let Me Find Out.* San Francisco: Davidson Films, 1972.

Snow, C. P. *"The Two Cultures: and a Second Look."* Cambridge, England: Cambridge University Press, 1969.

3

Understanding Students

The purpose of chapter 3 is to help you understand students in the context of teaching science in the classroom. You will learn how theories about motivation, learning, and development can be applied to teaching science. Model science lessons illustrate how to use these practical applications. Finally you will find a discussion of special students, and learn how to help both handicapped and gifted students.

KEY CONCEPTS AND IMPORTANT IDEAS
☐ motivation ☐ Maslow's hierarchy of needs ☐ physiological needs ☐ love and belongingness needs ☐ self-esteem needs ☐ self-actualization needs ☐ student learning ☐ reinforcement theory ☐ classical conditioning ☐ operant conditioning ☐ reinforcement schedules ☐ discovery learning ☐ humanistic learning ☐ physical development ☐ social and emotional development ☐ intellectual development ☐ exceptional students

Effective teachers understand and apply principles of motivation, learning, and development. As students enter elementary and middle school classrooms, they are full of interest, curiosity, and energy. Terms such as these are frequently used to describe students. In this chapter we discuss some of the basic psychological principles that are important for teaching in general. But we go one step further; we show how many of these principles can be applied through elementary and middle school science teaching. So, in several places we have included activities to illustrate how various psychological principles might be applied through instruction in science. Helping you to understand students in the context of science instruction is the goal of this chapter. You may already be familiar with some of the psychological theories, but now we ask that you try to understand how students are motivated, and how they learn and develop in the context of science instruction.

The last section of the chapter addresses special students and their needs. It is impossible to describe all handicapping conditions and educational remedies but we will provide you with an introduction to the challenges that handicapped students face. We will also suggest strategies that will help these students function better in the classroom. Let us start, then, with a view of motivation.

STUDENT MOTIVATION

In an imaginary observation of a science lesson, the behaviors of four students were observed. Maria was restless and fidgeting during the lesson; something seemed to be competing for her attention. Two desks away, Robert sat quietly, head on his desk, sleeping. Karen paid close attention to the teacher. She was careful to record the important points of the lesson and then started her assignment. When she completed her work and handed it in, the teacher gave her the recognition she had earned and deserved. Martin came into the class and immediately started clowning around, climbing on desks, and causing a commotion. He was also given recognition; it, too, was earned and deserved. However, the teacher had to make a great effort to get him to behave appropriately.

The motivation of these students could be discussed from different perspectives. The perspectives are revealed through the teacher's questions: "Why did the students behave as they did? How can the different motivations of students be explained?" Asking and answering questions about motivation is important to effective teaching. This section presents the psychological dimensions of motivation, and the importance of one theory of motivation for science education.

A tacit assumption about most science curriculum materials and teaching techniques is that they are designed to motivate individuals to learn more about themselves and the physical world. With this assumption, judgments about student behavior can be made. For example, in the imaginary classroom above, one could conclude that three of the students were poorly motivated. Teachers then ask, "How can I motivate these students?" Implicit in this question is the assumption that motivation results from external stimulation.

External motivation is associated with behavioral psychology. Except for basic drives, incentives and reinforcements existing outside the individual are the sources of motivation. In a science classroom extrinsic motivation might include teacher approval, grades, and other positive reinforcers associated with learning and applying science. Motivation can also be discussed from the students' point of view by asking questions such as, "What mo-

Many factors affect student motivation; some of these factors are internal while others are external.

tivated the behavior of the students in the opening vignette?" It is probably true that several factors such as level of maturation, past experience, and external stimulation were involved in motivating the behavior of the students. While these factors are important, teachers should also recognize the power and range that personal needs have on motivation.

Another way to describe individual motivation is in the context of Abraham Maslow's theory (1970) which posits a hierarchy of motivational needs that influence behavior (see figure 3–1). The **hierarchy of needs** is a con-

tinuum including physiological needs, physical needs, psychological needs, and at the highest level, needs that are creative, ethical, intellectual, and aesthetic.

The Hierarchy of Needs: Physiological

Physiological needs include food, water, air, and sleep. These needs have the greatest motivational force in the hierarchy. An individual's motivation, if totally deprived in a chronic way, would be toward fulfillment of physiological needs. In the introduction, there were

MASLOW'S HIERARCHY OF NEEDS

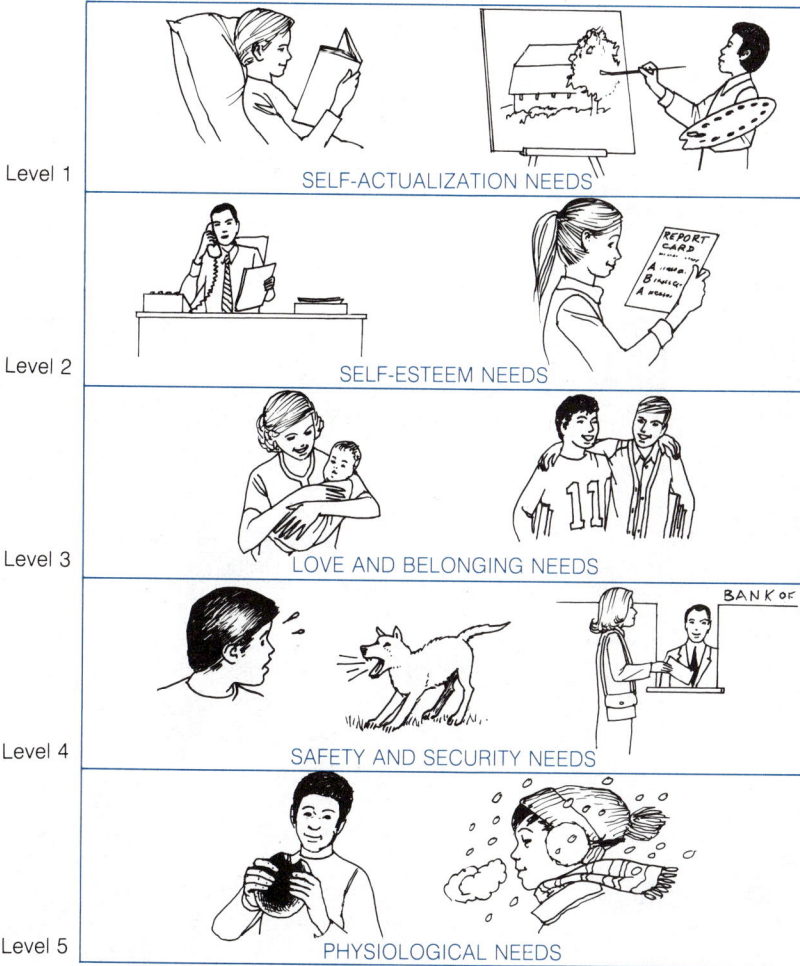

Level 1 — SELF-ACTUALIZATION NEEDS

Level 2 — SELF-ESTEEM NEEDS

Level 3 — LOVE AND BELONGING NEEDS

Level 4 — SAFETY AND SECURITY NEEDS

Level 5 — PHYSIOLOGICAL NEEDS

FIGURE 3–1
Abraham Maslow suggested that motivational needs can be thought of as hierarchical, with some needs such as those for food, security, and love taking precedence over others such as self-esteem and creativity.

examples of a restless girl, Maria, and a sleeping boy, Robert. Later that morning Maria was eating something, and it was learned that Robert had an early morning paper route. It then became more apparent that these students' behavior was at least in part motivated by physiological needs. To the degree that physiological needs no longer dominate behavior, new needs at the safety and security level may emerge.

The Hierarchy of Needs: Safety and Security

These needs arise from a need for order, structure, stability, security, and freedom from chaos and fear. In students, the indicators of safety needs can be clearly observed. Students generally prefer an undisrupted routine. They perceive their environment as secure when it has regular patterns and when people

Abraham H. Maslow (1908–1970)

Abraham H. Maslow is the acknowledged father of the humanistic movement in American psychology. The humanistic psychology Maslow developed was his attempt to expand the scope of psychology by combining elements of behaviorism, Freudian theory, and existentialism.

Abraham Maslow was born in New York City on April 1, 1908. His parents had emigrated to the United States from Kiev, Russia. As a youth Maslow was shy, skinny, and introverted. He was the first of seven children, and did not receive very much attention from his parents. Maslow was very intelligent (his IQ was later measured by E. L. Thorndike at 195) and spent much of his time in libraries reading. Despite his own family situation, there were always aunts, uncles, and teachers who took an interest in him. Early in life Maslow developed a strong desire to make a contribution to humanity.

After graduation from high school Maslow enrolled in New York City College for the academic year 1926–27. He transferred to Cornell but left after a semester. In the fall of 1928 Maslow went to the University of Wisconsin where he eventually was awarded all three of his degrees.

Classical behaviorism was Maslow's first introduction to psychology; his early education as a psychologist also included Freud's psychoanalytic theory. Upon leaving Wisconsin in 1935, Maslow returned to New York and worked for E. L. Thorndike, who was a disciple of John Watson, the founder of behaviorism in this country. At that time, Maslow's research turned to theories of dominance, self-esteem, and the authoritarian personality—theories not normally studied by the behaviorist. This research represents the beginning of Mas-

are consistent and dependable. In short, students need an orderly world. The open classroom, a recent trend in education, has overlooked these needs. The zeal with which some teachers "opened" the classroom and allowed total freedom left their students with a world they perceived as chaotic, unmanageable, and unorganized. This idea applies to a variety of approaches that might be used in science teaching. Many teachers have firsthand experience with students needing some security, limits, and organization.

low's defection from the behavioristic tradition, a defection which finally resulted in his leaving Thorndike's lab at Columbia and teaching at Brooklyn College. In 1951 he took a job at Brandeis where he was chairman of the psychology department until shortly before his death.

From 1935 to 1951, New York City was an intellectually exciting place for a young psychologist such as Maslow. He cultivated friendships with numerous individuals who had left a troubled Europe. Among his friends were people such as Max Wertheimer, Kurt Kaffka, and Kurt Lewin, founders of Gestalt psychology; Karen Horney, Alfred Adler, and Erich Fromm, all neo-Freudian psychologists; Kurt Goldstein, an organismic psychologist; and Ruth Benedict, an anthropologist. It was Maslow's genius to synthesize many of these great thinkers' ideas with his own ideas as he formed the new humanistic psychology. Maslow's ideas began developing with his decision to pursue the problem of personal growth and psychological health.

Maslow theorized that basic needs motivate human behavior. Some aspects of behaviorism, Freudian psychology, and Gestalt theory were incorporated into Maslow's theory as he attempted to strengthen weaknesses in psychological theory. The addition of existentialism to Maslow's theory completes the underlying structure. From existentialism Maslow incorporated identity as fundamental to human nature. Maslow can be credited with initiating a revolution in psychological theory and practice.

During his career Maslow served as president of the Massachusetts State Psychological Association and president of the New England Psychological Association. He was also president of the Division of Personality and Social Psychology, president of the Division of Aesthetics and in 1967–68 was the elected president of the American Psychological Association. Maslow's books include *The Psychology of Science* (1966), *Toward A Psychology of Being* (1968), *Motivation and Personality* (1970), and *The Farther Reaches of Human Nature* (1971).

In 1969 Maslow left Brandeis University and became Resident Fellow at the W. Price Laughlin Charitable Foundation. His intention was to devote himself to writing. After a year of work Abraham Maslow died of a heart attack on June 8, 1970. Maslow's contributions were many—the hierarchy of basic needs, self-actualization, and peak experiences—to name a few. Probably his most important contribution was opening the doors to new realms of human understanding.

The Hierarchy of Needs: Belongingness

Belongingness needs are characterized by many words associated with today's society: acceptance, togetherness, friendship, aliena-tion, loneliness, and rejection. The strivings of the individual are to give and receive affection, to have friends, to belong to groups, and to find a place with people and society.

Belongingness needs can be seen in these statements by Stuart, a four-year-old. The excerpt, a dialogue with his teacher, is about the need to give affection.

Stuart: Would you like somebody to take care of you?

Teacher: I can do it pretty well.

Stuart: I will take care of you. You come live with me. You can sleep in my bed. My mommy will take care of you. Or maybe I could come live with you (seriously and practically). But it doesn't matter where you live or what you do, you still have your insides! (Ends on a gleeful, impish note; goes off.) (L. Joseph Stone, ed., Stuart: A Four-Year-Old Views The World. Mimeographed material [Wilkes-Barre, PA: Vassar College, 1966], p. 40.)

Most teachers are aware of students who feel rejected, don't belong to groups, who need affection and positive regard from their teacher, and who are generally frustrated in the fulfillment of belongingness needs. Belongingness and esteem, the next level, seem to be important motivational needs in contemporary society. The lack of fulfillment of these needs is at the center of many problems that children have in school, and the implications for teachers seem clear—students should have a sense of belonging in the classroom and should feel that they are cared for.

The Hierarchy of Needs: Self-Esteem

The fourth level of needs is derived from a desire for a stable, firmly based positive perception of self. The need for personal adequacy is divided into mastery needs and prestige needs. The first, mastery, is the individual's need for achievement, competence, and confidence in carrying out duties and obligations. The second, prestige, is of a more social psychological nature. It is the need for others to show a respectful attitude toward the individual. Status, recognition, importance, and dignity are defining characteristics of need fulfillment at the prestige level.

Frustration of self-esteem needs produces insecurity, inferiority, and anxiety about personal worth. The resulting behavior of the student can be compensatory, which for the teacher, can cause discipline problems depending upon the degree of frustration and the availability of fulfillment through acceptable means. Two students in the opening story, Karen and Martin, were probably motivated by self-esteem needs. Karen was fulfilled in both the mastery and prestige needs, while Martin probably only fulfilled the need for prestige from his peer group, not the teacher. Martin is an example of self-esteem achieved negatively, a behavior teachers encounter daily.

Gratification of esteem needs results in perceptions of personal adequacy, individual integrity, and self-worth. Healthy self-esteem is based on mastery and prestige that is deserved and earned rather than on unwarranted adulation. Self-esteem fulfilled through an individual's distorted perceptions of self or through false imposition of perceptions of self from others, such as teachers, can cause personal difficulty. The maintenance of false esteem in the face of reality is a consuming exercise.

The Hierarchy of Needs: Self-Actualization

Motivation through the various levels of the hierarchy is toward the full development of an individual's potential. Self-actualization can be described as the full use of talents, capacities, and potentialities.

Such people seem to be fulfilling themselves and to be doing the best that they are capable of doing, reminding us of Nietzsche's exhortation, "Become what thou art!" They are people who have developed or are developing to the full stature of which they are capable. These potentialities may be either idiosyncratic or species wide. (Abraham Maslow, *Motivation and Personality* [New York: Harper and Row, 1970], p. 150)

From his studies of healthy people, Maslow synthesized characteristics of individual self-actualization. These include: creativity; a clear perception of reality; a continuing freshness of appreciation; a quality of detachment and need for privacy; acceptance of self, others, and nature; a problem-centered attitude; a philosophical, unhostile sense of humor; a democratic character structure; identification with humanity; and personal values that transcend the culture. Self-actualization is sometimes diagrammed as having different levels, such as aesthetic needs and the need to know. Although these are characteristics of self-actualization, Maslow did not indicate that these were levels of motivation. The concept of full self-actualization is a goal, for few adults (and probably no children) achieve this level except in occasional moments of heightened awareness and insight. During these times of peak experience individuals gain a glimpse of their own self-actualization. People reporting these experiences often use terms associated with values such as justice, truth,

beauty, goodness, and so forth. These are the heart of Maslow's theory and his argument for human values as a part of basic human nature.

To summarize, the motivational referents of Maslow's theory compose the interaction between internal needs and external satisfaction, forming a hierarchy that can be summarized as physiological, physical, and psychological. The higher levels of motivation form, to use Maslow's phrase, "the farther reaches of human nature."

Examining two Health Science activities can demonstrate the application of Maslow's theory. One activity is called The Four Food Families. All of us are motivated by the need for food. Though most of us never experience deep and chronic hunger, we all experience hunger if we miss a meal or two. Students should have this common background and subsequent interest, or need to know, about food since it is aligned with something they have experienced. The second activity, Communicating Emotions, is a slightly different application of Maslow's theory. If this activity is completed as suggested, the teacher should have some ideas of motivation needs of the students. The reason is rather simple; students will often express an emotion aligned with their needs. Safety needs might be expressed as fear and belongingness needs might be expressed as loneliness.

HEALTH SCIENCE ACTIVITY

The Four Food Families
(Kindergarten–Grade 2)

MAJOR CONCEPT

Developing eating patterns that result in a balanced diet contributes to good health.

OBJECTIVES

☐ To classify food words and/or pictures into the four basic food groups.

MATERIALS

Tagboard 12″ x 18″ or one manila file folder, tagboard cards for food words and pictures, food pictures from magazines, paste and scissors.

Optional: "Guide to Good Eating" poster of the four food groups. (This poster may be obtained from the National Dairy Council. Contact the office serving your area or write to: National Dairy Council, 6300 North River Road, Rosemont, Illinois 60018.)

VOCABULARY

dairy products   balanced diet
meat and protein   healthy
fruits and vegetables   menu
grain products   nutrition

PROCEDURES

A. *To make the Food Family Game,* take an envelope and paste it on one side of a piece of tagboard or file folder (Kaplan et al). Fill the envelope with food pictures and/or words. Next to the envelope, write the names of the four food groups or illustrate them with pictures. Direct the students to take food family members from the envelope and match them with the correct food family. You may want to make an answer key on the back of the tagboard (or file folder).

B. *To introduce the game,* discuss with the children the need for a balanced diet to insure proper growth and good health. Using a bulletin board or chart, display food pictures from the four basic food groups. Explain that a balanced diet should include some foods from each group every day. Invite the children to identify the foods pictured in each group. Point out how the same food can appear in many different forms (for example, tomatoes, ketchup, or spaghetti sauce).

EXTENDING THE ACTIVITY

C. Invite the children to draw pictures of their favorite foods. Display the pictures on a bulletin board.

D. Invite the students to cut out and classify magazine food advertisements into the four food groups.

E. Invite the students to make food games to take home for their families to play.

F. Invite the children to construct simple graphs categorizing foods eaten at breakfast or lunch into the four food groups. Use the number of servings of the foods to quantify the amount of each "food family" eaten.

HEALTH SCIENCE ACTIVITY

Communicating Emotions
(Kindergarten–Grade 2)

MAJOR CONCEPT

Emotional stability is influenced by an awareness of personal feelings and reactions.

OBJECTIVES

☐ Identify some common emotions experienced by primary-age children.

MATERIALS

None.

VOCABULARY

emotions

PROCEDURES

A. To introduce and explore the concept of emotions, have a class discussion using these guidelines.
   1. Ask the entire class to *show* how they really feel right now without using any words.
   2. Ask for two or three volunteers to *tell* in words how they feel right now.
   3. Ask the children to *show* you how they feel when eating an ice cream cone.
   4. Ask the children to *tell* you a word that describes how they feel when they eat an ice cream cone.
   5. Write the children's descriptive words on the chalkboard and tell them that these words are called *emotions*.
   6. Ask the children to *show* you how they feel when they have to stop playing with their favorite friend because it is time to go to bed.
   7. Ask the children to *tell* you, using just one word, how they feel in the above situation, and again write the words on the chalkboard.
   8. Ask the children what the words on the board are called. (emotions)
   9. Ask the children to *show* you how they feel when they listen to a very, very scary story in the dark.
   10. Ask the children to *tell* you all the words they can think of that describe how they feel in the above situation.

EXTENDING THE ACTIVITY

B. Lead class discussions in which children consider the following:
   1. When is it appropriate to "act out" emotions? When is it better to merely describe your feelings?
   2. Why is it important for you to know how you feel?
   3. Is it possible to have a feeling and not let anyone else know about it?

Adapted, with permission, from *My Body Is Mine, My Emotions Are Mine, My Environment Is Mine: A Health Education Curriculum.* Albany, New York: Albany Unified School District, 1978.

Implications

There are some specific and brief implications concerning the motivational needs of students. As an elementary or middle school teacher, you can infer the spirit of the implication and then, based on your understanding of Maslow's theory, your school, your students, and yourself, decide about the educational application.

1. Recognize the physiological and physical needs of students. Factors such as hunger, thirst, room temperature, and need for rest all contribute to a student's motivation. There are often overt behaviors that indicate needs at this level; try to be aware of these behaviors and respond as best you can within the school environment.

2. Maintain a physically and psychologically safe classroom. Help students with their "little" problems. Avoid situations that are threatening or frightening to the students. If routines are being changed, for example, when an open approach needs to be implemented, inform the students of the change and try to make a gradual transition.

3. Show your students that they "belong" in your room. Addressing them by name, recognizing their achievements (both in and out of your classroom), talking with them about topics of common interest, and listening to their ideas all can contribute to a feeling of acceptance and belongingness in your classroom.

4. Arrange learning experiences so all students will gain some esteem in your class. Give acknowledgment and praise when appropriate and avoid harsh criticism, ridicule, or humiliation. Rather, help students with difficult problems, concepts, and choices.

5. Provide opportunities for individuals to experience success, both individually and in a group. First, this means teaching in terms of an individual student's educational needs and not the demands of a text, lesson plan, or curriculum. Second, this means encouraging cooperation among students on tasks and projects as opposed to a continued competition among students.

6. Discuss and clarify personal values as a part of the educational program. Techniques such as class meetings, decision making, and value clarification are helpful in achieving this goal.

7. Arrange experiences that will enhance the choices for personal growth by the students. Sometimes this can be accomplished through ambiguous, though directed, assignments. When need be, suggest a possible activity that is challenging but attainable for individual students.

8. Provide for some time when students are allowed to make choices and learn in their own way. The most accurate interpretation of this goal is a freedom within limits—not unlimited freedom.

9. Become familiar with your students' behaviors, and try to recognize the needs they are expressing. To confirm your perceptions of their needs, offer an affirming recognition such as, "You look tired today, Robert; are you?" "Maria, are you as anxious for lunch as I am?" and "Martin, how would you like to help me plan a (skit, contest, and so on)?"

Maslow's theory can provide valuable understanding concerning your students. Further, and of equal importance, teachers are motivated by the same needs as their students. Teaching science can fulfill both student needs and teacher needs. Unfortunately the needs of students and teachers can also be in conflict. By understanding your own needs and the needs of others, serious conflicts can be avoided.

STUDENT LEARNING

Once you have begun to understand what motivates students' behavior, you will also want to know how students learn and what you can do to help them learn. After all, that is one of the reasons why most people become teachers. In this section, student learning is discussed. Here the focus is on several theories that explain how learning takes place, and on the kinds of situations in which these theories can be applied to teaching science in the elementary and middle school.

LEARNING THROUGH REINFORCEMENT

This section is primarily concerned with the behaviorist view in psychology. The behaviorist school maintains that observable behavior is the essential basis of study in psychology. Behaviorism further emphasizes the importance of **reinforcement,** or environmental influences, on behavior as opposed to internal conditions such as states of mind or emotions. There are *stimuli,* or events in environment, that cause *responses* by the organism. For this reason, some refer to this as stimulus-response (S-R) psychology. The work of three persons who made important contributions to behaviorism will be reviewed: Ivan Pavlov, Edward L. Thorndike, and B. F. Skinner. Before turning to their contributions we must define learning.

What do we mean by learning? The behaviorist school has provided educators with an answer to this question. Learning is defined as observable changes in behavior due to experience. Learned behavior is usually thought to be relatively permanent and consistent. This allows one to differentiate learned behavior from performance or motivated behavior. Personal thoughts, feelings, and desires are usually not considered learned behavior according to this definition.

Classical Conditioning. Classical or respondent **conditioning** refers to the ideas of Ivan Pavlov, a Russian physiologist who, in 1904, received a Nobel prize in medicine for his studies of digestive processes. Pavlov's theory of respondent conditioning was originally completed with dogs. When Pavlov presented food to dogs, the reflexive behavior was salivation. This is termed an unconditioned response (UCR); in this case the food is called an unconditioned stimulus (UCS). Let us say that another stimulus, a small bell, is presented simultaneously with the food. Initially, the bell would not cause the dog to salivate. However, if a bell is rung repeatedly as food is presented, the dog eventually learns to respond by salivating when only the bell is rung. The sound of the bell is then a condi-

tioned stimulus (CS). The dog's salivation to the bell's sound is referred to as a conditioned response (CR). It can be said that the dog has learned to respond to the bell. Classical conditioning is summarized in figure 3–2.

Associationism. Edward L. Thorndike, the father of educational psychology, further developed Pavlov's ideas. In one important experiment, Thorndike placed a hungry cat in a box that would open if the cat activated a release mechanism. Initially, the cat simply jumped around until it finally hit the release mechanism by accident. With time and additional trials, the cat narrowed its efforts to specific parts of the cage and eventually to the release mechanism itself. The cat learned how to release itself from the cage through a process of trial and success. In this case, the entire box was a stimulus and the cat had numerous behavioral responses of which only one was correct. Thorndike (1932) developed **associationism,** the idea that it was the association between the stimulus and the response that was important and that repetition of the correct response was essential to learning because it strengthened the association. These ideas were the basis for Thorndike's fa-

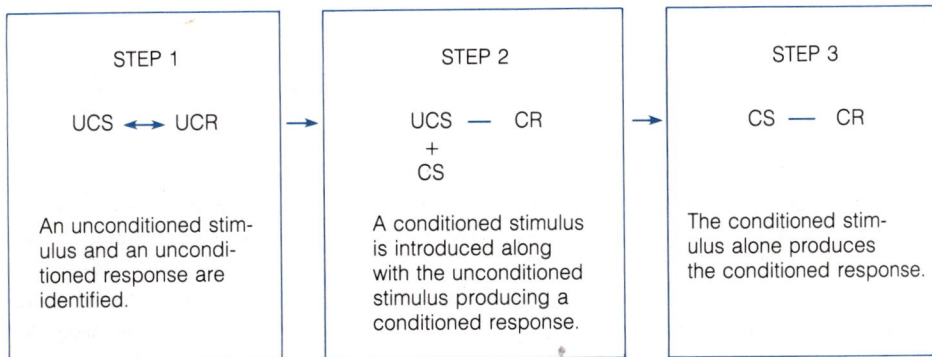

STEP 1	STEP 2	STEP 3
UCS ⟷ UCR	UCS — CR + CS	CS — CR
An unconditioned stimulus and an unconditioned response are identified.	A conditioned stimulus is introduced along with the unconditioned stimulus producing a conditioned response.	The conditioned stimulus alone produces the conditioned response.

FIGURE 3–2
The Process of Classical Conditioning

Courtesy B. F. Skinner

B. F. Skinner (b. 1904)

B. F. Skinner is both one of the most renowned and controversial figures in contemporary psychology. His contributions easily rank with those of Sigmund Freud and Jean Piaget. And, if one considers his

role in American psychology, he is surpassed by none. For over fifty years Skinner has contributed to our understanding of human behavior. During this time he has thoroughly researched and carefully applied his psychology—"radical behaviorism"—to psychotherapy, linguistics, social problems, and education. A 1958 award of the American Psychological Association for Distinguished Scientific Contributions described this renowned psychologist as "an imaginative and creative scientist, characterized by great objectivity in scientific matters and by warmth and enthusiasm in personal contact." Skinner has also been the subject of controversy because of his outspoken advocacy for using behaviorial principles to change human behavior.

Burrhus Frederic Skinner was born on March 4, 1904 in Susquehanna, Pennsylvania. His father was a lawyer and his mother a homemaker. Skinner has described his early years as "warm and stable." He was always interested in mechanics and gadgets. His inquisitiveness and handiness were later to be combined in scientific research and development.

Upon graduation from high school he enrolled at Hamilton College, majoring in

mous "law of effect." Stated simply, Thorndike's principle is that *successful behavior is likely to recur in similar situations.*

Operant Conditioning.   Operant conditioning is probably of greater interest to elementary and middle school teachers. In this form of behaviorism, an environmental consequence is paired with a behavior. The positive or negative consequence of a behavior (or lack of a consequence) influences the likelihood of that particular behavior recurring. B. F. Skinner (1953; 1968; 1971) is the psychologist most often associated with behaviorism and **operant conditioning.** The term "operant" indicates that the organism operates or acts on the environment to receive reinforcement. Let's look at an experiment

English with a minor in classical and romantic languages. Skinner had literary ambitions and at one time his literary career was even encouraged by Robert Frost. He received the baccalaureate degree in 1926. For a year after graduation he wrote fiction; however, he had to abandon hopes for a career as a professional writer.

About this same period, Skinner was influenced by Bertrand Russell's writing on behaviorism and John Watson's book, *Behaviorism.* As a result, in 1928 Skinner enrolled in psychology at Harvard University. Two years later he was awarded the M.A. and in 1931 the Ph.D. As a psychologist, Skinner never veered from his intention to develop a science of behavior that was free of mentalistic concepts such as "mind" or "free will." Instead, Skinner concentrated his research on observable behavior with the intention of understanding the laws governing behavior. Prediction and control of behavior were his long-range goals. He soon became known as a radical behaviorist. Over the years his research and theories have become the central model of American psychology.

Though Skinner may not have achieved a career in writing fiction, he has attained prominence for his writing in psychology. In 1948 he wrote a utopian novel, *Walden Two,* outlining the application of his behavioral ideas to a small community. Other books written by Skinner include *Science and Human Behavior* (1953), *Schedules of Reinforcement* (1957), and *Verbal Behavior* (1957). In 1971 he published his most controversial book, *Beyond Freedom and Dignity.*

Skinner used his interest and talents to develop teaching machines and programmed instruction. He also applied the basic principles of reinforcement to the learning of school subjects. Books entitled *A Program for Self-Instruction* (1961) and *The Technology of Teaching* (1968) detail many of his educational ideas.

B. F. Skinner is now retired from Harvard University where he spent most of his professional career. During that career he has received numerous awards, including the National Medal of Science, the Joseph P. Kennedy Jr. Foundation Award, and the Gold Medal of the American Psychological Association. Though retired, Skinner continues to work toward the application of his theories to advance humane goals.

to see how learning occurs through operant conditioning.

A hungry rat is placed in a small cage that contains a door with a bar. When the rat accidentally pushes this bar, the door opens and a small morsel of food drops into a tray. Through the process of trial and error, the rat eventually learns that pressing the bar releases other morsels of food. Once this behavior becomes part of the rat's repertoire of responses to hunger, you can establish some conditions for the release of food. For example, food pellets can be supplied under some conditions and not under others. Suppose you want the rat to push the bar when the bell is sounded. With appropriate reward, soon the rat will learn to push the bar at the sound of a bell. The rate of bar pushing would

By studying the way rats and other small animals learn, scientists further our understanding of human learning. (Larry Hamill)

increase when the bell was rung; it would decrease when there was no bell.

Several important concepts of behaviorism are demonstrated in this typical Skinnerian experiment. The rat learned a specific behavior due to *reinforcement*. When the behavior was not reinforced, *extinction* occurred; that is, the likelihood of the behavior occurring decreased. At first there was a *generalized* response, in that the rat pushed the bar as much when there was no bell as it did when there was a bell sound. However, with time the rat *discriminated* among stimuli and pushed the bar more often with the sound of a bell.

How is this experiment different from Pavlov's classical conditioning? In Pavlov's study the response had to be elicited from the dogs. Skinner's rats emitted a behavior that was later reinforced to increase the probability of

its recurring. Also, the rat operated the bar for the reinforcement.

Several important ideas for learning science are worth noting. First, rewarded behavior increases the *probability* of that behavior recurring. This does not mean however, that the behavior will always occur. Second, *extinction* can occur if the behavior is not reinforced. Third, it *takes time* for individuals to learn behaviors even though they have been reinforced. We should examine reinforcements closely because they are very important to this theory of learning.

Reinforcing Behaviors. Reinforcement increases the probability of a behavior recurring. There are two types of reinforcement—positive and negative. In positive reinforcement, some desirable or pleasant reward is presented after a behavior. For example, a student may feel pleased with himself or herself upon seeing an experiment work; a teacher may smile at a student who answers a question correctly or give praise at his or her completing a laboratory activity. Negative reinforcement is the removal of some noxious or aversive stimulus after a desired behavior occurs. So for example, a student in science class may always be looking around and not paying attention. The teacher continually frowns at and reprimands the student. Upon appropriate classroom behavior (listening, attending, responding, and so on) the teacher stops giving the student unpleasant looks. The student soon learns to avoid or escape the teacher's unpleasant looks by behaving appropriately.

Punishment is often confused with negative reinforcement. The two processes do differ and have different results. Punishment *decreases* the probability of a behavior occurring while reinforcement (both positive and negative) *increases* the probability of a behavior. Punishment results when an aversive stimulus is presented or a pleasure-producing stimulus is withheld. When a child fails to do his or her chores, he or she can be spanked (aversive stimulus) or he or she can be deprived of an allowance (pleasure-producing stimulus). Table 3–1 is a convenient way to re-member the differences between reinforcement and punishment.

When science teachers select reinforcers, they should be sure to use something that is important to the student whose behavior they wish to modify. A psychologist named David Premack provides an easy way of identifying a reinforcer that is directly applicable for classroom teachers. Premack's (1965) principle is that *any natural behavior that occurs relatively often can be used to reinforce another behavior that occurs less often.* "You must pick up materials and clean your laboratory area before recess." This is a simple example of the Premack principle. In order to apply this principle, the teacher must observe and note the types of activities preferred by the student, and then make the more desirable behaviors contingent on other, less desirable behaviors.

Scheduling Reinforcements. When applying the principle of operant conditioning in the classroom, it is best to use scheduled reinforcement. Rewards and punishments are given in classrooms every day. Unfortunately, most of the time, rewards for productive behavior are distributed randomly while punishment is aligned with some form of disruptive behavior. In addition, some things that teach-

TABLE 3–1
Reinforcement and Punishment

	Present Stimulus	Remove Stimulus
Pleasant Stimulus	Positive reinforcement and Increase in behavior	Punishment and Decrease in behavior
Aversive Stimulus	Punishment and Decrease in behavior	Negative reinforcement and Increase in behavior

ers think are reinforcers may not be for the student, and many things that are very powerful reinforcers may be only seldom used. As a teacher, you can use a much more organized and consistent approach to reinforcement. There are two basic schedules of reinforcement: ratio and interval. In ratio schedules, reinforcement is given after a specific number of responses. The number of responses can be *fixed*, for example, at every five, ten, or fifteen problems. Or the reinforcement can be *variable*; for example, an average of every five problems completed would be rewarded, but it may be the sixth, the eighth, and or the fifteenth problem.

Interval schedules require that an amount of time elapses before the next reward is given. Again, when the teacher reinforces the student after a uniform amount of time has elapsed, the teacher is using a *fixed* interval schedule of reinforcement; for example, reinforcement is every five minutes. Like the variable ratio schedule, on the *variable* interval schedule, rewards are dispensed around an average amount of time. On the average, a reward is given every five minutes but the rewards may be at two, six, and sixteen minutes.

Knowing about reinforcement schedules provides the opportunity for consistently rewarding desired behavior. It also helps the teacher who is trying to change disruptive behavior. If the teacher provides intermittent reinforcement to disruptive behavior, the behavior will continue longer than in cases where reinforcement has been constant and then ceases. Consider this example. There is a student who talks incessantly to other students and you wish to change this behavior. The best solution is one that usually does not happen. Usually the student receives occasional attention from the teacher and peers for talking. This is intermittent reinforcement for talk-

ing and results in the most difficult behavior pattern to change.

Most teachers use punishment or negative reinforcement as they attempt to change behavior. Since these methods are used, it is important that you know their negative consequences. There is the possibility of unintended conditioned emotional responses. For example, a student who is punished in science class may be conditioned to fear repeating that behavior but also may be simultaneously conditioned to fear other aspects of the science classroom. Another result is avoidance situations which produce the pain or anxiety associated with punishment. So, a student may skip school or not go to science class. Finally, one student's punishment may be another's reinforcement. If a student needs attention, receiving punishment is one way of getting attention, albeit a painful way. Punishment can result in emotional responses that are more enduring than the immediate, aversive stimuli. In general, one might first use positive reinforcement for desired behavior, and second, combine positive and negative reinforcement and punishment to increase desired behavior and reduce undesired behavior.

Two science activities can be used to illustrate the practical applications of learning theory. The first, an Earth Science activity, is intended to establish a way of describing objects and organisms. As such, the goal is to introduce words that will become part of the student's vocabulary. Notice that the procedures contain a series of questions that have specific responses, for example, the words, object and organism. As the students respond correctly to the questions, the teacher should reinforce their use of the correct words and extinguish (through lack of attention) their use of incorrect words. If the principles of learning theory are applied consistently, in a short time students will be using the proper terms. Yes, this is a simple example. But teaching can be much more efficient and effective with the proper application of learning principles.

In the second lesson, a Physical Science activity, there is an excellent example of a built-in reinforcer. Notice that the first part of the activity (Task Card #1) had the students try to light a bulb. When they complete the task correctly, they will be reinforced immediately; the bulb will light. The lesson continues by building concepts about electricity and circuits in a step-wise fashion. This is very typical of the approach suggested by learning theorists such as B. F. Skinner. As the students continue designing circuits, they continue to be reinforced if they complete their tasks correctly. All of the activities are later related to the idea of operational definitions, which are themselves also an application of behaviorist psychology. When objectives are stated in behavioral terms, what is expected is clearer to the learner. "The student should be able to construct an electrical circuit" is much clearer than an objective such as "know electrical circuits."

EARTH SCIENCE ACTIVITY

Describing the Environment: Objects, Organisms, and Colors
(Kindergarten–Grade 2)

MAJOR CONCEPTS

All nonliving things in the environment can be referred to as objects. Living things in the environment can be referred to as organisms. Living and nonliving things have identities based on their properties. All objects and organisms can be described by their properties. The property of color (or lack of color) can be used to describe objects.

OBJECTIVES

☐ Use the word *object* when referring to nonliving things in the environment.

☐ Identify objects using the property of color.

☐ Identify the colors red, green, blue, orange, yellow, purple, black, and white.

MATERIALS

Objects in the classroom.

VOCABULARY

object	blue
red	purple
green	white
orange	black
yellow	properties

PROCEDURES

A. To introduce this activity, have the following class discussion.

   1. Point to an object in the classroom and ask: *What is this object?* Repeat this procedure for several objects in

the room. The children will respond with answers such as a desk, the blackboard, a pencil, a floor, a door, and so forth.

2. Hold up a piece of paper and ask: *What is this object?* Emphasize the word *object*. As soon as the children respond, point out the fact that you have referred to all the things as objects.

3. Reverse the questions used earlier. *What is the door? pencil? desk?* All are objects. Point out that *object* is just another word the child will use when talking about things in his or her environment.

4. Repeat these procedures (2 and 3) for the word *organism*. Ask: *How do you know objects/organisms are different?* Find some organisms in the room. Have the children tell about each one. Choose items that include the following colors: red, orange, yellow, green, blue, and purple. At this time accept all responses that describe properties of the organism.

5. Tell the children: *The reason you know objects/organisms are different is because they have different properties.* The color of an object is one property.

6. Place *I Know My Color Words* on the board or a chart. Review the objects described before. As the children use a color to describe an object, print the color word on the board.

7. Review the color words. You find an object and ask the children for the *property of color.* Point to a color word and have the children find an object with that color.

8. Repeat this procedure for objects that are black and white. Note: You should respond to the level of the child. If colors are known, progress faster; in other cases, more review and a slower progression may be required. Continually use the word *objects/organisms* when referring to things in the classroom. Likewise, use the word *property*. The children will not use the word immediately. Encourage your students to use new words, so that they will become part of their vocabulary. These words are *not* spelling words. They are intended to be incorporated into the child's vocabulary. Be sure the child has directly experienced the word. In other words, you are operationally defining the words.

EXTENDING THE ACTIVITY

B. Bring several pieces of earth materials (rock, leaf, water) and have the class describe their color properties. Play the game "I'm thinking of an object/organism" and have the students guess what it is, based on its properties.

PHYSICAL SCIENCE ACTIVITY

Bulbs and Energy Paths*
(Grades 6–8)

MAJOR CONCEPT

A completed circuit will have an energy source, a path, and an indicator.

OBJECTIVES

☐ Construct an electrical circuit.

☐ Distinguish an operationally-stated definition from a non-operationally-stated definition.

MATERIALS

Flashlight batteries and bulbs; wire (insulated); battery holders; switches; miscellaneous materials, e.g., nail, cork, screw, string, paper clip, chalk, sugar, salt, aluminum foil, plastic spoon, penny, cloth, scissors, book, water, and rubber band.

VOCABULARY

circuit
operational definition

PROCEDURES

A. Light the Bulb
   1. Start this task by giving each pair of students in the group a plastic bag that contains a flashlight battery, bulb, and two pieces of wire about two decimeters (about eight inches) long. Have each pair complete Task Card #1.
   2. Now collect their diagrams and circulate them to others. It will help if they

also record on their task sheet the results of their testing of other's pictures.
   3. After they have completed their drawing of four ways to light the bulb, let the students share their results. You may want to award special honors for the most creative strategy.
   4. This will be an opportunity for your students to note first that a working plan is one in which the bulb lights and, second, that there may be more than one way to do the job.

B. Switches
   1. Give each group of three or four students a flashlight dry cell in a holder, a flashlight bulb in a holder, a switch, and three ten-centimeter (about four inches) lengths of insulated wire. Ask the groups to connect all of these things together in such a way that pressing or releasing the switch can turn the light on or off.

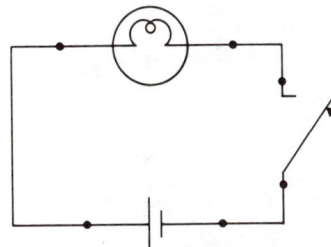

---

*"Science—A Process Approach," *Commentary for Teachers*, pp. 166–67 and 169–70. AAAS Pub. 68–7, 1968. Copyright © 1968 by the American Association for the Advancement of Science. Used with permission.

TASK CARD #1

Light the Bulb

Materials needed: This task sheet, plastic bag and its contents (battery, bulb, and two pieces of wire).

In your bag you will find four objects.

Show me that you can light the bulb.

Now draw a picture of what you did to make the bulb light.

Please keep a record of your results when you tried other plans.

Whose Plan Did You Try?	Did the Plan Work?
1.	
2.	
3.	
4.	
5.	
6.	

How did you decide if a plan worked? _____

_____

_____

Show at least four different ways you have now found to light the bulb.

1.	2.
3.	4.

Switch

As a group, use the materials your teacher supplies to construct a circuit.

If you had never seen this circuit, or one like it, which of the three definitions would most help you identify the switch? Circle your choice.

1. A switch is part of an electric circuit.
2. A switch is something that turns a bulb on and off.
3. A switch is something in an electric circuit that turns on a bulb when it is pressed.

Why is the definition the most useful? _____

_____

Does it specify what to do and what to observe? _____

A definition that specifies what to do and what to observe is an *operational definition*.

Circuits

Use these operational definitions to identify the circuits in the room. Put the letter or letters for one or more circuits beside the operational definition of the circuit.

_____1. *Series circuit:* Any number of lamps, a cell, a switch, and wires connected so that all the lamps glow when the switch is closed. If any one lamp is unscrewed, all the other lamps go out.

_____2. *Parallel circuit:* Any number of lamps, a cell, a switch, and wires connected so that all the lamps glow when the switch is closed. If any one lamp is unscrewed, the other lamps all still glow.

_____3. *Closed circuit:* Any number of lamps, cells, a switch, and wires arranged so that at least one of the lamps glows when the switch is closed.

_____4. *Open circuit:* Any number of lamps, cells, a switch, and wires arranged so that none of the lamps glow when the switch is closed.

_____5. *Series-parallel circuit:* Three lamps, a cell, a switch, and wires arranged so that all the lamps glow when the switch is closed. Two lamps are arranged so that if either is unscrewed the other will go out. There is one lamp in the circuit that can be unscrewed without causing the others to go out.

2. When they have completed this task, have them review Task Card #2. You may find it helpful to have them compare their definition of a switch with that in a dictionary.

C. Circuits

Now set up the five demonstrations shown below around the room. Using the Task Card #3, have the students decide which definition best fits each demonstration. Remember, each demonstration should have a label, but not the name of the circuit.

Do you agree that the student's choices should be

1.	Series circuit	E
2.	Parallel circuit	D
3.	Closed circuit	A, B
		D, E
4.	Open circuit	C
5.	Series-parallel circuit	A

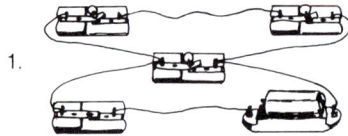

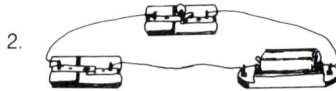

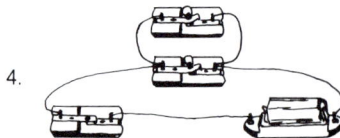

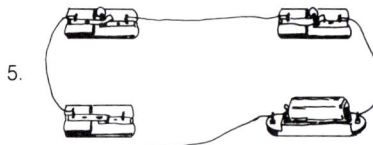

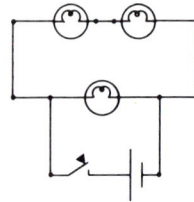

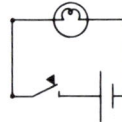

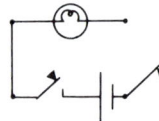

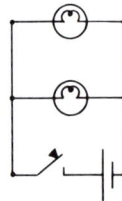

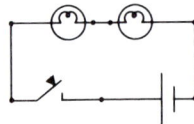

Lamp in Mounting

Lamp

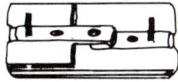

Switch

Switch

Cell

Cell

D. Conducting Through Energy Paths
  1. In this activity, your students will choose the most useful alternative definition in identifying conductors and nonconductors.
  2. They will need to recall that an operational definition specifies what must be done and what must be observed in order to identify or construct what is being defined.
  3. In order to be able to interpret the drawings that appear on the Task Card #4, the students must be able to name the parts represented by the symbols. On the chalkboard, sketch these symbols and the names shown above.
  4. Give each group of three or four students the equipment necessary to construct the circuit and the additional exploratory "junk" materials:

___Nail	___Paper
___Cork	___Plastic spoon
___Screw	___Penny
___String	___Cloth
___Paper clip	___Scissors

___Chalk	___Book
___Sugar	___Water
___Salt	___Your finger
___Aluminum foil	___Rubber band

  5. As they complete their tests have them record their results on a class chart:

		Group			
Item	A	B	C	D	E
Nail	con	con	con		

  6. Now have them select which of these statements are the best operational definitions.
    a. Conductor
      _____1 A conductor is a wire.
      _____2 A conductor is anything that makes a lamp light in an electric circuit.
      _____3 Connect a lamp, three wires, and a cell as shown in the diagram. A conductor is any object that will cause the lamp to light when it is connected across points A and B.

TASK CARD #4

Circuits Again

Here are operational definitions of two kinds of circuits.

Series circuit: Two or more lamps, one or more dry cells, a switch, and wires connected so that all the lamps glow when the switch is closed. If any one lamp is unscrewed all the other lamps go out.

Parallel circuit: Two or more lamps, one or more dry cells, a switch, and wires connected so that all the lamps glow when the switch is closed. If any one lamp is unscrewed, all the other lamps still glow.

Construct the circuit diagrammed below and then answer the questions listed.

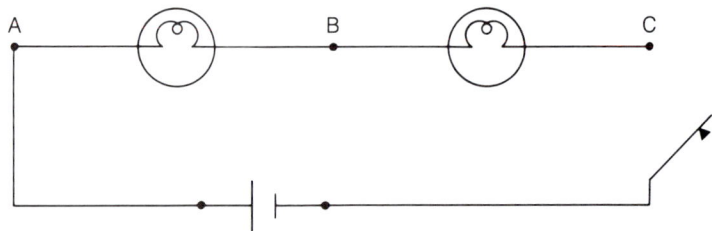

1. Using the operational definitions, what type of circuit is this?

   _____

2. a. What happens when you connect a wire from A to B? _____

   b. What happens when you connect a wire from B to C? _____

   c. What happens when you connect a wire from A to C? _____

   d. What happens when you insert another lamp so that the circuit is

      still the same type? _____

   e. What happens when you put another dry cell in the circuit con-

      nected as follows? _____

   f. What happens when you reverse one of the dry cells so they are

      connected as follows? _____

b. Nonconductor

_____1 Connect a lamp, three wires and a cell, as shown in the diagram. A nonconductor is any object that will not cause the lamp to light when it is connected across points A and B.

_____2 A nonconductor is a piece of chalk.

_____3 A nonconductor is something that will not cause a lamp to light when it is connected in a closed circuit.

E. More on Circuits

After your students have completed the activity described in Task Card #4, have them compare and share results such as these:

1. A series circuit
   a. Lamp 1 goes out and lamp 2 glows more brightly.
   b. Lamp 2 goes out and lamp 1 glows more brightly.
   c. Both lamps go out.
   d. Three lamps glow less brightly than two lamps.
   e. The lamps glow more brightly.
   f. The lamps do not glow.

2. A parallel circuit
   a. Both lamps go out.
   b. Three lamps glow less brightly.
   c. The lamps glow more brightly.
   d. The lamps do not glow.
   e. The lamps glow with the same brightness as they will with one dry cell.

3. You will find it useful to have them also discuss who needs to know

about circuits or operational definitions.

EXTENDING THE ACTIVITY

F. Short Circuits
   1. Give each group of students a piece of Christmas tree tinsel, an index card, and two clips arranged as shown below. Also, give them a bulb and holder, a battery and holder, a switch, and four ten-centimeter pieces of covered wire.

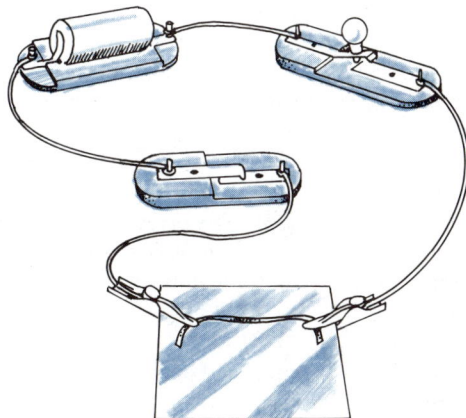

   2. Which items change the bulb?
   3. What is the meaning of "short circuit"?

G. Liquid Circuits
   1. This is a demonstration for you to do with the group. Using a grapefruit or weak acid solution, connect wires to two strips of material (for example, aluminum or zinc) and a meter to register the flow of current. As you make the comparison between the two kinds of strips specifically focus the discussion on what observations support their inferences.

Implications

What do you do when trying to apply the ideas of operant conditioning in classroom situations? There are some straightforward procedures that we can recommend.

1. *Identify the problems.* What is wrong? What are the specific behaviors that you wish to change? This step is especially important if you wish to change unnecessary or disruptive behaviors. (If you are only concerned with learning science concepts, this step may not be applicable.)

2. *Clarify the behaviors you desire.* Your expectations should be specific. "I want you to watch this film closely," or "You should finish your assignment in ten minutes," or "Today you should stay in your seat and not talk." One approach here is to use behavioral objectives. In this case, the terminal behavior is defined: the student should be able to sit in his or her seat. The student should be able to finish his or her work.

3. *Identify the available reinforcers.* Do not presume that candy or a smile is a reinforcer for the student. Spend some time identifying things that are reinforcing for the student. Use the Premack principle, if possible.

4. *Determine the most effective reinforcement schedule.* Since situations and behaviors vary, so should schedules of reinforcement. If you have a student that is very disruptive, you will be best served by selecting (1) reinforcers that are important to the student, (2) schedules that allow for immediate reinforcement of desired behaviors, and (3) continuous reinforcement of behavior. Less extreme situations, such as work on lessons and science concepts, require less extreme schedules.

5. *Establish a baseline of behaviors.* Once you have identified behaviors you wish to change, take a few days to observe how often these behaviors actually do or do not occur. It is helpful if these observations are graphed. This is a step most teachers do not follow. Yet it is important to have a standard of comparison so you will know how your reinforcement program is working.

6. *Determine successive approximations of the be-*

havior. In most cases, the desired behavior is more than can be achieved by the student, and one should not deliver a reinforcement until the desired behavior is demonstrated. You will have to *shape* the desired behavior. **Shaping behavior** is done by establishing successive approximations of the final behavior and then reinforcing those approximations until the full behavior pattern is established.

Reinforcement really works. Systematic application of the ideas presented in this section can decrease classroom problems and increase your effectiveness as a science teacher. Reinforcement occurs in every classroom, so it is not a question of whether or not to use reinforcement schedules. The question is one of your awareness of consistency and of direction. Are you, as a science teacher, consistent in reinforcement? Do you have goals and objectives that indicate a direction of learning for reinforcement programs? We hope you will.

LEARNING THROUGH DISCOVERY

Another way that learning takes place is through discovery. A chimpanzee is placed in a large cage with several different objects for his entertainment. Before long, the chimpanzee discovers that he can use a stick to pull things toward him if he does not feel like walking across the cage. One day a long stick and a banana are left outside the cage. This presents a problem. The chimpanzee wants the banana but cannot reach it with a shorter stick that is in the cage. The chimpanzee discovers this when he tries to rake in the banana with this stick. The frustration upsets the chimp and he retires to one corner of the cage and sits brooding. Suddenly the chimpanzee jumps up, runs across the cage, and seizes the short stick. He uses the short stick to rake in the longer stick and then uses the long stick to rake the banana into the cage.

We have just described a famous experi-

Wolfgang Köhler designed experiments to illustrate the role that insight plays in solving problems.

ment conducted by Wolfgang Köhler (1925), a Gestalt psychologist. In this experiment, the chimpanzee was first presented with a problem. In order to solve the problem, several old patterns of thought and behavior had to be combined into a new pattern, a process which we call **discovery learning.** How is this approach to learning different from B. F. Skinner's theory of reinforcement and operant conditioning? The chimpanzee (his name was Sultan) was not specifically conditioned or reinforced for any of his activities. Some might argue that the chimpanzee was indeed conditioned to take things and reinforced for this activity, even though the reinforcements were not consciously scheduled. However, the situation, problem, and solution seem unique enough to suggest that old behaviors

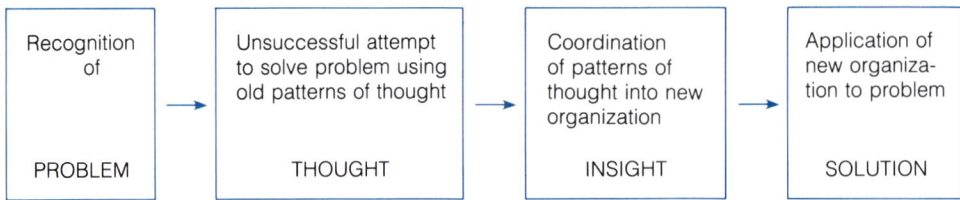

FIGURE 3–3
Learning Through Discovery

could not have been generalized to the new patterns necessary to solve the problem. Rather, it seems that the essence of learning in this case was the coordination of thought patterns into a new organization. See Figure 3–3 for a summary of the problem-solving process. The basis for this learning was a problem situation. In order to solve the problem, the chimpanzee had to reorganize patterns of thought. Insight was discovering the pattern that would solve the problem. Finally, the insight is applied and the problem is solved.

A second Earth Science activity, Population Growth, provides an application of learning through discovery. Note that the activity starts with a riddle that is to be completed as homework. With any luck, the students will be engaged by the problem as they think about the riddle (see activity). The insight (answer) is in realizing that the pond will be half full on the day before it is completely full. In the last seconds of the film "World Population" the students should be struck by the dramatic population increase. Next, graphing the growth of snails should establish the ideas (and problems) of exponential growth. The entire lesson is designed so the students discover this concept through the various activities of the lesson.

EARTH SCIENCE ACTIVITY

Population Growth
(Grades 6–8)

MAJOR CONCEPT

The environment has limits in its capacity for change. If the environment is to conserve its identity, changes must not exceed these limitations.

OBJECTIVES

☐ Construct and interpret graphs, predict and form hypotheses.

☐ Combine ideas in new ways; seek alternative points of view.

☐ Describe situations that require decisions.

MATERIALS

Film: *World Population* (ordered from Southern Illinois University, Carbondale, Illinois), graph paper, pencils.

VOCABULARY

population	exponential
growth	linear

PROCEDURES

A. This activity has one major goal: to lead the student to an understanding of the exponential growth of a population. Later lessons can deal with birth and death rates and problems of overpopulation.

Start with what is now a famous French riddle. Just before the end of class give the students the riddle:

There is a lily pond that has a single leaf. Each day the number of leaves doubles. On the second day there are two leaves. On the third day there are four leaves. On the fourth day there are eight leaves. On the thirtieth day the pond is full. When was the pond half full?

Leave the students with this riddle as homework. (The answer is the twenty-ninth day.)

B. On the next day start the class with the film, *World Population*. This is a four-minute silent film that illustrates world population growth extraordinarily well. Do not say anything about the film; simply start the projector. The students will be amazed at the rate of population growth depicted in the last seconds of the film. They will then have a first conceptualization of the problems of exponential growth. Discussing the film and the riddle should provide an excellent introduction to population growth rates.

C. Ask the students to graph an exponential growth rate. You can use a simulation of two snails that each have two offspring per month. And these, in turn, have two offspring a month and so on. Tell the students to construct a table describing the growth of the snail population for a school year of nine months. (A completed version follows.)

Simulated Growth of Snails for a School Year

$2 \times 2 = 4$	month 1 (September)
$4 \times 2 = 8$	month 2 (October)
$8 \times 2 = 16$	month 3 (November)
$16 \times 2 = 32$	month 4 (December)
$32 \times 2 = 64$	month 5 (January)
$64 \times 2 = 128$	month 6 (February)
$128 \times 2 = 256$	month 7 (March)
$256 \times 2 = 512$	month 8 (April)
$512 \times 2 = 1024$	month 9 (May)

4. Have the students graph the results. Graphing will clearly demonstrate the slow, then sudden population increase depicted by the J-shaped curve.

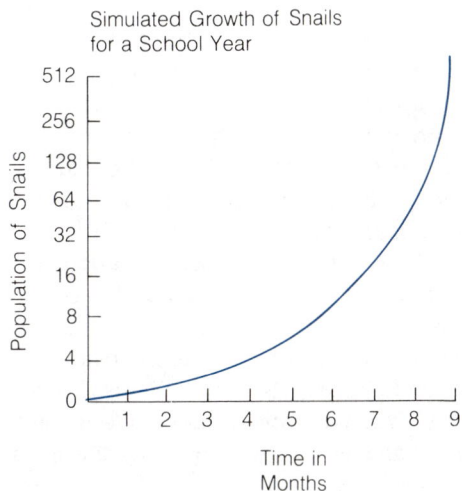

Simulated Growth of Snails for a School Year

5. Discuss the results of the students' activity and apply their findings to world population growth.

EXTENDING THE ACTIVITY

D. Find the present rate of world population growth. Determine the rate of population growth for your city or area. Contrast the geometric growth rate of the population with the arithmetic rate of food production. Read about Thomas Malthus and his theory about world populations.

Learning through discovery is an approach that is not only concerned with overt behavior, but also with the perceptions, motivations, and thought processes that influence behavior. Psychologists who developed this approach are referred to as cognitive field theorists. Learning amounts to rearranging or reconstructing old thought patterns into new ones. Compare this approach to the behaviorist who determines which behaviors are desired, changes the environment, and reinforces the desired behaviors. In this sense, the behaviorist controls behavior. The cognitive field theorist also arranges the environment, but the arrangement encourages learning through problem situations that result in perceptions of unique relationships among objects and events.

When these two theories are applied to teaching, very different methods and techniques are implied. The primary methods of the S-R approach are those of carefully organized programmed learning: the use of teaching machines and class monitoring of student behaviors. The discovery or problem-solving approach is the closest application of the cognitive field theory. In the discovery approach, you design problem situations that result in new insights by your students. As the teacher, you present the problems and provide the means and opportunities for students to solve the problems. Your role includes questioning, assisting, and giving clues or new directions toward solving the problem.

A cognitive field approach has been emphasized in recent years by Jerome Bruner (1960) in *The Process of Education*. Recall that the Gestalt and cognitive field theorists are interested in the individual's perceptions of the larger situation, the field or life space. Bruner's application of this idea to science education is to have students understand the structure of a discipline. The concept of discovery is emphasized by focusing on intuitive approaches to learning. Since students cannot learn all knowledge in a scientific discipline, their science education will of necessity be a sampling of physics, chemistry, biology, geology, astronomy, health, and so on. Presenting knowledge aligned with the structure of a discipline establishes the basis for the addition of knowledge in the future. Intuition is one of the informal intellectual processes of formulating possible solutions to problems without going through the formalized steps by which scientists determine the validity of knowledge. This is the insight portion of Bruner's position.

The learning cycle from an elementary program entitled Science Curriculum Improvement Study (SCIS) provides one of the clearest applications of the different approaches to the teaching and learning of science that we have discussed. According to the authors of SCIS, there are three steps in the learning cycle—exploration, invention, and discovery (see figure 3–4).

In the *exploration* stage, students are given materials that are intrinsically interesting and

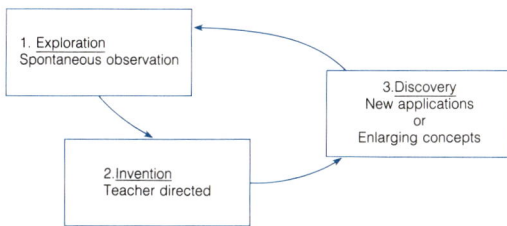

FIGURE 3–4
The SCIS Learning Cycle

related to science concepts that are to be developed. Children are encouraged to spontaneously explore the materials. They are given minimal guidance for the activity in this section of the learning cycle. This stage approaches learning from the child's own interest and natural tendency to inquire. Some questions will occur naturally in the child's mind, and with active manipulation of materials, the child will undoubtedly formulate some solutions to their questions through intuition, discovery, and insights.

The next stage is *invention*. What children learn from the exploration phase is limited by their prior knowledge and understanding of the natural world. There is a need for more formalized definitions, clear description of concepts, or clarification of principles. So, there is an invention section of the learning cycle. The approach in this phase is formal, clear, and directed toward objectives that have been established by the teacher. Though the invention phase is not programmed instruction, the model of education outlined for invention approximates the structure and organization of a behaviorist approach. Let's say you are defining the concept of kinetic energy. You might ask the children to write *kinetic energy is the energy of a moving object* on the chalkboard. The definition could be repeated. Then ask each child to find an example of ki-

netic energy. As they respond, they can be reinforced for their answer. It is possible to place several questions and multiple choice answers on the overhead projector and go through the activity in a semi-programmed manner, reinforcing their correct responses. So, because of the formal presentation, the emphasis on factual information, and the possibility for immediately reinforcing desired behavioral responses, this section of the learning cycle is closely aligned with the behaviorist model of teaching.

Discovery is the last stage in the SCIS learning cycle. In this stage, the children apply the concept learned in the earlier stages to new situations. The situations can be planned, or they may rely on the child's own ideas or investigations to apply the concept. The discovery of new applications and enlargement of the concepts is directly related to the cognitive field idea of learning through developing perceptions of new relationships. At the same time, the act of learning new applications through replication of the concept in different situations is also a means of reinforcing the concept.

Implications

In applying the cognitive field or discovery approach, there are several ideas to keep in mind:

1. *Establish an atmosphere conductive to exploration.* Students must feel free to explore problems. If they feel confined, the opportunities to learn are likewise restricted. Ideas concerning this suggestion are discussed in more detail in the next section.

2. *Present issues and problems that engage the students.* Science teaching is ideal for problems that grasp and hold the attention of students. Once you have their attention, teaching is fairly easy.

Jerome S. Bruner (b. 1915)

When the curriculum movement of the 1960s was beginning, there was need for an idea around which to organize the new curricula. Jerome S. Bruner's small but powerful book, *The Process of Education* (1960), set forth the needed conceptual organization. Bruner's idea that "the structure of discipline" should be the core of the curriculum had a substantial impact on curriculum reform and American education. Bruner is one of a small number of individuals whose ideas have directly influenced a generation of educational curricula and instruction.

Born in New York City on October 1, 1915, Bruner spent his undergraduate years at Duke University. There was a period of searching in which Bruner was not too enthusiastic about any field of study. He was initially in a pre-law program because his family expected him to eventually enter law, but he soon became interested in psychology. Bruner had an interesting entry to this field. At one time he was threatened with expulsion from Duke for cutting chapel. William McDougall, his psychology professor, intervened and saved Bruner from expulsion. After this incident he continued working in McDougall's laboratory and subsequently developed an in-

3. *Encourage perceptions of the whole situation.* Ask questions such as, "How does this problem seem to you? and "What are all the possible ways you might solve the problem?"

4. *Allow students the opportunity for insights.* This is sometimes hard for science teachers because we already have the answers and the solution seems so simple. The easiest way is to tell the students the answer. But it is possible that you are denying the students an opportunity to learn through their own insight. One of the most important applications of discovery approach to teaching is allowing students time to live with their questions.

LEARNING THROUGH PERSONAL MEANING

Psychologists and educators have identified different "domains" of learning. The cognitive, affective, and psychomotor domains are examples. Such divisions are important, even essential for clarifying and discussing different aspects of learning and for doing educational research. However, students enter our science classes as more than "cognitive domains," "affective responses," or "psychomotor structures." Their motivation, learning, develop-

terest in psychology. From Duke, Bruner went to Harvard University and studied psychology, completing his M.A. in 1939 and Ph.D. in 1941.

After completing his advanced degrees he remained at Harvard and eventually became Director of the Center for Cognitive Studies. The work of Bruner and his colleagues at the Center has advanced our understanding of child development and learning. Bruner is now a member of the Department of Experimental Psychology at Oxford University.

Before World War II Bruner studied learning and perception. During the war he analyzed propaganda and public opinion, and he developed an interest in the nature of the processes underlying opinion. Studies in perception, thought, learning, and language resulted from this interest. Eventually he became interested in child development and the process of education. Bruner's contributions to education can be thought of as being on two sides of the same coin: how children learn and how educators can facilitate learning.

As early as 1970 Bruner noted that education was in a state of crisis because it had failed to respond to changes in society. No doubt this great educator would include science education in the present educational crisis and changes in science and technology as important social changes.

In 1962 Jerome Bruner was awarded the Distinguished Scientific Contribution Award by the American Psychological Association. In 1964 he served as president of the same organization. Among his other affiliations are the National Academy of Education, the American Academy of Arts and Sciences, and the American Association for the Advancement of Science.

Throughout his career Bruner has been a prolific writer. His books include *On Knowing: Essays for the Left Hand* (1962), *Studies in Cognitive Growth* (1966), *Toward a Theory of Instruction* (1966), *The Relevance of Education* (1971), *Beyond the Information Given* (1973), and *Play: Its Role in Evolution and Development* (1976).

ment, and behaviors are a combination of intellectual, emotional, physical, and social responses. Student behavior, whether directed toward learning science concepts or creating classroom chaos, is a result of numerous personal and environmental variables. Science teachers often forget these simple ideas, and they forget that categories such as "cognitive domain" are created for study, communication, and evaluation. Students are more than the categories we create. The term *humanistic* reminds us that our students are, after all, whole human beings. In its simplest

form, this is what humanistic education means. The implication is that science teachers should recognize the importance of personal qualities of students and interpersonal dynamics in classrooms as these relate to learning science. Education in science contributes to the student's development in ways unique to the discipline; that is, the knowledge, methods, and values of science are central to science teaching. At the same time, we cannot ignore the personal development and interpersonal skills of the student. The personal, humanistic dimension of teaching is in-

Motivation for learning stems from internal sources such as a sense of belongingness and self-esteem; it is also triggered by enthusiastic teachers and fascinating materials. (Strix Pix)

evitable if one is to be an effective science teacher.

In the following pages, we shall look at the dimensions of **humanistic learning.** One humanistic psychologist's views will be emphasized, those of Carl R. Rogers, as expressed in his book, *Freedom to Learn for the 80's.*

Personal Meaning and Learning

For Rogers (1983), there are two types of learning: learning that is important, significant, and meaningful *to the individual,* and learning that is unimportant, insignificant, and meaningless. The *individual* is the final authority on the appropriateness and importance of the learning experience. Rogers goes on to describe the components of significant learning. *The learning is initiated by the individual.* The

motivation to learn, to engage in an activity, originates within the individual, not solely in the environment. The second aspect of significant learning is *personal involvement.* The learner is actually involved with activities, materials, and environments that lead to new understandings about himself or herself and the physical world. Self-initiation and personal involvement do not necessarily mean that students should be left to learn on their own, though we all know they often do learn on their own. Teachers have the crucial role of designing a learning environment that will maximize the opportunities for learning and will facilitate the process of learning once it starts.

In the humanistic model, learning is summarized as: *Learning = Cognition + Meaning.* Note that this is somewhat similar to earlier models, with the additional stress on meaning. Cognition is used here in the usual definition: the process of acquiring knowledge or that which comes to be known. Usually, cognition refers to knowledge about the physical world. Humanistic theory includes personal knowledge or cognition the individual has about himself or herself. Notice that the amount of learning is equal to cognition *plus* the meaning those cognitions have for the individual. Meaning is more than reinforcement as described by behaviorists or discoveries described by cognitive field theorists. The meaning of any particular knowledge is a result of the individual's perceptions of the importance of the knowledge. In this view, a reinforcement (let us say food, a smile, or a grade) may or may not have meaning depending on individual perceptions.

Personal meaning is a result of psychological closeness (self-initiation and motivation), and/or physical closeness (involvement of the individual with actual materials). Let us say a student has a question about the different colored "things" (minerals) in rocks. Ideally, the elementary science teacher would let the student examine and describe the "things," discuss their properties, and see if the same "things" can be found in different rocks. In this example, learning was self-initiated, followed by personal involvement with the actual materials. The reverse is also possible. Students can engage in activities with materials that originally have meaning due to their physical closeness. If the materials are interesting, the next step is predictable—students will ask why, how, what, and when—they will initiate their own learning. So, the critical component of personal meaning can work either way: from self-initiation to involvement, or from physical involvement to interest and questions. But personal meaning must be there. To the degree that there is personal meaning, learning is increased.

According to Rogers, the teacher's goal should be to facilitate student learning as opposed to concept presentation. The emphasis of this goal is only subtly different from the goals of other theorists but important to understand. The key words are "facilitate"—assist, make easy, direct; and "present"—introduce, offer, display. Both are critical to learning. Emphasizing the facilitation of learning directs the elementary and middle school teacher toward the more personal aspects of teaching and meaningful learning. Facilitating learning includes the traditional goals of understanding science concepts and methods and places these ideas in a context of personal meaning for the student.

The Environmental Biology activity, Recording Changes in Our Gardens, uses personal meaning as an important component of learning. Children are naturally curious and interested in the growth of seeds. By providing students with *their own* garden, the activity acquires a personal meaning which becomes a central aspect of the lesson. This interest will make students more eager to learn about other concepts related to the activity.

ENVIRONMENTAL BIOLOGY

Recording Changes in Our Gardens
(Grades 3–5)

MAJOR CONCEPT
Living things change when they interact with other things (living and nonliving) in the environment.

OBJECTIVES
☐ Record changes in plants.
☐ Observe, compare, measure.
☐ Handle and care for plants under supervision; handle nontoxic substances
☐ Value the health of plants
☐ Select alternatives in our behavior and accept the consequences.

MATERIALS
Seeds (peas, beans, squash, and radishes are fast growing), soil mix, measuring cup, containers (milk cartons, pressed peat moss cups, and plastic tumblers with holes punched in their bottoms with a hot ice

pick), trays to hold containers, paper for drawing.

VOCABULARY

seeds (by name)	germinate
soil mix	sprout
containers	seedling
water	leaves
measure	flowers
grow	fruit
develop	

Note: This list is an auditory vocabulary.

PROCEDURES

A. On the backs of the seed packages, you will find detailed descriptions of how to prepare soil for gardening and how to plant seeds.

B. Arrange a time when all students can plant their seeds at once. Encourage students to work in teams at tables so that they can share their experiences and enthusiasm with each other. You may want to arrange for two parents, volunteers, or upper grade tutors to come assist the children during this thirty to forty-five minute planting activity.

C. After the planting activity is over and students are waiting for their seeds to sprout or germinate, introduce the idea of a picture book that tells the story about their plants. Begin by showing them a "blank" book: five to ten pages of drawing or ditto paper, covered by two sheets of colored paper, stapled or held together by brads.

Open the blank book and explain how each student will have a book and keep a record to show how his or her plants grow. Every week they can draw a new picture to show how much their plants grow each week, and tell or write a short story about their plant. The story can be about how much water they gave the plant, what new features (leaves/flowers) it has developed, and whether the container was in a window sill, or on their desk, or outside in the sun, rain, and wind.

Ask students to begin their books before the seeds sprout, with the first page showing a picture of what seeds they planted and telling how they started their container garden.

D. Hold brief informal daily observation/discussions of about five to ten minutes, giving students a chance to see what changes their plants have made overnight. These periods are also a good time to have students water their plants and share their experiences with each other.

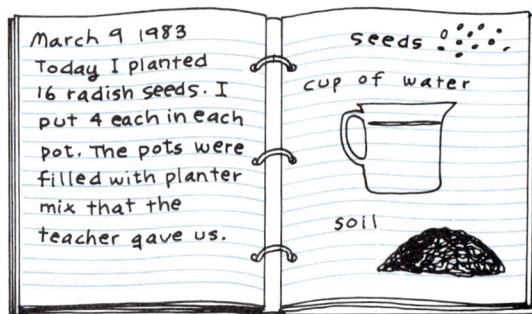

Implications

How does a science teacher achieve the goal suggested by Carl Rogers? What does an elementary or middle school teacher actually do to facilitate learning in science? Being an effec-

tive facilitator requires an understanding of scientific knowledge and the ability to plan lessons, organize classroom procedures, and use a variety of teaching methods. In addition, there is also the essential attitude of all teaching, namely, personal interaction with your students in the learning process. Rogerian theory is especially important for the interpersonal aspects of facilitating learning in science. The following are some personal qualities essential to facilitating learning:

1. *Effective facilitators are first and foremost themselves.* They teach without a distorting or artificial facade. They exhibit the quality of being real, genuine, and natural. Being authentic includes expressing emotions such as anger, disappointment, or personal inadequacy. It also includes being happy, excited, and proud.

2. *Effective facilitators care for the learner.* The caring is directed toward students simply because they are people. Caring is not based on qualities, actions, or expectations of the other. While teachers admit they can't love all their children, they are encouraged to strive toward this goal of respecting and caring for the individual; it is essential for effective science teaching.

3. *Effective facilitators accept students as they are.* This is closely related to caring. Accepting does not mean setting up expectations that must be met before acceptance is shown. Accepting the students does not mean condoning all their behavior. Teachers can accept and respect individuals and not condone disruptive behavior.

4. *Effective facilitators develop an empathy toward students.* This attribute refers to the ability to understand students from their frame of reference, to see problems as students see them, and to feel the same ways that students do. Frustration at not finding the answer to a problem, uneasiness about a question, or anxiety about scientific ideas are three examples where empathy will help the teacher-student interaction and, as a consequence, perhaps overcome the problem and facilitate learning.

Carl Rogers' ideas provide valuable insights concerning the dynamics teachers should use to gain an accurate understanding of the student's perceptions. Then, based on a clearer understanding of the student's learning and development, the teacher can present challenges and opportunities that will further learning. The preceding implications outlined personal qualities of the teacher. We can go one step further and discuss the interpersonal qualities that are important for effective elementary and middle school science teaching.

1. *Developing a rapport.* In order to understand the learner, there must be a warm and friendly rapport between the teacher and student. Developing a rapport with students communicates a desire to help and nurtures the reciprocal process, the desire to be helped. The initial burden for establishing and developing a helping relationship is on the teacher. With time, the student should communicate the desire to be helped.

2. *Attending to the student.* Observing and listening to the learner communicates the desire to understand. It is the active process of trying to comprehend just how the student views the situation. Active attention is personal involvement in the learner's perception of the problem, concept, or activity.

3. *Responding to the student.* The next important part of understanding the learner is communicating your perceptions of the situation. Reflecting and clarifying what the student has expressed establishes better communication, and helps the student gain new insights, and makes him or her feel understood. Importantly, this communication gives you a clearer view of the student's understanding.

4. *Questioning for understanding.* Your questions can be a key to further understanding. Ask the student to justify his or her answer. "How did you come to this conclusion?" "Can you tell me why your decision is the best one?" After you have asked the question—wait. When the student answers—listen.

5. *Facilitating the student's learning.* The next step is suggesting or initiating activities, projects, reading, and so on that will enable the student to move a step further in his or her learning. By this time you should have a clear understanding of what the student hopes to accomplish and

what he or she understands in science. Then it is up to you to assist, direct, guide, and make easy the next steps in the learning process.

STUDENT DEVELOPMENT

The playground scene below suggests some of the topics covered in this section. The characteristics of children and adolescents between the ages of five and fourteen years will be described in three dimensions: physical development, social-emotional development, and intellectual development.

Scene: (on a playground)

First Child: Come on. I've got the ball. Let's play Four Square.

Second Child: I like to play ball.

First Child: Here, you get in the square. I'll start.

Second Child: I want the ball.

First Child: Hey, wait a minute! You're supposed to stand there, in the square......And let go of the ball.

Second Child: Let me have the ball!

First Child: O.K. I'll get in the square and you can start with the ball....Hey! Where are you going with the ball? Come on back here!

Question: From this description, what can you tell about the developmental levels of the two children?

Analysis: If you guessed that the two children are of different ages, you were correct. What are the clues? First, their use of language is quite different. One uses a more elaborate language than the other. Another clue is in their different reactions to the ball: one sees the potential of playing a game so that the two can use the same ball, while the other sees it as an object to be played with alone. A third

clue is that the first child has mastered the rules of a game which the second child does not understand. And finally, the first child is aware of the second child's point of view, and is willing to switch roles or turns so that the game can begin. The second child, ignoring this courtesy, goes off with the ball unaware of the game's importance. How might an age-equal of the older student have replied if he or she didn't care for Four Square? He or she would have suggested an alternative game or activity.

To give you a sense of the developmental changes that take place during the elementary and middle school years, students are described in age groups. Children between the ages of five and seven are grouped under primary grades. Children between the ages of eight and ten are usually in the third through fifth grades and are described under intermediate grades. Children who are becoming adolescents (ages eleven to fourteen) are represented by the middle grades which include sixth through eighth grades. From these descriptions, you will begin to understand some of the physical, social-emotional, and intellectual limits of your students and anticipate some of the changes that will take place in your students as they grow and develop during the school year. However, keep in mind that some students will be well ahead or behind the majority of their classmates in any of these characteristics.

Physical Development

The first impression you will have of your students will be based on their physical appearance. Later you will form impressions about their social-emotional and intellectual quali-

Boys and girls are nearly equal in size during the intermediate grades (ages 8 to 10 years) and have developed fine eye-hand coordination. (Donald Birdd)

ties. Students' impressions of themselves and their classmates follow a similar course: first they become aware of their physical characteristics, and later form impressions of identities based on behavior. During their formative years, students are especially aware of their physical appearance, and their reactions to themselves and others. To a large extent, they are governed by these impressions. Thus, physical development is a logical place to begin a discussion of the general development of children and adolescents.

What should you know about the physical development of your students? First, it will be helpful to think of physical development as composed of two processes: growth, or sim-

ply an increase in size, weight, and height; and morphogenesis (from the Greek words *form* and *to be born*) meaning the appearance of new form or shape such as the changes that occur during puberty.

Growth and morphogenesis lead to other characteristics of students that will interest you. If you were to observe three students, one in the primary grades, another from the intermediate grades, and a third from the middle grades, you would soon become aware of differences in their activity levels. You also would become aware of differences in their ability to coordinate their muscles and their five senses. These four characteristics—growth, morphogenesis, activity level, and coordination—will be described for each of the age groups or grade levels, and summarized in table 3–2.

Primary Grades. Between the ages of five and seven, children range in weight from an average of 15 to 25 kilograms (approximately 35 to 45 pounds) and range in height from an average of 100–120 centimeters (about 40 to 50 inches) or approximately 60–75 percent of their adult height. Nutrition plays an important part in their rate of growth but both weight and height are also governed by genetic factors. Differences in rate of growth or morphogenetic change are not apparent at this stage of physical development.

During the primary grades, children are very active and require frequent rests. When they become overly tired, quarrels and crying are frequent. An educational plan that alternates moderate physical activity with rests or quiet activity, or that is structured so that students move from teacher-directed to student-directed activity, seems to best fit the physical needs of primary children.

The amount of time that primary children are able to sit and attend to one task requiring their attention varies considerably among children and with different kinds of tasks. The more intellectually taxing or the more demanding in psychomotor skills (muscular activity coordinated with the senses and thought processes) a task is for individual children, the shorter the amount of time they can attend to the task. Thus, an average attention span for this age group might be fifteen to thirty minutes. Deviations from this average, however, are certain since children will vary widely in their development and experience.

Running, skipping, jumping, and balancing are typical activities at this age when children are improving control of their large muscles. Eye-hand coordination is just beginning to develop for the youngest children; tasks which

require the use of simple tools such as a hammer, ruler, or brush, or tasks which require students to pour liquids are probably helpful in the development of eye-hand coordination.

Intermediate Grades. In general, between the ages of eight and ten or eleven, children enter a stabilizing period of physical development. Boys and girls are about equal in size and rate of change until near the end of this period. Then some girls begin their growth spurt and enter puberty. Weights range from an average of 20 to 35 kilograms (about 45 to 75 pounds) and heights range from an average of 115–140 centimeters (about 45 to 55 inches) or approximately 70–90 percent of their adult height.

Students are still active and often require rest after exertion at recess. In the classroom, they may be able to sit and attend to a single task for twenty to forty minutes, but their ability to concentrate on a task is related not only to their need for physical activity but also to success with the task. Changes of pace are helpful; for example, a science activity might include making observations of some kind for fifteen minutes and recording observations for fifteen minutes.

Intermediate grade children find eye-hand coordination much easier than they did in the primary grades and are ready for tasks that involve more advanced psychomotor skills such as using hand lenses, thermometers, or simple balances; constructing simple projects without supervision; or handling plants and animals without supervision.

Middle Grades. Students in the middle grades undergo the greatest physical changes, and teachers are likely to observe more physical variation among different students. By the beginning of this period, most girls reach puberty and become aware of their developing breasts and hips accompanying the onset of

TABLE 3–2
Summary of Developmental Characteristics of Children and Adolescents
Aged Five Through Fourteen Years

	Primary Grades (K–2)	Intermediate Grades (3–5)	Middle Grades (6–8)
Physical Characteristics	*Growth and Development:* Height 100–120 cm./40–48 in. (60–75 percent of adult height); weight 15–25 km./35–50 lbs. Boys and girls about equal in size per same age	*Growth and Development:* Height 115–140 cm./46–56 in. (70–90 percent of adult height); weight 20–30 km./47–75 lbs. Boys and girls about equal in size per same age	*Growth and Development:* Height 135–165 cm./54–65 in. (85–99 percent of adult height); weight 30–50 km/70–110 lbs. Girls begin to surpass boys in size/development per same age
	Activity Level: Very high; need frequent rests. Able to sit for fifteen to thirty minutes between activity breaks	*Activity Level:* Active; may need rest after recess. Able to sit twenty to forty minutes between activity breaks	*Activity Level:* Moderately active when given the chance. Able to sit thirty to sixty minutes at task with change of pace in task
	Coordination: Large muscles develop in running, tumbling, balancing; eye-hand coordination begins to develop with written tasks and manipulative activity	*Coordination:* Refinements in eye-hand coordination; good control of large muscles	*Coordination:* Growth spurts cause awkwardness; continued refinement of eye-hand coordination
Social/Emotional Characteristics	*Dependence:* Asks for help from adults frequently	*Dependence:* May ask for adult help on some tasks/decisions	*Dependence:* Rarely asks for adult help on tasks/decisions
	Friendship bonds: Weak; changes friends often; plays in twos and threes	*Friendship bonds:* Stable; plays in threes and fours and on teams, occasionally with conflict	*Friendship bonds:* Very strong; shares in threes and fours; enjoys team activities

TABLE 3–2, *continued*

	Primary Grades (K–2)	Intermediate Grades (3–5)	Middle Grades (6–8)
	Authority: Readily accepts; begins to value rules and fair play at end of primary grades	*Authority:* Generally accepts, but begins to question; emergence of peer group norms; strict adherence to rules and fair play	*Authority:* Accepts authority but frequently rebels in preference to group norms; follows rules/fair play, yet accepts exceptions
	Goals: Sets unrealistic goals; often cries when frustrated	*Goals:* Becomes more realistic; occasionally cries when frustrated	*Goals:* Unstable; self-expectations fluctuate from excessively high to low; rarely cries in front of others
Intellectual Characteristics	*Language:* Highly imaginative; changes from egocentric (unaware of another's point of view) to socialized speech	*Language:* Socialized speech develops further; imaginative but without fantasy; enjoys secret of hidden codes and meanings; begins to use antisocial speech late in intermediate grades	*Language:* Uses elaborated speech for classroom but restricted speech for close friends; frequent use of antisocial speech with some students
	Use of symbols: Struggles with word and number concepts first two years but on the average acquires fluency by third year; exceptions become frustrated	*Use of symbols:* Increased fluency with word and number concepts; anticipates meaning will be ascribed to symbols	*Use of symbols:* Highly developed sense of word and number concepts; manipulates familiar concepts with ease
	Reasoning: Focus limited to one variable at a time (length, width); learns to seriate and classify by end of third year	*Reasoning:* Able to focus on more than one variable at a time; can describe rules and relations of objects when concrete examples are available	*Reasoning:* Able to focus on several variables at once; begins to describe rules and relations of objects applied to hypothetical (real but not visible) situations

Note: Physical characteristics based on data from *Development and Classroom Learning: An Introduction to Educational Psychology* by J. M. Stephens and E. D. Evans. Copyright © 1973 by Holt, Rinehart and Winston, Inc. Reprinted by permission on Holt, Rinehart and Winston, CBS College Publishing.

During the intermediate and middle school years, students are less dependent upon adults for direct assistance and enjoy working on their own or with peers but they still need and depend upon feedback from adults. (Michael Hayman, Corn's Photo Service)

menstruation. Among those boys who also reach puberty near the end of this time period (eleven to fourteen), there is considerable self-consciousness over the uncontrolled changes of voice and the awkwardness associated with getting used to their new weight and height. However, it is generally more difficult for those boys who show no sign of such growth or development and find themselves among the smallest students in the class.

Weights may range from an average of 30 to 50 kilograms (70–110 pounds) and heights can range from an average of 135–165 centimeters (about 55 to 65 inches) or approximately 85–99 percent of their adult height. Activity levels are moderate at this stage, although the wide variation in students' rates of growth will make more apparent differences in their energy levels and needs for rest and nutrition. As a teacher, you will need to re-

spond to these differences and help students understand the need for exceptions to general rules. Students are able to sit and attend to a task for thirty to sixty minutes, but changes of pace are helpful for most students and so tasks that include a combination of reading, writing, talking, watching films, or physical activity in the classroom or outdoors enables students to attend to learning tasks for longer periods of time.

Coordination of large muscles can be awkward for those students experiencing growth spurts. They find that skills they had previously mastered are momentarily difficult again until they become accustomed to their new height and weight. Eye-hand coordination is finely developed by this time and students are able to use science tools such as microscopes, cameras, and equal arm balances that require fine discriminations. And students find great satisfaction in constructing projects or handling plants and animals, all without supervision.

To help you and your students focus attention on the characteristic physical development, consider the activity below which also appears in chapter 9, "Environmental Biology." The activity is also designed to appeal to and be appropriate for younger students (ages five to seven) but could be used with older students as well.

ENVIRONMENTAL BIOLOGY ACTIVITY

Describing Myself and Others
(Kindergarten–Grade 2)

MAJOR CONCEPT

Living things have identities based on their characteristics.

OBJECTIVES

☐ Describe oneself and another person in such a way that the identity of the individual is recognized by others.

☐ Observe and measure.

☐ Use tools.

☐ Value the uniqueness of the individual.

☐ Recognize personal decisions that involve self.

MATERIALS

Mirror, tape measure, bathroom scales, color chart, crayons, paper, and pencils. If a floor length mirror is available, secure it against the wall temporarily and attach the tape measure to the mirror.

VOCABULARY

color words for hair, skin, eyes, clothing
texture words for clothing
size words for height, weight, and hair length
shape words for articles of clothing

PROCEDURE

A. Set up a learning center with the required materials. Demonstrate how children are to use the learning center, perhaps by using it yourself. Point out the special features that give each person an identity: appearance in height; weight; hair color, length, style; color of eyes; clothing; and so forth.

Then invite children to describe themselves, first by looking in the mirror and recording what they see, using crayons and paper, and later by telling someone else. Provide informal assistance to individual children who need to improve their skills of observation and description. Make a temporary display of the completed pictures.

EXTENDING THE ACTIVITY

B. On another day, invite the children to choose a partner to describe. Each pair of children should use the learning center together. Pair less articulate children with partners who have more advanced verbal skills, and ask each pair to work together as a team writing descriptions of each other. Guide the students in writing their descriptions by saying:

1. Try to describe your partner so that we (members of the class) can guess who it is.
2. Use words to describe your partner's color of hair, eyes, and clothing.
3. Tell how tall your partner is and how much he or she weighs.

Use yourself as an example; write on the chalkboard a description of yourself. Then ask someone to read the description aloud with you standing nearby and ask the class who in the room the description fits best.

As students complete their descriptions, have them print the name of the person they have described on the back of the page. Later in the week, children may use these pictures and verbal descriptions to play "Guess Who."

C. To assist children in understanding the concept of an individual's identity, have two students who look somewhat alike stand up together, and ask the class how they can tell these two classmates apart. Help students focus on the characteristics that differ between the two children. Repeat this activity several times until children's descriptions are clear in their distinctions between pairs of students.

Social and Emotional Development

If you were to spend a day going from a primary school classroom to an intermediate grade classroom and, finally, to a classroom in the middle grades, you would notice several ways that students in these three age groups differ in their social-emotional development. First, they would differ in terms of their dependence upon the adult in the classroom and how much contact and support they seek from that adult. Second, you would notice great differences in how students function with others their own age: how well they get along, how they deal with conflicts or handle competition, and how well they are able to cooperate or function as a member of a team. Third, you would also notice differences in student's reliance on and acceptance of authority, standards, codes of behavior, and socially-defined roles. And finally, if your observations were to continue over several weeks, you would detect differences in the kinds of goals students set for themselves and how they react to success or frustration in their pursuit of those goals. These four characteristics will be described for each of the age groups and summarized in table 3–2.

Primary Grades. Children between the ages of five and seven years are very dependent upon the adults in the classroom for a wide variety of needs: projection, approval, guidance, physical contact and affection, nutrition, and assistance with most tasks that are new to them. Primary classrooms are generally places where a great deal of helping and nurturing goes on, and most children appear to have very positive attitudes toward school and adults.

In this social setting, primary-aged children are usually open to guidance in how to get along with others in a group and, although quarrels are frequent, they usually end quickly. Children begin to form friendships by the end of this time period, but the bonds are temporary among younger children and frequent changes of "best friend" are common. Authority of adults is accepted. Gradually, children develop a sense of fair play and expectations about following rules. Adults are expected to enforce rules and reprimand offenders. Role-playing is overt and children engage freely in portraying their perceptions of social roles.

Goals set by children in the primary grades differ from those set by adults. Because the sense of time, in relation to recurring or expected events, develops slowly, children lack the capacity to work systematically toward long-range goals. Moreover, because they lack experience, children's short-term goals or expectations often are unrealistic. For example, many kindergarteners are surprised to find that they cannot be chosen first every time or cannot always be permitted to use some favorite piece of equipment. When their short-term goals or expectations are not satisfied, they become frustrated and frequently cry. But primary school children can just as easily be comforted by a hug or an empathetic explanation.

During the primary grades, children's unrealistic expectations change as they begin to see themselves as members of a larger group. In science activities, learning to set realistic expectations and taking turns talking or using equipment can become a part of any science lesson. Children also need the opportunity to explore on their own in situations that are planned by the teacher to insure safety and learning.

Intermediate Grades. By the time children reach the intermediate grades, they are less dependent upon adults in the classroom. Although they still seek help on various tasks, they begin to take pride in being able to do things for themselves. Nevertheless, they need and depend upon feedback and try to earn approval and recognition from adults. Rarely do students in the intermediate grades seek the amount of physical contact and affection from adults that is characteristic of primary grade students.

During the third through fifth grades, students reach a new level of socialized behavior and willingly participate in group and team activities. However, they freely express their distaste at being assigned to a group not to their liking, and frequently engage in quarrels over leadership, the interpretation of rules, and fair play, or the task assigned to their group. Friendships stabilize and loyalty bonds are sometimes tested.

Authority is generally accepted during these years, but most students will begin to question authority as they approach puberty, a time when peer group norms start to emerge. During these intermediate years, it is helpful to introduce students to the concept and value of *evidence* as students consider various sources of authority.

Goals set by students between the ages of eight and ten or eleven years become more realistic. For example, students enjoy earning specific privileges or developing the skills that are required for particular sports, scholastic activities, or elected offices. Clubs are especially popular at this age if they offer concrete rewards or symbolic badges that accord recognition of accomplishments. As students begin to develop and express their independence, they also learn to deal with frustration; some prefer verbal support to the consoling hug of the teacher.

Middle Grades. The early adolescent years are a time when students begin their overt move away from dependence upon adults and toward equality with adults. This movement produces friction as students challenge adults and expect them to explain or defend their reasons for particular rules, standards, assignments, or behaviors. Unless teachers and parents understand this natural stage in adolescent social-emotional development, they may mistake it for personal rejection. It is a time when students themselves experience a sense of loneliness, isolation, and vulnerability as they begin to separate themselves from dependence on specific adults. Thus, it is crucial that adults accept students' needs for separation and individuation and assist them in their struggle.

As logical reasoning skills develop, students enjoy trying to "prove" their answers to questions teachers pose. (Strix Pix)

As dependence upon adults decreases, reliance upon peers increases. Friendship bonds are very strong at this stage and students generally prefer the company of their own friends to that of adults. Group norms grow in importance as does peer approval. Team activity and group projects are valuable ways for teachers to take advantage of students' natural preferences. Authority is questioned as students search for new and reliable sources of information to restructure their perceptions of the world and their newly emerging set of values. Evidence of this search is apparent in the challenge of authorities nearest them. They test rules or limits of tolerance and the consequences of exceeding those limits. More distant authorities such as textbooks or advertisements are not usually questioned until late adolescence. Evidence of the adolescent emerging set of values is also apparent in

their ability to make exceptions for others in following rules or interpreting standards. For example, a less mature classmate may not be expected to shoulder the same responsibility, have the same understanding, or meet the same physical requirements as a more mature classmate.

Goals can become very unstable during this period of development. As students attempt to define themselves in new ways, their self-expectations fluctuate, swinging excessively from highs to lows. At one moment, a student might aspire to be a great singer and the next moment will want to live alone in a cave because of his or her despair over a changing voice. Adults are frequently unaware of such fluctuations because students tend to share private feelings with their peers more than with adults. But evidence of mood changes is readily apparent in every middle school classroom. Most middle school teachers attempt to minimize the effect of students' mood changes by offering stabilizing support and feedback. Crying is rarely associated with frustration or failure.

The Health Science activity, Peer Pressure and Health Choices, uses the social-emotional development of students as both the motivation for and the content of the lesson. Early adolescents are naturally curious about the changes they experience. Rather than ignore their problems of separation-individuation, or identity and choice of peers, why not start with their questions and build upon this interest? This activity is an example of how this can be done and, in this case, it is done with an emphasis on spelling.

HEALTH SCIENCE ACTIVITY

Peer Pressure and Health Choices
(Grades 3–5)

MAJOR CONCEPT
Life styles, peer groups, and individual families influence health choices.

OBJECTIVES
☐ Classify group pressure in terms of harmfulness and helpfulness.
☐ Identify pressure groups that affect intermediate-level students.

MATERIALS
Speller from the students' current grade level; speller from one grade level above the students; chalkboard; drawing paper, crayons, and pens.

VOCABULARY
group pressure
peer group

PROCEDURES
A. Invite the students to participate in a competitive spelling bee. Divide the class into teams of equal numbers. Select one or more students to act as scorekeepers if it is necessary to equalize the number of players on each team. Encourage the students to think of names for their teams.
B. Explain the game rules to the students:
  1. The game consists of two rounds. During each round, every member of the team has one turn.
  2. When it is your turn, stand in front of your team.
  3. You may ask the teacher to give you a word from one of three different levels: Level One, Level Two, or Super Speller Level Three.
    [Note: Words selected from a speller of the students' current grade level may be used for Level One. Words from a speller of the next grade level may be used for Level Two. Save the most difficult words from this speller for Level Three.]

4. If you spell a word from Level One correctly, your team gains one point. A word from Level Two spelled correctly earns three points. A word from Super Speller Level Three, if spelled correctly, earns six points.
5. You cannot change your mind and ask for another level word after the word has been given to you. However, you may decide on a different level for your round-two turn.
6. A misspelled word earns no points. Once you have misspelled a word, you do not have a second opportunity to spell it.

C. After round one, ask the scorekeepers to add up each team's points and record the totals on the chalkboard. This subtotal is posted to encourage competition between the teams, increasing group pressure.

D. At the end of round two, ask the scorekeepers to calculate round-two scores for each team and final scores for both rounds one and two. The team with the highest number of points wins the game.

E. Following the game, lead the class into a discussion of the results. You might ask the following questions:
1. During the game, which of you chose a higher level word than you really wanted to? Are you glad you did? Did that decision earn more points?
2. Which of you decided to ask for a lower-level word, even though you

really wanted to try a harder word? Why didn't you?
3. Were the round two scores higher than the scores from round one? If so, why do you suppose this happened?

F. Next, introduce and define the terms *group pressure* and *peer group*. Encourage the students to discuss the following questions. List and classify students responses on the chalkboard.
1. When can group pressure be harmful?
2. When can group pressure be helpful?
3. Can you think of some different pressure groups that might influence the decisions you make about the kind of bike you want or the type of clothes you like to wear? (Look for answers such as family, classmates, neighborhood friends, and club members, perhaps scouts.)
4. Can you remember and describe a situation when you, or someone you know, had to make a decision in the face of group pressure, like right in front of your pals? Can you explain some of the possible consequences that could have occurred had you decided differently?

Throughout the elementary and middle school grades, teachers frequently organize discussions of classroom rules and expected social behaviors in order to provide the structure students need for their social-emotional development.

From *Health Activities Project,* Trial Edition. Berkeley, CA: University of California, Lawrence Hall of Science, 1976, 1977, 1978.

Intellectual Development

Our understanding of how the human brain works and how learning takes place changes almost daily based upon new discoveries reported by neuroscientists and cognitive scientists. For example, as we write this chapter scientists are conducting experiments to learn how the brain develops and

changes from birth to adulthood.

By the time you were six years of age, your brain reached its present approximate weight: between 1,000 and 1,400 grams, or just about 3 pounds. But there are major differences between your brain's activity then and now. When you were six years old, your brain's activity was much slower and more ir-

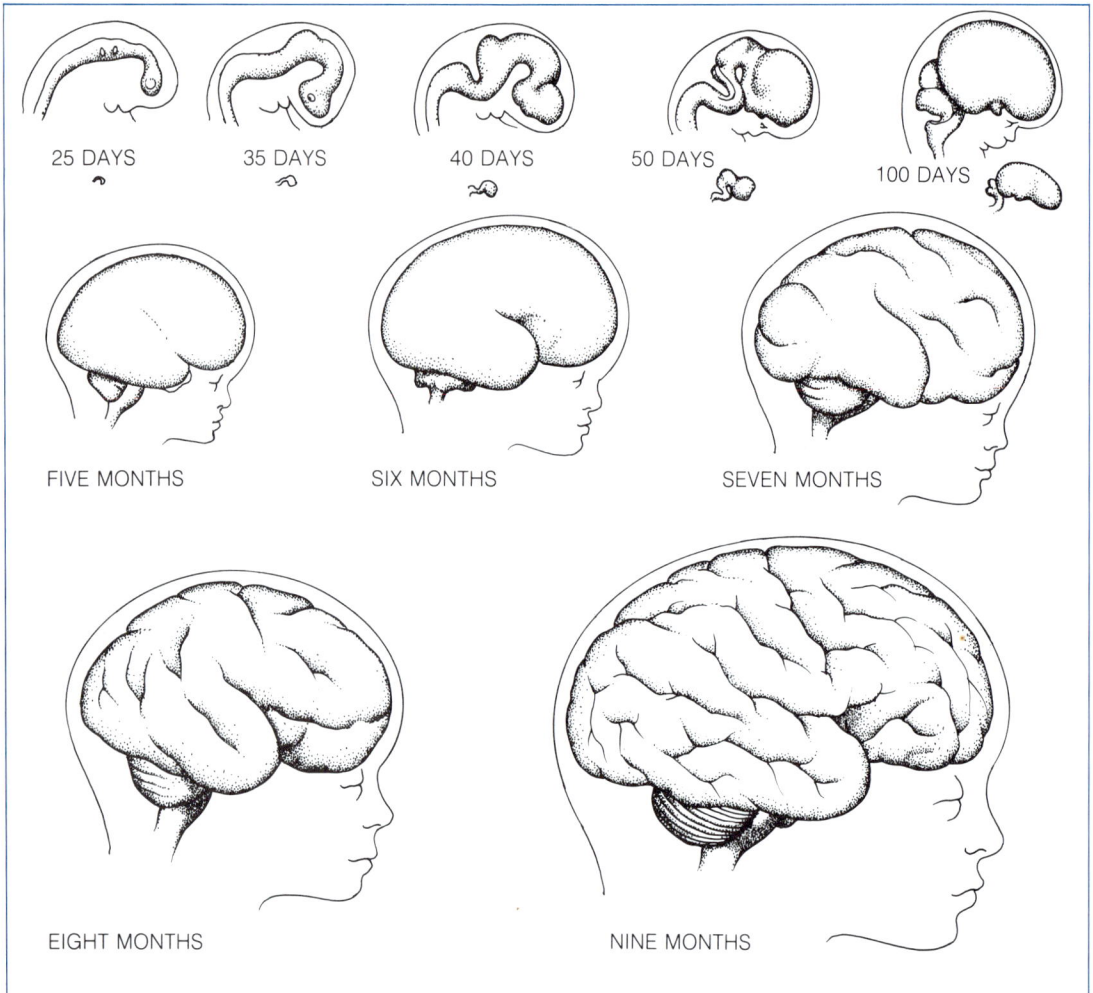

25 DAYS   35 DAYS   40 DAYS   50 DAYS   100 DAYS

FIVE MONTHS   SIX MONTHS   SEVEN MONTHS

EIGHT MONTHS   NINE MONTHS

FIGURE 3–5
The views of the developing human brain depict the change from a relatively simple organ at twenty-five days to one with characteristic convolutions and invaginations on the surface and about 100 billion neurons at birth.

regular, if you were to compare it with the nine to twelve waves per second an electroencephalograph (EEG) would show for you now. Moreover, your brain had more connections between its cells, interconnections called synapses, when you were six than you will possess at any other time in your life.

The circumference of your head probably showed three growth spurts during your school years; at the approximate ages of seven, eleven, and fifteen years. And interestingly, two of these increases and possibly all three increases in circumference were correlated positively with major spurts of your brain growth and mental capabilities. Ages seven and eleven or twelve years are clearly ages of major growth in intellectual development and there is some evidence that a spurt occurs at age fifteen as well.

What happens as intellectual capacities develop during the years individuals attend school? This topic has been of interest to neuroscientists, psychologists, and educators for the past five decades. Most of their research has dealt with changes in the complexity of children's and adolescents' thought processes rather than with their accumulated knowledge and perceptions. Thus, the next section describes the general characteristics of intellectual development and patterns of thought.

Observing and talking with several students in the primary grades, the intermediate grades, and the middle grades will quickly reveal differences in their use of language and written symbols. Given a few days to talk with them individually, you will also become familiar with differences in their patterns of reasoning. These three areas of intellectual development—spoken language, use of written symbols, and reasoning ability—will be characterized for the three age groups. Intellectual development is summarized in table 3–2.

Primary Grades.   Between the ages of five and seven, the language of primary school children undergoes a substantial change. Evidence of the change is found not only in speaking but also in the way they use language in various social settings. The need to develop new capabilities in language occurs for most children when they enter kindergarten although those who have attended nursery school or preschool have begun the transformation earlier. Children find out that the language they used with their family is not completely satisfactory in the new setting of the school. People outside of their families have not shared the same histories and therefore do not understand the same abbreviated or personalized forms of language that are understood and accepted in their homes.

The complexity of sentences, including such factors as verb tense (I go/goed/went), syntax (why the dog won't eat/ why won't the dog eat?), plurals (the kitty's feets/feet), the length of sentences, and the number of syllables used, are all areas that show gradual transformation during the primary grades.

A second major change takes place in language development during these years. Children move from egocentric speech, wherein they are unaware of another's point of view, to socialized speech which reflects this awareness. You might observe, for example, that kindergarteners respond to a story from a single, personal frame of reference. Talking activities such as "Show and Tell" are actually a series of testimonials that are independent from one another and the audience. Even if you, as a teacher, solicit sharing and telling experiences on a specific topic to develop a thread, volunteers are likely to veer off the topic as some non-key word reminds them of an experience that is of special interst to them.

In the primary grades, the major focus of the curriculum is to introduce children to the use of written symbols in reading, writing, and arithmetic. As children first encounter the letters of the alphabet and the numerals, there is a sense of excitement and accomplishment. But suddenly these written symbols are used in various combinations in order to form printed words and to learn mathematical processes such as addition and subtraction. Then the use of symbols becomes difficult for many children. By the end of the second grade, most children understand that letters make up words, which in turn form sentences, and that numerals represent a language for counting and describing changes that take place in sets of numbers or objects. However, the range of

The Bettman Archive, Inc.

Jean Piaget (1896-1980)

On September 16, 1980, Jean Piaget died at the age of eighty-four, leaving over sixty years of professional work. Piaget was recognized for his contributions to biology,

psychology, philosophy, sociology, education, and even literature. The quantity and quality of his work would truly rank him among the great intellectuals of our age, if not of all history.

Jean Piaget was born in the small Swiss town of Neuchatel on August 9, 1896. His father, who was a scholar of medieval literature at the local university, had a critical mind and approached matters systematically. His mother was a homemaker—intelligent and energetic. Both his father's logical approach and his mother's mental outlook had an influence on Piaget's life and work. He was to study the development of logical thought and avoid the study of psychotherapy.

Early in his life Piaget observed nature and expressed his ideas in written form. At age ten he sighted a partly albino sparrow. He wrote about his observation and submitted it to the natural history journal of Neuchatel. The paper was published and Piaget's career began.

During his undergraduate and graduate education Piaget also considered philosophical and psychological questions. Piaget's undergraduate degree, however,

differences in children's abilities to understand and use written symbols has already begun and usually the gap widens during the next six years.

In addition to changes in children's use of spoken language and written symbols, a major change also occurs in their ability to use logical thought during the primary grades. For example, if children are given a pan of water and several objects and asked to explain why

some objects float while others sink, kindergarteners will typically place the objects in the water and tell you which ones sink and which ones float. They do not generalize about the objects by saying these float because....In contrast, second graders will generally perform the same action but, after placing the objects in the water, they can usually observe that the heavy objects sink while the light ones float.

was in biology, which he received at the age of eighteen. He began his graduate studies immediately, also in biology, and was awarded a doctorate at the age of twenty-one.

After completing his degree, Piaget went to Zurich to work in a psychological laboratory. This experience provided some background in clinical techniques and psychodynamics. Piaget then went to Paris and spent some time working at a laboratory school standardizing reasoning tests. From there Piaget returned to Geneva and developed his own studies on the development of thought in children. For the next thirty years Piaget studied and wrote about the cognitive development of young children.

About 1950 Piaget began extending his ideas to other areas of interest. He returned to philosophy, biology, and education and continued his studies of cognitive development by examing adolescent thinking.

Throughout his career, Piaget was recognized for his professional achievements. Honorary degrees from Harvard (1936), the Sorbonne (1946), and the University of Brussels (1949) are a few examples of the recognition he received. In 1969 the American Psychological Association gave long overdue honor to Piaget for his distinguished scientific contribution to psychology. Among the many books that Piaget wrote, there are several that are important for elementary educators: *The Child's Conception of the World* (1929); *Play, Dreams and Imitations in Childhood* (1951), *The Origins of Intelligence in Children* (1952), *The Growth of Logical Thinking from Childhood to Adolescence* (1958), *Language and Thought of the Child* (1962), *To Understand Is to Invent* (1973), and *The Development of Thought* (1977).

Jean Piaget's professional career was long, varied, and productive. There is little doubt that scholars will evaluate Piaget's work, find errors, and suggest revisions. This is even as Piaget's theories would predict. Still, Piaget's contributions must rank with Sigmund Freud and B. F. Skinner. Piaget's life and work provide a model for all. Throughout his career he never failed to find uniqueness in common observations, He never lost sight of his larger task, and he never ceased to find new problems to study.

Intermediate Grades. By the time students reach the intermediate grades, their speech, use of symbols, and ability to reason logically take on some adult characteristics. For example, their speech reflects an awareness of another person's point of view, they become somewhat fluent in the use of written symbols, and they can apply rules to concrete situations in order to classify events. Yet in conversations, one is constantly reminded of the differences in experience associated with age.

During the third and fifth grades, students expand vocabulary, and seek and interpret information from written pages. This search for meaning can be contrasted with the guidance that primary school children usually need to interpret written material. At the ages of eight to eleven years, secret codes and meanings are popular, and antisocial speech begins to appear among older students. The ability to use logic in reasoning about concrete situations is distinctive. In the previous example using a pan of water and several objects, intermediate grade students not only generalize about heavy and light objects, but are able to perform simple tests to "prove" that the weight of an object alone does not account for its floating or sinking. Toward the end of this period, students begin to look for the reasons behind rules and test the limits of rules. Their search for understanding how and why things work can be the basis for many science lessons.

Middle Grades.   As students approach adolescence their speech, use of symbols, and ability to reason logically take on further characteristics of adult thinking and speech. Students differentiate between the language expected in the classroom and the one they use with friends; and occasionally they alternate between the two for shock effect or for attention. By this time, students depend upon written symbols to express ideas that are important to them. Keeping intricate records or private journals and using mathematics to calculate amounts of time or money are evidence of their dependence upon symbols to represent ideas.

During this period of early adolescence, students begin to apply logical thought to hypothetical or imagined events as well as to "real" or concrete events. For teaching science, the implication of this new ability to reason abstractly is that students can be guided to form hypotheses about cause-effect events in nature. Using the example of the pan of water and several objects that either float or sink, an adolescent could "prove" that the material an object is made of determines whether or not the object will float. An adolescent might suggest using three imaginary cubes of the same size, with one composed of wood, another of metal, and a third of

sponge, in order to illustrate the effect of density in determining whether objects would float or sink.

Using table 3–2, trace the changes in physical, social, emotional, and intellectual development of students as they progress from kindergarten through middle school. Then select a single age group and try to imagine how you would plan a science activity to teach them that plants and animals change in form as they grow and develop.

UNDERSTANDING EXCEPTIONAL STUDENTS

So far our discussion has focused on "average" students. Now we introduce characteristics of exceptional students. Students in elementary and middle schools represent a wide range of social, economic, and educational situations. As a teacher you will have to understand the motivation, learning, and development of students who, for a variety of reasons, are exceptional. This section begins with a rationale for having exceptional students in regular classrooms. Then we provide some general and science-related suggestions and recommendations for helping exceptional students.

Exceptional Students in Education: A Rationale

Of the many issues that educators will have to face during the 1980s, perhaps one of the most encompassing is that of a "right to education" for all students. In the late 1970s at-

Portions of this section on exceptional students are revisions of the article, "Helping the Special Student Fit In" by Rodger W. Bybee. Reprinted with permission from *The Science Teacher* (Volume 46, Number 7, October 1979), published by the National Science Teachers Association. (This section appeared as chapter 6 of L. Trowbridge, R. Bybee, and R. Sund, *Becoming a Secondary School Science Teacher.* Columbus, Ohio: Charles E. Merrill Publishing Company, 1981. Reprinted with permission.)

tention was focused on the educational rights of students who were traditionally placed in restricted special education programs. One result of this movement is the recognition of individual differences and the conclusion that has been clear to many teachers for a long time—*all students are* exceptional.

We all stand to gain from having special students in the regular science classroom. Although it is only natural to expect some initial hesitation, frustration, or fear on the part of students and teachers alike, once this period passes, the gains are clear: Special students encounter a whole new range of educational opportunities; regular students learn that in terms of basic human needs and wants, special students are not very different from themselves; and teachers become more sensitive to the realities of different learning styles, subtleties of instruction, and of modifying curriculum to meet students' personal needs. In the end, we all find out more about what it means to be human.

There is another reason for including exceptional students in the classroom. We have a responsibility to provide the best science program for *all* our students. This course has not always been followed. Teachers know that students have unique needs that are not fulfilled by curriculum materials alone. The task is to accommodate our programs and teaching methods to the needs of students, not to make students adapt to our science programs and teaching strategies.

Exceptional Students in Science Programs: The Law

Appeals to personal and professional benefit and to justice have not completely convinced teachers of the need to include special students in the science classroom. The most immediate and forceful argument seems to be the law. We have a *legal responsibility* to include special students in the mainstream of school programs.

One of the first laws that included protection of the rights of special students was the Rehabilitation Act of 1973, Public Law 93–112, Section 504 of which states:

> No otherwise qualified handicapped individual in the United States . . . shall, solely by reason of his handicap, be excluded from participation in, be denied the benefits of, or be subjected to discrimination under any program of activity receiving federal financial assistance.

Since most, if not all, school systems receive federal financial assistance under this law, special students must be allowed to participate, receive the benefits, and have open access to educational programs.

Another piece of federal legislation that included safeguards concerning the rights of special students was Public Law 93–380, the Education Amendments of 1974. This law mandated due process procedures at the state and local levels for placement of special students, assured placement of special students in the least restrictive environment, and set a goal of providing full educational opportunities for all handicapped students within each state. Public Law 94–142, the Education for All Handicapped Children Act of 1975, the regulation with which most U.S. school personnel are probably familiar, requires that special students be integrated into regular classrooms whenever possible:

> It is the purpose of this Act to assure that all handicapped children have available to them . . . a free appropriate public education which emphasizes special education and related services designed to meet their unique needs, to assure that the rights of handicapped children and their parents or guardians are protected, to assist states and localities to provide for the education of all handicapped children, and to as-

sess and assure the effectiveness of efforts to educate handicapped children.

Specifically, P. L. 94–192 requires the following of school personnel:

1. Zero Rejection. No student may be rejected from a free public education and related services. Court cases have resulted in a legal commitment to the public schools for the education of all school-age students. That all students have a "right to education, regardless of their present level of functioning," results in a principle of zero rejection. It should also be noted that education is defined as the development of students from their present level to the next appropriate level. In brief, the assumption is *all* students are educable.

2. Classification of Placement. Evaluation of students shall be nondiscriminatory. Diagnostic and assessment procedures are to be established by each state to ensure that cultural and racial bias are not evident in the system used for identifying special students. Tests shall be a fair evaluation of the student's strengths and weaknesses.

3. Appropriate Education. This stipulation is a requirement for an Individualized Education Program (IEP). An IEP should have statements concerning the student's present level of performance, how he or she will participate in the regular educational program, the type of special services needed, the date special services were initiated, and the expected length of services. In addition, the IEP should set short- and long-term minimum standards, measures of achievement, and an evaluation of educational progress which includes a conference between school personnel, parents, and the special student.

4. Least Restrictive Placement. To the maximum extent possible, handicapped students will be educated with nonhandicapped students. "Least restrictive placement" means that handicapped students should be educated in the "mainstream," the regular educational environment. They can be educated in special programs when the nature or severity of their handicap requires such treatment.

5. Due Process. The handicapped student (usually through parents or a guardian) has a right to question testing and placement. That is, special students are guaranteed procedural safeguards in the placement and provision of special services.

6. Parental Participation. Parents of the handicapped student have the right to be present for their child's evaluation, placement, and development of an IEP.

Public Laws 93–112, 93–380, and 94–142 are based on fundamental principles guaranteed in the Constitution. The handicapped have been systematically excluded from educational programs, which, in essence, has been a violation of the constitutional rights of approximately 35 million Americans. The Fourteenth Amendment guarantees equal protection under the law for all Americans. Recall that the *Brown* v. *Board of Education of Topeka,* 347 U.S. 483 (1954) overturned the earlier "separate but equal" ruling of *Plessy* v. *Ferguson,* 163, U.S. 537 (1896). Separate educational facilities for some students are, by definition, unequal; thus, the handicapped have been deprived of the equal protection of the laws guaranteed by our Constitution.

Of the three laws, P.L. 94–142 was probably the most significant piece of educational legislation of the 1970s and its effect will be felt throughout the 1980s. There are several reasons for this fact. First, P.L. 94–142 incorporates parts of the other laws and clarifies the fundamental right of all students to an education. Second, because it is a federal law, it establishes the "right to education" as a national priority. Third, P.L. 94–142 commits us to the recognition of individual differences and to appropriate educational programs for special students because it is permanent legislation with *no* expiration date. This fact demonstrates the gravity with which Congress approached the legislation.

Exceptional Students in the Science Class: Some Guidelines

Teachers' concerns are not in understanding why special students ought to be in science classrooms but rather, in dealing with the fact that they are in classrooms. And so the problems may be stated, "What can be done to provide the best science education program possible?" "What are the first steps?" "What should I do now?" The next sections are addressed to these questions.

General Guidelines for Helping Handicapped Students

1. Obtain and read all the background information available on the student.
2. Spend time educating yourself on the physical and/or psychological nature of the handicap

and how it affects the student's potential for learning.

3. Determine whether or not special help can be made available to you through the resources of a special education expert.
4. Determine any special equipment needed by the student.
5. Talk with the student about limitations due to his or her handicap and about particular needs in the science class.
6. Use resource teachers and aides to assist you.
7. Establish a team of fellow teachers (including resource teachers and aides) to share information and ideas about the special students. A team approach is helpful in overcoming initial fears and the sense of aloneness in dealing with the situation. You may need to take responsibility for contacting appropriate school personnel and establishing the team; if so, take courage and do it.
8. Other students are often willing to help special students. Encourage them to do so.
9. Be aware of barriers, both physical and psychological, to the fullest possible functioning of the special student.

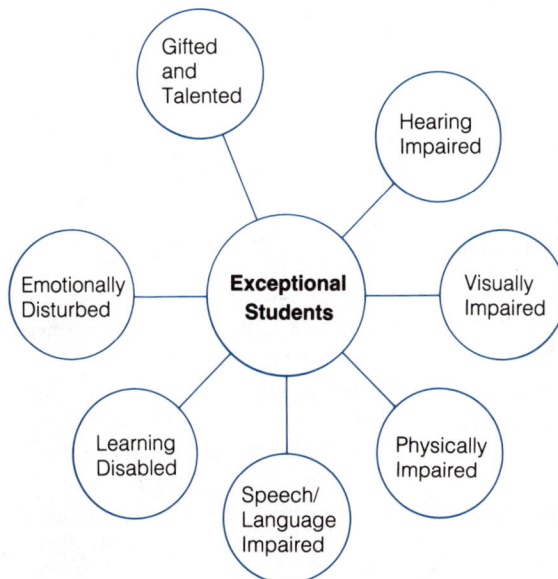

The recent focus on special education has helped to clarify the nature of handicaps, giftedness, and individual differences.

10. Consider how to modify or adapt curriculum materials and teaching strategies for the special student without sacrificing content, processes, or activities.

11. Do not underestimate the capabilities of the special student. Teachers' perceptions of a student's abilities have a way of becoming self-fulfilling prophecies. If these perceptions are negative, they may detrimentally affect the student and your ability to create new options for him or her.

12. Use the same standards of grading and discipline for the special student as you do for the rest of the class.

13. Develop a trusting relationship with the special student.

14. Educate the other students about handicaps in general, as well as specific handicaps of students in their class.

Hearing Impaired Students

From an early age most children learn through listening. And there is every indication that most teaching is telling. Hearing impairment is defined as an auditory problem adversely affecting the student's educational performance. Students with hearing impairments often have developmental delays in speech and language. These delays have obvious effects on the ability to communicate. Hearing-impaired students will not necessarily have problems acquiring science concepts, although they may have difficulty learning the written or oral language to communicate their understanding.

Helping Hearing-Impaired Students

1. The hearing-impaired depend heavily on visual perception. Therefore, seat the student for optimal viewing.

2. Determine whether an interpreter will be needed and the nature of the child's speech language problems.

3. Learn the child's most effective way of communicating.

4. Find the student a "listening helper."

Visually Impaired Students

Like the hearing-impaired, the visually impaired are those students whose vision is limited enough to require educational adaptions. Students who can read material with the use of magnifying devices and/or enlarged print are classified as partially seeing. Students who require braille or taped materials are classified as educationally blind.

Helping Visually Impaired Students

1. Visually impaired students learn through sensory channels other than vision, primarily hearing. Therefore, seat students for optimal listening.

2. Determine from the student what constitutes the best lighting.

3. Change the room arrangement whenever necessary but always make a special effort, formally and informally, to reorient the student.

4. Allow the student to manipulate tangible materials and models. Do not unduly "protect" students from materials.

5. Speak aloud what you have written on the chalkboard and charts.

6. Use the student's name: otherwise, the student may not know when he or she is being addressed.

7. Since smiles and facial gestures might not be seen, touching is the most effective means of reinforcing the student's work.

8. Be aware of student eye fatigue. This fatigue can be overcome by varying activities, using good lighting, and providing close visual work.

9. Have the student use his/her visual capacity when possible (unless otherwise directed).

Physically Impaired Students

Students with physical and health impairments represent a diverse group of special needs, for this category includes students with allergies, asthma, arthritis, amputations, diabetes, epilepsy, cerebral palsy, spina bifida, and muscular dystrophy. Some are mobile and others are confined to wheelchairs; some have good use of their limbs and others do not. Some have a single crippling condition and some have multiple handicaps.

Helping Physically Impaired Students

1. Eliminate architectural barriers.
2. Become familiar with the basic mechanics and maintenance of braces, prostheses, and wheelchairs.
3. Understand the effects of medication on students and know the prescribed dosage.
4. Obtain special devices such as pencil holders or reading aids for students who need them.
5. Learn about the symptoms of special health problems and appropriate responses.

Speech- and Language-Impaired Students

Until recently, classroom teachers had more contact with speech- and language-impaired students than any others with handicapping conditions. This situation may still be true in most schools, but learning disabilities programs are growing rapidly. Speech and language handicaps that you might encounter are articulation (the most common problem), dyslexia, delayed speech, voice problems, and stuttering. In addition, students with other handicaps such as cleft palate, cerebral palsy, and hearing loss may have speech and language problems.

Helping Speech- and Language-Impaired Students

1. Help the student become aware of his or her problem; students must be able to hear their own errors.
2. Incorporate and draw attention to newly learned sounds in familiar words.
3. Know what to listen for and match appropriate remedial exercises with the student's problem.
4. Be sure your speech is articulate; students often develop speech and language patterns through modeling.

Learning Disabled and Mildly Mentally Impaired Students

There is a distinction between learning disabilities and mild mental handicaps. The difference is much too technical to summarize. Students with mild mental handicaps should be identified only through the use of multiple criteria. Classroom teachers may observe indications of mental handicaps in a student's social interaction, general intelligence, emotional maturity, and academic achievement. In contrast, students with learning disabilities show significant discrepancy between their achievements and the apparent ability to achieve. The problem is manifest as a disorder of learning and not mental ability. Teachers may observe learning disabilities in the areas of arithmetic, listening, reading, spelling, logical thinking, speaking, and writing.

Helping Learning-Disabled and Mentally Impaired Students

1. Listen closely so you can understand the student's perception and understanding of concepts and procedures.
2. Use an individualized approach based on the student's learning style, level of understanding, and readiness.
3. Use multisensory approaches to learning: visual, auditory, kinesthetic, and tactile.

4. Find and use the student's most refined sensory mode to aid in development of mental capacities.

5. Make use of the student's strengths and work on diminishing his or her deficiencies.

6. Reduce or control interruptions since many special students have short attention spans.

7. Stay within the student's limits of frustration. Rely on your judgment, not the level of curriculum materials.

8. Start conceptual development at a sensory-motor or concrete level and work toward more abstract levels.

9. Work on speech and language development.

10. Help special students to develop self-esteem: a good, firmly grounded self-concept is essential to their continued development.

Emotionally Disturbed Students

These students probably cause the greatest concern and frustration for teachers. As it turns out, they are also the ones who have been in classrooms all along! Some emotionally disturbed and disruptive students show behavior that ranges from mild, attention-getting "pranks" to violent assault. They may also be mildly withdrawn to clinically depressed and suicidal. Other examples of behavior that teachers might identify as disturbed or disruptive are regression, fears and phobias, chronic complaints of pains and illness, aggressiveness, overdependence, social isolation, perfectionism, excessive dieting, obesity, chemical dependency, defiance, and vandalism.

Helping Emotionally Disturbed Students

1. Spend time with the student when he or she is not being disruptive.

2. Make rules reasonable and clear.

3. Provide realistic, reasonable, and appropriate consequences if rules are broken.

4. Disruptive behavior ranges from low levels at which a student may merely be looking for attention or recognition through a spectrum that ends in rage, tantrums, or complete withdrawal. Try always to be alert to behaviors that, though minimally disruptive, could become more serious problems.

5. Avoid personal confrontations or situations that provoke troubled students.

6. Make directions for assignments, classwork, and laboratory procedures direct, clear, and complete.

7. Be aware of and prepare for transitional times in the classroom.

8. Provide troubled students with success experiences.

9. Resolve conflicts by talking about specific behaviors, reasoning, and involving the student in the problem-solving process. Once a course toward aggressive or uncontrolled behavior is started, it is hard to stop.

10. Convey your intention to help resolve the problem mutually, "*We* have a problem here and *we* are going to resolve it."

11. If behavior problems escalate, try to talk about the process while providing solutions to the problem. For example, "We are both getting angry; can't we settle this calmly," or "I see you are upset; let's try to solve the problem."

12. Avoid using comparision, embarassment, ridicule, and unwarranted threats to change behavior.

13. Never use physical punishment for rule violation.

Gifted and Talented Students

Definitions of giftedness vary. Most, however, are paraphrased from the congressional report submitted by past Commissioner of Education Sidney Marland in *Education of the Gifted and Talented*. Gifted and talented students are those identified by professionals who, by virtue of their abilities, are capable of high achievement. These students require educational programs beyond those normally provided, to fulfill their personal potentials and encourage their contribution to society. In a less cumbersome definition: gifted students have superior academic abilities. Talented students have special aptitudes in specific areas. The difference between giftedness and talent is sometimes not distinct, since most gifted students have talent. Talented students may only be gifted in one or two areas and may perform poorly in other areas.

Characteristics of the Gifted and Talented in Science Class

1. Enjoys solving scientific problems.
2. Solves problems easily and logically.
3. Demonstrates advanced ethical, cognitive, and aesthetic development.
4. Learns science faster than other students.
5. Understands scientific concepts quickly.
6. Asks many questions about science.
7. Shows an awareness of science far beyond that of other students.
8. Is motivated to read and study science.
9. Demonstrates unique abilities in designing laboratory equipment to solve problems.
10. Is highly creative.
11. Shows normal social adjustment.

In addition, there are a few negative behaviors such as boredom, frustration, and complaints that you may observe. This list gives a subjective and preliminary means of identifying gifted and talented students. If you think you have such a student, it is best to consult the school counselor who can administer appropriate tests to confirm your initial impressions.

Adapting school programs for the gifted can be achieved in many ways. Businesses, industries, colleges, and universities often have programs for students showing special abilities. There are special classes, programs, and schools. Gifted students can work on advanced placement courses, accelerated schedules, take extra classes, and work on part-time projects where they can develop their talents. You can easily find many options for the gifted students in your school.

Helping Gifted and Talented Students

1. Use questions, problems, and projects that will facilitate higher levels of cognitive, affective, and psychomotor development.
2. Develop independent study programs.
3. Have special "honors seminars."
4. Initiate extracurricular science activities such as science fairs or opportunities to teach elementary science classes.
5. Assign special projects.
6. Use inquiry and problem solving.
7. Emphasize scientific methods.
8. Individualize a program based on the student's interests.

Science teaching in the 1980s will, due to legal mandates, include greater recognition of the unique handicaps, gifts, and talents of students. Handicapped students will be mainstreamed in regular classrooms, and the gifted will also receive special attention. Although each special student, whether handicapped or gifted, presents a unique case, there are some guides and suggestions that can help the science teacher meet the unique needs of students.

Education has recognized the needs of students at either end of the continuum—handicapped to gifted. The process has clarified individual differences in general and has emphasized the theme of this chapter. Understanding students

COMMENTARY

is essential to effective teaching and a prerequisite to deciding what teaching methods to use. In the next chapter you will find a variety of teaching methods described. They are presented to enable you to reach your educational objectives while meeting students' needs.

BIBLIOGRAPHY   Albany Unified School District. *My Body is Mine, My Emotions are Mine, My Environment is Mine: A Health Education Curriculum.* Alameda County,CA: Office of Education, 1978.

Ausubel, David. *Educational Psychology: A Cognitive View.* New York: Holt, Rinehart, & Winston, 1968.

Axelrod, Saul. *Behavior Modification for the Classroom Teacher.* New York: McGraw-Hill, 1977.

Biehler, Robert and Snowran, Jack. *Psychology Applied to Teaching.* 4th ed. Boston: Houghton Mifflin, 1982.

Bigge, Morris. *Learning Theories for Teachers.* New York: Harper & Row, 1976.

Blackhurst, Edward A., and Berdine, William H., eds. *Introduction to Special Education.* Boston: Little, Brown & Co., 1981.

Blake, Evelyn. *Educating Exceptional Pupils.* Reading, MA: Addison-Wesley Publishing Co., 1981.

Bruner, Jerome. *Toward a Theory of Instruction.* New York: W W Norton & Co., 1966.

Bruner, Jerome. *The Process of Education,* New York: The Vantage Books, 1960.

Bybee, Rodger W. "Helping Special Students Fit In." *The Science Teacher* 46 (October 1979): 22–24.

Bybee, Rodger W. and Sund, Robert B. *Piaget for Educators.* Columbus, Ohio: Charles E. Merrill Publishing Company, 1982.

Erickson, Erik, *Childhood and Society,* New York: W W Norton & Co., 1963.

———. *Identity and the Life Cycle.* New York: W W Norton & Co., 1980.

Gagne, Robert. *The Conditions of Learning.* New York: Holt, Rinehart & Winston, 1980.

Heward, William and Orlansky, Michael. *Exceptional Children.* 2nd ed. Columbus, Ohio: Charles E. Merrill Publishing Company, 1984.

Kaplan, Sandra N.; Kaplan, Jo Ann B.; Madsen, Sheila K., and Taylor, Bette K. *Change for Children: Ideas and Activities for Individualizing Learning.* Pacific Palisades, CA: Goodyear Publishing Company, 1973.

Köhler, Wolfgang. *The Mentality of Apes.* New York: Harcourt Brace & World, 1925.

Marland, Sidney. *Education of the Gifted and Talented.* Washington, D.C.: U.S. Government Printing Office, 1972.

Maslow, Abraham. *Motivation and Personality.* New York: Harper & Row, 1970.

Pavlov, Ivan. *Conditioned Reflexes.* New York: Dover Publications, 1960.

Piaget, Jean, and Inhelder, Barbel. *The Psychology of the Child.* New York: Basic Books, 1969.

Premack, David. "Reinforcement Theory." In *Nebraska Symposium on Motivation.* edited by D. Levine. Lincoln, NE: University of Nebraska Press, 1965.

Rogers, Carl. *Freedom to Learn for the 80's.* Columbus, Ohio: Charles E. Merrill Publishing Company, 1983.

Skinner, B. F. *Beyond Freedom and Dignity.* New York: Knopf Publishers, 1971.

_____. *Science and Human Behavior.* New York: Macmillan, 1953.

_____. *The Technology of Teaching.* New York: Appleton-Century-Crofts, 1968.

Stephens, J. M., and Evans, E. D. *Development and Classroom Learning.* New York: Holt, Rinehart & Winston, 1973.

Stone, Joseph, ed. *Stuart-A Four-Year-Old Views the World.* Vassar College, 1966. Mimeographed.

Thorndike, Edward L. *The Fundamentals of Learning.* New York: Teachers College Press, Columbia University, 1932.

Trowbridge, Leslie W.; Bybee, Rodger W.; and Sund, Robert B. *Becoming a Secondary School Science Teacher.* Columbus, Ohio: Charles E. Merrill Publishing Company, 1981.

4

Taxonomy of Common Teaching Methods

The purpose of chapter 4 is to provide brief descriptions of various teaching methods so that beginning teachers will be aware of the many alternative methods that are available for teaching science.

KEY CONCEPTS AND IMPORTANT IDEAS

☐ taxonomy ☐ teaching methods
☐ auditory learning ☐ listening-speaking methods ☐ visual learning ☐ reading-writing methods ☐ kinesthetic learning
☐ watching-doing methods ☐ degree of teacher dominance in teaching ☐ appropriateness of methods for groups of various sizes

Every time teachers write something on a chalkboard, show a film, give a lecture, or ask students to read, answer questions, write, explore, or conduct an experiment, they are guiding the way students receive information. In doing so, teachers choose one teaching method over others, whether they are aware of it or not. Some teachers use only one or two methods to teach, and they defend their approach by saying that it fits their teaching style or the subject matter. Others alternate among a variety of methods as they match their teaching objectives with students' needs, with the materials that are available, and with the environment they want to create. Because students, school systems, and instructional objectives vary so widely, it is important that teachers consciously choose the methods that best help their students achieve the objectives of instruction. But knowing which method to choose for particular desired outcomes requires knowledge of the rich variety of alternative methods of teaching that exist and a willingness to experiment with various methods. Those concerned with teacher education serve beginning teachers best if they avoid advocacy of a single method or approach to teaching, and instead encourage experimentation with a variety of methods. This will help beginning teachers to find those methods which meet their needs and which meet the needs of their students and schools.

The purpose of this taxonomy of common teaching methods is to provide an overview of teaching methodologies and brief descriptions of the many teaching methods that exist so that beginning teachers will be aware of the rich alternative methods of teaching. Thirty-five methods are described and organized as a taxonomy that reflects the similarities and differences among methods. The taxonomy is intended to encourage readers to try a variety of teaching methods and thereby develop some measure of versatility and style in teaching.

What is a taxonomy? A **taxonomy** is a system of classifying things to reflect the totality of their similarities and differences so that users can then recognize the individual things described within the system and the clusters or hierarchies that organize the system. Taxonomies can be organized in many ways; there is no one "right" way. To help readers use this taxonomy, a definition of teaching methods is offered, followed by an explanation of the organization of this taxonomy so that readers can find what they need.

TEACHING METHODS

For purposes of definition, teaching methods refer here to actions initiated by teachers to guide instruction and learning. They are by nature transactions between teachers and students. The methods described here include actions that focus on the teacher, such as lecturing, leading field trips, or doing demonstrations. Methods also include actions initiated by the teacher which focus on students' responses, such as giving oral reports, solving problems in teams, or reading alone.

Teaching methods involve the use of materials but they are distinct from the materials themselves. For example, a teacher may choose to have students read from a textbook, build a model described in a textbook, or develop note-taking skills by outlining a passage in the textbook. Although teaching methods are distinct from materials, we have sometimes used the name of the material to identify a method (for example, the chalkboard method) because a more appropriate identification could not be found.

Teaching methods require skill to become effective but they are distinctive from the skill itself. One teacher may be very exciting as a lecturer or storyteller while another is boring, and still another confuses students.

Teaching methods are generally chosen to enhance learning but they cannot guarantee that learning takes place. Because students vary in their abilities, interests, and previous experience, many experienced teachers use a variety of methods to maximize learning for the largest number of students. Yet even if one were to individualize teaching methods so that every student was presented information in the way he or she learned best, it would be impossible to guarantee that learning would take place. This is because learning processes are controlled by many factors in addition to the method of teaching.

What can be said, however, is that teachers who use a full repertoire of teaching methods and who are aware of their students' needs, somehow communicate a sense of excitement about learning and create an atmosphere that is conducive to learning.

ORGANIZATION OF THE TAXONOMY

This taxonomy is organized into three sections around three common kinds of transactions that occur between teachers and their students as they engage in instruction:

- ☐ **Listening-Speaking**
- ☐ **Reading-Writing**
- ☐ **Watching-Doing**

These three kinds of transactions coincide with the three predominant ways students learn in schools—through auditory, visual, and kinesthetic experiences. Good teachers try to provide experiences for students in all three of these areas to ensure maximum learning.

Table 4–1 illustrates how all of the methods in the taxonomy are organized into sections and are related by their common forms of transaction. For example, under Listening-Speaking, such common methods as *lecture, discussion,* and *debate* are described. Within each section on the table, the methods are further organized according to the degree of **teacher dominance** in the transaction and the **usefulness** of the method for groups of various sizes. For example, the *lecture method* is nearly 100 percent *teacher dominated* and is typically used with the entire class, often a group of thirty students, although lectures might also be given to larger or smaller groups. Consider one more example: *Oral reports* by students are almost lacking in teacher-dominance (probably less than 5 percent) and may be presented to the entire class of thirty students or smaller groups of fifteen to five students. Thus, table 4–1 shows the relationship between various methods in terms of the nature of the transaction (Listening-Speaking, and so on), the degree of teacher dominance, and the usefulness of the method with groups of various sizes.

Many methods could be placed in more than one section. For example, the *drill method* could be used in Listening-Speaking transactions or in Reading-Writing transactions between teachers and students. But each method is described only once; if a method can be used with more than one type of transaction, that information is given in the text and is also noted in table 4–1.

One last point needs to be made about the names of the teaching methods. It is important to recognize that individual teaching methods, like species of wild flowers, are known by many different common names depending upon the locale where they are found or used. As of now, there is no standardized or universally accepted nomenclature to describe individual teaching methods. Therefore, this taxonomy represents an attempt to describe a variety of commonly used teaching methods. The common names chosen for various methods reflect predominant usage in the United States in the mid 1980s. In a few cases, names have been created for methods

TABLE 4–1
Taxonomy of Common Teaching Methods

Nature of Transaction and Method	Amount of Teacher Dominance in Method			Usefulness of Method with Groups of Various Sizes (in numbers of students)				
	High	Medium	Low	1	5	10	15	30
Listening-Speaking Methods								
Lecture Method	*					*	*	*
√Giving Instructions	*			*	*	*	*	*
Recitation Method		*			*	*	*	*
√Drill Method		*		*	*	*	*	*
Review Method		*			*	*	*	*
√Questioning Method		*		*	*	*	*	*
Oral Exam Method		*		*	(Repeated with entire class)			
Discussion Method		*			*	*	*	*
Film Analysis Method			*			*	*	*
Debate Method			*				*	*
Oral Report Method			*		*	*	*	*
Brainstorming Method			*		*	*	*	*
Reading-Writing Methods								
Textbook Method	*			*	*	*	*	*
Workbook Method	*			*	*	*	*	*
Chalkboard Method	*				*	*	*	
Bulletin Board Method	*							*
√Problem-Solving Method		*		*	*	*	*	*
Laboratory Report Method		*				*	*	*
√Team Learning Method			*	*				
				(Repeated with entire class)				
Peer Review Method			*	*	(Repeated with entire class)			
√Peer Tutoring Method			*	*	(Repeated with entire class)			
Programmed Instruction			*	*	*	*	*	*
√Individualized Instruction			*	*	(Repeated with entire class)			
Note-Taking Method			*	*		*	*	*
Journal-Keeping Method			*	*	*	*	*	*
Watching-Doing Methods								
Demonstration Method	*					*	*	*
Field Trip Method	*							*
√Contract Method		*		*	(Repeated with entire class)			
"Hands-On"/Lab Method		*			*	*	*	*
√Inquiry Method		*		*	*	*	*	*
√Learning Center Method			*	*	(Repeated with entire class)			
Projects Method			*	*	*			
				(Repeated with entire class)				
Simulation Method			*			*	*	*
Games Method			*	*	*	*	*	*
√Exploration-Discovery Method			*	*	*	*	*	*

NOTE: Checkmark (√) indicates that the method is used in more than one kind of transaction: Listening-Speaking, Reading-Writing, and Watching-Doing.

where names were uncommon or nonexistent.

Many teaching methods are enhanced by combining more than one kind of transaction. To help readers make fruitful combinations, cross-references are used. For example, in the *demonstration method,* the primary transaction is Watching-Doing, but the teacher also talks while students listen; so a method called *giving instructions* is cross-referenced under *demonstration method.*

Within the body of the taxonomy, several kinds of information are given with each method described:

1. The common name of the method, the degree of teacher dominance, and the usefulness of the method for groups of various sizes;

2. The general purpose of the method;

3. The nature of the transaction: What the teacher does and what students do;

4. Conditions that may affect the use or effectiveness of the method; and

5. Other methods used in conjunction with or instead of the method described.

SECTION 1:
LISTENING-SPEAKING TEACHING METHODS

LECTURE METHOD

GIVING INSTRUCTIONS

RECITATION METHOD

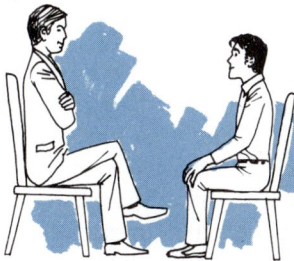

ORAL EXAM METHOD

DISCUSSION METHOD

FILM ANALYSIS METHOD

DEBATE METHOD

ORAL REPORT METHOD

BRAINSTORMING METHOD

Listening-Speaking teaching methods are used to enhance students' auditory learning.

1   LECTURE METHOD

Teacher Dominance: High
Group Size: 10–30 Students

The purpose of the *lecture method* is to present orally a body of information that has been organized in advance. The lecture method is more frequently used in high schools and colleges than in elementary and middle or junior high schools. Younger students seem to need a greater variety of information sources (visual and kinesthetic as well as auditory).

Teachers may add variety to lectures by using slides, an overhead projector, the chalkboard, or other techniques to "show" ideas or information described in the lecture. They need to be alert to the level of attention given by students during lectures: it is easy to assume that students are listening when they actually are not. By stopping every few moments and directing questions to specific individuals, teachers can check students' attention. But more than attention is needed during lectures. Teachers need to find out if students understand the material as well; questions asked by the teacher during the lecture are an excellent technique for assessing students' understanding.

During the lecture, students may take notes or ask questions; they may participate in a discussion at the end of the lecture. However, they generally need to learn how to take lecture notes. Teachers can help by putting an outline on the chalkboard, by reviewing students' lecture notes individually, and by offering suggestions for improvement.

The room should be arranged so that everyone in the room can see and hear the lecturer. The speaker should talk in a voice that is audible to everyone and should convey a sense of enthusiasm for the topic.

The lecture method resembles the textbook method in that the teacher is presumed to be the source of authority on the topic. Teachers need to evaluate carefully students' learning achievement when using the lecture method, and compare the results with other teaching methods.

2   GIVING INSTRUCTIONS

Teacher-Dominance: High
Group Size: 1–30 Students

The purpose of *giving instructions* is to provide direction for students who are about to begin a task. Teachers spend a great deal of time giving instructions. Instructions are generally a part of every teaching method teachers use and a part of every activity teachers introduce. Just as important, perhaps, is the fact that students need to learn how to follow instructions at increasing levels of complexity as they proceed through school. Thus, it is important for teachers to provide clear, succinct instructions before learning can proceed, and they need to serve as models for presenting clear instructions. This method is also useful in Reading-Writing transactions.

There are a few simple guidelines for giving instructions that result in clear, easy-to-follow instructions:

1. Write the instructions for yourself first. This will ensure that you have covered all of the important steps and have them in proper sequence.
2. Reduce the instructions to a few steps (four to six steps, if possible). If instructions take more than four to six steps, it is probably better to give a demonstration along with the instructions, or to divide the learning task into seg-

ments and give the instructions in segments with each part of the task.

3. Convert the instructions to the language level used by the students, and state them as simple, direct commands.

4. Ask the class for their full attention before you begin. Do not begin until they are quiet and listening.

5. Give the instructions orally and put them in writing (on the chalkboard or in a ditto). Provide an oral or visual example with each step, if appropriate.

6. Check to see how well students understand the instructions. This can be done by asking if there are any questions, or by asking one student what is the first thing to do, asking a second student what to do next, and so forth through the steps. Explain that you will handle any additional questions individually after the class begins work on the assignment. If a student asks a question that is likely to be of interest to all students, interrupt the work of the class to explain the point, but interruptions should be kept to a minimum.

One way to improve giving instructions is to analyze the problems students encounter after you have given instructions. People tend to make the same errors rather consistently. Teachers who find students having difficulty following their instructions, and who have tried to improve their method of giving instructions, have nearly always seen improvement.

3 | RECITATION METHOD

Teacher-Dominance: Medium
Group Size: 5–30 Students

The *recitation method* is used to assess what students know or understand on a regular, informal, sometimes daily basis and to reinforce correct or accurate understandings. Generally, recitation involves recalling information that has just been presented or learned, and stating or repeating it in the exact wording or paraphrasing it in the students' own words. Often students are asked to read or recite a particular passage or state a specific rule from memory when paraphrasing is not complete enough. At other times, recitation is used by the teacher to summarize what was learned the day before in order to set the stage for the new day's learning.

In the recitation method the teacher asks questions and students provide the answers. Students are usually permitted to use their books, reports, notes, or other material unless they have been asked to have memorized the material.

Many teachers use the recitation method as a way of dealing with great differences in students' ability to read the textbook; those who have been able to read and understand become the paraphrasers for students who have not understood the text. Thus, learning occurs in auditory form for less able readers. Teachers also use recitation to improve students' critical reading abilities, by asking students to paraphrase passages in the text and then to look at the text closely to check the accuracy of their understanding.

Other methods that are often used in conjunction with the *recitation method* are *drill* and *review*, though both differ from *recitation* in important ways.

4 | DRILL METHOD

Teacher-Dominance: Medium
Group Size: 1–30 Students

The purpose of the *drill method* is to provide students with enough practice and reinforcement for specific learning so that they can repeat the information more or less automatically when asked to do so. Most teachers think there is value in having students memorize certain kinds of information that are used so often that it saves time to have the information at one's "fingertips" rather than to have to look it up in a book.

The essential question teachers must ask themselves about drill is whether the information is really important to know in an exact form (such as the correct spelling of words, knowing arithmetic facts, or recognizing words by sight) or whether other forms of knowing are more appropriate (such as being able to apply information to real life situations, solving problems on one's own, or explaining things in one's own words). This method is also useful in Reading-Writing transactions.

Once teachers decide which form of learning is appropriate, a learning drill has the following characteristics: (1) practice in understanding visual and auditory information so that students learn to discuss and write about what they have learned; (2) immediate feedback on their responses so that correct responses are reinforced (and acknowledged), and incorrect responses are analyzed for errors; (3) frequent and spaced practice at regular intervals at first and later at irregular intervals.

Because drill involves repetition, it can become boring. Variety can alleviate boredom. For example, games and contests introduce competition and challenge some students. For noncompetitive students, variety can be added to drill by having students write on unusual surfaces (steamed windows, paper towels, wet sand, magic slates, and so on), or having them repeat the information orally in unusual places (while swinging, bouncing balls, or lying on the floor). Many people can assist students in their drill: the teacher, class-

room aide, other students in the classroom (peer tutors), or family members at home.

The *drill method* is generally used in conjunction with *written* and *oral exam* methods, and with *workbook, games,* and *programmed instruction* methods.

5 | REVIEW METHOD

Teacher-Dominance: Medium Group Size: 5–30 Students

The purpose of the *review method* is to have students look again at something they have learned before. Review is not to be confused with drill or recitation. Rather, review is reviewing or re-examining what has been learned. It implies that one can see things in a new light or in a broader context. Although the teacher cannot guarantee that students will come to a new understanding through review, it is an essential step in the consolidation of information and the use of applications.

To use the review method, teachers first ask students to recall previous learning. As students remember what they have learned, teachers need to listen for misconceptions and try to correct them. If the original learning involves a large number of facts, concepts, or generalizations, teachers may help to emphasize key ideas or point out relationships among main ideas, thereby leading students to see or discover the new application or relationship on their own. During the review process, students listen to the teacher and to each other, and try to add their own knowledge when asked. They may refer to the textbook, their workbook, lecture notes, or other written reports to assist them in recalling and seeing new relationships.

In order for review to be of benefit, the original learning must have been understood. For example, if students learn several formulas for chemical reactions, they must know what actually occurs in specific chemical reactions before they can generalize about the reactions and discover a new relationship among them. An effective review lesson generates enthusiasm because it enables students to see new value, association, or application of the knowledge they have gained. Review that is limited to recall of information, without any reanalysis, can be boring. It can be even worse if students are asked to repeat information they do not understand.

Several methods are used in conjunction with the review method. Since the purpose of review is to re-view what has been learned, *inquiry* and *discussion* methods are frequently used, along with *questioning,* as the teacher conducts a review session.

6 | QUESTIONING METHOD

Teacher Dominance: Medium Group Size: 1–30 Students

The purpose of the *questioning method* is to stimulate thinking. Asking questions is a part of nearly every instructional transaction between teachers and students; yet strategies for asking questions are not well understood and many teachers ask only one or two kinds of questions. This method can also be used in Reading-Writing transactions.

Most of the questions teachers ask, regardless of subject matter, generally have one of six outcomes. Questions may require students to (1) recall specific facts, principles, or generalizations; (2) demonstrate comprehension by rephrasing something in their own words; (3) apply information to a new situation; (4) analyze information by seeking out new relationships, underlying assumptions, or implications; (5) synthesize information by combining it in new ways; or (6) evaluate information by using some criteria to make judgments.

Once the teacher has asked the question, the student needs to be given enough time to compose a suitable answer. Recall and comprehension questions require less time to answer than questions which ask students to apply, analyze, synthesize, or evaluate information. Teachers who have experimented with giving students more time to answer questions have found that the quality of students' responses improves as the amount of time to compose the answer increases. Although some teachers limit their questions to recall and comprehension, with practice they can learn to ask higher order questions and increase the amount of time they wait for answers.

Students generally try to learn information they think is important, and what they think is important is reflected by teachers' questions. When teachers ask only one or two kinds of questions, other critical thinking skills are neglected. If students are to develop their abilities to think critically, they need to have experience answering different kinds of questions.

Another teaching method which focuses on questions is called the *Socratic method.* The purpose of the Socratic method is to guide students to new conclusions by asking them a series of easily answered questions which lead to logical and (to students) unforeseen conclusions. It is a method of teaching in which the teacher, like Socrates, professes to impart no information, but draws information from the students, demonstrating to them that they already possess the knowledge but only need help drawing it out.

7 | ORAL EXAM METHOD

Teacher Dominance: Medium
Group Size: 1:1, Repeated with
Entire Class

The purpose of the *oral exam method* is to provide students with an opportunity to demonstrate what they have learned without having to express it in writing. Oral exams usually follow a set of questions designed by the teacher, but they also enable students to demonstrate their knowledge in areas that go beyond the questions asked. Oral exams take place between the teacher and one student at a time, and can be especially useful when students need to take make-up exams, or when students are unable to demonstrate what they know on written exams. Because of their more personal nature, oral exams offer the opportunity for teachers to become more involved with the students' understanding of the subject matter. Such exams need not be formal but can more nearly resemble "talks" or "chats" if students are overly anxious. The teacher asks questions; the student answers them. When the student's answer indicates a misconception or lack of understanding, the teacher can provide the answer, thereby teaching during the exam. Teachers also can ask students to tell them what other things they have learned, in addition to the questions asked by the teacher during the oral exam.

One condition that affects the success of oral exams is the way teachers deal with grading. If the oral exam is seen as a form parallel to the written exam taken by other class members, then general scoring (pass/fail) of the oral exam is simpler than precise scoring (percentage points), since the two forms (written or oral) are not really parallel.

Other methods used in conjunction with oral exams are *questioning*, *contract*, and *projects*.

8 | DISCUSSION METHOD

Teacher Dominance: Medium
Group Size: 5–30 Students

The purpose of the *discussion method* is to promote the free exchange of ideas among several people on a given topic. Classroom discussions, unlike discussions which occur spontaneously in social settings, need careful planning if they are to stimulate students' interest and guide them to new understandings.

Teachers usually initiate a discussion by announcing the topic. For example, "Today let's talk about weather and the way it affects how people feel and behave," announces the topic and the fact that several classmates or everyone will be participating (as opposed to a question addressed to a single student). Frequently discussions begin with a question. "What will we need to conduct the experiment?" and "How can we improve the environment here at our school?" are examples of questions used to initiate a problem-solving discussion. When discussions are used to solve problems, it is important for students to understand that they are not being asked to *brainstorm*, another teaching method described in the taxonomy. In *discussion*, each student is expected to contribute to the discussion, and to have others in the group react to his or her ideas. In other words, ideas are evaluated as they are expressed during discussion. (In *brainstorming*, evaluation is delayed until the entire activity is completed.)

Teachers responsible for planning a discussion need to (1) plan in advance all or most of the key issues to be discussed within the topic; (2) make certain everyone has a chance to express a view or to ask questions; (3) encourage the objective evaluation of ideas as they are expressed; and (4) summarize the main ideas or conclusions from the discussion, or have a member of the group summarize them. A good discussion is characterized by lively give and take among students. No one dominates the discussion and the teacher need not control the exchange by calling on speakers unless it is needed to keep the discussion moving. The teacher helps to keep students on the topic by redirecting attention to the topic, by restating the original question, and by moving to the next question. Students interact with each other so long as only one student speaks at a time and everyone can hear. If possible, the classroom should be arranged so that speakers can see one another.

The discussion method is one of the most commonly used methods of teaching; it is especially useful in conjunction with *textbook, field trip, film analysis,* and *oral report* methods.

9 | FILM ANALYSIS METHOD

Teacher Dominance: Low
Group Size: 10–30 Students

The purpose of *film analysis* as a method of teaching is to guide students toward new ideas or generalizations. Films can be more dramatic and extensive than written materials, lectures, pictures, or firsthand experience. The major single advantage of films over firsthand experience is that films can condense or expand time and space, thus heightening the impact of events by providing information that is not observable with normal vision, within a single lifespan, or within given environments. Any film generally arouses interest but teachers need to guide students' attention in order to attain instructional objectives.

The teacher usually prepares students for the film through introductory remarks which point out what will be seen on the film. Some teachers write key questions on the chalkboard, while others have the same information on ditto sheets; but most teachers limit their introduction to oral remarks. During the film, students should be discouraged from taking notes and focus their full attention on the film. Likewise, the teacher should avoid distracting students by talking in competition with the film narrative. When the film is over, the teacher guides the discussion to determine how well the key points were understood and to correct any misconceptions. Students are encouraged to ask questions and students' interpretations need to be encouraged. Showing the film a second time is an excellent technique for handling discrepancies or differences of opinion. Ultimately, the teacher leads students to link the new information from the film to their previous knowledge and experience so that new generalizations can be derived.

Conditions that affect the success of film showing and film analysis relate to scheduling and the teacher's skill in helping students derive generalizations. Scheduling is important; the film should be scheduled to fit in with related learning. Yet experienced teachers always plan an alternative activity for the day the film is to be shown in the event that it does not arrive on schedule. Sometimes they must provide more introductory background before showing a film that arrives too early or too late.

Several methods that are used in conjunction with the film analysis method are *review, discussion, questioning* and, later in a sequence, *field trips, projects,* or *experiments.*

10 DEBATE METHOD

Teacher-Dominance: Low
Group Size: 20–30 Students

The purpose of the *debate method* of teaching is to enable students to discuss opposing sides of a question or issue. Because of its competitive nature, it is an effective technique for getting students involved with subject matter. Debate also sharpens students' abilities to analyze information and to deal with issues of immediate importance to them.

The teacher generally selects the problem or issue to be debated. Questions which deal with factual information in relation to human attitudes or values are best for debate. For example, the question "Should nuclear reactors be used for peacetime purposes?" allows students to argue the advantages and disadvantages of the development of nuclear power to individuals and the environment. The teacher and students together need to arrive at a clear and concise statement of the two alternative positions to be debated. The teacher then sets the guidelines for the debate and is generally responsible for enforcing them; however, once a class has been introduced to debate, other students can assume the role of referee. Guidelines usually include: (1) setting time limits for speakers to prepare and present their positions; (2) setting time limits for speakers to prepare and present their rebuttals and make their closing statements; and (3) enforcing a "No interruptions" rule while any speaker is talking.

Debate teams typically are composed of three to four members per side of the issue to be discussed. While the team members take turns presenting their arguments to support their views, the rest of the class listens. The class, or "audience," can become more involved with the issue by voting before and after the debate by secret ballot. Often a debate can be continued for several days. The original speakers make their opening statement on one day, have a day or two to prepare their rebuttals, and make their closing statements on a third day. If issues are more broadly defined, several debate teams can deal with various aspects of the issue or problem.

Other methods often used in conjunction with the *debate* method are *note-taking, research,* and *outlining.*

11 ORAL REPORT METHOD

Teacher Dominance: Low
Group Size: 5–30 Students

The purpose of *oral reporting* as a method of teaching is to provide students with an opportunity to demonstrate what they have learned. Like the teacher's *lecture* method, *oral reporting* allows students to tell about a topic they have studied and organized for themselves. During the oral report, students may show materials they have collected or produced, or illustrate ideas on charts, the chalkboard, and through pictures. As each student gives an oral report, the others listen. Frequently listeners are expected to ask questions at the end of the report. The teacher has responsibility for assuring that the information given in the oral report is accurate so that other students do not form misconceptions; this is usually accomplished by talking with the speaker before the oral report. The teacher also helps the student keep track of time during the report, and may assist in arranging for or setting up any special equipment the student needs in order to present visual material.

Students need to learn how to give oral reports. In primary grades, students give brief reports of two to three minutes before small groups of classmates, sometimes without any preparation. The teacher can help develop their oral reporting skills by writing questions on the chalkboard to provide structure: What did you do? How do you do it? What did you find out? Older students appreciate written guidelines for oral reports. They usually enjoy making oral reports of five to fifteen minutes—if they are permitted to use notes and are supported by the teacher and class. Evaluation of oral reports is generally a private communication between the teacher and student because many students are somewhat insecure about giving oral reports until they have had several successful experiences. The classroom should be arranged so that everyone can see and hear the speaker.

Other methods that are often used in conjunction with oral reporting are *projects, notetaking, individualized,* and *contract* methods.

12 | BRAINSTORMING METHOD

Teacher Dominance: Low
Group Size: 5–30 Students

The purpose of the *brainstorming method* is to encourage students to use their creative imagination. Brainstorming sessions are structured to permit the free and spontaneous expression of ideas. Its overall objective is to find new ideas and combinations of ideas, rather than to analyze existing ideas or judge each idea at the time it is expressed. By temporarily delaying evaluation of the ideas and focusing on the generation of as many ideas as possible, group brainstorming sessions can unharness much creative thought. This is accomplished by building a momentum for acceptance of any serious idea expressed.

All brainstorming sessions have a focus. Creative thinking is usually directed toward solving some problem, finding new ways to act in familiar settings, discovering new activities to explore, or eliciting a wide variety of interpretations of an event. The teacher sets the stage by describing the purpose of brainstorming and the problem or event that will be the focus of the session. Then rules for brainstorming are stated, usually in writing if the students are unfamiliar with the process. Rules generally include: (1) all ideas are accepted; (2) criticism of ideas is not allowed; (3) as many ideas as possible are needed; (4) spontaneity is important; and (5) "piggy-backing" on one anothers' ideas is welcomed. Once the topic is clearly understood and the rules have been explained, students begin to generate ideas. The teacher or some students act as scribes, writing down on the chalkboard or on a chart all of the ideas as they are expressed.

When the flow of ideas begins to wane, the teacher usually congratulates the students on their creative activity and calls the session to a close. Several options can follow. Students may be asked to think about the topic overnight, allowing incubation time for any additional ideas to be added to the list. Incubated ideas are usually of a different quality than those expressed during brainstorming, and they can be equally creative or of even greater value to the group as a whole. If incubation time is to be an option, it is important to postpone evaluation of the ideas generated during the brainstorming session and assure students that the rules of brainstorming still apply.

Another option following brainstorming is to sort through the list of ideas to identify those with the greatest potential for solving the problem, providing new insights, or whatever the purpose of the activity was. During evaluation, the teacher actively encourages students to look for combinations of ideas that will best work together. Refinement of the ideas is a secondary process, and students need to see it as a natural follow-up of brainstorming.

Several conditions affect the success of brainstorming. Teachers need to exhibit enthusiasm for the expression of creative thought. At the same time, they need to act as referees, enforcing rules that ensure freedom of expression without criticism. And finally, teachers need to find meaningful use for the refined ideas created; otherwise brainstorming is pointless.

The brainstorming method is frequently used with the *problem-solving* method.

SECTION 2:
READING-WRITING TEACHING METHODS

TEXTBOOK METHOD

WORKBOOK METHOD

CHALKBOARD METHOD

BULLETIN BOARD METHOD

PROBLEM-SOLVING METHOD

PEER TUTORING METHOD

PROGRAMMED INSTRUCTION METHOD

NOTE-TAKING METHOD

JOURNAL-KEEPING METHOD

Reading-Writing teaching methods are used to enhance students' visual learning.

1 | TEXTBOOK METHOD

Teacher Dominance: High
Group Size: 1–30 Students

The purpose of the *textbook method* is to assure uniformity in learning through reading. It is an effective method of teaching when students can read and understand information presented in written and pictorial form. But if some students' reading abilities limit their understanding of the material, and if teachers use the textbook exclusively, it can be an ineffective method of teaching.

To use the textbook method, teachers (1) select the textbook that best fits students' reading abilities and their own teaching objectives; (2) assign reading in conjunction with other activities that are related to the topic; and (3) conduct activities that involve other methods (discussion, panels, debates, laboratory, workbooks, outlining, and review) in order to assess students' understanding of the material and to provide opportunities to apply the ideas presented in the textbook.

The effectiveness of the textbook method can be increased by combining it with other techniques. First, many experienced teachers use a variety of textbooks for each subject. They use different science textbooks for advanced readers, average readers, and slow readers. The class is typically divided into groups, based on the textbook used, and then each group proceeds through the textbook as it is sequenced by the author. However, some teachers prefer to keep the class together and use multilevel textbooks that introduce unifying themes; they ask students to report what their various textbooks have to say on each topic. Most experienced teachers introduce one or more probing questions before students begin their reading in order to give focus, interest, and advance organization to students' reading.

Because students' learning styles vary, it is important for the information presented in textbooks to be presented in auditory or kinesthetic forms as well, and for teachers to assess all students' understanding of the material as they proceed. If the textbook method is ineffective for some students, other teaching methods such as *recitation* and *discussion* can be used to meet various needs.

2 | WORKBOOK METHOD

Teacher Dominance: High
Group Size: 1–30 Students

The purpose of the *workbook method* is to provide practice of skills and application. The workbook method varies in its usefulness depending on (1) how closely the workbook matches students' readiness for handling the material; (2) how closely the purpose of the workbook activity correlates with the teacher's objectives; and (3) how the workbook is actually used by the teacher.

Teachers usually (1) select workbooks that fit their own instructional objectives and fit students' levels of ability; (2) assign pages in the workbook to coincide with learning from other sources (textbooks, lectures, discussions, labs, and demonstrations); (3) arrange for "grading" the workbooks; and (4) provide students with feedback on their performance in the workbooks. Students usually try to do the work in workbooks and ask for help when they have difficulty combining or applying information from other sources to answer questions in the workbook.

The workbook method can be used in several ways. First, teachers can alternate between grading workbooks themselves and sharing that responsibility with others (classroom aides, students, or cross-age tutors) thereby freeing teachers to spend more time discussing students' performance. Second, teachers can use workbooks for "extra credit" activity, with pairs of students testing each other, or as items for oral "spelling-bee" activity between teams. Sometimes workbooks are used to reinforce what has been read in textbooks. Workbooks can be used as homework or cut apart and used for materials in learning centers, contracts, or individualized instruction.

Other methods used in conjunction with the workbook method are *individualized instruction, contracts, learning centers, drill, review, laboratory report,* and *programmed instruction.*

3 | CHALKBOARD METHOD

Teacher Dominance: High
Group Size: 5–30 Students

The purpose of the *chalkboard method* is to emphasize specific ideas or information in writing or graphic form for students to see. The chalkboard has such varied uses and values that a method for using it is described here. What distinguishes the chalkboard from all other teaching materials is its constantly renewable surface: information written or drawn on it can be more easily changed on the chalkboard than on any other surface. Students who gain most from the chalkboard are those who need to see things to remember them.

To use chalkboards effectively, teachers usually: (1) write legibly and large enough for students at the back of the room to see; (2) stand aside after writing the information so that students can see or take notes while the information is being discussed; (3) show how ideas are related or organized (imagine that students learn the information in the form that you present it on the chalkboard); and (4) encourage students to ask questions about the information presented on the chalkboard. Naturally, the information on the chalkboard needs to be related to learning that has preceded and follows it.

There are many valuable uses teachers have found for the chalkboard: (1) to list daily assignments, instructions, or schedules; (2) to outline key points or questions to be discussed, read, or seen in films or in experiments; (3) to present maps, drawings, or diagrams of something students will see; (4) to present a written exam or quiz; (5) to record students' responses during brainstorming or discussion; and (6) to drill students, either by having them work at the board in teams or by having them respond to information written by the teacher for drill.

Thus, the chalkboard is used in conjunction with *lecturing, discussion, brainstorming, demonstration,* and *giving instructions.*

4 | BULLETIN BOARD METHOD

Teacher Dominance: High
Group Size: 30 Students

The bulletin board can be used by itself for teaching or with other teaching methods to make them more effective. The purpose of the *bulletin board method* is to display information so that it will attract attention and be available on a sustained basis until the display is changed. It frequently involves one-of-a-kind materials.

Teachers or students who prepare bulletin boards can improve the effectiveness of learning if they: (1) decide on a clear, unambiguous purpose for the learning (what is to be learned can be conveyed through the title or a question); (2) present a self-contained lesson on the bulletin board (students should be able to "get the message" from everything on the bulletin board); (3) show how the ideas are related through color-coding, arrows, or other techniques; (4) select colorful interesting materials that are easily understood (photos, pictures, reading levels, language, and so on); and (5) relate the information on the bulletin board to other learning in progress, usually through discussion.

There are several uses teachers make of bulletin boards in addition to presenting "a lesson" as described by the steps above. Teachers use bulletin boards to motivate students as they introduce a unit or topic of study, to display informal self-test quizzes ("Question-of-the-Week" topics), and to display the work of students (tests, drawings, and so forth).

Bulletin boards are used in conjunction with several teaching methods: *lecture, projects, problem-solving, inquiry, field trips, laboratory,* and *textbook* methods.

5 | PROBLEM-SOLVING METHOD

Teacher Dominance: Medium Group Size: 1–30 Students

It is generally assumed that the purpose of the *problem-solving method* is to provide stu-

dents with practice and support while they learn to solve problems. The strategies they learn will ostensibly, transfer to real life when they are adults. But there is no single problem-solving method: there are as many methods as there are kinds of problems to solve. Problems in math, science, and social studies vary greatly. Students also vary in their learning styles, previous experience, and capacities to solve problems, all of which makes problem solving a challenge to teach. Therefore, the following list of strategies will be more useful with some students than others, and totally unnecessary with still others; they will be appropriate for some problems and not for others. This method can be used with Listening-Speaking and Watching-Doing transactions.

First, problems are, by definition, situations that require some action not immediately obvious to the observer. Second, there is a general framework for solving some problems that is most frequently presented in schools. It includes: (1) understanding the problem statement, that is, being able to say in one's own words what is expected or being asked; (2) generating many possible ways to solve the problem by naming as many actions as one can think of which might work or be useful; (3) selecting the most reasonable way to solve the problem or trying out several ways to solve the problem; and (4) evaluating the solution(s) for fitness. Naturally this generalized framework does not begin to account for all of the difficulties students encounter when they are trying to solve problems. Teachers need to look closely at each of these steps to locate the specific sources of student difficulty. A few examples are described below.

1. Understanding the problem. When students have difficulty understanding what is being asked of them in written problems, they are often confused by the context. Some are helped

by hearing someone else read the problem or by saying it in their own words (auditory learners). Others have to imagine themselves or others carrying out the actions suggested by the problem. Still others actually have to act out the problem the first time they encounter it (kinesthetic learners). Some students are helped by sketching a picture of the problem. Students who have been given guidance at this stage of problem-solving (for example, exploring various strategies), usually are able to repeat strategies on their own with similar problems in an effort to understand the problem. Interestingly, in science labs or "hands-on" activities, some students have less difficulty understanding the problem because the materials are in front of them.

2. Generating many ways to solve the problem. Some students are helped by trying to recall a similar situation which they faced successfully before. Others are helped if they try to picture how to solve small pieces of the problem. (It's easier to light one bulb with a battery and wires than it is to light two or three bulbs.) Others imagine solving a similar but simpler problem. (It's easier to divide one apple among four friends than to divide a dozen among thirty people; once the process is recognized, the problem is easier.) Still other students are helped if they visualize turning the problem over to someone else they know and imagining how that person would behave. Typically, most students need to list the various actions they could take to solve a new problem; afterward, they begin to examine the context clues to recognize which action will work best.

3. Selecting the most reasonable action to take in solving a problem. Some students are helped by picturing "in their heads" or on paper what would happen if each of the actions or solutions were tried. Others need to actually carry out each approach. Still others cannot carry out any of the actions without help because they lack experience with the actions themselves. It's difficult to solve math problems requiring multiplication if you still have difficulty with the operation itself.

4. Evaluate the solution to a problem. Students have to compare their estimate or action with some alternative "answers" and with the original problem statement. To develop skill in evaluating alternative solutions or actions, most students need models. The teacher can show how various actions or solutions compare by showing how some are closer estimates than others. A bathroom scale is not as accurate as a postal scale for weighing small objects. The teacher can show how some are faster solutions than others. Multiplication is faster than a series of additions. And teachers can show how some solutions are socially beneficial to greater numbers of people. Cafeteria lines are more efficient than having everyone try to get through the door at once.

In each of the steps of the generalized framework for solving problems, some students feel more relaxed when they are able to share solving the problem with peers. Team or group learning is an efficient way for teachers to provide the support students need when they are learning to solve problems. Not all students need such support, but it should be available to those who do need it until they are successful on their own.

The problem-solving methods described here are often used in conjunction with *laboratory, individualized instruction, contract, questioning,* and *inquiry* methods.

6 | LABORATORY REPORT METHOD

Teacher Dominance: Medium Group Size: 5–30 Students

The purpose of the *laboratory report method* is to stimulate students' thinking and help them to generalize from their laboratory experience. Lab reports are usually written after the laboratory activity is completed or while it is in progress. Teachers are responsible for (1) providing a meaningful purpose for lab reports; (2) describing (orally and in writing) the procedures for writing the report; (3) providing assistance while students write reports; (4) evaluating students' work in lab reports; and (5) correcting misconceptions and reinforcing learning that has taken place. Students are responsible for following the guidelines for writing the report and asking for assistance when needed.

There are several ways to increase the learning value of the laboratory report method. First, avoid using lab reports as "busy work" or as a way to control behavior. Whether workbooks or teacher-made outlines are used for lab reports, students need to see the connection between the content of their laboratory report and issues or applications of interest to them. (For example, how can the result of an evaporation experiment be applied to the loss of water at the local swimming pool?) Second, many students enjoy working in teams of three to five persons and sharing responsibility for lab activity and reports. As teachers circulate around the room, they can ask probing questions to stimulate students' thinking. They can ask students to reflect their joint thinking in the lab report and thereby foster the exchange of ideas. Finally, many teachers have found students are most motivated to write lab reports when the results on their individual or team lab reports are pooled and compared with others. For example, when all the results of the lab reports are entered on a master chart or matrix on the chalkboard, students are eager to compare the results of their work with the results of others.

Other methods used in conjunction with the *laboratory* method include *peer review*, *team learning*, *problem solving*, *questioning*, *workbook*, and *discussion*.

7 | TEAM LEARNING METHOD

Teacher Dominance: Low
Group Size: Multiples of 5 or 6

The purpose of the *team learning method* is to foster a spirit of sharing and cooperation while students learn, by grouping students in such a way that their different skills and interests complement each other while they work on assignments as a team. This method can also be included in Listening-Speaking and Watching-Doing transactions.

Teachers form teams after careful observation of students' behavior. Typically, teams are most successful when there is a balance among the members of the team in several qualities: (1) ability to read and follow instructions; (2) ability to organize people and to act as a catalyst in getting things started; (3) ability to follow through on tasks and keep plodding along when others give up; (4) capacity to generate lots of ideas; and (5) willingness to accept difficult or less popular assignments. Many teachers form temporary groups and present them with games of various sorts to make assessments of students' skills and interests. Then they switch students until they find the best balance for the teams. They suggest coordinators or team leaders, and rotate leadership within teams depending on the special interest or skill needed.

Team learning can be used effectively in many subject areas. It is a method of teaching that lends itself to classes where students' instructional levels vary widely; there the method serves as a built-in tutoring technique. But team learning is equally effective with students of similar abilities, where it serves as a device for promoting cooperative, enjoyable learning.

Other teaching methods that are used in conjunction with team learning include *laboratory, lab report, problem solving, projects, oral report,* and *inquiry.*

8 | PEER REVIEW METHOD

Teacher Dominance: Low
Group Size: 1:1, Repeated with Entire Class

The purpose of the *peer review method* of teaching is to enhance students' learning by having them apply high standards provided by the teacher to evaluate each other's work. By evaluating the work of their peers on a task identical to the one they have done, students generally improve their own work.

To use the peer review method, teachers should (1) explain the purpose of peer review; (2) guarantee anonymity if students are insecure about having others read and evaluate what they have written; (3) distribute and discuss the criteria to be used in reviewing the papers; (4) arrange for the exchange of papers; (5) monitor the peer review process, answering questions as they arise; (6) evaluate students' first efforts at peer review and add their own comments to those of the peer reviewers' (these may sometimes contradict those of the peer reviewer); and (7) discuss with the class as a whole ways in which they can improve their peer reviewing process.

One condition that affects the success of the peer review method is the age of the students. Older students rather than younger ones are better able to use this process. Another condition is a climate of cooperation that assures all students of the primary objective of peer review: to improve everyone's skills as much as possible. Thus, to be successful, peer review should be followed by an opportunity for students to revise their papers. Finally, as a cooperative atmosphere is associated with peer review and students no longer feel the need to keep their identity (authorship) unknown, students begin to act as reviewers for each other and talk about how their peer can better communicate what he or she is trying to say in writing. Then, peer review takes on the form of *peer tutoring,* a new level of interaction.

The peer review method is useful in conjunction with any activities in which students have to express themselves in writing, but especially with material written in paragraph form.

9 | PEER TUTORING METHOD

Teacher Dominance: Low
Group Size: 1:1, Repeated with Entire Class

The purpose of the *peer tutoring method* is to have one student help another to learn. There are many reasons why teachers use peer tutoring. Teachers find they can provide more assistance to students who need it. Others use peer tutors to provide a qualitatively different kind of help for students—one that comes from someone closer to the student's point of view and therefore, someone who may explain things differently. Still others occasionally use peer tutoring to provide incentive and practice for students who have had difficulty learning material themselves. They do so by helping even slower students with the same material. This method could also be included with Listening-Speaking and Watching-Doing transactions.

The teacher's responsibility in peer tutoring is to (1) explain to both students (the tutor and the tutored) what is expected of them; (2) provide guidelines for the tutor which suggest whether the tutor should help the other student in just reading problems, by actually helping to solve the problems, or by only pointing out which problems are incorrect and explaining why they are incorrect; and (3) evaluate the effectiveness of the tutor-tutored pair by talking privately with each student.

Teachers who use peer tutoring as a part of their regular classroom instruction say it helps to create a cooperative and friendly atmosphere in their classroom. These teachers train tutors at the beginning of the school year and give every student an opportunity to tutor someone else in their own area of expertise or to tutor younger students in other classrooms. The best time to use tutoring is when students are given a block of time to work on written assignments or projects of their own, and the teacher circulates around the room helping students. Then students can also be helped by tutors and have the feeling they are being helped sooner than otherwise would be possible. Some teachers use peer tutoring in the context of *team learning*.

Peer tutoring lends itself to many other teaching methods: *problem solving, workbooks, drill, games,* and *peer review.*

10 | PROGRAMMED INSTRUCTION METHOD

Teacher Dominance: Low
Group Size: 1–30 Students

Programmed instruction is a method of teaching that is often associated with computer technology. In order to program the instruction, one takes a "chunk" of knowledge (for example, some rules of grammar or animal groups in the Animal Kingdom), one breaks it into small "bites," and then one sequences the "bites," and separates each "bite" with a test that the learner must take in order to find out whether to proceed or to reexamine previous learning. Typically, students proceed at their own pace, and several or all students in class learn from the same material.

In programmed instruction, the teacher usually selects the programmed material and assigns specific portions to students after diagnosing their needs. The materials may involve books, boxes, or kits with an assortment of cards and booklets. Some may need to be used with a small classroom computer. Most are usually produced commercially, rather than teacher-made. Students generally manage the materials themselves if the teacher's diagnosis is accurate and the materials match the learner's readiness. Teachers periodically check students' progress. Many programmed instructional materials are exceptionally attractive to students, but not all students like working within the narrow framework of programmed instruction. Programmed instructional materials are a mainstay for teachers who wish to individualize instruction because of the thoroughness, organization, and completeness of the knowledge that is covered by the materials.

Other methods used in conjunction with programmed instruction include *individualized instruction* and *learning centers.* It is important for teachers to balance programmed instruction with the creative and open-ended activities derived by using *brainstorming, exploration-discovery,* and *inquiry* methods.

11 | INDIVIDUALIZED INSTRUCTION METHOD

Teacher Dominance: Low
Group Size: 1:1, Repeated with Entire Class

The purposes of the *individualized instruction method* are to present each student with learning materials that have been especially selected for him or her, to fit specific learning needs, and to promote greatest progress for each student. This method can also be included with other Listening-Speaking and Watching-Doing transactions.

Students who use individualized instructional materials may work alone on materials that are unlike those of any other classmate, or they may work with several other students who have nearly identical instructional needs and materials. The teacher's job in individualizing instruction is to match learning materials with individual students' needs and abilities. It is very much like teaching thirty classes, each consisting of one student. Yet, by gradually phasing individualized instruction into two or three subjects in the curriculum, most teachers find themselves more satisfied with the results of their own efforts and those of their students. With the advent of mainstreaming handicapped students into regular classrooms, individualized instructional materials are more often prescribed by special education teachers than they were before mainstreaming, and a richer array of materials is available at all levels.

Individualized instruction can be an effective method of teaching, especially when it is used in conjunction with other teaching methods that enhance sharing and cooperation among students: *team learning, peer review, peer tutoring,* and *discussion.*

12 | NOTE-TAKING METHOD

Teacher Dominance: Low
Group Size: 1–30 Students

The purpose of the *note-taking method* is to enable students to record information in an organized and abbreviated form, so that it can be referred to at a later time. Because audiotapes, videotapes, and cassettes are available, some teachers might overlook the need to teach students to take notes. But it is an essential skill since students still need to use note-taking to summarize what they have read. Note-taking is certainly more common in high schools and colleges, but elementary and middle or junior high school students also need this skill for book and lab reports and to improve their auditory and listening skills.

Teachers can teach or improve note-taking skills by following four strategies. Students should anticipate the structure of the information. For written material, the structure can usually be found by looking at the table of contents for the main ideas, looking at bold-faced type, or looking for key sentences in each paragraph. For lectures, structure is not always easy to recognize unless the teacher puts an outline of the lecture on the chalkboard. However, students can always take abbreviated notes and restructure them later. Students need to be shown how and what to abbreviate; the subject matter itself usually dictates this, but teachers need to illustrate the process by using concrete examples. Students need to be encouraged to ask questions about information they do not understand. This is especially true if they cannot differentiate between major and minor points. Students should ask for feedback on their note-taking skill. Teachers need to provide students with specific feedback on improving their note-taking skills or have peer tutors do so.

Note-taking is used in conjunction with *lecture, demonstration, project, field trip,* and *review* methods. It is a method of learning that is essential for visual and auditory learners.

13 | JOURNAL-KEEPING METHOD

Teacher Dominance: Low
Group Size: 1–30 Students

The purpose of the *journal-keeping method* is to encourage students to make keen observations and to record their observations in some systematic way. For some purposes, the style of writing may be quite open and free-flowing but the observations are recorded on a regular basis, as in personal diaries. Or the amount and kind of information may be detailed and systematic but the frequency of recording observations might be quite irregular. This would be true of students who describe specific details about seasonal changes in the environment as they occur, or students who write poetry.

To introduce journal-keeping, the teacher needs to: (1) explain the purpose of keeping the journal (for example, science journals are quite different from personal journals); (2) provide concrete examples of journals such as Leopold's *Sand County Almanac,* Samuel Pepys' diary, or those written by other students, if they are not too personal; (3) demonstrate the procedure students should follow with regard to form and frequency of entry; (4) arrange times for writing in or sharing the contents of the journals, unless they are confidential; and (5) provide students with the journal, booklet, or other material to be used for their observations.

Journal-keeping is a method of teaching that is often used in conjunction with *project, field trip,* and *individualized* methods of instruction.

SECTION 3:
WATCHING-DOING TEACHING METHODS

DEMONSTRATION METHOD

FIELD TRIP METHOD

"HANDS-ON" METHOD

INQUIRY METHOD

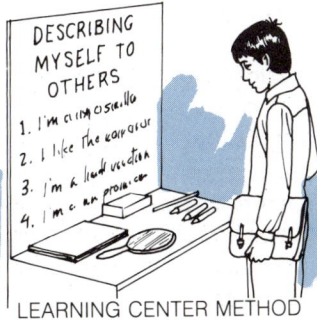

LEARNING CENTER METHOD

PROJECT METHOD

SIMULATION METHOD

GAMES METHOD

EXPLORATION-DISCOVERY METHOD

Watching-Doing teaching methods are used to enhance students' kinesthetic learning.

1 | DEMONSTRATION METHOD

Teacher Dominance: High
Group Size: 5–30 Students

The purpose of the *demonstration method* is to enable students to see and follow a series of events or steps, or to witness a phenomenon or single event. A demonstration is sometimes used to prepare students to conduct their own similar activity or experiment. The demonstration prepares them to use some tool or piece of equipment, to anticipate a result or difficulty, or to use a particular technique or sequence. The demonstration method is also used as a substitute for students acting on their own, in which case students become spectators while the teacher controls the learning by showing what is to be learned and by pointing out key features or steps in the sequence. Demonstrations are frequently used instead of the *laboratory* method to cut costs, to save time, or to safeguard students when the materials or process may be dangerous for them to handle. The teacher is responsible for knowing the subject matter and educational value of the demonstration, and for preparing the materials for the demonstration. The teacher presents the event, describes it in easily understood terms, and provides an opportunity for students to ask questions. The students observe and listen and may take notes or make drawings and diagrams. The classroom should be set up so that everyone can see the event and hear the teacher. Large groups can usually be accommodated if students sitting nearest the demonstration remain seated while those farther away are allowed to stand. For small groups of students, standing near the demonstration is often most interesting. It is important for teachers to conduct practice dem-

onstrations so that unforeseen difficulties may be dealt with or prevented and the length of time needed can be estimated for the final demonstration.

Several methods are frequently used in conjunction with the demonstration method: *giving instructions, lecture, diagram/drawing, questioning,* and *reviewing* are just a few.

2 | FIELD TRIP METHOD

Teacher Dominance: High
Group Size: 30 Students

The purpose of the *field trip method* is to provide students with experiences that cannot take place in the classroom or place of normal instruction. The destination of the field trip may be in- or out-of-doors. The most common examples are field trips to museums, zoos, theaters, parks, wilderness areas, neighborhoods, supermarkets, stores, public businesses, agencies, or offices.

The teacher prepares students for the event by telling them the objectives of the trip, explaining the codes of dress and behavior that are expected, and describing some of the experiences they may anticipate. The teacher generally leads the field trip or turns over leadership to a guide at the site. Likewise, the teacher usually assumes full responsibility for arranging transportation, admission or right to enter, permission from parents of students to go on the field trip, and knowledge of the educational value of the site itself. If additional adults are needed for supervision of students, the teacher also arranges for their presence. On the other hand, students' responsibilities are to come prepared in appropriate clothing, with lunch or lunch money if that option exists, and with any equipment that has been recommended by the teacher. In order to keep track of students, they may be paired, grouped by teams, or assigned to specific supervising adults. When the field trip is over, the teacher and students usually summarize the event and may begin follow-up activities that grow out of the field trip.

Several conditions affect the success and value of any field trip. First, a field trip is more effective if the teacher goes on the trip shortly before the trip is scheduled for the class. This "dry run" provides important information: the time it takes to get to the site; the time it takes to see everything; the location of key places such as restrooms, lunch sites, bus stops, and so on; the quality of the on-site leaders; the weather or seasonal fluctuations in natural history; and the presence of any conditions that require special precaution for students' safety. Secondly, if maps to the site are available, they add to the effectiveness of the field trip. Brochures, pictures, or other printed material at the site can make the experience more enriching. If the field trip is held in an outdoor setting, a battery-powered microphone can be useful for calling students who are out of normal hearing range. It is always useful to have one extra vehicle at the site of the field trip to handle irregularities or emergencies for field trips that are more than a few minutes from school.

The field trip method is usually combined with brief *lecture, discussion, questioning,* and *review methods.* Many teachers also use *journal-keeping* or *note-taking* methods either during or after a field trip. There are few substitutes for the field trip method—for being there to see, hear, smell, touch, taste, or have other experiences that can never be brought to the classroom. Perhaps *film showing* comes the closest to the field trip method for providing a similar experience.

3 | CONTRACT METHOD

Teacher Dominance: Medium to High
Group Size: 1:1, Repeated with Entire Class

The purpose of the *contract method* of teaching is to guide students toward independence in learning by setting objectives for individuals or groups of students, engaging them in planning how to reach those objectives, and packaging materials as a series of learning tasks intended to help them reach the objectives. As learning contracts are commonly used in schools, there is less that is negotiated by the students and more that is specified by the teachers; more joint decision making might be desirable. This method would also be used in Listening-Speaking and Reading-Writing transactions.

To prepare contracts, teachers typically use a set of behavioral objectives that specify what the students should be able to do after completing the assignments. Then teachers put together one to six week's work in a packet for a given subject. For example, a contract in science might be organized to guide students' learning about the circulatory system in the human body. Included in the packet would be instructions for reading a section of a textbook, watching a filmstrip, doing an experiment with a stethoscope, drawing a diagram, writing a report, and taking a five-minute oral exam with the teacher. In this example, the students might be instructed to do any five of the six activities, as the negotiable element of the contract. Teachers usually try to individualize contracts to some extent, especially in choices of reading materials. Contract learning is best liked by students who enjoy independence. Age and previous experience are important factors in determining the ease or difficulty of contract learning.

The contract method of learning is often used in conjunction with *individualized instruction, programmed instruction, projects, workbooks* (frequently cannibalized and doled out with each week's packet), *oral reports,* and *drill* methods.

| 4 | "HANDS-ON" OR LABORATORY METHOD |

Teacher Dominance: Medium Group Size: 5–30 Students

The purpose of the *"hands-on"* or *laboratory method* of teaching is to provide students with an opportunity to solve problems or apply conceptual knowledge to concrete problems or situations. "Hands-on" activities predominate in the primary grades in art, cooking, social studies, and science. Laboratories in the upper grades follow conceptual learning presented either in lectures, textbooks, or in both; and typically involve the use of equipment and materials needed to solve problems or illustrate concepts. For example, labs in science often present problems that require a formula to solve or illustrate a natural phenomenon to be observed and described. However, labs in the upper grades also can be used to introduce concepts rather than follow such introductions. The essence of laboratories is that students have their hands on materials of some kind: hence the term "hands-on."

The laboratory method familiar to junior high and high school teachers typically involves rooms furnished with water, heat sources, and basic components such as tables and counters. Elementary school classrooms have "hands-on" activity, but teachers simply do not refer to them as a lab.

To provide for laboratory or "hands-on" learning activity, the teacher is responsible for: (1) selecting problems or situations that illustrate the concept or skill they have taught or wish to introduce, many of which are found in teacher guides or lab manuals; (2) arranging the classroom with appropriate materials and equipment, often with student help; (3) providing written and oral instructions for the activity, experiment, or problem using ditto sheets or workbooks; (4) circulating around the room to ensure safety and learning while students are engaged in the work; and (5) evaluating students' work, formally or informally, by having them tell what they have learned.

There are several things that contribute to successful laboratory experiences for students. First, the teachers' own enthusiasm for learning the results of students' lab activities is contagious. Second, because students vary in their ability to conduct experiments on their own and in their ability to see the connection between the concept and its application, it is often helpful to group students so that stronger students can help weaker students. Third, when students work in teams, the teacher can ask each team to put the results of their activity or experiment on the chalkboard or a chart, showing how many teams' results compare. Other techniques that add to success of a lab include having a brief exploratory-discovery period at the beginning of the lab or "hands-on" activity so that students can become familiar with the materials and not become distracted during the time instructions are given. Another technique is to have the teacher or a student do a demonstration before the lab activity begins. Each of these techniques has merit for the effect it is designed to achieve.

The laboratory method is frequently linked to several other teaching methods besides *exploration-discovery* and *demonstration*: *inquiry, problem solving, projects, field trips, lab report*, and *discussion*.

5 | INQUIRY METHOD

Teacher-Dominance: Medium Group Size: 1–30 Students

The purpose of the *inquiry method* is to enable students to gain experience and confidence in their ability to find answers to questions and problems that interest them or that must be solved, and to do so at the same time they are acquiring subject matter knowledge. For example, when the inquiry method of teaching is used to help students learn about natural phenomena in the world around them, not only is knowledge about the phenomenon in question important, but the process of learning itself is important. In essence, students learn to engage in reflective thought. This method is also used in Listening-Speaking and Reading-Writing transactions.

Teachers use a general framework which involves selecting some problem, puzzle, question, or situation which has a mystery or contradiction in it. Beginning teachers usually contrive these events by planning for their occurrence, but with experience, teachers can use spontaneously occurring events to promote students' inquiry training. Teachers then identify the essence of the problem, question, or contradiction to be investigated. They help students to decide what information is needed, have the necessary materials available, and stand by while students gather the information. Finally teachers encourage students to compare their interpretation of the information with those of others interested in the same question or problem.

There are many approaches to use of the inquiry method. Some involve having students discover things for themselves with responsibility for their own chain of reasoning; others involve the entire class in a group process of reasoning. Some teachers introduce inquiry activities by "talking through" the process as it occurs; other teachers allow students to discover the inquiry process after they have been using it for awhile. What is ultimately essential is that students become aware of and responsible for directing their own search of knowledge to find answers to specific questions or problems. It is important to keep in mind that students develop these skills gradually, and younger students especially may need more help or "talking through" the process than older students. Because many students have learned to depend upon teachers and textbooks as infallible sources of knowledge, they must be guided to rely on their own ability to find answers, to have confidence in information they gather from their senses, and to look for firsthand evidence to support their conclusions. Teachers can help students by temporarily assuming a role of non-expert or a friendly "devil's advocate."

Other methods used in conjunction with the inquiry method are *questioning* and *discovery-exploration* methods.

6 | LEARNING CENTER METHOD

Teacher Dominance: Low
Group Size: 1:1, Repeated with Entire Class

Learning centers are really a facility, but the use of learning centers has become so popular that a method for using them is described here. The purpose of learning centers is to promote independent learning among students. Learning centers are usually located at a table, booth, shelf, or in a corner, and they offer students a varied and extensive collection of materials related to some topic to be learned. This method also involves Listening-Speaking and Reading-Writing transactions.

The materials are organized in such a way that students can easily find what they need to perform some task or engage in an activity of their choice. Some rules for behavior or use of the center or materials are usually discussed in class before learning centers are introduced so that independent learning can proceed with a minimum of disruption. Usually, one or two students use a learning center at a time. Some classrooms have the entire classroom and curriculum organized for learning centers so that students move from one center to another.

The teacher needs to have a clear notion of what learning is to occur at the learning center. The next step is to assemble the materials that allow students to learn by themselves. That means the teacher creates simple directions to follow, and plans for giving assistance when students who are not fully independent need help (a classmate or aide can often help). The teacher further plans for evaluation of what has been learned, usually by means of a product.

Learning centers are usually set up so that there is something for everyone, when each member of the class takes a turn at the center. Thus, it is important for the materials to be varied so that students with differing abilities can find tasks, problems, or activities that fit their level of understanding and ability. This variety is also a factor in motivation; if students have the opportunity to choose from several activities, they find learning more interesting. Coding cards or instruction sheets can guide students to materials at their level of ability and still provide them with options.

Most teachers set up their own learning centers but many commercial materials are available to help teachers in this effort. Experienced teachers have students help them prepare the centers by bringing in materials and duplicating or decorating them. The most successful learning centers are usually colorful, have materials that are inviting to handle, have questions or tasks that stimulate curiosity, and offer sufficient variety so the slowest and the brightest students are challenged. Some teachers pair students so that strong and weak readers use the center at the same time, each doing a different kind of activity. The strong reader can help the weaker reader understand the instructions for her or his task.

Two other teaching methods often used in conjunction with learning centers are *exploration-discovery* and *games*.

7 | PROJECT METHOD

Teacher Dominance: Low
Group Size: 1–5, Repeated with Entire Class

The purpose of the *project method* is to increase students' involvement with conceptual ideas and to develop skills that might not otherwise be developed by reading textbooks, listening to lectures, or any other method. Projects usually require students to acquire new knowledge or require them to apply knowledge they have gained to lifelike situations, problem solving or in the creation of products.

The teacher's responsibility is to provide students with the opportunity to select among alternative topics for their projects, to provide written guidelines (except in primary grades), and to explain orally for students how to follow guidelines throughout their projects. Guidelines usually include information about (1) the purpose of the project; (2) the nature of the final product; (3) any special procedures to be followed; and (4) the time the final product is due.

The students' responsibility is to follow the guidelines and ask for assistance when needed. Students usually need assistance in locating resources for their projects. The teacher can provide this assistance or guide students to seek help from others such as classroom aides, librarians, people who work in local offices, agencies or stores, and from parents or classmates. Students vary in their abilities to conduct projects on their own and many prefer to share responsibility in team projects. Having team and individual projects in the same classroom or assignment rarely causes difficulty if guidelines are developed beforehand. Team projects are usually broader or more comprehensive than individual projects.

Many factors affect the success of projects; students' abilities, the clarity of the purpose of the project, and the availability of resources are strong influences. For these reasons, it is important for teachers to support students' efforts in their projects and to recognize the need for multiple standards in evaluating projects.

Other teaching methods related to projects are *problem solving, oral reporting, written reports,* and *note-taking.*

8 SIMULATION METHOD

Teacher Dominance: Low
Group Size: 10–30 Students

The purpose of the *simulation method* is to increase students' abilities to apply concepts to the solution of problems, to resolve social issues, and to see other points of view. Simulations are, by definition, not real situations but are calculated to represent reality in condensed, abbreviated, scaled-down, or simplified form so that problems or issues become more manageable.

The teacher has several responsibilities: (1) to select a problem or issue which deals with a conflict of interest or opposing views; (2) to ensure that the key elements of the problem or issue are contained within the simulation; (3) to identify the concept that is to be applied to the problem or issue; and (4) to identify the procedure to be followed during the simulation. Procedures include how students are to behave, the roles they are to play, and any rules that govern their actions. Students participate in simulation by assuming the roles of particular individuals or interest groups, following the rules or procedures identified by the teacher, and then spontaneously "living" the situation by behaving as the social controls permit. For example, in a simulation of conflicts of interest in land use, a miniature "lake" measuring 4' x 6' x 6" was actually created on a school ground. Lots surrounding the artificial lake were auctioned off and students had to decide how commercial and recreational properties were to be developed and governed, and how the environment was to be protected by the conservation of natural areas.

Simulations are generally more successful when lifelike materials are used. Simulated money, governing boards or offices, and votes taken, are all good examples. The teacher must be ready to help students apply the concept that is needed to resolve the issue, conflict, or problem; otherwise students may become "wrapped up" in the simulation and lose sight of the educational objective. When the simulation is over, a discussion of the feelings, points of view, and difficulties in resolving conflicts is essential if students are to benefit from the experience.

Simulation is also used in conjunction with *textbook, discussion* and *questioning* methods.

9 | GAMES METHOD

Teacher Dominance: Low
Group Size: 1–30 Students

The purpose of *games* in the classroom is to add enjoyment to learning. Games have properties that make learning a pleasant experience. Perhaps their most appealing characteristic is the element of chance. It relieves the boredom associated with other nongame forms of drill or repetition. Competition is sometimes associated with games and with regular classroom assignments. But in games, students control the level of competition by how long they play and with whom they play the game, and there are never any grades for games.

Teachers who make their own games, adapt commercial games to their own subject matter, or have their students prepare the materials necessary for adapting commercial games to classroom subjects, are better able to use games for instructional advantage than are teachers who depend strictly on commercial games. Although such games are often quite limited in their application to subject matter, they can be adapted easily to a wide variety of subjects. For young students, Bingo and other simple games that include the use of spinners or cards can have reading or science vocabulary or math combinations substituted for commercially printed words by using self-adhering labels. Older students enjoy games like Chess and Monopoly. Chess can be adapted to reinforce learning about ecological food chains, for example, if chessmen are replaced by pieces representing six species of plants and animals in a food chain. Monopoly can become a game that illustrates the problems and issues associated with environmental planning and protection when the Chance and Community Chest cards are replaced by cards that place restrictions on land use and development.

Factors which affect the success of games in the classroom include how easy or difficult the games are, how closely they match the subject or teacher's instructional objectives, and how often games are used. A game, as experienced teachers know, can become as boring as any other activity. Games are used as an excellent substitute for, or in conjunction with *drill*.

10 | EXPLORATION-DISCOVERY

Teacher Dominance: Low
Group Size: 1–30 Students

The term *exploration-discovery* is used here to describe a method of teaching intended to promote students' self-guided learning. Teachers do no more than to provide specific materials and allow students to pursue their own interests with the materials. The kind of learning which occurs when self-guided exploration and discovery are permitted appears to provide essential and fundamental knowledge about materials that are new to students. They find out what something is, how it works, what some story or article is about, and how they can use it. Exploration-discovery is a method designed to promote students' curiosity, desire to know, and freedom to explore—all of which contribute to their motivation to learn and discover new knowledge. This method is also suited to Listening-Speaking and Reading-Writing transactions.

Teachers select materials which have characteristics that invite self-guided exploration and discovery: (1) materials which are new to students and that have the potential to be important for them to know about and (2) materials which permit simple and more complex investigation so that students may choose their level of interest. Materials such as special one-of-a-kind books, new plants or animals in the classroom, or new pieces of equipment or tools are examples of the kinds of materials that lend themselves to the exploration-discovery method of teaching. Then teachers need to allot time for free exploration of the materials without dictating outcomes that are expected, and specify a place for exploration to take place so that distraction to others is avoided. For example, when a new piece of equipment is introduced to the entire class the teacher might say, "See what you can find out about the . . . during the next five minutes." Or when several new books are brought in and placed on a special table,

the teacher might say simply, "I have found some new books about . . . that several of you might be interested in. When you finish your regular assignment, feel free to go back and look at them." These might even be checked out. When an interesting animal is brought in, the teacher might say nothing but place it in a conspicuous place and wait for students to comment. This leads to a discussion of what the students have observed or what they already know about the animal.

Exploration-discovery is spontaneous learning, but it is not without accountability. Once teachers have provided materials and a time and space for exploration, they may follow students' interest in these materials by asking how students found them interesting and what they learned from the materials. Implicit in the teacher's questions must be the fact that any level of interest is acceptable.

The term *guided discovery* is often used to refer to a method where the teacher has a more or less specific discovery in mind that students are to encounter. For example, if the teacher wanted students to find out what the effect of the length of a pendulum was on the speed of its swing, the teacher might ask them to see if they could find out "What makes a pendulum go faster or slower?" The teacher is guiding students' discovery rather than permitting them to explore the equipment on their own. Of course, both methods have value. Other methods of teaching that are similar to exploration-discovery are *inquiry, problem solving* and *games,* but all of these involve more teacher dominance. Exploration-discovery is the most open-ended of the teaching methods described in this taxonomy.

The essence of artful, competent teaching is found among those rare individuals who so nimbly trade one teaching method for another from their rich, pedagogical repertoire that their students hardly realize they are being taught.

COMMENTARY

In this chapter we have opened the door and invited you to explore the many methods used by teachers. As you learn about procedures for planning science lessons, units, and programs in the next chapter, keep these methods in mind.

BIBLIOGRAPHY

Carkhuff, Robert; Berenson, David; and Pierce, Richard. *The Skills of Teaching: Interpersonal Skills.* Amherst, MA: Human Resource Development Press, 1977.

Cronquist, Arthur. "Taxonomy." *Journal of College Science Teaching* (November 1979): 76–79.

Esler, William. *Teaching Elementary Science.* Belmont, CA: Wadsworth Publishing Co., 1979.

Hoover, Kenneth, and Hollingsworth, Paul. *A Handbook for Elementary School Teachers.* Boston, MA: Allyn & Bacon, 1973.

Joyce, Bruce, and Weil, Marsha. *Models of Teaching.* Englewood Cliffs, NJ: Prentice-Hall, 1980.

Rowe, Mary B. *Teaching Science as Continuous Inquiry.* New York: McGraw-Hill, 1973.

5

Planning to Teach

The focus in chapter 5 is on planning a science program, science units, and individual science lessons. The chapter begins with a Planning Checklist that will help you coordinate critical steps in planning. You will find suggestions for selecting a science program; you will learn how to make planning charts; and how to order, maintain, and store science materials. Finally, you will find ideas for scheduling science and integrating it with other subjects in the curriculum.

KEY CONCEPTS AND IMPORTANT IDEAS

☐ planning checklist ☐ establishing broad goals for the year ☐ assessment of students' capabilities ☐ selecting a science program ☐ preparing planning charts ☐ ordering materials ☐ reserving films and buses ☐ getting permission and clearance ☐ recruiting volunteers ☐ integrating science with other subjects

Once you have been hired for your first teaching job and your science methods course is behind you, you will be faced with planning science for your first year. Whether you will be teaching science along with many other subjects in a self-contained classroom or teaching several sections of science to different groups of middle school or junior high school students, you will need to use the material contained in this chapter to help you plan your first science program.

USING A PLANNING CHECKLIST

When experienced teachers plan a science program for their students, their decisions usually are made in stages that more or less follow the yearly school calendar. The **Planning Checklist** in figure 5–1 illustrates a sequence of planning decisions for a science program that is based on conversations with many experienced teachers.

To get a picture of the planning process in your head, think of a process that begins with "seeing the big picture" of science during the coming school year. It is much like seeing an entire city at a glance as your plane flies over it. The first three steps in the Planning Checklist focus on this big picture. Then the planning process shifts to looking at three-month periods of the school year; it is somewhat like using a city map to look at sections of the city. Steps four through six show how chunk planning takes place. Finally, in step seven of the Planning Checklist, planning gets down to the weekly and daily schedules.

The Planning Checklist describes steps you will find easy to follow once you have completed chapters 5 and 6. When should you begin planning your science program? It is a good idea to start planning before the school year begins. Most experienced teachers find

long-range planning easier to do before the students arrive in September. As we noted in figure 5–1, the Planning Checklist, many teachers complete Steps 1–5 before school starts, if they have access to official school records that show their new students' capabilities in reading, mathematics, arithmetic, and language. In cases where teachers do complete Steps 1–5 before school starts, they make final adjustments in the selection of the program, materials, and units of activity once they have assessed students' areas of interest in science.

Sometimes teachers don't know what classes, subjects, or grade levels they will be assigned to teach until the first week of school. This is especially true in school districts where enrollments are changing or where classes have to be reorganized after students have arrived and tested for assessment of their capabilities. If you find yourself in this situation, it is advantageous to begin the first two weeks of science with activities that are easy to prepare and easy to accomplish with minimal reading skills. Yet these activities must offer challenge to inventive minds and arouse your own enthusiasm. Many such activities appear in part two. Then, during the weeks that follow, go back to Steps 1–5 of the Planning Checklist. Let's take a closer look at the seven steps in the Planning Checklist.

STEP 1: ESTABLISH BROAD GOALS FOR THE YEAR

Before the school year begins, think about the goals you would like your students to achieve by the end of the year in science and write them down. Look back at some of the suggestions in chapter 2 for setting goals and objectives, and then talk to other teachers at

The Big Picture	August	Before school starts	Step 1:	Set broad goals for year. Tentatively select a science program or become familiar with the program adopted by the school.
	September	First two weeks of school	Step 2:	Assess students' capabilities in reading, math, and language, and their interests in areas of science.
		Third week of school	Step 3:	Become familiar with the district's program or make the final selection of a science program to fit students' capabilities and interests. A published science program is easiest to follow the first year.
Chunks		Fourth week of school	Step 4:	Prepare planning charts to show sequence of units for the year.
			Step 5:	Order materials for September through December. Reserve buses for any field trips using approximate dates and destinations. Also reserve and schedule films.
	October	Fifth week of school	Step 6:	Recruit volunteers for tasks through December.
Weekly	October through December	Fifth through twelfth weeks of school	Step 7:	Once each week, plan details for science activities for the coming week based on planning charts from Step 4. Estimate and reserve preparation time needed for each day's activity.

NOTE: Steps 1–5 can be partially completed before school starts if school records for Step 2 are available. Steps 5–7 are repeated in December through February for the twelfth through twenty-fourth weeks and are repeated in March through May for twenty-fourth through thirty-sixth weeks. These dates and sequences are suggestions based on conversations with teachers.

FIGURE 5–1
Planning Checklist for Science

your school. If they are not available, look at the goals for science listed in curriculum libraries found in the district or county offices. After comparing various sources, make a short list of your own goals for the coming year.

Next, look at the science programs that are available in your school or district office. Decide which programs will best help you to meet your goals. Do not hesitate to use two or more science programs if you think they will do the job better than a single program. Think, at the same time, about the rest of your curriculum and how science might fit into your overall goals for the year. Consider your choice of a science program to be tentative until you have had a chance to assess your students' capabilities and interests.

STEP 2: ASSESS STUDENTS' CAPABILITIES

Become familiar with your students' capabilities in reading, mathematics, and language, since these are crucial skills for reading about science, solving simple science problems involving math, and writing about science experiences. Check the reading level of the science materials you have chosen tentatively for your program. If you have students reading below the level of the material you have chosen, develop an alternative plan for meeting the needs of these students. You need not necessarily eliminate reading from the science program, although many teachers do substitute "hands-on" activities for students or even entire classes designed for students with lower-than-average reading ability. An easy solution might be to have students read in pairs, consisting of a strong and a weak reader. This same strategy is an effective way to compensate for weaknesses in math or language. One teacher we know has her stu-

dents work in teams of five: one strong reader, one student strong in math, one student strong in language, and others less able in these areas. Occasionally her teams are composed of four students weak in these basic skills working with one student who reads, computes, and writes "on grade level" and who acts as the interpreter of the material. Every team member performs at grade level in activities involving firsthand experience; each member can tell about his or her experience with science-related concepts in solving practical problems in science or applying science to daily actions. Thus, the lack of basic skills need not prevent this teacher's students from enjoying science and the teacher can use the science program of his or her choice. This example, however, is the exception; most teachers are able to match their science program with their students' capabilities.

There are other factors to look for as you assess your students' capabilities and interests in science. For example, there may be gifted as well as handicapped students in your class. Be sure to reread the section of chapter 3, Understanding Exceptional Students, which deals with the identification and teaching of gifted and handicapped students.

STEP 3: SELECT A SCIENCE PROGRAM

Select the science program you think will best meet the needs of the students in your class, based on the assessment of their capabilities and interests that you made in Step 2. There are literally dozens of science programs to choose from if you consider all of the published material available in textbook series, kits or boxes, and occasional one-of-a-kind idea books for science activities prepared for teachers. Table 5–1 identifies many of the published science programs available. Use

MAGNETS

Team learning is an effective way to compensate for weaknesses in math or language when teams include some students with strengths in these areas and others with weaknesses. (Donald Birdd)

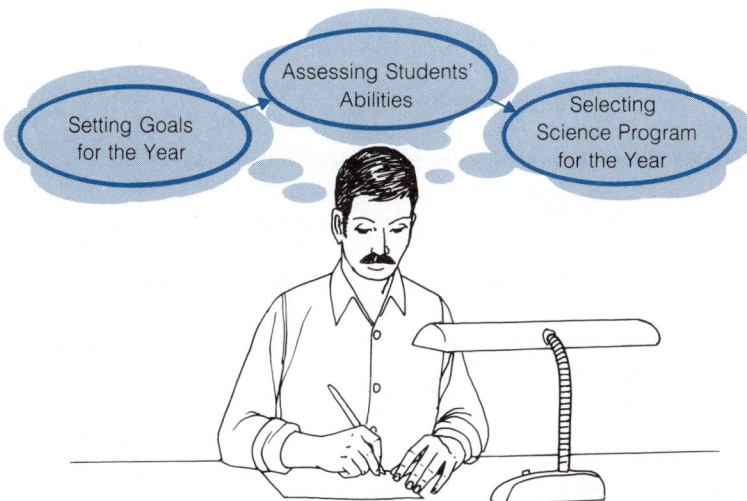

Setting Goals for the Year

Assessing Students' Abilities

Selecting Science Program for the Year

The science goals to be reached and the capabilities of the students should determine what kind of science program you choose.

TABLE 5–1
Published Science Programs

Kindergarten

Accent on Science (Merrill, 1983)
Modular Activities Program in Science (Houghton Mifflin, 1982)
Pictures That Teach (Silver Burdett, 1984)
Science: A Process Approach (Ginn, 1984) (Spanish version, 1984)

Grade 1

Accent on Science (Merrill, 1983)
Concepts in Science (Harcourt Brace Jovanovich, 1982)
Investigating in Science (American Book, 1984)
Learning Science (Macmillan, 1984)
Modular Activities Program in Science (Houghton Mifflin, 1982)
Now You Know About Animals (EBE, 1984)
Science: A Process Approach (Ginn, 1984) (Spanish version, 1984)
Understanding Your Environment (Silver Burdett, 1982) (Spanish version, 1982)

Grade 2

Accent on Science (Merrill, 1983)
Concepts in Science (Harcourt Brace Jovanovich, 1982)
Investigating in Science (American Book, 1984)
Learning Science (Macmillan, 1984)
Modular Activities Program in Science (Houghton Mifflin, 1982)
Our Environment: Choices & Actions (Harcourt Brace Jovanovich, 1984)
Science: A Process Approach (Ginn, 1984) (Spanish version, 1984)
Understanding Your Environment (Silver Burdett, 1982); (Spanish version, 1982)

Grade 3

Accent on Science (Merrill, 1983)
Concepts in Science (Harcourt Brace Jovanovich, 1982)
Investigating in Science (American Book, 1984)
Learning Science (Macmillan, 1984)
Modular Activities Program in Science (Houghton Mifflin, 1982)
Nature Series (Bowmar, 1982)
Science: A Process Approach (Ginn, 1984) (Spanish version, 1984)
Understanding Your Environment (Silver Burdett, 1982) (Spanish version, 1982)

Grade 4

Accent on Science (Merrill, 1983)
Concepts in Science (Harcourt Brace Jovanovich, 1982)
Investigating in Science (American Book, 1984)
Learning Science (Macmillan, 1984)
Little Nature Books (EBE, 1984)
Modular Activities Program In Science (Houghton Mifflin, 1982)
Science: A Process Approach (Ginn, 1984)
Understanding Your Environment (Silver Burdett, 1982) (Spanish version, 1982)

TABLE 5–1, *continued*
Published Science Programs

Grade 5

Accent on Science (Merrill, 1983)
California Gray Whale/Polar Bear/Salt Water Otter (Husco Films (SYSTEM),
    1984)
Concepts in Science (Harcourt Brace Jovanovich, 1982)
Do Your Project Right! (Barr Films, 1984)
Investigating in Science (American Book, 1984)
Learning Science (Macmillan, 1984)
Modular Activities Program in Science (Houghton Mifflin, 1982)
Science: A Process Approach (Ginn, 1984)
The Web Of Living Things (Barr Films, 1984)
This Earth Series (Guidance Associates, 1982)
Understanding Your Environment (Silver Burdett, 1982) (Spanish version,
    1982)

Grade 6

Accent on Science (Merrill, 1983)
Concepts in Science (Harcourt Brace Jovanovich, 1982)
Earth & Moon Geology Unit/Chemical Weathering/Hypothesis Machine (Educ. Materials
    and Equipment, 1984)
Focus on Life Science (Merrill, 1984)
Investigating in Science (American Book, 1984)
Learning Science (Macmillan, 1984)
Modular Activities Program in Science (Houghton Mifflin, 1982)
Science: A Process Approach (Ginn, 1984)
Self-Paced Investigations for Elementary Science (Silver Burdett, 1982)

Grade 7

Earth Science: Patterns in Our Environment (Prentice-Hall, 1982)
Focus On Earth Science (Merrill, 1984)
Life: A Biological Science/Matter: An Earth Science (Harcourt Brace Jovanovich (SYSTEM)
    1982)
Life Science/Physical Science (Ginn (SYSTEM)), 1982)
Modular Activities Program In Science (Houghton Mifflin, 1982)
Principles of Science, Books One and Two (Merrill, 1983)
Spaceship Earth: Life Science/Physical Science (Houghton, Mifflin (SYSTEM), 1984)
You and The Environment (Houghton Mifflin, 1984)

Grade 8

Exploring Science/Life Science/Earth Science (Allyn & Bacon, (SYSTEM), 1984)
Focus on Physical Science (Merrill, 1984)
Idea Science Program (Steck-Vaughn, 1984)
Ideas and Investigations in Science: Physical Science/Life Science/Earth Science (Prentice-
    Hall (SYSTEM) 1984)
Interaction of Earth & Time/Man & the Biosphere/Curriculum Project (Rand McNally (SYS-
    TEM), 1984)
Investigating The Earth (Houghton Mifflin, 1984)

TABLE 5–1, *continued*
Published Science Programs

Investigations in Science Series (Wiley & Sons, 1984)
Modular Activities Program in Science (Houghton Mifflin, 1982)
Natural World 1 & 2 (Silver Burdett (SYSTEM), 1984)
Physical Science: Energy, Matter & Change (Scott, Foresman, 1982)

Multigraded

Asimov On Physics (Doubleday, 1984) (6–8) (7–8)
Earth Science: The World We Live In (American Book, 1984) (7–8)
Elementary Science Sets 1–4; Spanish set 1 (ACI, 1984) (5–7)
17 Miscellaneous Titles (Educational Design, 1984) (6–8)
Our Changing Earth (Coronet, 1984) (4–6)

Non-Graded

Climate In 3-D (Activity Resources, 1982)
Discovering Natural Science (EBE, 1982)
Ecology (EBE, 1982)
Evaluation Program for ESS (McGraw-Hill, 1982)
5 Software Packages (Activity Resources, 1984)
Let's Discover (Guidance Associates, 1982)
Making Small Things Look Bigger/Substances & Mixtures & Powders & Liquids (Educ. Sci.
  Consult., 1984)
Puzzles, Problems, Challenges (Selective Educ, 1982) Kitchen Physics (Selective Educ.
  1982)
Science of Good Earthkeeping (Barr Films, 1984)
Science Readers (Reader's Digest, 1982)
Tales of Endangered Wildlife; The Human Machine; Things That Live and Grow (Coronet,
  1984)

this list as a guide to see how many of the programs are available for inspection at your school, district, or county office.

Every year, you have an opportunity to see a large collection of science teaching materials published if you are able to travel to the annual national or regional meetings of the National Science Teachers Association. (For the location and dates of these meetings, write the National Science Teachers Association, 1742 Connecticut Avenue, NW, Washington, D.C. 20550.) Commercial publishers display their science programs and associated mate-

rials including films, science equipment, and of course, teachers' guides, student workbooks, and textbooks. Frequently a single publisher will have several science programs, each of which appeals to a different kind of learner or teacher.

Whenever you find science programs and materials, use an organized set of criteria to evaluate and compare various programs. Some teachers develop their own criteria for choosing science materials, while others prefer to rely on guidelines developed by others. Figure 5–2, General Guide for Selecting Sci-

High / Medium / Low		
○ ○ ○	Content accuracy	Is the content accurate? Check with a local expert if you are in doubt; ask a science consultant, or high school, or college teacher.
○ ○ ○	Content coverage	Are most of the important topics covered? Will content be wasted? Will you be able to learn science, too?
○ ○ ○	Content up-to-date	Examine an area like space exploration to determine if the science information is outdated. Is the content new to you or to others around you?
○ ○ ○	Content depth	Is there sufficient information about each topic? Can you and your students probe at different levels in the book or material?
○ ○ ○	Motivational appeal	Is the material attractive to students? Don't guess; let students tell you their favorites. Are there enough illustrations?
○ ○ ○	Illustrations	Are materials interesting, clear, colorful, and accurate in detail?
○ ○ ○	Organization	Is there a logical sequence to the ideas? Is it easy to find what you want?
○ ○ ○	Readability	Does the text fit the reading level of most of the students? Does it fit their math and language capabilities?
○ ○ ○	Freedom from bias	Are all groups in society treated with equal respect and accorded equal treatment in the text?
○ ○ ○	Teacher support	Are there enough aids for teachers: vocabulary lists, an index, suggestions for films and resource materials, and sections with science background for the teacher?

NOTE: Fill in the circle that matches your rating. These ten points cover *reading materials only*. Be sure to fill out a separate form of this kind for each program you evaluate. (Permission is granted by the publisher to reproduce this figure.) Then compare your rating sheets for each program.

ence Textbooks and Other Reading Materials, and figure 5–3, General Guide for Selecting Science Textbooks, Equipment, and "Hands-on" Materials, include criteria from several different sources. Look over the guides and se-

lect those criteria that seem important for your materials. The faculties in many schools make a joint selection of a science program for their school. Thus, students can move sequentially through the series of texts or other learning

FIGURE 5–3
General Guide for Selecting Science Textbooks, Equipment, and "Hands-On" Materials

High Medium Low				
○ ○ ○	Purpose/Goals	Are the goals of the program made clear? Will your teaching goals be met? Is the purpose of the equipment or other incidental material clearly related to objectives?
○ ○ ○	Evaluation	Does the material include a way to evaluate students' performance? If not, try to picture how you could evaluate student performance.
○ ○ ○	Versatility	Can the program or material be used easily in many ways? Can it be used for group or individualized instruction? Can science be related easily to other subjects in students' experience? Can different purposes or ability levels be served?
○ ○ ○	Provocativeness	Does the material create in the reader a desire to act, to ask questions, and to seek answers?
○ ○ ○	Cost	Given the number of other strengths and weaknesses of material, is this program or material a good buy?
○ ○ ○	Safety	How much supervision is required for students to use the material; can they use it alone or in small groups without close supervision? Can students be hurt by disassembling the material or by assembling it in new (or "wrong") ways?
○ ○ ○	Maintenance	Will it be easy to store, care for, repair, or replace this material? Can you find a volunteer to be responsible for its maintenance?

NOTE: Fill in the circle that matches your rating. These seven points cover textbooks, equipment, and "hands-on" materials. Be sure to fill out a separate form of this kind for each program you evaluate. (Permission is granted by the publisher to reproduce this figure.) Then compare your rating sheets for each program.

materials as they progress through the primary, intermediate, and middle grades. Selections typically take place every five years or so.

Whether you will select a science program on your own or with a team of teachers from your school or district, try to present your final candidates for adoption (those programs you think will best fit your students' needs) to the students themselves and ask for their reactions and preferences. Research has shown that teachers and librarians frequently misjudge the appeal of books for students.

When you use a set of criteria or a rating scale to help you select a science program or materials, remember these points.

☐ Identify the criteria that are most important to you, and weigh them more heavily than the others. Don't be afraid to eliminate some criteria. For example, the cost of a program may be critical to one school district and relatively unimportant to another.

☐ Every good program usually has some drawbacks. As you compare the strengths of various programs, see which have the least serious drawbacks. For example, the amount of time required to prepare for "hands-on" science activities may be a drawback, but the activity may be so motivating to students that it pays to recruit parent or student help for preparation and clean-up.

☐ Include students in the decision-making process. Once you have narrowed the choice to three or four programs, get students' reactions to the choices. Invite students to look at, read, and talk about the materials, and perhaps try an activity. Students frequently react differently than do adults to illustrations, the size and density of print, and the style of language. Since students are the ultimate consumers, it is valuable to get their input.

We suggested using published science programs during the first year of teaching, but this is only to ease the burden of planning. Even for experienced teachers, planning thirty hours of new instruction every week, regardless of the subjects taught (six hours per day, five days per week) is a very demanding responsibility. Especially during the first year of teaching, the task of creating every lesson or science unit "from scratch" would be overwhelming by about the third month. However, that is not to say you should not try to create your own science program; it is a great deal of fun! Summers are an excellent time to engage in this creative activity. When you are ready to teach your first science program, you can decide whether to use one of the published science programs, create your own, or use the activities provided in part two of this book as a basis for your science program.

STEP 4: PREPARE PLANNING CHARTS

Once you have chosen the book, series of textbooks, program-in-a-box-or-kit, or collection of assorted materials that will constitute your science program, the next step is to map out the program in segments of instruction for the school year.

If you have chosen a program or materials that already includes **planning charts,** you are ahead of the game—unless, of course, you decide to revise them. Begin by asking yourself, "Do I want to follow the book as it is or make some changes?" If you decide to make changes or if you have chosen materials that have no planning guide, you will ask, "When will I teach what? What comes first?" Be sure to begin with a series of activities that really excites you! Getting off to a good start is important.

Curricular segments corresponding to fall, winter, and spring terms are easy to work with as you begin to think about what kinds of activities go best with different seasons.

A creative teacher considers many factors when planning the course of science study for the year.

Published Science Programs

Grading Periods and Seasons of the Year

Relating Science to Other Subjects

Gardening or growing things, for example, is often better studied in the spring than in the winter. You might want to intersperse the study of weather throughout the year rather teach it during a single season. As you peruse the material, try various arrangements of content. Then decide on a tentative arrangement you think makes most sense. You can always change your mind later—so plunge in!

At this stage of planning, look for groups of activities or sequences of lessons that fit well together. Simply list them by titles or descriptive phrases that convey a sense of what students will be doing. Here are a few: describing groups of plants and animals around us, starting a nature collection, learning how scientists organize nature, and making an identification key for a nature collection. Many teachers plan instruction so that it will fit easily into grading or report card periods. These sections are frequently called science units, and can be planned for two to four weeks, six to eight weeks, or ten to twelve weeks. Their main virtue is that units have the appearance of intact, cohesive, and related learning content. Teaching and learning are more meaningful be-

cause units highlight the relationship of ideas. Most of the teachers we know organize their science teaching into units, and fit some number of them into fall, winter, and spring terms. Since there are thirty-six weeks in the school year, and about twelve weeks per quarter, the number of science units teachers fit into that schedule depends entirely upon how many weeks their science units take.

Now, make your planning chart. Chart 5–1 (pp. 166–67) is a sample planning chart for middle or junior high school grades from the Environmental Biology unit. It includes the names of activities and the objectives to be covered by these activities. The "Teaching methods" column is another way of stipulating the kind of activity the students will experience, for example, guided discovery, reading, individual projects, and so on. A final section indicates other areas of the curriculum to which the science activity can be related.

When you are ready to make your own planning chart, we suggest you look at various examples before adopting or designing your own. As illustrated for Environmental Biology in figure 5–4, monthly calendars are

SEPTEMBER

S	M	T	W	T	F	S
	NOTES TO MYSELF • Remember my goals for science • Plan attentions for kids who need them • Try to integrate science w/ other subjects			1 2:15-3:00 "What is an Environment?" Prep: Color 3×5 slides, pictures, maps. PP. 207-210	2 "Planning Field Trip to Compare Environment" Prep: Ditto Outline for Chalkboard PP. 211-214	3
4	5 LABOR DAY	6 Continue "Planning for Field Trip" Prep: Phone parents from Committees	7 Continue "Planning Field Trip" Prep: Maps/Charts for Committees	8 "Taking a Field Trip to Compare Environments" Prep: Field Equipment PP. 215-223	9 "Taking a Field Trip to Compare Environments" No Prep.	10
11	12 PP. 220-224 Follow-up of Field Trip Prep: Chart paper	13 Follow-up of Field Trip Prep: Outline for Reports	14 PP. 224-235 "Other Ways to Compare Environments" Prep: Discussion	15 PP. 226-230 "Field Trip to Museum of Natural History" Prep: Parent volunteers	16 PP. 230-232 "Talking to Old Timers about Changes in the Environment" Prep: Tape recorders	17
18	19 PP. 233-235 "Reading Old Newspapers about Environment" Prep: Library	20 Continue "Reading Old Newspapers" to find changes in Environment Prep: Reports	21 PP. 236-240 Discussion of all activities to study environment, plan for displays No Prep.	22 PP. 241-245 Prepare science displays for Open House Prep: Art materials	23 Open House "Studies of Our Environment"	24
25	26 PP. 246-248 "Exploring Limits of Environmental Change" Prep: Discussion	27 PP. 249-251 "Creating a Plan to Improve our School Environment" Prep: Outline for Chalkboard	28 PP. 252-257 "Implementing Plan to Improve School Environment" Prep: Parent Volunteers	29 Continue/Improvement of School Environment Prep: Work Groups	30 Continue/Improvement of School Environment No Prep.	

FIGURE 5–4
Science Monthly Calendar

CHART 5-1
Planning Chart for Environmental Biology (Grades 6–8)

ACTIVITY	OBJECTIVES					TEACHING METHODS	RELATED CURRICULUM
	Conceptual Knowledge	*Learning Processes*	*Psychomotor Skills*	*Attitudes and Values*	*Decision-Making Skills*		
1. What is an environment?	Environments have identities based on physical and biological characteristics.	Observe, compare.	. . .	Explore and ask questions.		Discussion, film analysis	Social Studies
2. Planning a field trip to compare environments	Environments have identities based on physical and biological characteristics.	. . .	. . .	Compare alternative points of view. Assume responsibility for health and welfare of others.	Consider consequences of alternatives before making decisions. Make decisions based on consequences to self and others.	Chalkboard, discussion	Social Studies
3. Taking a field trip to compare environments	Environments have identities based on physical and biological characteristics.	Observe compare, measure, classify, gather and organize information.	Use tools that require fine adjustment and discrimination.	Assume responsibility for health and welfare of others. Assume responsibility for wise use of natural resources.	. . .	Field trip	Social Studies
4. Other ways to compare environments	Environments have identities based on physical and biological characteristics.	Gather and organize information, infer, describe relation among variables.	. . .	Actively explore and ask questions. Compare alternative points of view.	. . .	Visiting expert, pen pals	Social Studies
5. Summarizing differences between environments	Environments have identities based on physical and biological characteristics.	Organize information, describe various parts of environment to whole.	Construct complex project with some guidance.	. . .	. . .	Class project	Social Studies
6. Our environment today, one hundred, and two hundred years ago	Environments change as living and nonliving matter interact.	Gather and organize information, infer, describe relation among variables.	Use equipment that requires fine adjustment and discrimination.	Actively explore and ask questions. Seek alternative points of view and multiple sources of evidence.	. . .	Student projects	Social Studies
7. Talking to "old-timers," reading old newspapers	Environments change as living and nonliving matter interact.	Gather and organize information, infer, describe relation among variables.	Use equipment that requires fine adjustment and discrimination.	Actively explore and ask questions. Seek alternative points of view and multiple sources of evidence.	. . .	Student projects	Social Studies
8. Field trips to museums of history, natural history, and visits to city planners	Environments change as living and nonliving matter interact.	Gather and organize information, infer, describe relation among variables.	Use equipment that requires fine adjustment and discrimination.	Actively explore and ask questions. Seek alternative points of view and multiple sources of evidence.	. . .	Student projects	Social Studies

						Student projects	Social Studies
9. Cooking with native plants	Environments change as living and nonliving matter interact.	Gather and organize information, infer, describe relation's among variables.	Use equipment that requires fine adjustment and discrimination.	Actively explore and ask questions. Seek alternative points of view and multiple sources of evidence.	...		...
10. Making insect and plant collections	Environments change as living and nonliving matter interact.	Gather and organize information, infer, describe relation's among variables.	Use equipment that requires fine adjustment and discrimination.	Actively explore and ask questions. Seek alternative points of view and multiple sources of evidence.	...	Student Projects	...
11. Exploring the limits of environmental change	An environment can change and still conserve its identity; but when its capacity to change is exceeded, it loses its identity.	Observe, compare, classify, form hypotheses, predict infer.	Handle plants and animals; use and care for tools and equipment.	Actively explore and ask questions.	...	Field trip	...
12. Creating a plan to improve our school environment	Environments can change and still conserve their identities.	Gather and organize needed information, seek alternative points of view.	...	Assume responsibility for the wise use of natural resources, and for contributing to the health and welfare of others.	Describe decisions that involve self and others. Differentiate between facts, attitudes, and values. Formulate alternatives which reflect facts, attitudes, and values.	Brainstorming, discussion	
13. Implementing a plan to improve the school environment	Environments can change and still conserve their identities.	Observe, compare, measure, gather and organize information.	Use and care for tools and equipment; handle and care for plants in new situations.	Combine objects and materials in new ways. Demonstrate responsibility for health/welfare of self and others. Use natural resources wisely.	Describe immediate and long-range consequences of various alternatives in changing the environment.	Field projects, discussion	...
14. Surveying world environments that have been changed	When the environment's capacity to change is exceeded, it loses its identity, even though matter and energy are conserved.	Gather and organize information, describe relations among variables.	...	Seek alternative points of view and multiple sources of evidence.	Describe immediate and long-range consequences of various alternatives in changing the environment.	Team learning, oral reports, simulation	...
15. Understanding the consequences of changing environment	When the environment's capacity to change is exceeded, it loses its identity, even though matter and energy are conserved.	Gather and organize information, describe relations among variables.	...	Demonstrate the need to use natural resources wisely, and contribute to the health and welfare of society.	Describe immediate and long-range consequences of various alternatives in changing the environment.	Debate, discussion	...

preferred by many teachers. They make six to eight copies of each month's calendar and use a separate calendar page for each subject: science, math, reading, and so forth, or alternatively, they use a different page for each science class in junior high school.

STEP 5: ORDER MATERIALS AND RESERVE FILMS AND BUSES

After your planning chart is sketched in, you are ready to order materials and reserve films or buses for field trips. The easiest way to do this is to review the activities or lessons you have listed on your calendar or planning chart and make a master list of equipment and other materials you should order and those you should "check-out" of the supply room. This is the time to look through the film catalog for films and other audio-visual material that will enrich the science program; and it is the time to arrange transportation for field trips. Don't be overly concerned about specific dates at this time; estimates will be fine. Then as your science program gets underway, you will begin to see how close your first estimates were and you can adjust other reservations accordingly.

Many teachers find it easiest to order materials and reserve films and buses in intervals of about three months. However, you may find yourself in a school where all requests must be turned in at the beginning of the school year; inquire before the first week of school so that you will have enough time to prepare the list.

Where do you order science materials? There are many sources; a few of them are:

☐ The district office or warehouse
☐ The county office: curriculum center or library
☐ Local stores

☐ From advertisements in professional magazines like *Science and Children* and *The Science Teacher*
☐ Catalogs usually available in the county office

How much equipment will you need? That depends upon the activities themselves and how many students will be using the equipment at one time. Many teachers group children in order to share equipment. Other teachers put specialized, one-of-a-kind equipment in a learning center and have students take turns using the center. Still other teachers demonstrate the use of equipment themselves and do not allow students to use it at all. There are some circumstances in which such demonstrations are especially warranted, for example those involving sharp knives and special tools. But too many demonstrations and too few "hands-on" activities for students can dampen their enthusiasm for science. Teachers tend to buy equipment for each member of the class if the equipment is versatile or if it can be obtained at low cost. For example, plastic hand lenses are inexpensive and are often provided for every student in the primary grades. Microscopes, on the other hand, can be moderately priced or very expensive and hence may be provided one for every two, five, or thirty students at the junior high level, depending on cost. Since microscopes are reasonably durable and used for many different activities, science teachers usually have to make such investments once every five years or so.

Find out what your yearly school budget is for science. Then consider the cost of equipment and other materials (for example, seeds, potting soil, fish food, and so on) and begin to adjust your plans to fit the budget. If the budget is far below what you need, don't give up! Ask other teachers how you might acquire some of the things on your list. Most beginning teachers are surprised to find the resourcefulness of their colleagues in getting the science materials they need. Incidentally, high school science teachers can be helpful in this regard, too.

Finally, in ordering equipment, we encourage you to have a *wish list* ready at hand which names pieces of equipment you would like to have for the students. It should list the costs and all information necessary for materials purchase. Quite often, and on the spur of the moment, teachers are told that there is some money left in the budget for materials. Teachers who have their lists ready are usually the first in line! It is useful to have several things of different prices on your list so that you can fit your request to the situation. Your wish list might also include special books, reference books, science picture books, or even subscriptions to special science magazines for students. And your list should reflect certain criteria for items to be purchased. When making up your list, be sure to consider such factors as:

- ☐ cost
- ☐ upkeep
- ☐ consumable or nonconsumable
- ☐ versatility (e.g., bar magnets instead of horseshoe magnets)
- ☐ durability
- ☐ safety
- ☐ reliability (or accuracy of instruments)
- ☐ storage (space and security)

What else do you need to know about equipment and materials for teaching science? Here are a few suggestions from teachers:

- ☐ If you are going to have students bring materials from home, have them bring the material *at least a week* before you plan to use it in class.

- ☐ If you wish to design simple science equipment yourself, two good resources are: *The Everyday Science Sourcebook* by Lawrence F. Lowery and *Outdoor Biology Instructional Strategies* (modules available from Delta Education, 1980).

- ☐ The more you use the natural environment, the more meaningful it is for students—and the less it costs.

- ☐ When an activity is over, have a student check the equipment and make sure it is in good condition. If not, see who you can get to repair it before storage.

- ☐ When you are reserving the bus for field trips, be sure to ask the office about permission slips for the students.

Safety

Safety is one of the most important considerations regarding science materials and equipment. Precautions are important when using all types of materials. General precautions include:

- ☐ Check materials and equipment for defects or hazards before allowing the students to use them.
- ☐ Supervise the students constantly.
- ☐ Make sure the workspace is large enough for students to work.
- ☐ Have students work in small groups to avoid the accidents that large groups can create.
- ☐ Give the students enough time to carry out experiments; rushing can cause accidents.
- ☐ Have a first-aid kit available.

Other precautions are necessary when students are working with specific variables or sources.

Heat

- ☐ Know the school's fire regulations and procedures. Make sure that you and the students follow general safety guidelines when using heat.

☐ Keep volatile or flammable materials away from heat.

☐ Use pot holders or other such safeguards when handling hot objects.

☐ Have insulated pads on which to set hot objects.

Electricity

☐ Tell students not to experiment with electricity on their own.

☐ Don't let students handle electrical devices with wet hands.

☐ Show students how to remove electrical plugs from sockets by pulling the plug and not the cord.

Glass

☐ Make sure all bottles are clearly labeled to show contents and precautions.

☐ Tell students that only heat-treated glass should be heated.

☐ Tape sharp edges of glass.

☐ Never pick up broken glass with your hands; use dustpans and whisk brooms.

Animals

☐ Provide proper surroundings and food for animals.

☐ All mammals should have rabies inoculations.

☐ Caution students about teasing animals or poking their cages.

☐ Students should handle animals only when necessary.

FIGURE 5–5
Permission Slip

Request for Permission to Go on a Field Trip

Description of the Field Trip:

Destination _____

Purpose _____

_____

Location _____

Time of departure_____Time of return_____

Form of transportation _____

Special Arrangements:

Clothing _____

Lunch _____

Equipment _____

(Student's Name) _____

has my permission to go on this field trip.

(Legal Guardian) _____

☐ Students should wash their hands after handling animals.

Plants

☐ Help children learn to identify poisonous plants.
☐ Caution children against parts of plants that may be poisonous.

These are basic precautions; however, more detail on safety precautions is available. A particularly comprehensive source is *Safety in the Secondary Classroom,* published by the National Science Teachers Association (1978).

Permission and Clearance

Who must give their permission before you take the students out of the classroom? Your principal is the best source of information on both school and district policies. If you need to secure written permission from parents, be sure to follow the expected policies and procedures of your school. Figure 5–5 illustrates a typical school permission slip.

As you plan segments of your science curriculum, remember to consider ways for others to assist with learning. Many people in the field enjoy volunteering their assistance. (© C. Quinlan)

STEP 6: RECRUIT VOLUNTEERS

As you look back over your planning chart and the list of materials you have ordered, think about the kinds of tasks associated with the science activities you have planned for the first twelve weeks. Which activities will require assistance? Remember, just because you are the teacher doesn't mean that you have to do everything for the students. Identify the big tasks and plan to get help. You may want help with the field trip, for example. A guest speaker may be an important source of special information. Someone responsible for cleaning up once a week or feeding the fish each day can be a helpful aid, too. This is the time to recruit volunteers who will share responsibility for some of the learning that takes place in science. Chapter 6 will provide many suggestions for how and when to recruit volunteers and how to delegate responsibility. The point to keep in mind here is that it is easiest to recruit volunteers for segments of the year. That way the volunteers will know their responsibility lasts for a specific period of time.

STEP 7: PLAN THE DETAIL FOR LESSONS

How you make the transition from planning in segments—whether they are twelve-week periods or science units for smaller numbers of weeks—to planning the final details of each day's activity is an individual choice. Some teachers like to spend time once a week planning the details for the coming week; others plan the final details on a daily basis, still others do both. How much detail should you plan? The best way to answer this question is to introduce a science activity, "You and Your Teeth," similar to activities found in part two of this book. The activity is planned for students in the primary grades, although you can see that the activity would be quite suitable for students through junior high school. Notice that the activity lists the *Major Concept* to be learned, *Objectives*, *Materials* needed, *Vocabulary* used, and *Procedures* for Day One, Day Two, and *Extending the Activity*.

If you are following a teacher's guide which has activities preplanned for you, the amount

There are many resource people available to teachers, including students, parents, other school personnel, and community health and social service volunteers.

HEALTH SCIENCE ACTIVITY

You and Your Teeth
(Kindergarten—Grade 2)

MAJOR CONCEPT

Developing dental hygiene practices contributes to health.

OBJECTIVE

☐ Interpret color change as evidence of areas of potential tooth decay.

MATERIAL

For the class: one tube of toothpaste. For each student: paper towel, personal toothbrush, one red-dye disclosure tablet (available without prescription from any pharmacy), small mirror, diagram of upper and lower teeth, piece of apple, one paper napkin, one stick of sugarless gum. Optional: Dental health packet from the American Dental Association.

VOCABULARY

incisors	dental floss
cuspids (or canines)	dental hygienist
biscuspids	dentist
molars	grind
bacteria	plaque
bite off	tear
crush	

PROCEDURES

A. Day One
1. Introduce a discussion of teeth and how they function.
2. Distribute mirrors and diagrams of teeth to the students. Have the students use the mirrors to compare their own teeth with the diagram, noting loose or missing teeth.
3. *Distribute apple pieces.* Invite the students to bite down on the apple. Ask them to tell which teeth they used to "bite off" a chunk. Now identify the incisors and explain their cutting function. Ask the children to locate the incisors on their diagrams and color or label them.
4. *Distribute paper towels.* Invite the children to tear the towels with their teeth. Identify the canines, and explain their tearing function. Ask the children to locate the canines on their diagrams and color or label them.
5. *Distribute pieces of chewing gum.* Invite the children to chew on a piece of gum. Ask the children to tell which teeth they use to "grind" the gum. Identify the molars, and explain their grinding function. Ask the children to locate the molars on their diagrams and color or label them.

B. Day Two
Ask the students to bring their toothbrushes to school for the next day's activity.

1. Define *plaque* in simple terms for the students. Tell the children that they are going to locate plaque areas on their teeth by using some red dye that turns plaque areas a dark, pink color.
2. Give each student a mirror, a red pencil, an empty paper cup, and a red-dye disclosure tablet. Instruct the students to chew the tablets vigorously and, using the tongue and saliva, to swish the dye throughout the mouth for thirty seconds. Students then expectorate the dye into the paper cup, discard it, and rinse their mouths twice with plain water. With supervision, children could complete the steps, in this procedure, one at a time, at a drinking faucet or classroom sink.
3. The children return to their desks. Using the mirrors, the students can inspect their teeth. Tell the children to locate plaque areas, indicated by dark pink, and mark these areas on their diagrams with the red pencils.
4. Have the children brush their teeth to remove the dye.
5. Invite the children to share their findings.
6. Ask them:
   "What kinds of foods stick to the teeth and lead to plaque?"
   "Where does the food accumulate in the mouth?"
   "How does eating between meals contribute to plaque?"
   "What can you do to control plaque?"
7. Invite the children to role-play as dental professionals instructing others on the need and practices for proper tooth care.

EXTENDING THE ACTIVITY

C. Demonstrate or invite a dental hygienist to demonstrate to the class the proper techniques to brush and, for older children, to floss.
D. Have some students who have recently lost teeth bring them into class. Invite the class to do an experiment that will determine the effects of sugar on tooth enamel. Place one tooth in a cola drink and the same kind of tooth in plain water. Observe the teeth at regular intervals for one week. At the end of the observation period, test each tooth for hardness.

From *Health Activities Project,* Trial Edition. Berkeley, CA: University of California, Lawrence Hall of Science, 1976, 1977, 1978.

of time required for planning detail is minimal. However, do take time to read through each activity, conduct the activity yourself, and see how long it takes you to get out the materials and put them away. That way you can begin learning how to estimate the time it takes to conduct each kind of activity.

The final part of Step 7 in planning is to learn how to estimate the amount of time it takes to accomplish the various tasks you propose to accomplish in teaching. It is difficult for beginning teachers to predict how much time will be required for each science activity. For that reason, you should find out how much time it takes *you* to do each activity and record the results. From your observations, you can estimate how long it might take twenty to thirty students to complete the activity. But how close is your time requirement going to match that of your students? After you complete the task with your students, you can reflect back on the **lesson plan** and make more appropriate time estimates for the future. Repeated estimates and appraisals of your estimates should improve your skill at estimating the time required.

Time Plan	Task	Actual Time
3 minutes	Set up task for the day—find pictures in magazines that are symmetrical or non-symmetrical.	2 minutes
2 minutes	Distribute scissors, paste, chart paper, and magazines.	10 minutes
15 minutes	Children work in pairs to complete chart.	20 minutes
5 minutes	Display charts and have each pair tell how they decided on which chart to put pictures.	(time ran out)
5 minutes	Clean up.	

FIGURE 5–6

Lesson Plan for *Symmetry* for Second Graders
(Time: 30 minutes)

In the example lesson plan above, what changes would you suggest for the student teacher?

When it became apparent during the course of the lesson that students would not be able to finish the activity, what would you have done: urged the students to hurry; "borrowed" time from the next subject or class; or postponed completion of the project until the next day?

INTEGRATING SCIENCE WITH OTHER SUBJECTS

The world your students experience does not come in neat bundles called reading, mathematics, spelling, or science. The curriculum has an interconnectedness that we, as teachers, tend to forget. There are two key ways to make science a part of the "natural" learning of your students.

First, help students extract data or skills obtained from science activities so that they may use this information in other subjects. One example is in mathematics. Suppose your objective in a mathematics lesson is to have students practice measurement skills. You can do this by having them keep charts of their heights or maintain daily records of plant growth. If you are introducing the concept of area in mathematics, for example, compare how many squirrel foot prints it takes to cover an elephant foot print (kindergarten–grade 2) or how many square feet there are in the floor of the gymnasium (grades 3–8).

Second, use other subject areas as reinforcements of science skills. After introducing the concept of the distinction between observations and inferences in science, reinforce that concept in the reading group by having students make observations and inferences about the characters in a story. After introducing classification skills with seeds in science, have your students continue to make and describe sets in their mathematics activities.

A simple and effective strategy is to look at your objectives in reading, mathematics or social studies and ask yourself what kind of an activity would help students practice their science skills.

Theme Integration

Integrating science with other subjects by means of transferring newly acquired skills to another subject area is not the only approach to subject matter integration. A process of planning around a central theme rather than around a specific subject has some interesting

THEME: UNDERSTANDING OUR TOWN/ CITY

☐ Science:
Study the weather of the town or city on a year round basis. Survey local plants, animals, and sources of water. Learn about changes that have occurred in the environment over the past one hundred years.

☐ Social Studies:
Study the people in your area: find out where their ancestors came from, and ask them what jobs they hold. Study how the government works. Discover what services, recreation, and kinds of transportation are available.

☐ Art:
Experience the rich variety of art, music, and dance forms that the population of the town or city creates or enjoys. Portray the town or city in poetry and in pictures.

☐ Language Arts:
Write stories about life today and life as it was one hundred years ago in the city or town. Publish a small school newspaper containing articles about people, business, recreation, and the natural environment of the town or city.

potential advantages for teachers in self-contained (nondepartmentalized) classrooms.

Lessons planned from this perspective tend to be woven together in a practical and seemingly natural way, potentially increasing student motivation. Additionally, recent research findings suggest strongly that effective learning makes use of both the right and left hemispheres of the brain. More traditional subject matter planning often results in emphasis on the verbal, logical, and linear learning associated with the left part of the brain. Theme planning requires more creativity in the teacher planning phase and has the potential of demanding more right brain creativity, spatial reasoning, and artistically-related right brain capabilities from the student. The integrative, multisensory approach to teaching has strong support in learning theory and developmental research.

Two points of caution regarding theme integration of subjects need to be mentioned. Teachers often find that more time and thought are required in the pre-planning phase for theme planning. It is also necessary to make sure that the concepts the teacher identifies as important are actually being learned during theme-integrated lessons. The potential for blurred goals—and consequent, diminished student learning—is a possibility to be kept in mind when considering theme planning as a form of subject matter integration. Potential limitations can be avoided by giving careful thought to evaluation of expected outcomes.

COMMENTARY

Planning to teach science, the focus of this chapter, involves considerations not found in some subject areas taught in elementary, middle, or junior high school. The use of manipulative materials that are either central or supplemental to each science lesson requires special planning on the teacher's part. Although quality planning is the key to all fine teaching, using a multimedia approach demands different planning from that required with a textbook approach. It is necessary to plan the purchasing or design of equipment. This in turn requires maintenance and storage planning. Special procedures with regard to safety require a level of planning not necessary when students are using textbooks, pencils, and paper. Active involvement with heat, electrical sources, glassware, animals, and plants necessitates specific techniques that may not be common to the average classroom teacher. But, like preparation for art and physical education activities, the time invested in planning for "hands-on" or laboratory science activity pays off well in terms of student motivation and learning.

Once your planning is underway, it is time to focus on implementation of your plans. In the next chapter, we will focus on instructional management, and you will find many suggestions for achieving your objectives in the classroom.

BIBLIOGRAPHY

Aylesworth, Thomas. *Planning for Effective Science Teaching.* Columbus, Ohio: American Education Publications, 1963.

Block, J. H. *Mastery Learning: Theory and Practice.* New York: Holt, Rinehart & Winston, 1971.

Brandwein, Paul. "A General Theory of Instruction." *Science Education* 63:3 (1979) 288–293.

Bruner, Jerome. *Toward a Theory of Instruction.* New York: W W Norton & Co., 1968.

Cooper, James et al. *Classroom Teaching Skills: A Handbook.* Lexingtn, MA: D. C. Heath & Co., 1977.

Curtis, Thomas, and Bidwell, Wilma. *Curriculum and Instruction for Emerging Adolescents.* Reading, MA: Addison-Wesley Publishing, 1977.

Dean, Robert A.; Dean, Melanie Messer; and Motz, LaMoine L. *Safety in the Secondary Science Classroom.* Washington, D.C.: National Science Teachers Association, 1978.

DeVito, Alfred, and Krockover, Gerald H. *Creative Sciencing* (2 vols.). Boston: Little, Brown & Co., 1976.

Gronlund, Norman. *Stating Behavioral Objectives for Classroom Instuction.* New York: Macmillan, 1970.

Hannah, Larry S., and Michaelis, John U. *A Comprehensive Framework for Instructional Objectives.* Reading, MA: Addison-Wesley, 1977.

Johnson, David, and Johnson, Roger. *Learning Together and Alone: Cooperation, Competition, and Individualization.* Englewood Cliffs, NJ: Prentice-Hall, 1975.

Kauchak, Donald, and Eggen, Paul. *Exploring Science in the Elementary Schools.* Chicago: Rand McNally, 1980.

Lowery. Lawrence F. *The Everyday Science Sourcebook: Ideas for Teaching in the Elementary and Middle School* (Abridged ed.). Boston: Allyn & Bacon, 1978.

Nerbovig, Marcella. *Unit Planning: A Model for Curriculum Development.* Worthington, Ohio: Charles A. Jones Publishing Company, 1970.

Piltz, Albert, and Sund, Robert. *Creative Teaching of Science in the Elementary School.* Boston: Allyn & Bacon, 1968.

Triezenberg, Henry J., ed. *Individualized Science. Like It Is.* Washington, D.C.: National Science Teachers Association, 1972.

6

Instructional Management

The purpose of chapter 6, which focuses on instructional management, is to offer a pot-pourri of suggestions to make things in the classroom run smoothly. You will find ideas for being the kind of manager you prefer and for establishing the kind of classroom climate you want: you will be given guidelines for forming groups within the classroom, recruiting volunteers, and delegating responsibility; for distributing and collecting materials; for guiding discussions and asking questions; and for handling disruptions and resolving conflicts. Finally you will find suggestions for setting limits for safety in the classroom.

KEY CONCEPTS AND IMPORTANT IDEAS

☐ instructional management ☐ directive management ☐ participative management ☐ permissive management ☐ classroom environment ☐ forming groups ☐ distributing and collecting materials ☐ delegating responsibility ☐ asking questions ☐ handling distractions ☐ resolving conflicts ☐ setting limits for safety

What is **instructional management?** In its most general sense, instructional management is being responsible for everything that goes on in the classroom, including all of those administrative tasks like taking attendance and collecting lunch tickets. But here we will focus on those aspects of instructional management that have a clear connection with teaching science.

INSTRUCTIONAL MANAGEMENT

It is two o'clock on Wednesday afternoon, and the last period of the day at Lincoln School is science. Wander down the hall and picture for a few minutes three different classrooms at the end of the hall. After you take this mental picture, we will define instructional management and describe some of the factors that contribute to effective instructional management in classrooms. Watch as the studnts enter the first classroom.

> The students enter class and go immediately to their seats. The teacher tells the students exactly how to set up and complete the science activity. Worksheets are provided with specific step-by-step instructions. The teacher assigns student groups, indicates who should obtain the materials, how long each section of the activity should take, and when the written report is due. The students work within this structure, following directions closely for the remainder of the period; the report is due at class time the next day.

What's happening in the classroom next door? Take a quick look.

> The students enter class and go to their seats. The teacher tells the students that they are going to try to solve a problem. The teacher describes the problem and asks the students how it might be solved. Several suggestions are offered by the students. The suggestions are summarized

into a science activity. The teacher clarifies different aspects of the activity and informs the students of the available materials. They are then told to form groups, obtain materials, and set up and complete the activity. The students go to work for the remainder of the period. The next day there is some time to finish laboratory work, time to have small-group discussions, time to complete the report, and have a post-activity class discussion.

There is one more classroom at the end of the hall. Let's see how the third teacher approaches the teaching of science.

> The students enter class and wander around looking at various bulletin boards and displays. The teacher calls for their attention and presents a problem, informs them of the location of materials, and tells the students to try to solve the problem any way they can and to expect little help from the teacher.

Interesting. In each of these three classrooms you saw a different style of classroom management: the first teacher was **directive,** the second teacher was **participative,** and the third teacher was **permissive.** As a science teacher, you may use all three approaches at one time or another. However, one style will tend to represent how you work with students most of the time unless you consciously alter and vary your management style.

Obviously, teacher and student roles differ in the three styles of management described. In directive management the teacher becomes the active focus of attention, the one who provides direction, information, and synthesis of experience. In participatory management the teacher shares in decision making with regard to what is to be learned and how it is to be learned. In permissive management, the teacher may provide the major goals but subsequently allows students to secure the information and develop the synthesis themselves.

While teachers may have a preferred style of management, experience has shown that in teaching science, one is required to use different management styles at different times. The effective teacher has insight, is flexible, and chooses when to direct, when to participate, and when to permit. This flexibility is based on the teacher's perception of the classroom environment, educational goals, and awareness of group and individual needs.

The following considerations may be of use in evaluating the advantages and disadvantages of the three management styles:

1. *Directive Management.* Care must be taken to make sure that students do not feel powerless and at a loss in terms of control over their work and learning. Because students are in a more passive role in this type of management style, the teacher needs to take special effort to insure that students are understanding the material. Beginning teachers may be vulnerable to more disruption when leading a class discussion than during a science demonstration.

2. *Participatory Management.* Care must be taken to ensure that there is an equal sharing of responsibility for communicating directions, securing information, and searching for meaning. In this way, problem solving and learning comes primarily from group learning. The teacher must be sure that the structure of the student groups allows maximum participation for all students. Beginning teachers may be more vulnerable to disruption when they share decision making with the students.

3. *Permissive Management.* Care must be taken to ensure that students successfully understand the learning goals and how to reach them because the teacher exhibits minimal leadership in this classroom role. The greater demands placed on students for responsibility require the teacher to be particularly aware of students who are unable to identify a clear sense of purpose and direction. Beginning teachers may be more vulnerable to disruption when they delegate responsibility to the students for developing goals and securing information independently.

Before we leave Lincoln School, let's stop in the teachers' lounge for a few moments and talk about instructional management. So far we have discussed one important part of instructional management, the management style. Now we will look at other teacher activities that affect the smoothness and quality of instructional management.

CLASSROOM ENVIRONMENT

Let us begin by considering the classroom itself. There are many ways to create an environment that is conducive to learning science. The furniture, color, pictures and charts, plants, and materials available to explore, all convey an atmosphere that is either formal or informal, inviting or uninviting, and organized or disorganized. More than anyone else, you as the teacher are responsible for the classroom environment or climate. And whether you teach science as one of several subjects, or teach mainly science to several classes, you can create a classroom environment that is conducive to learning.

You may have little choice in the kind of furniture in the classroom, although we know of teachers who were able to exchange furnishings in their classrooms for other furniture in the district warehouse. Tables and chairs are especially useful for a variety of science activities, because they lend themselves to multiple arrangements.

Colorful pictures, posters, and charts that relate to science lessons invite learning and contribute to an interesting environment. Another way to invite informal learning in the classroom is to place materials and equipment that require little or no supervision on cupboard tops or shelves. For example, a mi-

The environment you create in the classroom is your nonverbal message to students, it says how you feel about learning. A classroom that is rich with materials to explore says you invite learning; one that is sterile or boring offers less incentive to learn. (Donald Birdd)

croscope or equal arm balance that has been "retired" from use in a local high school or hospital can intrigue elementary and middle or junior high school students and invite hours of informal learning. Just supply a variety of material to look at or to weigh. Some teachers like to combine such materials with bulletin board directions and have a *Question-of-the-Week* display. This might invite students to search through one or two books on display, to do some easily managed experiment, or to find the answer to the question on the bulletin board. Placing students' answers in a box (ballot-box style) adds to the excitment; students wait until the end of the week to find out whose answer came closest or how many students found the answer to the science puzzle, problem, or question. Ideas for curiosity-arousing questions or problems are plentiful in science teacher magazines like *Science and Children* and *The Science Teacher.*

Plants of many kinds can be grown in the classroom and help to create the feeling of bringing the outdoors inside. Some plants that are easily maintained in the classroom are spider plants, jade plants, piggybacks, and creeping charlies. The same is true for having small animals like fish and turtles in the classroom.

Interesting books, magazines, and science-related pamphlets can be borrowed from the library and exchanged for new ones every few weeks. Change reading and picture materials often and leave one or two open to stimulating pictures. Reading materials become an invitation to informal learning and require no direction or instruction from you.

Some teachers are concerned about having everything "go together" in the classroom environment. For example they want to have materials that match the theme of their science unit on space travel, on African animals, or whatever. It is interesting to go into a classroom in which the bulletin boards, pictures, charts, library books, and lab materials or displays are all related to a particular theme or current science unit. But variety is equally interesting and easier to assemble, whether you or someone else collects the materials.

The best time to arrange the physical environment of the classroom is in the fall before school starts. That way, the students sense that they can expect to learn many interesting things in your classroom environment during the coming year. Later, if you prefer, you can

Informal classroom learning can be encouraged through creative, colorful displays, allowing students the opportunity to solve a problem or puzzle, research a question, or observe nature.

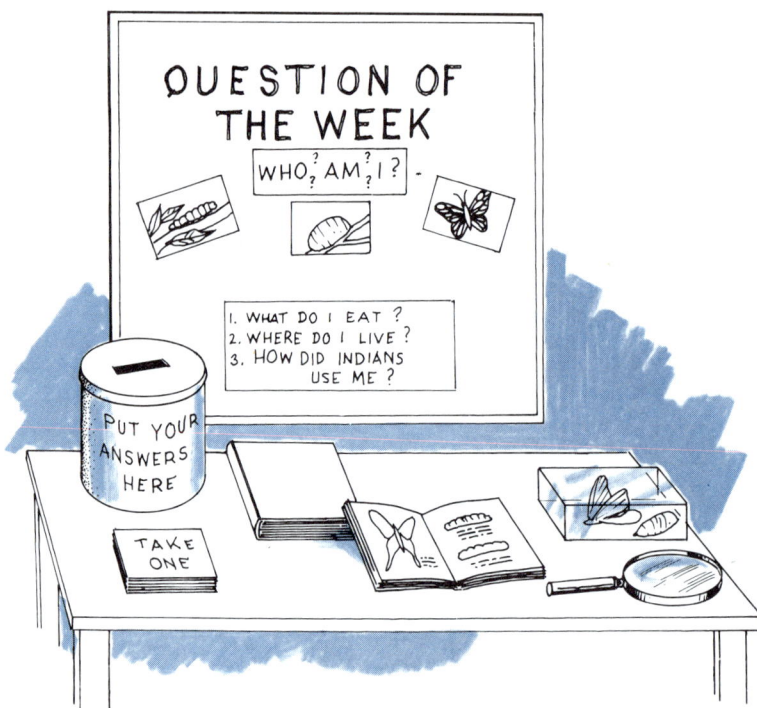

ask students (or parents, if you teach in the primary grades) to help you to change the physical environment. All you need to do is make lists of the things you need or note the theme you want to follow, and you can usually find many who are willing to help plan or implement your ideas. Of course, for many teachers the task of arranging the physical environment is an opportunity for self-expression, and a responsibility they prefer not to share. What seems to be most important in the long run is not whether the teacher or someone else arranges the environment, but the fact that the environment changes and offers continuously new opportunities and stimulation for learning.

There is more to classroom climate than the physical environment. We refer to less tangible factors such as respect, trust, cohe-

siveness, good interpersonal relationships, and caring among and between students and the teacher. These interactions contribute to a positive or negative learning environment. Where there is *respect,* teachers and students are treated with dignity, and classroom conflicts are resolved with respect for the rights of everyone involved.

Trust in the classroom is evident when students feel that the teacher is on their side and will not let them down. Here all members of the class contribute their share and are dependable and honest. Trust is evident when procedures are consistent and predictable, and when rules are not changed suddenly or applied unfairly.

Cohesiveness exists between members of the science class when students feel that they belong to and are part of the group. It is im-

In every classroom there are one or more students who value the opportunity to care for plants and animals. (Harvey R. Phillips/Phillips Photo Illustrators)

portant for students to feel that others care about them, and that their health and welfare are of some concern to the group and to the teacher.

FORMING GROUPS

We learn while doing things, whether we are on our own or with others. Meaningful science learning may, in other words, be experienced by students alone, in pairs, with several others who are doing the same task, or with the whole class. It is your responsibility to decide how to facilitate learning for the entire science class. A key to successful teaching is the grouping of learners based on their immediate interests and needs.

The Science Class as a Group

A group consists of two or more persons who interact and influence one another. With this definition we point out that students are placed in classrooms based on criteria such as age, ability, or scheduling compatibility. This is to say that students in your science classroom will not necessarily be a group. Let us look at two educational situations to clarify the concept of a group. In the first situation, the science class is totally individualized. Student tests and aptitude scores are forwarded to the science teacher, and the teacher, in turn, designs totally individualized programs. Individualized programs require reading the science textbook, doing selected units of programmed instruction, completing designated laboratory activities, and taking tests. All of this is done individually, with an occasional meeting with the teacher. In contrast, a second approach to science teaching involves teacher presentation to the entire class; laboratory work involving two, three, or four students working together; individual study time; discussions with the teacher; and discussions among the students themselves.

The first situation may in fact be good educational practice, but from the teacher and student points of view, there is probably no classroom group. The second situation can be considered a group because it has some immediately recognizable group properties: interaction, structure, goals and norms, and cohesiveness. We will look briefly at each of these properties.

Interaction

Interaction is an important property of groups through which students affect and communicate with one another. The interaction among individuals in the science class contributes to the development of other group characteristics such as cohesiveness, structure, goals, and behavioral norms. Many factors affect group interaction in science classrooms:

- ☐ The science teacher facilitates the type of interaction.
- ☐ Frequent interaction affects students' attitudes toward the class.
- ☐ Seating arrangements can be conducive to the desired type of interaction.
- ☐ Status and authority assigned to members can affect interaction.

Structure

As groups develop, structures emerge. Some students will be perceived by and located differently with respect to other members of the group; this is the group's structure. Structure

within a group can be influenced by the physical setting of the classroom, the "neighborhood" or micro-environment, and other factors such as students' race, sex, and socioeconomic status. Needless to say, there are structures that facilitate groups and structures that debilitate groups. Often you must guard against detrimental group structures.

Goals

Group goals are an important property. Different types of goals can coexist in a group: student goals, teacher goals, educational goals, and so on. Discussion, laboratory activity, projects, and field trips are group activities that involve different goals. Groups that formulate or clarify their own goals tend to have more interaction, clearer structures, and greater cohesiveness.

Norms

Groups have normative beliefs and behavioral expectations. What should individual students do in certain situations? What is expected of individuals in the group? When you assign a laboratory activity and groups start to work, norms of behavior are often clearly seen in student comments such as "Let's work together," or "If we each do our share, we should be able to get this laboratory assignment done."

Cohesiveness

Cohesiveness is the degree of attachment members of a group have for one another. Cohesiveness can be expressed as the number of friendships within the group, the degree to which group members function together effectively, the bonds that bind members to the group, or the degree of similarity among ideas, behavior, and so on. For science, a test of cohesiveness is the degree to which students will speak of *our* science project, *our* group assignment, and *our* results. Figure 6–1 is an inventory designed to assist teachers in analyzing the characteristics of groups in their classes.

DISTRIBUTING AND COLLECTING MATERIALS

The materials used to teach science can facilitate or inhibit learning. Materials may be made available at the wrong time or may be in such short supply that students become frustrated over trying to share them. Below you will find six keys that are helpful to remember:

☐ *Key One:* For each part of the day, what materials will be needed? Even if it is only a pencil, note that a pencil is essential.

☐ *Key Two:* Where are the needed materials stored? Is there an adequate supply available to the students, and do they know that if needed, they may use these supplies?

☐ *Key Three:* Who uses the materials? Some may be used only by you. If so, it will help if you put them "off limits" and, if possible, out of sight.

☐ *Key Four:* When do materials need to be available? It may be that you need a large collection of brightly colored balls for the science les-

---

Group Structure in the Science Classroom:
An Analysis

To determine the group structure in a science class, answer the following questions. Use the 1 to 7 continuum to indicate the degree to which each group characteristic is present.

	Low	Medium	High
To what degree do the students have *solidarity* of opinion, purpose, and interests when they work in your class?	1 2 3 4 5 6 7		
To what degree are the students *satisfied* with their class?	1 2 3 4 5 6 7		
To what degree are members of other classes *attracted* to this class?	1 2 3 4 5 6 7		
To what degree are members of this class *attracted* to the class?	1 2 3 4 5 6 7		
To what degree do children in the class group feel they *belong* in the class?	1 2 3 4 5 6 7		
To what degree (both qualitatively and quantitatively) do students *interact* through various types of communication?	1 2 3 4 5 6 7		
To what degree is the group *structure* beneficial to achieving goals?	1 2 3 4 5 6 7		
To what degree are the *norms* of behavior adhered to by group members?	1 2 3 4 5 6 7		

FIGURE 6–1
Group Structure Inventory

son. If you do not need them until the fifth period, will that large container of balls be a springboard for their curiosity or a distractor from other areas of study?

☐ *Key Five:*   Where are the materials to be returned? When a group is finished with a task, they should return them to a clearly established place of return.

☐ *Key Six:*   Who's responsible? Materials wear out, get used, or get lost. Who is supposed to see that all the crayons are returned to the box and are ready to be used again or that all the pieces to the puzzle are returned?

RECRUITING VOLUNTEERS AND DELEGATING RESPONSIBILITY

One of the most useful things a teacher can know is when and how to recruit volunteers and delegate responsibility. Delegating or sharing responsibility in the science classroom has two advantages: it provides student or adult volunteers with a sense of ownership for what happens and it helps get the job done. Yet because teachers are paid for teaching, their requests for voluntary assistance must be handled professionally and thoughtfully. One must avoid asking others to do the work teachers are paid to do, and avoid appearing to make such a request.

When clean-up is a shared responsibility, everyone benefits. (Donald Birdd)

When Should You Ask for Assistance?

If you follow two simple rules, you will avoid most difficulties. First, ask for assistance *only* for tasks where there is a clear and obvious connection between the outcome of the task and some improved opportunity for students to learn about science. That means, for example, that you can ask a parent to help you prepare for a field trip—an obvious benefit to students. However, you cannot ask a parent to pick up your dry cleaning while you grade papers. This rule may be ambiguous because if you asked a parent to grade weekly lab papers or help to plan a science program, those are tasks that most people think you are paid to do. But if, for example, you want to individualize a part of your science program so that each student has a chance to do activities that fit his or her level of ability or interest, it would be appropriate to provide a grading key and to ask the students to grade each other's work. In fact, many teachers have students grade their own work because it helps to build trust and provides students with immediate feedback on how well they understood the material. Spot-checking for students' accuracy in self-grading is usually sufficient for small daily assignments.

The second rule is *never* cajole or beg for assistance; instead, state simply *what* assistance you need, *why* you need the assistance, and show the direct instructional *benefit* to students. Consider the following example: You know that a field trip to a local weather station would improve students' understanding of weather prediction (the benefit). So you ask for parent volunteers who will act as chaperons (what assistance is needed). The assistance is needed in order to increase the level of supervision while students are travel-ing away from school (why the assistance is needed). If no one volunteered, you would simply proceed with the textbook and class-room activities on weather prediction and forego the field trip. Obviously you would not ask students to beg their parents to volunteer, since other responsibilities may have made it impossible for them to offer assistance.

How Do You Get People to Volunteer?

If you understand *why* someone is likely to volunteer, you are more likely to get the assistance you need.

☐ People volunteer for tasks because it provides them with visibility or the chance to be the center of attention. This is sometimes true for students and adults.

☐ People volunteer for tasks because the task itself provides an opportunity to learn new information or a new skill. This is true for both students and adults.

☐ People volunteer for tasks because there is an obvious and direct benefit to students' learning. This is especially true for parents whose own children are learning.

☐ People volunteer for tasks because they like to feel needed or to feel that they are contributing something important to society. This is especially true for senior citizens and for grandparents of students in the class.

As you think about each task that you wish to delegate, think about the reasons various people might have for volunteering their assistance. Realize that each time you delegate responsibility, you not only get the work done but you also contribute to the volunteer's development or feeling of self-worth.

How Do You Delegate Responsibility?

The key to success here is thinking the task through—what needs to be done—before selecting the person(s) to do the task. You are then able to match the volunteer's skills and needs with the requirements of the task. If you wanted to ask a particular student to do a classroom task because it would help to raise his or her self-esteem, it would be important to give the student a task at which he or she could succeed. By first thinking the task through, you will better ensure that the task is matched to the student's ability. To get started, try the following three-step exercise.

1. Make a list of several tasks which clearly would improve students' learning in science, and which would require some assistance. Organize your list to show daily, weekly, or occasional tasks.

Example

LIST OF TASKS FOR SCIENCE

Daily Tasks
Feed the fish and chipmunk.
Help to set up and clean up science activities.

Weekly Tasks
Water the plants.
Clean the chipmunk's cage.
Run dittos for the coming week.

Occasional or Monthly Tasks
Pick up, operate, and return the film projector.
Record results of students' experiments on chart.
Change bulletin boards and hall wall exhibit.
Share special experience through photo slide show.
Demonstrate special skill or technique to class.

2. On 3″ × 5″ cards, list the materials needed for several of these tasks, and in three to five steps describe how to carry out each task. Be sure to note special precautions that must be taken.

Example

FEEDING THE FISH

Materials
Fish food is in a small round container beside the aquarium.

Steps
1. Feed fish each morning.
2. Shake container gently three times above the water. Too much food (four or five shakes) makes the water cloudy and is unhealthy for the fish.
3. When you feed the fish, check to be sure all of the fish are active—one sign that they are healthy. If any fish seem inactive, watch them during the day. If you are worried, tell the teacher.

These instructions are written for students at the middle school level or for parents. For younger students, these same instructions might read:

Example

FEEDING THE FISH

Materials
Fish food is in *can* beside aquarium.

Steps
1. Feed fish *one* time each day.
2. Shake can *three* times only!
3. Tell the teacher if any of the fish don't move or swim.

3. To help volunteers find materials they need, place self-adhesive labels on drawers and cupboard doors.

Empty Containers: Glass/Plastic

Chart Materials: Rulers, Chartboard, Paint, Brushes, Pencils

Plant and Animal Care Materials

GUIDING DISCUSSION

How do you keep a science discussion going? It seems like it should be easy; after all, there are twenty-five of you in the classroom! Yet many beginning teachers have difficulty when they first try to hold discussions in science classes. If you were to observe a discussion in their rooms, you would probably discover that the key ingredients to good discussions were missing. For discussions to be productive, they need to have many of the following characteristics: enthusiasm, purpose, spontaneity, focus, relevance, and group involvement.

Enthusiasm. The teacher *must* feel and communicate genuine interest in the topic and enthusiasm for it.

Purpose. There needs to be a clear purpose for discussion, and that purpose must be made explicit by the teacher. Sometimes it is helpful to put the topic on the chalkboard and list the questions, terms, or ideas that will be discussed.

Spontaneity. Discussions, however, can become boring if students know in advance everything that will be said or who will say it.

Avoid putting on the chalkboard everything you expect students to say. There are several ways you can add spontaneity to discussions; some involve taking side explorations.

☐ Pick out a part of a student's answer and ask a new question.

☐ Ask students to extend their answers, showing a surge of new interest yourself by saying, "Tell me/us more about it."

☐ Ask students to justify their answers by saying, "What is your evidence?"

☐ Ask the same question of another student, saying, "Do you agree with everything (name) said?"

Focus. If you encourage spontaneity and exploration of unexpected possibilities, it is sometimes easy to lose the thread of the discussion. Therefore, you also need to help students refocus on the purpose of the discussion. To help refocus, try the following:

☐ Ask students to summarize. "Who can summarize what we have said so far in relation to _____." Point to the topic listed on the chalkboard.

☐ Ask a single student how his or her answer relates to the topic. If you have already tried to bring the class back to the topic and they seem to be going astray again, this may help.

☐ Look at the clock or your watch, and then ask how the class can refocus on the topic.

☐ Ask any one student how another student's comments are related to the topic.

Relevance. If students are to volunteer or contribute to discussions, the topic must be related to their experience, or related to material they can read. One reason many teachers use "hands-on" activities in their classes along with a science text is that some students are unable to get the full meaning of concepts when they read the science text. A firsthand experience allows less able readers to participate fully in a discussion of ideas. Obviously, not all discussions can be based on "hands-on" experience. When you do provide such kinesthetic experiences, make a special effort to bring less able readers into the discussion.

Group Involvement. Most students in a class need to have an opportunity to offer their ideas in order for the discussion to hold their interest. The most common way to hold students' interest is to ask them questions or to call on them by name. One teacher tells his students a discussion is like a basketball game: everyone is expected to take part in and pay attention to what is going on so that they can be ready to play the ball that is thrown to them, or take the ball away from anyone that "goofs up." He actually divides the class into two teams and keeps score.

ASKING QUESTIONS

Recently a group of teachers (grades 1–8) talked about why they ask their students questions. Here are some of their comments:

☐ I am usually looking for agreement. I ask questions to emphasize agreement and right answers.

☐ Not me. I ask questions to make students think. I like questions that make them probe for reasons rather than mere recall.

☐ So do I. I want my questions to help students gain insights; so I am always asking my kids to clarify or justify their answers.

☐ That's fine for older students but my little guys can't do that. It's all they can do to keep their minds on the topic! I have to ask questions that keep them on the subject—lots of repeating what they just heard.

☐ Yes, I agree with that. It really helps students learn if you ask them to repeat, and ask them

for specific answers. You know, like 'What are you supposed to do next in the experiment?'

☐ My students are in seventh grade, but they still have to hear things over and over again! It's not just the little guys!

☐ Many times I ask a particular student a question because I know he or she knows the answer—one of the few times that student will be able to answer a question. So you could say I ask questions to help build kids' confidence or self-esteem. Of course, I also ask questions for most of the reasons given by everybody else here. I'd have to say this, though: I have taught first grade, third grade, and sixth grade. Right now I'm subbing in an eighth grade science class. And the kids are all the same when it comes to asking questions. You really have to make the question fit whatever you are trying to accomplish: drilling, keeping their attention on the subject, making them think, or giving them a chance to show what they know. I think you can get first graders to think by asking them thought-provoking questions, too.

Important points were made by all of the teachers. There are many purposes for asking questions and many ways to ask them. It is also helpful to think about questions in terms of their being *open* or *closed*.

Open questions are questions which encourage students to respond to a single question with diverse responses. For example, following a field trip, an experiment, or a film, you might ask, "What did you learn?" Most students will have something to contribute. From their diverse answers, you may be able to weave a coherent summary or have one of the more able students summarize what he or she has heard the class say. Open questions tap students' enthusiasm for describing what they have learned.

Closed questions are questions which focus attention on specific, acceptable answers and reinforce those "correct" answers. Reviewing the rules to be followed on a field trip, learning the differences between toxic and nontoxic substances, or repeating the steps to be followed in an experiment are all examples of closed questions.

Mary Budd Rowe (1973) has studied the effect of teachers' questions on students' answering behavior. She and her colleagues found that the amount of time a teacher waits for students to answer questions before calling on someone else or asking a new question will dramatically affect the kind and the length of student answers. One implication of Rowe's research is that if you want to encourage students to probe, think deeply, and gain insights from your questions, you need to pause after your questions and give students an opportunity to think about their answers; don't call on someone else immediately. Rowe (1968) found that teachers tend to wait longer for bright youngsters to answer questions and to give slower students less time to respond. Ordinarily, teachers ask a question about every thirty seconds during discussions. It is difficult enough for children to answer questions easily without being bombarded by one question after another.

Many teachers ask open, open-ended, or thought-provoking questions—and direct the question to the entire class. Then they wait until most students have formed their answers before calling on a student. In this way, they allow many students to give their answers. Waiting for students to form answers is also important in asking closed questions. Remember that students *want* to give the "right" answer, and it sometimes takes more than a few seconds for them to sort it out. Although some students can always be ready with an appropriate answer in less than five seconds, you deny the opportunity to others who take a little longer if you always call on the first person who is ready.

The work of Benjamin Bloom (1980) is a good source for teachers to use in learning about the relationship between the form of a question and the level of thinking required by students to answer it.

HANDLING DISTRACTIONS

Be aware of things that will distract students during discussions. For example, having students sit too closely together on the floor is a natural distractor in the primary grades; leaving interesting equipment on desks during discussions becomes a natural distractor for intermediate and middle grade students; an open window that blows or ruffles papers is a natural distractor for students of any age. You cannot always prevent distractors, but often you can control students' reactions to them in one of several ways:

- ☐ Recognize the distractor and make it part of the lesson. A joke or a mistake on your part can be useful if you use it. This strategy requires you to be on your toes and to see ways to use the unexpected. For example, you are having a discussion on gravity, and you accidentally drop a book, saying, "See how fast things fall to the ground?"

- ☐ Recognize the distractor and ignore it, if you know it will go away. This strategy works for distractors of short duration, such as a sudden slam of a door or the sound of someone walking down the hall.

- ☐ Recognize the distraction and do nothing because you don't know what to do. Honesty helps when you are faced with a situation you don't know how to handle. Your students can be helpful here if you stop, tell them what you think is a distractor, and get their suggestions as to what to do next. For example, "I know you're really hot; but how can we finish our discussion before the bell rings?"

- ☐ Remove distractors; put away the interesting materials, move the stack of papers that are

Open questions are asked to invite diverse responses. They require more discussion time than closed questions because they usually involve more students. Closed questions are asked when a single "correct" answer is needed to reinforce learning.

blown by the wind, or move students who are sitting too closely together and kicking each other.

RESOLVING CONFLICTS

As a teacher you will be responsible for the management of discipline problems in your classroom. A useful way to picture such prob-

lems is to think of them in terms of conflict resolution. You cannot hope to have a conflict-free classroom. You will have to resolve conflicts between different students and between yourself and students.

A conflict exists when the activities of two or more people are incompatible, resulting in less than a satisfactory opportunity for all concerned to achieve their goals. It is sometimes

Good discussions don't just happen by accident or luck; they are planned by teachers who know how to keep students involved. (Paul Conklin)

helpful in understanding conflicts to recall Maslow's discussion of motivation (chapter 3). Students' behavior may be motivated by needs such as thirst, hunger, or lack of sleep, needs for security, self-esteem, or self-actualization. Such needs can trigger behavior that leads to conflict.

Conflicts in the science classroom disrupt the normal flow of activities and require your attention for resolution. Sometimes you may be tempted to spend far too much time trying to establish blame for problems and then realize you have not resolved the problem at all.

Think about the following description of a classroom conflict. This conflict is unique because it has been described by both the student and the teacher.

The Student's View

I was sitting at a table in the science laboratory with my friends. We had finished our work and were just talking. The teacher came over to our table and told us to get some work out, saying, "You can't just sit there and talk."

I ignored her and, after she left we continued our conversation. She came back and repeated her previous order. This time I told her I must be responsible enough to know when and how to do my schoolwork, since I was as big as she was. "So," I said, "I don't need you to supervise my study habits." I reminded her we weren't being noisy, just talking amongst ourselves, and that we weren't bothering anyone.

She became angry. She told us we were bothering others and said there was a rule of "no talking" in the science laboratory. *She* was making a big scene and just wanted to push us around. When she started treating us like little kids, I finally lost my temper. Who wouldn't? She then kicked me out of science for the remainder of the day.

The Teacher's View

It was about half way through the period when I noted three boys sitting at a table talking. I went over to the table and told them that this was a science class and they should finish their science laboratory. I said they should do something besides talk.

I went about my business until, a few minutes later, I noticed the boys were still talking. So, I went to the table again and explained that they should be working. I hadn't even finished what I was saying when one of the boys said he was smart enough to take care of himself so I could just leave him alone.

At this I told the boys that a few students do not have the right to talk and disturb other students who want to work. The boy got angry and used very strong language, totally inappropriate! At this, I asked the boy to leave the room and he did.

What would you have done if you were the teacher? Did you notice the similarities to other classroom conflicts? The conflict was very brief, lasting only four to five minutes. Apparently there was a rule governing conduct for the situation. The perceptions of both the student and teacher in this conflict were different. Did you notice how communication was somewhat similar at the beginning but deteriorated as the situation continued to the point of threats, name-calling, and assertion of power? The characteristics of a trusting attitude were lacking. Both the student and teacher thought they were correct, and thus

the problem belonged to the other person. When the conflict ended, it was not resolved. There is every reason to suspect that future problems would occur between the student and the teacher.

All conflicts are not the same. Table 6–1 outlines levels of student behaviors that may result in class conflict. Understanding student (and teacher) behaviors that lead to conflict situations prevents conflict and gives direction and guidelines toward intervention and resolution. Also, it clarifies how conflicts escalate; each party moves to a higher level, hoping the other party will succumb. In some cases, simple conflicts escalate to violent and destructive episodes.

Disruption of the class is obviously unacceptable. Although disruptive behavior can be harmless (talking or laughing) or dangerous (playing with sharp objects), the main objective in dealing with misbehavior is ending it, getting the students back to work, and letting the flow of classroom activity continue uninterrupted. It is not enough to stop the disruption; one must get the disruptive student quickly back into work so there is less opportunity of a relapse. One example of a direct, misbehavior-ending strategy of this kind is one developed by Kounin and Gump (1958) in which a boy who is bothering a girl is asked by the teacher to stop and either work on a project or read a book. Such a strategy is not time-consuming; it encourages students to continue work and it lets all students know what the teacher objects to and what is expected. This "ripple effect" can be seen to affect the rest of the class, as well as the misbehaving student (Tanner, 1978).

The way in which a teacher deals with disruption should conform to both educational and developmental goals. Discipline should not include tactics that discourage students from wanting to learn or instill in them unacceptable means of dealing with situations.

TABLE 6–1
Disruptive Behaviors in
Classroom Conflicts

Level	Description of Behavior
I	*Behaviors seeking personal affirmation.* This is the first level of behaviors that can be potentially disruptive. Behaviors seeking affirmation are subtle and often overlooked in the classroom. Examples include seeking recognition from the teacher or other students due to new clothes, outstanding work, or mild disruptive statements. Also, there are those who are detached and mildly withdrawn, or who show off and are mischievous in ways that do not cause teacher alarm.
II	*Behaviors showing personal assertion.* At this level the student makes it very clear to a large number of individuals (including the teacher) that others *must* pay attention. This stage is slightly more intense than the earlier stage of affirmation. Examples from this stage are being stubborn, defying authority, being withdrawn or depressed, refusing to work, or talking back. In general, this stage is differentiated from the earlier one in that the teacher or other students *must* pay attention.
III	*Behaviors demonstrating personal aggression.* The aggressive student attempts to take over and control the immediate situation. These students are openly and maliciously disobedient or have disassociated and withdrawn from the classroom. There is no longer any working relationship between student and teacher. The student is attempting to dominate the classroom environment by actively moving forward and taking power or digressing totally into a withdrawn state either mentally or physically.
IV	*Behaviors that are violent.* This is the stage of physical force, the demonstration of rage and temper tantrums. In the withdrawn inward form, this may be a flight from reality through chemicals, psychological withdrawal, and attempts at self-mutilation or self-destruction.

Overloading a student with work or reacting with angry retribution does not enhance the educational or moral development of the student.

Behavior modification can sometimes be used in conjunction with ignoring disruptive behavior; an example is praising good behavior while ignoring disruptive behavior. Behavior modification will not work, of course, with dangerous misbehavior. Immediate means are necessary for dealing with dangerous behavior.

In some cases, actual punishment must be administered. Punishment may mean withdrawal of privileges, verbal reprimand, scolding, or even just a frown or a shake of the head. The punishment must be one that is considered punishment by the student, not just by the teacher. Punishment is effective when the behavior that provoked it ceases. Making the student leave the class should be reserved for extreme, repeated cases, and it should be handled matter-of-factly rather than as an angry retribution. However disruptive behavior is handled, the most important consideration is returning the student to work as quickly as possible and preventing relapses.

When conflicts between you and your students occur, the following recommendations may help to avoid serious conflicts and to bring about constructive consequences.

1. Try to recognize consistent patterns of behavior that result in conflicts.
2. Clarify and reiterate daily classroom rules.
3. Clarify each person's perceptions of the conflict situation: "How does this situation seem to you?"
4. Maintain communication. You should be able to keep the lines of communication open for several minutes. This means one should avoid personal insults or threats.
5. Define the conflict as a mutual problem. "Look, you would like to visit and I would like it quiet so other students can work. How can *we* resolve this?"
6. Avoid using physical power to resolve the conflict. This can escalate the conflict and/or end it without resolutions.

Several specific techniques for helping teachers resolve disruptive behavior have been developed that utilize particular communication skills. Sources for these techniques are included in the bibliography at the end of this chapter.

SETTING LIMITS FOR SAFETY

When survival or success is at stake, most people are interested in knowing what the limits are, and they are usually willing to live within them. This generalization is certainly true for students in the classroom. You will discover that students want to know limits, and want to have freedom to act within these boundaries. You also may have observed that younger students need more assistance from adults than do older students in order to define the limits of their activities. That is one reason why there are frequently more adults

in kindergarten classrooms than there are in junior high school classrooms. A significant aspect of your science teaching is the responsibility for defining and enforcing limits with regard to the personal safety of students in your class and the school. First, look over figure 6–2, the Safety Checklist, to see how many of the questions you can answer appropriately.

How do you make students aware of limits and how do you enforce rules? Most teachers find the job manageable by making lists of safety rules that apply to different situations. For example, there are rules about what to do during a fire drill, an earthquake, or a tornado. These rules all have one thing in common: they pertain to personal safety during a serious or major disaster. Rules of this kind are not limited to teaching science; while school authorities are responsible for setting the rules, you are responsible for helping students practice these rules.

There are yet other kinds of rules you must make that will help to set limits for students' behavior during science. These rules relate to the use of toxic substances, the use of potentially injurious tools, and the behavior of students when they are on field trips or when they are handling animals in the classroom. Usually teachers prepare these shorter lists of rules, one for each topic, and present them to the students just prior to the time students will first encounter the situation to which they are relevant. Other safety considerations appear in chapter 5.

To emphasize the importance of rules, teachers usually put them on the chalkboard or in writing on a chart, and then go over the rules with the class. Discussion of what the rules mean and the reason for each rule generally follows. Teachers usually end the discussion by noting what will happen to someone who breaks the rules; comments may be couched in terms of what harm could happen

FIGURE 6–2
Safety Checklist

1. Do you know the location of fire extinguishers and first aid kits in your room? Are they in working order?
2. In case of fire, do you know the appropriate evacuation procedures?
3. Do you already have proper disposal plans for broken glass, burned materials, and toxic substances?
4. Do you have a safe, locked place for storage of potentially hazardous chemicals?
5. Are all your potentially hazardous materials fully labeled?
6. Do you know the materials you are using well enough to recognize any that might be harmful to a student if spilled or eaten?
7. Do you know how to communicate to your students the responsible use of materials without making them fearful of those materials?
8. Do you plan enough time for the activity so that students will not have to rush?
9. Have you clearly set—and enforced—the rule that, in science, one *never* eats or drinks anything used in experiments?
10. Do your students know *who* is authorized to use *what* materials and *when?*
11. If there is a potential hazard, do you first show students the hazard potential and then explain why it is a hazard?
12. Are you with your students at all times when they are working with science laboratory activities?

to the offender or to classmates, or in terms of punishment. Once the potentially hazardous activity is under way, the teacher speaks up quickly when some student first violates one of the rules. The teacher reminds students of the reason for the rule and the possible danger if the rule is broken again. If students intentionally violate rules, there is an immediate need to remove the student until you think the behavior will be altered.

COMMENTARY

The importance of effective instructional management cannot be overemphasized. Student learning and teacher satisfaction are directly proportional to this teacher variable. Chapter 6 has attempted to identify those areas in which teachers make decisions that will affect the overall management of the classroom.

Although successful instructional management styles vary, each teacher makes daily decisions that determine the degree to which the classroom runs smoothly. This chapter suggested alternative methods to help the teacher make these decisions with confidence. Perhaps the guiding principle for each teacher should be selecting strategies that encourage the uninterrupted flow of classroom activity and the continued educational development of the students. In the next chapter, we explore ways to evaluate students' learning and their reactions to teaching, and how to evaluate science programs and your own teaching.

BIBLIOGRAPHY

Becker, Wesley; Engelmann, Siegfried; and Thomas, Don. *Teaching 1: Classroom Management.* Chicago: Science Research Associates, 1975.

Bloom, B. S.; Hastings, J. T.; and Madaus, G. F. *Taxonomy of Educational Objectives: Handbook 1, The Cognitive Domain.* New York: David McKay, 1980.

Bybee, Rodger W., and Gee, Gordon E. *Violence, Values, and Justice in the Schools.* Boston: Allyn & Bacon, 1982.

Chamberlain, Leslie. *Effective Instruction Through Dynamic Discipline.* Columbus, Ohio: Charles E. Merrill Publishing Company, 1971.

Combs, Arthur; Purkey, William; and Avila, Donald. *Introduction to the Helping Relationship.* Boston: Allyn & Bacon, 1978.

DeCecco, John P., and Richards, Arlene K. *Growing Pains: Uses of School Conflict.* New York: Aberdeen Press, 1974.

Deutsch, Morton. *The Resolution of Conflict.* New Haven: Yale University Press, 1973.

Dinkmeyer, Don, and Dinkmeyer, Don, Jr. "Logical Consequences: A Key to the Reduction of Disciplinary Problems." *Phi Delta Kappan.* 57 (1976): 85–86.

Dreikurs, Rudolf, and Cassel, Pearl. *Discipline Without Tears.* New York: Hawthorne Books, 1972.

Kounin, Jacob S., and Gump, Paul V. "The Ripple Effect in Discipline." *The Elementary School Journal* 59 (December 1958): 158–162.

National Education Association. *Classroom Group Management.* Washington, D.C., 1974.

———. *Discipline and Learning: An Inquiry into Student-Teacher Relationships.* Washington, D.C., 1975.

———. *Discipline in the Classroom* (rev. ed.). Washington, D.C., 1974.

———. *Discipline Techniques.* Washington, D.C., 1976.

Paul, James L., and Epanchin, Betty C. *Emotional Disturbance in Children: Theories and Methods for Teachers.* Columbus, Ohio: Charles E. Merrill Publishing Company, 1982.

Rowe, Mary Budd. *Teaching Science as Continuous Inquiry.* New York: McGraw-Hill, 1973.

———. "SCIS in the Inner City School." *SCIS Newsletter,* (11) (Winter 1968).

Skinner, B. F., "Contingency Management in the Classroom." *Education* 90 (November 1969): 10–15.

Tanner, Laurel N. *Classroom Discipline for Effective Teaching and Learning.* New York: Holt, Rinehart & Winston, 1978.

Trowbridge, Leslie; Bybee, Rodger; and Sund, Robert. *Becoming a Secondary School Science Teacher.* Columbus, Ohio: Charles E. Merrill Publishing Company, 1981.

7

Evaluation

The purpose of chapter 7 is to focus on evaluation. The chapter begins with descriptions of a variety of informal evaluation procedures and instruments as well as more traditional and formal techniques. You will become familiar with common grading practices, report cards, and parent conferences. You will learn to evaluate your own science program. Chapter 7 concludes with suggestions for self-evaluation of your own teaching. Throughout the chapter you will find examples of tests, surveys, informal checklists, and forms that are used currently in schools.

KEY CONCEPTS AND IMPORTANT IDEAS

☐ formal and informal evaluation ☐ assessing different kinds of learning ☐ conceptual knowledge ☐ learning processes ☐ attitudes and values ☐ psychomotor skills ☐ decision-making skills ☐ practical laboratory tests ☐ picture tests ☐ problem tests ☐ true-false tests ☐ multiple choice tests ☐ completion tests ☐ guidelines for using tests ☐ grades, reports, and conferences ☐ program evaluation ☐ teacher evaluation

BEGINNING YOUR SYSTEM OF EVALUATION

This chapter includes an overview of several current evaluation practices as well as examples of criteria used to evaluate science programs. The chapter concludes with techniques for evaluating and improving your effectiveness as an elementary science teacher. Note that while we have primarily addressed methods of student evaluation, we also include two other elements essential to effective science teaching in the elementary school—the program and the teacher. A good system of **evaluation** incorporates all three realms. Developing a coordinated system of evaluation is an important challenge for all teachers.

As you read about different types of evaluation techniques, you will undoubtedly find ideas and methods that you would like to include in your teaching. You will slowly build your own system of evaluation that is unique and based on an understanding of yourself, your students, your program, and your school. Our purpose is to help you begin conceptualizing that system of evaluation. Beginning your own system of evaluation includes the synthesis of strong points from different theories along with your own ideas about evaluation. Here are a few helpful guidelines.

Start by taking a few minutes to think about your own aims and objectives as an elementary, middle, or junior high school science teacher. Then read and study the different approaches to evaluation in this chapter. You might wish to make some notes about the ideas and modifications of techniques that you feel are important. Ask yourself, "What situations are most applicable for this type of evaluation?" "What are the strengths and weaknesses of this technique?" "Will these methods work for all students?" "How can the results of this type of evaluation be used to: improve my instruction, modify the program, bring about better student learning, and aid in parent conferences?"

As you progress in the initial development of your evaluation system, you will begin to identify advantages and disadvantages of different approaches. This analysis of different systems can prove helpful. As you begin to formulate a coherent system of evaluation, keep in mind that you want to have a system that (1) aligns with your goals and preferences as teacher; (2) gives direction to student learning, program revision, and your effectiveness; (3) includes evaluation for the range of achievement in your classroom; and (4) organizes your observations for improvement and reporting of the elementary, middle, or junior high school science program. In the next sections we have provided ideas, techniques, charts, and materials to help you begin—but this is only a beginning. You will have to modify these ideas in ways appropriate for your teaching style, your students, and your school.

Traditional, **formal evaluation** is exemplified by midterms, finals, and end-of-chapter tests. These results are supposed to show what one has learned and determine what instruction to present next. Teachers use these results of tests to certify student mastery of major cognitive objectives and to communicate pupil progress to pupils and others. For decades, school systems have selected students by using formal evaluation techniques that cull the least able. Occasionally, formal evaluation techniques are used to judge the overall effectiveness of a teacher or to compare curricula. Unfortunately, formal evaluation occurs after the teaching and learning have taken place. It is then too late to make instructional changes that will benefit current students.

A newer, broader, and more informal view of evaluation goes beyond merely classifying students as "good" or "poor" according to a single test score. Everything that happens to a student in school has an educational effect: a smile, a thank you, or learning about spontaneous combustion. As a teacher you will want not only to assess how much the child understands cognitively, but also how he or she feels about the learning that has occurred. You will want to discover what you did to facilitate and hinder learning and attitude development. **Informal evaluation** supplies constant feedback that can help you make decisions concerning the quality of student or group performance; the need for adjustment in the pace or style of your teaching; and the success of a particular activity, unit, or curriculum in relation to desired objectives.

During the 1980s many countries in the world will be educating at least 80 percent of their populations through the high school grades. With increasing numbers of people acquiring more schooling, we need broader, more humane, and more productive views on evaluation. Informal evaluation procedures are needed to supplement traditional, formal techniques in order to more appropriately assess the complexities involved in the teaching-learning interaction. Good teaching must include evaluation that starts before a lesson begins and continues as an ongoing process until learning culminates.

STUDENT EVALUATION

We shall begin with an examination of various approaches to the evaluation of student outcomes. Then we continue with program evaluations.

Formal and Informal Evaluation Approaches

The wide variety of teaching methods, learning objectives, and evaluation techniques require a teacher to match evaluation procedures to specific goals and teaching approaches. In other words, you must choose a mode of assessment that corresponds to both the concept or skill your students are learning and the method you are using to teach that concept or skill. For example, in some of your science objectives, your approach to evaluation tells students what is important to learn. The process of an inquiry-centered exercise in that case is more important than learning specific facts. Your assessment techniques must correlate with the learning processes that are to be taking place.

Depending on your objective, specific evaluations will range from the informal, **formative** approaches to the formal, **summative** techniques. The following list illustrates evaluation approaches ranging from informal to formal.

1. Observe students at work.
2. Ask an individual student to explain a particular response or action.
3. Listen to a discussion involving a small or large number of students.
4. Observe simulations or role-playing situations and note evidence of behaviors that indicate understanding of the teaching objective.
5. Observe children as they play lesson-related games.
6. Listen as students peer teach.
7. Evaluate, using specific criteria, written and oral projects.
8. Grade teacher-made tests given at the conclusion of a unit.
9. Correct problem sets at the end of a chapter.
10. Score standardized tests given at yearly intervals.

As an example of the multiplicity of evaluation techniques necessary to adequately evaluate the objectives of a series of related lessons, let us look at the Health Science activity titled "You and Your Vision," from chapter 12. The activities cover approximately three class periods as students participate in simple experiments to learn the functions of the various parts of the human eye. Table 7–1 suggests specific evaluation techniques to use with each of the five learning objectives. The column on the left lists the specific objectives of the activity; the middle column suggests one or more alternative evaluation techniques re-lated to this objective; the column on the right describes observable student behaviors that give evidence of student understanding regarding the specific objective.

Assessment of Different Kinds of Learning

Successful learning is more than remembering "right answers" at the end of a chapter. It is an active process that results in long-term

TABLE 7–1
Evaluation techniques for "You and Your Vision" activity

Objectives	Evaluation Techniques	Student Behaviors
Conceptual Knowledge Knowledge of the basic structure and function of the human eye encourages personal practices that promote and protect visual health.	Give each of the youngsters a simple outline of the eye. Have them make posters of the eye and label all parts with names and functions, including accessory parts.	Does the student include all names and functions of the components of the human eye that you have taught?
	Invite students to draw pictures illustrating one practice that promotes or protects visual health.	Does the student illustrate a practice that promotes or protects visual health?
Learning Processes Understand and use the names of the parts of the eye in describing their functions.	Invite students to teach what they have learned about the human eye to students in another class.	Does the student demonstrate understanding by the ability to answer questions from other students?
	Have students take a summative test in which they are required to use the names of the parts of the eye.	Do your students explain all of the parts of the eye that you have taught? Do summative test results demonstrate understanding?
Attitudes and Values Learner assumes responsibility for care of self and others in the environment.	Invite students to conduct surveys in which they evaluate home or community environments. In a report to the class, they are to suggest ways of improving these environments for optimal visual health.	Do students consider (1) proper lighting; (2) viewing distance from TV; (3) the use of sun glasses and safety goggles; and (4) other conditions that affect visual health?

TABLE 7–1, *continued*
Evaluation techniques for "You and Your Vision" activity

Objectives	Evaluation Techniques	Student Behaviors
Psychomotor Skills Student handles and uses simple tools with care.	Invite students to play a game in which they simulate the proper handling of simple tools. Fellow students try to guess which tasks they are simulating and what tools they are pretending to use. (Suggested demonstrations include pouring substances into containers or rinsing the eye with cool water.	Do students demonstrate care and safety in the way they handle tools?
Decision-Making Skills Students become aware of the consequences of various practices that affect the eye.	Invite children to simulate circumstances requiring them to decide upon a mode of behavior based upon probable consequences.	Do students select behaviors that promote or protect visual health?
	Conduct a class discussion to consider the consequences of these various behaviors.	Do children correlate probable outcomes with specific behavior choices that affect visual health?

growth and change within an individual in many areas. For example, in an activity involving planting and caring for seedlings, primary grade students may modify their conceptual understanding regarding the growth and identity of living things. They also have the opportunity to become more skilled in observing, comparing, and classifying. Simultaneously, the pupils may develop their psychomotor skills by handling simple equipment and substances and learn to accept guidance in caring for their plants. They also may learn more appreciation for living things. Students have the opportunity to develop decision-making skills as they consider the consequences of such alternative actions as watering their plants more or less often or exposing their plants to more or less sunlight.

As teachers, we are continually making decisions that set limits on the learning potential and possibilities inherent in a particular lesson. We need to ask: What is really important for our students to know? How do we want them to change? What role do we take in the process? And finally, how do we monitor the changes taking place in order to determine what has happened with respect to a given objective? For this last question, evaluation information supplies the data to help us make the necessary judgments.

Earlier, you were introduced to a new direction for the future of science education. Five areas of knowledge and skills were defined: conceptual knowledge, learning processes, psychomotor skills, attitudes and values, and decision-making processes. Because these areas differ, techniques for evaluating them vary. We shall clarify some evaluation procedures for these areas of knowledge and skill.

Conceptual Knowledge

Evaluating conceptual knowledge—that is, evaluating ideas about how things appear or are related and how events occur—utilizes both formal and informal approaches. An essential component to any technique is the opportunity to use both convergent and divergent responses. **Convergent evaluation** requires the student to answer a specific, directed, question. **Divergent evaluation** requires the student to answer a general, open, question. Asking the students to draw or tell you the correct way to place the wires, battery, and bulb in order for the bulb to light is an example of convergent evaluation. Following this with an opportunity for the students to explain the reasons for the specific answers given is an example of divergent evaluation.

It is also a good idea to double-check students' reasoning by asking them to evaluate and explain their reasoning with regard to a negative proposition. For example, you might ask the students what they would think of an arrangement in which the bulb touched the side of the battery and the wire touched each end of the battery. Why might that arrangement work or not work? It is possible for a student to give a "correct" convergent response (true/false, yes/no, this way or that way) and yet not understand the concept except at a very superficial level. Analyzing the students' divergent reasoning processes as an extension of a specific, convergent response gives the teacher an understanding of the adequacy of the students' logical understanding of the concept being learned.

Learning Processes

During the past two decades, a great deal of research has been conducted to identify basic learning processes, observe when they take place, and discover how people use or integrate these processes. Ten learning processes are presented below that range, in order, from simple to complex reasoning demands.

1. *Observing* usually includes using any one or a combination of the five senses.
2. *Comparing* begins with real objects or things when children are young but gradually develops to include the comparison of ideas or concepts.
3. *Measuring* refers to quantitative descriptions, made directly by the learner or by the use of a piece of equipment.
4. *Classifying* or *forming groups* begins with real objects for young children and becomes more advanced when older children can classify ideas.
5. *Gathering* and *organizing information* may be accomplished in either verbal form (oral or written) or in pictorial form.
6. *Constructing* and *interpreting graphs* generally includes numerical representation.
7. *Inferring* means suggesting more about a situation or condition than can be observed.
8. *Predicting* usually includes making estimates or reasonable guesses based on observations.
9. *Forming hypotheses* is the process of finding a conditional relationship and expressing it in the form of an if-then statement.
10. *Describing relationships among variables* occurs when one makes clear how a condition or event is similar to or different from other conditions or events.

The primary way to evaluate these learning processes is by setting up tasks for your students to accomplish independently and observing them as they work through the tasks. Examples at the primary grade level might be to have students illustrate their observations from a nature walk or a tour in the city, or to have them group objects (blocks, nature items, or pictures) according to similarities and differences. Intermediate students might gather data on food favorites of people from different ages or cultural groups. They might subsequently organize and graph the data, make inferences, and finally, based upon their observations, make predictions as to what other foods each group might prefer.

Psychomotor Skills

You have just seen that learning processes can be used in different ways: to explore the environment as a participant or spectator, and to explore verbal or numerical descriptions of the environment. When people explore as participants, they coordinate muscular activity with their intellect and senses to learn about the environment. The result is called psychomotor activity. This kind of activity has a special significance in science. Scientists, for example, might use tools and equipment to analyze substances, observe plants and animals, and construct models. These activities involve essential basic skills. You might wonder if it is necessary for students to have these skills since very few, if any, will actually become scientists. A major importance of psychomotor skills lies in the fact that they are important ways to gain new knowledge. It is not simply a matter of having students behave as scientists in order to appreciate how new knowledge is produced. Rather the conceptual knowledge that students gain through the combination of psychomotor activity and learning processes such as observing, comparing, and measuring is qualitatively different than the knowledge gained through second-

The evaluation of psychomotor skills can occur during laboratory or hands-on activities by simply observing students' skills as they complete their assignments. (© C. Quinlan)

hand sources such as reading and discussion. Unfortunately, few specific criteria for evaluating psychomotor skills have been developed. One method to assess your students' performance of typical laboratory techniques with regard to care and safety is to devise several simple exercises around the classroom or science area. Invite students to move from one task to another in two- or three-minute intervals. Each activity requires that students manipulate equipment, objects, plants, animals, and so on in order to find a solution to a problem. The results they submit will enable you to observe students' skill and success in applying the techniques.

Attitudes and Values

The development of attitudes and values regarding science, its methods, and its applications play an important role in the future of your students and our society. Necessary conditions for positive growth are classroom climate and circumstances that permit children to practice desired affective behaviors. This calls for an environment of recognizable patience and mutual respect for every student's ideas. Additionally, you might provide a variety of objects and materials for classroom use: time for independent and individual investigations as well as group activities, open-ended discussions, students' assignments for daily responsibilities, and opportunities and skills that equip your students to be involved in the health and welfare of themselves and others.

To assess the development of attitudes and values among your students, you might observe your students in light of the following questions or considerations similar to these:

☐ Do your students discuss their work with one another?

☐ Do they initiate any new activities in class?

☐ Are they resourceful in finding out what they desire to know?

☐ Are your students curious? Do they continue to explore unassigned topics when they are out of the classroom?

☐ Can your students cope objectively with differences of opinions? Can they express their opinions of right or wrong without letting peer pressure influence them?

☐ Do your students listen to one another?

☐ Do they critically question ideas to attain deeper understandings?

☐ Are they creative and flexible in problem solving?

☐ Are they self-motivated?

☐ Are they able to recognize contradictory evidence or opposing viewpoints?

Another organizer that may be helpful to you in evaluating affective involvement appears as table 7–2. It is arranged from the least positively involved student behaviors to the most positively involved student behaviors.

Evaluating change in affective behaviors—attitudes and values, emotions, feelings, interests, and appreciation—may be difficult. Although less direct means are used to determine the extent to which affective goals have been achieved, affective behavior cannot be ignored in the total evaluation picture of today's learner.

Decision Making

A part of our responsibility as educators is to develop the understanding needed to help future citizens participate in the democratic process. Decision making is thus an important goal of education. Some of the decisions that students, as future citizens, must make will be in the area of science and science-related, social issues. Decision making is a recent addition to the goals of science teaching. Subsequently, evaluation of decision-making processes is also new.

TABLE 7–2
Affective behavior evaluation

Receiving	Student attends class and does not sleep or interfere with learning of others.
Responding	Student reads assignments, answers questions, enters discussion if asked in class, and participates in lab activities when asked.
Valuing	Student voluntarily enters discussions in class or initiates them on his or her own with teacher or students. Student helps in class lab experiences and encourages others to participate.
Organizing	Student explains why material and experiences of discipline are of interest or of practical value in his or her work.
Characterizing	Student brings example or tells about uses and application of subject from his or her job, or from preparation outside class.

SOURCE: From *Taxonomy of Educational Objectives: Handbook II: Affective Domain* by David R. Krathwohl et al. Copyright © 1964 by Longman Inc. Reprinted by permission of Longman Inc., New York.

Basically, the decision-making process requires students to acquire knowledge and values that apply to a particular personal or social problem. They must evaluate the information available and rationally analyze, synthesize, and evaluate the short- and long-term consequences of following alternative courses of action or lack of action.

Evaluation of decision-making processes might consist of the following. Present students with a personal and/or social problem appropriate to their understanding. Many examples are provided by the activities encountered in any classroom. As the students are involved in the process of making a decision you can evaluate the degree to which they: (1) present appropriate knowledge; (2) use the best sources of information available; (3) evaluate the costs, risks, and consequences of different courses of action; and (4) define a specific course of action to resolve the problem.

DIFFERENT ASSESSMENTS OF LEARNING

The emphasis in the last section was on different kinds of learning. In this section discussion is directed toward different kinds of assessments used by science teachers.

Testing

Designing evaluation procedures begins with the teacher's review of objectives. Ideally, objectives should have been written as the lesson or unit was being developed. And, they should have been specific enough to be translated into test items. Examples of Science and Society objectives might be: After completing the unit on Science and Society the student should be able to: (1) define basic research and (2) differentiate between basic and applied research.

As you develop evaluation instruments it is important to consider whether you wish to use **criterion referenced tests** or **norm referenced tests.** Criterion referenced evaluation tests are used to identify learning difficulties of individual students. The students are then given extra time and instruction so they can learn material missed or not learned earlier. Student achievement is determined by an absolute level of mastery on each test, (for example, 80 percent correct), not a performance level relative to all other students in the class. The latter are norm referenced tests, (for example, grading on a curve). We generally recommend the use of criterion referenced tests. After all, the goal is to learn science, and this approach to evaluation directs the efforts of students and teachers toward that end.

Considerations for Different Elementary Science Tests

The first and strongest consideration is that the test be appropriate to the kind of learning that is the object of teaching. The second consideration is that the evaluation format reflect an understanding of the student's developmental levels, reading abilities, and prior experiences. Considerations such as these are left to the teacher's judgment; you are the person who will best know your objectives, students, and school. The next paragraphs are helpful hints for developing different tests that may be used in elementary and middle school science programs.

Practical Laboratory Tests. In this type of test the students are directed to complete a laboratory activity. As the students complete the activity the teacher observes and notes the different techniques and skills used by the students. It would be very easy to use an evaluation sheet, such as that in table 7–3, to systematize the recording process. As the students work on the activity the teacher checks their work under categories related to the activity.

Picture Tests. Many tests evaluate reading more than science. Use of picture tests is one way to evaluate students' understanding of science concepts. There is some evidence to indicate that many students do better on tests that consist mainly of pictures and that require only a check mark for correct responses (Finkelstein and Hammill, 1969). The use of picture tests requires the teacher to obtain pictures and ask questions related to the science content, processes, or skills represented.

Problem Tests. Problem solving and decision making are important objectives of science teaching. An appropriate means of evaluating these objectives is through the use of problem tests. Students are given a simple problem and asked to provide recommendations for action that will result in a solution. Begin by giving students information concerning the problem: There is a vacant lot near school that is designated to be used as a dump. Have the students (1) devise their own hypothesis about the positive and negative aspects of the problem; (2) find information about the problem; (3) collect and evaluate information; and (4) make recommendations.

True-False Tests. In order to accurately evaluate student understanding through the use of True-False tests, they should be long (about one hundred items) and carefully constructed. You may find the following guidelines helpful:

- ☐ Be sure the number of true and false items are about the same.
- ☐ Avoid trick statements.
- ☐ Use language that is familiar to the students.
- ☐ Avoid double negative statements.
- ☐ Avoid complex sentences.

TABLE 7–3
Recording student achievement in practical laboratory tests

Kinds of Learning	Students' Names														
CONCEPT KNOWLEDGE Knows concepts Applies rules															
LEARNING PROCESSES Observes Measures Infers															
PSYCHOMOTOR SKILLS Manipulates apparatus Constructs projects															
ATTITUDES AND VALUES Open to new information Respects others' views Organizes information															
DECISION MAKING Presents knowledge Uses best resources Evaluates costs and benefits															
OTHER															

NOTE: The five kinds of learning listed here are discussed in chapter 2. Examples of more specific objectives are also presented. Specific items should represent the objectives of the laboratory being used in the evaluation.

Multiple Choice Tests. Multiple choice tests are more appropriate for upper elementary and middle school students. Most teachers are familiar with multiple choice examinations. A question or statement is presented with three or more possible responses. In designing multiple choice items, keep in mind the following guidelines:

- ☐ All responses should be plausible.
- ☐ Answers should be grammatically consistent.
- ☐ Responses should be about the same length.
- ☐ Answers should be placed in a random order.

Completion Tests. Completion tests rely on a student's verbal ability and have a tendency to measure how well a student can memorize facts and definitions. One flaw of most completion tests is that they less effectively measure a student's understanding of science concepts, skills, and processes. If a teacher decides to use a completion test, he or she might want to follow these suggestions:

- ☐ Try to design some statements that require more than recall.
- ☐ Write clear and direct statements.
- ☐ Stress concepts, skills, and processes that students can reasonably be expected to know.

Guidelines for Using Tests in Elementary and Middle School Science

When preparing and administering tests remember that you are primarily concerned with student learning and secondarily with the determination of a grade. You want to use tests in the most appropriate, effective way in order to achieve your objectives. Some guidelines may help you with this task:

- ☐ Use a combination of formal and informal techniques to evaluate students' learning, attitudes, and perceptions.
- ☐ Use tests to diagnose and prescribe for student learning, not just to evaluate and decide on student grades.
- ☐ Do not use tests as punishment.
- ☐ Review each student's test results with the student.
- ☐ Use a variety of testing techniques to evaluate.
- ☐ Begin tests with easier items.
- ☐ Consider the total developmental levels of your students when constructing tests.
- ☐ Test for the different kinds of learning: concept knowledge, learning processes, psychomotor skills, attitudes and values, and decision-making processes.

Grades, Reports, and Conferences

Just as there are numerous methods, varying from formal to informal, used to evaluate your students, a variety of practices exists for reporting student progress to parents and administrators. A survey of reporting practices employed among urban, suburban, rural, and small town public school districts indicates that a variety of reporting styles are current among schools within typical American communities.

At the kindergarten level, teacher-parent conferences are most common. One school district surveyed requires that the teacher issue a kindergarten report card at the time of the conference (see figure 7–1). In this case, science is grouped with art and music. The teacher simply checks one of two comments describing the pupil's participation in science activities. Written comments that supplement grades and marks (or less commonly, comments based upon a teacher-parent conference) are requested of the teacher at all grade levels.

Student Progress Report

West Branch Public Schools
West Branch, Oklahoma

KINDERGARTEN

STUDENT	SCHOOL	YEAR ENDING
PRINCIPAL	TEACHER	

PROGRESS	1st Term		2nd Term	
	Making satisfactory progress	Needs to improve	Making satisfactory progress	Needs to improve

SOCIAL DEVELOPMENT
- Shares and takes turns
- Is developing self-control
- Contributes ideas to group discussion
- Respects the rights and properties of others
- Benefits from constructive comments
- Contributes to group discussions

WORK HABITS
- Attention span is adequate
- Works independently without disturbing others
- Completes work
- Accepts responsibility for cleanup
- Listens to and follows directions

LANGUAGE ARTS
- Expresses ideas verbally
- Speaks distinctly
- Prints first name correctly
- Can identify rhyming sounds
- Recognizes some words

ART, MUSIC, SCIENCE
- Participates in
- Art activities
- Music activities
- Science activities
- Group activities

PHYSICAL DEVELOPMENT AND HEALTH
- Good large muscle coordination
- Good small muscle coordination
- Chooses appropriate clothing for weather

ALPHABET

Recognizes these letters (underlined items have been presented, circled items indicate mastery).

A B C D E F G H I J K L M N O P Q R S T U V W X Y Z
a b c d e f g h i j k l m n o p q r s t u v w x y z

Can associate these letters with these sounds

a b c d e f g h i j k l m n o p q r s t u v w x y z

COLORS

Recognizes and names basic colors

NUMBERS

Can count from 1–10   1–20   1–30

1–5   1–10

Can count objects with understanding

Can identify and write these numerals

1   2   3   4   5   6   7   8   9   10

Can draw these shapes
○ ▢ ☐ △

Knows left and right hand

Writes own name

Knows address and telephone number

Knows date of birth

Knows the days of the week

Comments: 1st term   2nd Term

Parent's Signature

Parent's Signature

FIGURE 7–1
Student Progress Report for Kindergarten

Health and science are frequently grouped together at the primary grade levels. In some cases, the primary grade teacher is again asked simply to check one of two or three levels of performance, as shown in figure 7–2. Other school districts require that teachers utilize marks such as $+$, $\sqrt{}$, and $-$ to indicate student progress, as shown in figure 7–3.

Letter grades most commonly enter the picture at the intermediate and middle grade levels four through eight, as illustrated in figures 7–4 and 7–5. Computerized report cards (figure 7–6), used increasingly for middle and upper grades, simplify the science teacher's task because they allow letters, numbers, and symbols to indicate progress and comments. You might want to compare our samples with the reporting procedures practiced in the school where you student-teach. Inquiring among your peers, master teacher, or principal about alternative reporting styles will broaden your picture of evaluation reporting practices.

Regardless of the type of school you choose, teaching will probably require that you participate in parent conferences at some time. Indicating student progress and supporting your comments and concerns requires that you keep personal records. Overall class profiles enable you to compare and contrast one student's progress in a given subject with that of the class as a whole. (See figures 7–7 and 7–8.) An individual student profile that records, for example, a single student's attitudes, perceptions, and grasp of science concepts and processes is also helpful. If you practice ongoing, formative evaluation, you may need to use record keeping symbols that can be modified as students master given skills. For example, you might use an X to indicate partial understanding of a skill, then fill in the area when mastery is achieved. See figure 7–9 for one example.

Using your individual and class profile records to supply specific facts about each student's progress, you might devise a general outline to follow during conferences such as the one in figure 7–10 created by an elementary school teacher.

PROGRAM EVALUATION

Assessing the effectiveness of the curriculum and supplementary materials that you use in teaching science is another important aspect of evaluation. As a teacher you will have to decide whether to continue using the same program or to make planning changes for both the day-to-day implementation of the program and next year's science program. Only through careful and continual evaluation will discrepancies between anticipated and actual outcomes of your program become apparent. The criteria for evaluating a science program, both formatively and summatively, are similar to criteria considered in program selection. In one state (California), the state department of education recommends the guidelines listed in table 7–4 for evaluating the effectiveness of the program in use. You might check with your state department of education for such guidelines.

TEACHER EVALUATION

The crucial role that a teacher plays in the growth and change of his or her students is undisputed. Though as a beginning teacher you may rely largely upon specialists when selecting and preparing your curriculum, you alone are the one who interacts with your students. Given such responsibility, how can you continually measure and improve your effectiveness as their teacher? Several sources and alternatives for evaluating teaching exist within the school environment. One valuable judge of your teaching is you! Through the

Student Progress Report

GRADES ONE, TWO, and THREE

West Branch Public Schools
West Branch, Oklahoma

Student _____

Grade _____   Teacher _____

A check (✓) indicates level of achievement.

Academic Development

LANGUAGE ARTS

READING

	1st			2nd			3rd				4th		
	Very Well Progressing	Progressing	Experiencing Difficulty	Very Well Progressing	Progressing	Experiencing Difficulty	Very Well Progressing	Progressing	Experiencing Difficulty		Very Well Progressing	Progressing	Experiencing Difficulty
Phonics													
Word Structure													
Comprehension													
Vocabulary													
Oral Reading													

Reading Level (Circled)
Grade Level Kindergarten Readiness Preprimers Primers 1st Reader 2¹ 2² 3¹ 3²
Reading Levels 1–2 3 4–5–6 7–8 9–10 11–12 13–14 15–16 17–18
Comments: _____

HANDWRITING

SENTENCE STRUCTURE													
SPELLING													
LISTENING													
ORAL EXPRESSION													
DICTIONARY SKILLS													

Comments: _____

MATHEMATICS

Recognizes & Counts Numbers													
Writes Numerals													
Addition													
Subtraction													
Multiplication													
Problem Solving													
Work Level													

Comments: _____

	1st		2nd		3rd		4th	
	Progressing	Experiencing Difficulty	Progressing	Experiencing Difficulty	Progressing	Experiencing Difficulty	Progressing	Experiencing Difficulty
SCIENCE AND HEALTH								
MUSIC								
ART								
SOCIAL STUDIES								
PHYSICAL EDUCATION								
Participation								
Sportsmanship								

Social Development

WORK HABITS

Uses Time Well								
Completes Work								
Follows Directions								
Works Independently								
Works Carefully								
Keeps Work Orderly								

BEHAVIOR

Respects Rights and Feelings of Others								
Assumes Responsibilities								
Displays Self-Control								
Follows School Rules								

Comments: _____

ATTENDANCE	1	2	3	4
Days Absent				
Days Tardy				

PROMOTED TO GRADE _____

FIGURE 7–2
Student Progress Report: Grades One, Two, and Three

Hanover School District
PROGRESS REPORT 19____ — 19____
GRADES 1–3

STUDENT _____
TEACHER _____
SCHOOL _____ GRADE _____

EXPLANATION OF MARKS
+ AREA OF STRENGTH
√ DEMONSTRATES GROWTH
— IMPROVEMENT NEEDED

	1	2	3	4
READING LEVEL				
Comprehension				
Word Attack Skills				
Reads with Reasonable Speed				
Shows Interest in Leisure Reading				
MATHEMATICS LEVEL				
Number Facts				
Computation				
Reasoning				
Accuracy				
LANGUAGE				
Written Expression/Creativity				
Grammar/Usage				
Oral Expression/Discussion				
Punctuation and Capitalization				
SPELLING LEVEL				
Assigned Words				
Independent Application				
HANDWRITING				
Manuscript				
Cursive				
HEALTH/SCIENCE				
Basic Concepts				
Participation				
Interest				
SOCIAL STUDIES				
PHYSICAL EDUCATION				
ART				
MUSIC				
ATTENDANCE				
Days Absent				
Days Tardy				

	1	2	3	4
SCHOLASTIC RESPONSIBILITY				
Completes Assignments				
Listens Attentively				
Displays Initiative				
Follows Directions				
Neatness/Accuracy				
Contributes to Group Activities				
Works Independently				
CITIZENSHIP				
Displays a Positive Attitude				
Shows Consideration for Others				
Accepts Responsibility				
Follows Class and School Rules				
Respects Personal and Public Property				

1st REPORT COMMENTS

2nd REPORT COMMENTS

3rd REPORT COMMENTS

ASSIGNMENT FOR NEXT YEAR

FIGURE 7–3

Another Student Progress Report: Grades One, Two, and Three

Hanover School District
PROGRESS REPORT
GRADES 4–6
19 ___ ___ 19 ___

A = Outstanding   D = Below Expectations
B = Exceeds Expectations   F = Failure
C = Meets Expectations   I = Incomplete

SCHOOL _____
TEACHER _____
STUDENT _____
GRADE _____

Academic Achievement

	Term 1	Term 2	Term 3	Term 4
READING				
Grade Level				
Vocabulary				
Independent Reading				
Comprehension				
Reference Material Use				
LANGUAGE				
Grammar				
Mechanics				
Listening Skills				
Expresses Ideas Clearly				
Organizes Ideas when Writing				
Note Taking/Outlining				
SPELLING				
Assigned Words				
Application				
HANDWRITING				
Letter Formation				
Neatness and Clarity				
MATHEMATICS				
Grade Level				
Basic Concepts				
Number Facts				
Good Reasoning				
SCIENCE/HEALTH				
Concepts and Relationships				
SOCIAL STUDIES				
ART				
MUSIC				
ORCHESTRA/BAND				
PHYSICAL EDUCATION				
Skills				
Sportsmanship				

General Achievement

	Term 1	Term 2	Term 3	Term 4
RESPONSIBILITY				
Completes Work				
Listens & Follows Directions				
Works Independently				
Uses Time Well				
Seeks Help When Needed				
CITIZENSHIP				
Displays a Positive Attitude				
Is Considerate of Others				
Is Cooperative				
Follows Rules				
Respects Property				
Participates Constructively				

ATTENDANCE
Days Absent			
Days Tardy			

Parent's Signature

_____ Term 1

_____ Term 2

_____ Term 3

Comments:

_____ is promoted to _____ grade.

FIGURE 7–4
Student Progress Report: Intermediate Grades

STUDENT PROGRESS REPORT
Hanover School District
Hanover Middle School

Student _____   Grade _____ Year _____

Subject _____   Teacher _____

Semester 1 Quarter 1	Semester 1 Quarter 2	Semester 2 Quarter 1	Semester 2 Quarter 2
Academic Development:	Academic Development:	Academic Development:	Academic Development:

A   C+
A− C
B+ C−
B   D+
B− D
      D−
      F
      Inc.
	A C+
	A− C
	B+ C−
	B D+
	B− D
	      D−
	      F
	      Inc.
		A C+
		A− C
		B+ C−
		B D+
		B− D
		      D−
		      F
		      Inc.
			A C+
			A− C
			B+ C−
			B D+
			B− D
			      D−
			      F
			      Inc.

Pass Fail	Pass Fail	Pass Fail	Pass Fail
Conference Requested □	Conference Requested □	Conference Requested □	Conference Requested □
General Development:	General Development:	General Development:	General Development:
Satisfactory □	Satisfactory □	Satisfactory □	Satisfactory □
Needs Improvement □	Needs Improvement □	Needs Improvement □	Needs Improvement □
Unsatisfactory □	Unsatisfactory □	Unsatisfactory □	Unsatisfactory □
Comments: _____	Comments: _____	Comments: _____	Comments: _____

FIGURE 7–5
Middle School Student Progress Report

LITTLE CREEK CITY SCHOOLS

PROGRESS REPORT

School Name	School Term	Counselor	Gd.	Adv.	Student Name
LITTLE CREEK MIDDLE	4	HUMPHREYS	7		CLEARY, SANDRA J.

Per./Sect.	COURSE	MARKS				Com-ments	Qtr.* Abs.	TEACHER	Credit
		1	2	Sem					
3	DEVELOPMENTAL READING	A	B	A	G		2	DAUBENMIER, C.	4

PARENT TEACHER CONFERENCE DAY—NOV 16. CALL 131-8381 FOR APPT.

CURR GPA 3.34783 YTD GPA

EXPLANATION OF MARKS

A—Outstanding achievement
B—Good achievement
C—Satisfactory achievement
D—Minimum achievement
F—Failure due to unsatisfactory achievement
I—Incomplete due to justifiable absence

EXPLANATION OF COMMENTS

X—Excellent progress
G—Good attitude/conduct
1—Showing some improvement
2—Achievement is not up to apparent ability
3—Absences/tardiness affecting school work
4—Books/materials are not brought to class
5—Assignments are incomplete or unsatisfactory
6—Oral participation needed
7—Inattentive/wastes time/does not follow directions
8—Conduct in class is not satisfactory
9—Please contact teacher through counselor
*—This number indicates all absences from the class

FIGURE 7–6
Computer Progress Report

FIGURE 7–7
An Example of a Class Profile

FIGURE 7–8

Another Class Profile

Reprinted by permission from *Subsystems and Variables Evaluation Supplement,* written and published by the Science Curriculum Improvement Study. Copyright © 1972 by the Regents of the University of California.

EVALUATION

During these two weeks, in which the children extended their understanding of the process of identifying and controlling variables, the teacher took the opportunity to observe the children working at their problems.

In addition to assessing their level of understanding of isolating and controlling variables, she also noted their inventiveness, critical thinking, and persistence.

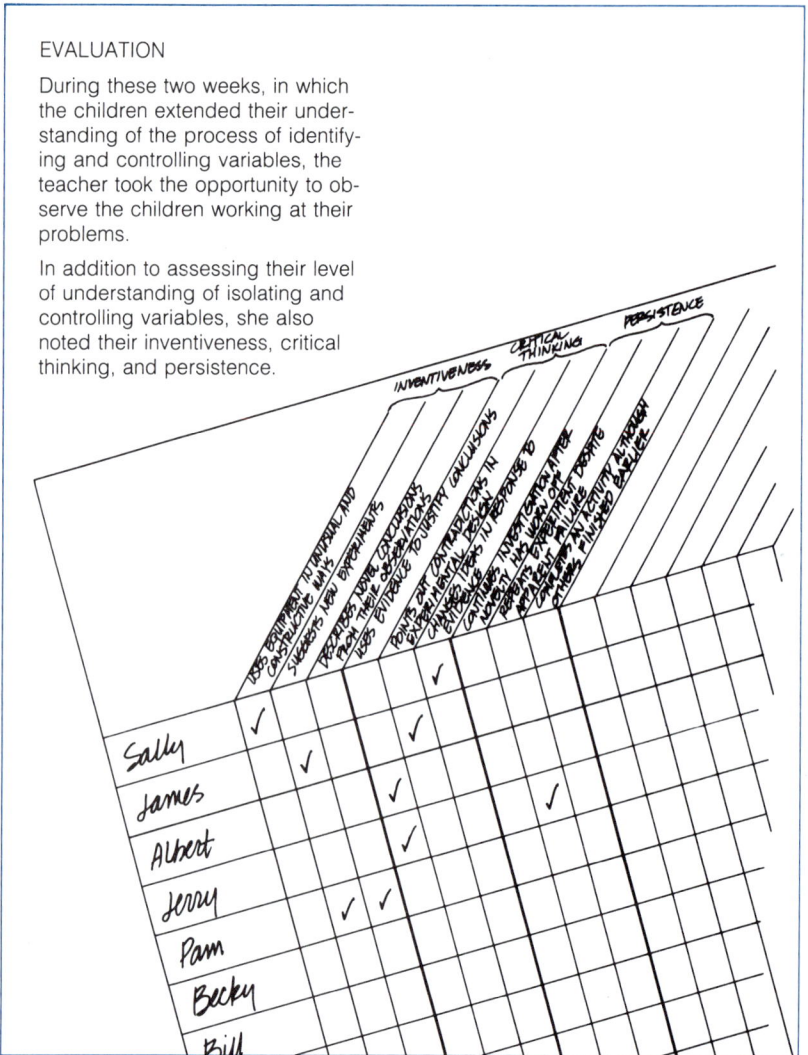

use of videotaped or audiotaped lessons, you can personally examine your presentation, pace, enthusiasm, and clarity. You can ask yourself: What do I like? What don't I like? What's missing? Keeping a journal of your own observations—what works and what doesn't—is also helpful. You might write up a more formal lesson plan, and add a critique after you perform the lesson. Rewrite or compare your lesson plan and conclusions with others and note these discussions in your personal file.

Another source of teaching appraisal is feedback from your peers. Four informal and nonthreatening ways to do this are offered here.

SCIENCE CURRICULUM IMPROVEMENT STUDY

INDIVIDUAL PROFILE

Interaction and Systems

Student   Grade   Date

Teacher   School

PHASE I Attitudes in Science (record the date or chapter of an observation)												
curiosity/interest												
inventiveness												
independent or critical thinking												
persistence												

PHASE II   Perception of Classroom Environment

Our Science Class		Picture no.				
Date	Activity	1	2	3	4	5

Faces					
Date	Question	⊗	⊗	☺	☺

PHASE III   Concept/Process Objectives

⬤ Activity 1   Identifying Interacting Objects

⊗ Activity 2   Keeping Track of a System

◯ Activity 3   Identifying Systems of Inter-acting Objects

◯ Activity 4   Describing Evidence of Interaction

◯ Activity 5   Identifying Interaction-at-a-Distance

◯ Activity 6   Writing Experiment Reports

FIGURE 7–9

An Example of an Individual Student Profile

TABLE 7–4
Guidelines for evaluation of a science program

The Appropriateness of Course Content and Curriculum

How Effective is Your School in Providing For Each of the Following:	Ineffective	Somewhat Effective	Effective	Very Effective
1 Having the program planners examine a wide variety of curriculum guides, textbooks, and other materials as part of the process of curriculum planning?				
2 Ensuring that the information in science textbooks and materials is valid, accurate, and up-to-date?				
3 Making certain that the course content being learned by students is consistent with the basic goals of the *Science Framework for California Public Schools:* attainment of positive attitudes toward science; attainment of manipulative and communicative skills in science; and attainment of scientific knowledge?				
4 Keeping teachers informed about new materials and significant new developments in science?				
5 Establishing at the elementary level a balance of instruction in life science, earth science, and physical science?				
6 Offering at the secondary level a wide variety of courses each year, including advanced courses, general courses, and minicourses in specialized areas of science?				
7 Ensuring that the sequence of the curriculum reflects students' general orientation to the world around them?				

Teaching Methods

How Effective is Your School in Providing For Each of the Following:	Ineffective	Somewhat Effective	Effective	Very Effective
1 Encouraging teachers of science to use a variety of teaching methods, such as demonstrations, group experiments, modeling, discussion, multimedia instruction, individual investigation or research, student tutoring, and questioning?				
2 Encouraging teachers to modify their teaching methods to take into account the needs, strengths, and interests of individual students or groups?				
3 Permitting high school students to choose among classes and teachers that offer a variety of instructional methods and teaching styles?				
4 Encouraging teachers to modify their teaching methods when evaluation data or direct observation indicate that these methods are not successful?				
5 Assessing student learning regularly through quizzes, tests, examinations, and reports?				

SOURCE: *Science Education for the 1980s: A Planning and Assessment Handbook for Kindergarten Through Grade Twelve* (Sacramento, CA: California State Department of Education, 1982), pp. 34-39. Used with permission.

How Effective is Your School in Providing For Each of the Following:	Ineffective	Somewhat Effective	Effective	Very Effective
6 Checking students continually to determine whether they are understanding what is being taught?				

Sustaining Student Involvement and Motivation in Science

How Effective is Your School in Providing For Each of the Following:	Ineffective	Somewhat Effective	Effective	Very Effective
1 Allowing time in classes for open discussion of science and related topics so that students feel free to ask questions?				
2 Encouraging teachers to modify lessons, when necessary or appropriate, to take students' needs and interests into account?				
3 Giving students opportunities to initiate scientific investigations and explorations in areas of special interest?				
4 Translating current events into interesting science activities?				
5 Encouraging students to explore science-related issues and to test hypotheses?				
6 Securing an adequate number of science materials for students who have limited proficiency in English and making certain that teachers use these materials sensitively and adequately?				
7 Ensuring that students who have limited proficiency in English are as involved in science classes as students who are fluent in English and that they are learning the same concepts as those which students who are fluent in English are learning?				
8 Giving handicapped students adequate opportunities to participate in science classes and providing handicapped students with adequate opportunities and special learning materials to become actively involved in laboratory activities? Ensuring that handicapped students are learning the same concepts that nonhandicapped students are learning?				
9 Giving gifted and talented students adequate opportunities to pursue science projects and interests in science independently, in groups, and in tutorials with teachers, older students, and volunteers?				
10 Providing science-oriented extracurricular activities and making students aware, on a regular basis, of such activities?				
11 Involving the staff of special programs (e.g., special education teachers, bilingual education teachers) in the planning of the science education program?				

The Amount and Quality of Time Spent on Science Education

How Effective is Your School in Providing For Each of the Following:	Ineffective	Somewhat Effective	Effective	Very Effective
1 Allocating a specific amount of time for science education for each student during each school week and making that a provision of the school-level plan?				
2 Encouraging students to become positively and actively involved in science classes?				
3 Varying science activities sufficiently to allow different kinds of student participation and engagement; e.g., combination of lectures, laboratory activities, discussions, small group activities, and field studies?				
4 Encouraging teachers to make homework assignments that review and reinforce concepts learned in class and letting students know they are expected to complete the assignments?				
5 Requesting that students complete reading assignments before class time?				
6 Contacting parents when students fail to complete homework assignments on time?				
7 Having well-known and uniformly applied consequences for not completing the homework assignments on time?				
8 Training parents to supervise and assist their children with homework?				

Standards and Expectations

How Effective is Your School in Providing For Each of the Following:	Ineffective	Somewhat Effective	Effective	Very Effective
1 Adopting schoolwide standards for homework, academic performance, classroom behavior, workmanship, and so forth, which are included in the written plan?				
2 Making students and parents aware of the adopted standards and the consequences for not meeting them?				
3 Applying the standards uniformly?				
4 Encouraging teachers, students, and parents to regard academic achievement as a highly positive attribute?				
5 Encouraging students to finish homework on time?				
6 Making certain that teachers and students are on time and ready to begin work when class begins?				
7 Making students and parents aware of the amount and quality of academic work necessary to receive a specified grade?				
8 Making students and parents aware of the standards of thoroughness, neatness, care, and precision expected of students' work?				

FIGURE 7–10
Outline for Conference

Conference Agenda

Reading

Grade level _____

Describe the materials that the student is reading.

Discuss the student's strengths and weaknesses.

Find out how much reading the student does at home.

Stress to parents that they should help the student develop his or her vocabulary.

Math

Grade level (_____)

Explain to parents how the student is progressing along the district skills chart.

Describe the student's basic fact knowledge.

Discuss the student's understanding of concepts.

Explain how well the student works during independent work time.

Make suggestions on how parents can help children work at home.

Science

Discuss the concepts that the student is currently studying.

Describe the processess that the student is working on.

Tell the parent what the student's attitude is toward classwork.

Give examples of student behavior that exemplify science-related attitudes (curiosity, persistence, inventiveness, critical thinking).

Spelling

Placement _____

Mention how parents can help students apply spelling principles to daily spelling.

Cursive Writing

Questions about Curriculum

Social Relationships

Discuss the student's playground behavior.

Describe how well the student gets along with classmates.

Describe the student's attitude, ability to listen, independent work habits, confidence, quality of work, and ability to keep up with the other students.

1. Invite an associate to attend your class during a lesson and ask him or her to observe your students. Are they attentive, restless, involved, and interested?

2. Simply talk over your concerns with your co-teachers. Experienced teachers utilize the teachers' lunchroom, faculty lounge, or faculty meetings to discuss and compare with colleagues of other grade levels the content matter and their expectations. As a beginning teacher, you may prefer to talk about your science program in a more private setting.

3. Team teaching also provides an opportunity for you and a colleague to informally observe and share teaching methods, ideas, and plans for the same subject matter.

4. Find out if your school district has identified Master Teachers or Mentor Teachers who assist new teachers. If so, seek their support, guidance, and feedback.

Feedback from your master teacher, supervisor, or principal can be particularly helpful in both informal and formal ways. One common method of evaluation, called **clinical evaluation,** proceeds like an informal interview. The experienced, supervising teacher asks the teacher trainee to assess his or her lesson and performance, and then a reciprocal dialogue takes place. Alternatively, your master teacher may be asked to complete a more formal, summative evaluation form that requests a rating and/or comments concerning your specific strengths and weaknesses in the classroom. One sample student teacher evaluation form, issued by a teacher-training college program, focuses on four major categories considered important in effective teaching: classroom effectiveness, relationships with students, planning for teaching, and personal characteristics. The master teacher is asked to rate the student teacher on a 5-point scale and then to comment on the student teacher's performance in a number of skills or behaviors in each category. Selected examples follow.

Classroom Effectiveness

☐ is attentive to the physical environment
☐ gives clear directions and examples
☐ presents material in logical sequence and at appropriate levels

Relationship with Students

☐ at ease with students in informal situations
☐ consistent and fair in expectations for students
☐ handles discipline problems tactfully and effectively

Planning for Teaching

☐ selects appropriate unit objectives and carries them through in the development of daily lessons
☐ plans for individual differences in students' abilities and interests
☐ uses a variety of instructional materials
☐ plans for a variety of teaching procedures, including "hands-on," teacher-directed, and peer-teaching

Personal Characteristics

☐ uses his or her voice effectively
☐ establishes good relationships with co-workers
☐ seeks suggestions and utilizes criticisms for improvement

Finally, do not overlook your students as a meaningful source for teaching assessment. Both informal and formal techniques can be employed. Here again, videotaping when possible enables you to carefully observe student behavior and success in response to your directions and explanations of an activity. Additionally, through simple, informal interviews you can learn how children perceive an activity. On a more formal level, the Science Curriculum Improvement Study (SCIS), developed through the University of California, Berkeley, in the 1970s provides written activ-

FIGURE 7–11

Example of Primary Students' Perceptions of the Classroom Environment (teacher's material page)

Reprinted by permission from *Populations Evaluation Supplement*, written and published by the Science Curriculum Improvement Study. Copyright © 1972 by the Regents of the University of California.

Perception of Classroom Environment

Our Science Class

This activity assesses your pupils' views of the classroom environment during science. Use it with the entire class whenever you wish to know how the children perceive the science class.
Time required: 5-10 minutes

Evaluation Materials

For each child:
    1 pencil or crayon*
    small piece of paper*

For the class:
    Our Science Class chart
*provided by the teacher

Our Science Class

Background Information

Our Science Class. The chart entitled *Our Science Class* has five pictures, each showing an activity that may take place during a science session. The five activities are working with materials, recording (writing or drawing), listening to the teacher, talking in a group, and reading. Your pupils' selections from these five pictures approximate what they think is happening in science class.

Evaluation Suggestions

Introducing *Our Science Class.* The first time you implement this activity, plan to use it mainly for practice—to explain the chart and teach your pupils how to use it.

Post the *Our Science Class* chart in a prominent place and tell your pupils that the pictures show what might go on during science class. To make sure they can interpret the pictures correctly, ask them to find the pictures that show children listening to the teacher (3), working with materials (1), reading (5), drawing or writing about their experiments (2), and talking about their experiments (4).

Using *Our Science Class.* When you are certain the children understand what each picture represents, distribute a small piece of paper to each child. Ask the children to choose the picture which shows what happened most during science class on that day and to write the number of the picture on the paper. Some children may wish to record more than one number. Permitting them to identify two pictures is reasonable, but listing more than two reduces the effectiveness of the activity.

Collect the papers and see how the children viewed the science class. Do not be concerned with individual responses, but rather with the consensus expressed by the class. Compare the results with your own impression of what happened. For example, if you find that most of the pupils identified the picture of children listening to the teacher (3) after a discussion session, they may feel they were not active participants in the discussion.

When most children perceive a different emphasis than you intended, you might consult the "Clues for the Teacher" section and the "Teaching Suggestions" sections in the teacher's guide to help you plan future science periods.

We suggest that you invite your pupils to respond to the *Our Science Class* chart several times during the *Organisms* unit.

You might also wish to use *Our Science Class* to assess the children's preference for specific activities. The children can identify the pictures which show activities they like best.

ities that assess pupils' perceptions of the classroom environment and students' feelings about their science activities. The tasks designed for primary grade students (below) could be adapted for more advanced grade levels.

☐ Our Science Class. This chart has five pictures, each showing an activity that may take place during a science lesson: working with materials, recording (writing or drawing), listening to the teacher, talking in a group, and reading. Your pupils' selections from these five pictures approximate what they think is happening in science class. (See figure 7–11 for greater detail.)

☐ Faces. The answer page with this title has three rows of faces. Each row includes a happy, a neutral, a sad, and an angry face. By circling a face, a pupil indicates how he or she feels about an activity that you have described. Sadness or anger on the part of most of the

FIGURE 7–12

Example of Students' Feelings about Their Science Activity (teacher's material page)

Reprinted by permission from *Life Cycles Evaluation Supplement,* written and published by the Science Curriculum Improvement Study. Copyright © 1972 by the Regents of the University of California.

Faces

This activity assesses your pupils' feelings about their science activities. Use it with the entire class occasionally during the unit. Time required: 5-10 minutes

Evaluation Materials

For each child:
    1 *Faces* page
    1 pencil or crayon (provided
       by the teacher)

Background Information

Faces. The answer page entitled *Faces* has three rows of faces. Each row includes a happy, a neutral, a sad, and an angry face. By circling a face, a pupil indicates how he feels about an activity you select. Sadness or anger on the part of most of the class in response to a science activity is a signal that you should consider changing the emphasis of the activities.

Advance Preparation

Duplicate the necessary number of answer pages from the ditto master.

Evaluation Suggestions

Introducing *Faces*. The first time you ask your pupils to respond on the *Faces* page, invite them to describe the faces. Point out that the answer page has three rows of faces and you will ask them three questions. By circling one face in each row, your pupils will indicate how they feel about three activities you describe.

Using *Faces*. Select three clearly different situations. Plan to use the first two questions to establish each pupil's range of feeling at the time he is using *Faces*. You might ask "How do you feel when it is time to play?" to establish his upper range and "How do you feel when you have had a fight with your best friend?" to establish his lower range. The third question may then be used to assess his feelings about a particular science activity (e.g., "How do you feel when you plant seeds?").

Collect the papers and review your pupils' responses. In this way you will be able to interpret each pupil's feelings about science in relation to his range of feelings on this occasion.

Another way to use this activity is to compare children's reactions to science in relation to other subject areas. Supply a specific example for each activity or subject area. These might be (1) reading a book, (2) taking care of guppies, and (3) drawing a picture.

class in response to a science activity is a signal that you should consider changing the emphasis of the activities or at least ask students to explain the reasons for their reactions. (See figure 7–12.)

At the intermediate level, students are asked to complete a Science Report Card in which pupils express their preference for, and estimate the time spent on, various science activities. (See figure 7–13 for further details.)

Perception of Classroom Environment

Science Report Card

When? *At the beginning of a class period two or three times during the unit*
Who? *The entire class*
What? *Pupils express their preference for, and estimate the time spent on, various science activities*
How long? *Five minutes*

Evaluation Materials

For each child:
1 *Science Report Card* page
(provided by the teacher)

Name _____

Science Report Card

1. What do you think of these?

    paper airplanes _____

    variables _____

    rolling spheres _____

2. How much of your time in science did you spend on each of these?

experimenting	much	some	a little	none
writing	much	some	a little	none
listening to the teacher	much	some	a little	none
discussing	much	some	a little	none
reading	much	some	a little	none

3. Circle the activity in question 2 that describes what you like best.

Background Information

The report card may have up to three sections. On one of these, your pupils express their opinions of various science topics, on the second they indicate the way their time has been spent, and on the third they express their preference for a type of activity. You may omit one or two sections to suit your purpose.

Advance Preparation

Design a *Science Report Card* similar to the example illustrated. On this page, list science topics that have been investigated during the past three or four weeks, and also list the five types of pupil participation. Duplicate the required number of pages.

Evaluation Suggestions

Introducing the *Science Report Card*. The first time you ask your pupils to respond on the report card, distribute the pages, read the text to the class, and explain the procedure. Each child should rate the topics in question 1 by writing his opinion on the line provided, either in words (interesting, very good, fair, boring) or by means of a letter grade. In question 2, each child should circle the word that best describes the amount of time he spent on each activity during the past two weeks. And for item 3, he should circle the activity he prefers of the five listed in question 2.

Collect the papers and review your students' ideas. Look for topics that aroused a great deal of interest and follow these up with additional optional activities. Look also for "weak" spots in the program, that is, topics that were not well liked by a large number of pupils. Try to identify the reasons why a topic did not interest your pupils. Invite them also to suggest ways of modifying the procedure or the topic to make it more interesting.

As you look at items 2 and 3, keep in mind that a balanced program includes all five types of activities, but that experimenting should generally be most prominent, and that reading and listening to the teacher should usually be less prominent than the others. If the children perceived a different emphasis than you had intended, you might consult the "Clues for the Teacher" and "Teaching Suggestions" in your teacher's guide.

Using the *Science Report Card*. If you found the feedback from your first use of the report card helpful, we suggest that you use all of it, or sections, at later times during the unit.

Letter

When? *About a week after the completion of the unit*
Who? *Entire class*
What? *Pupils describe their ideas concerning the unit in a letter*
How long? *Approximately fifteen minutes*

Evaluation Materials

For each child:
1 sheet of writing paper
(provided by the teacher)

Evaluation Suggestions

Invite volunteers to describe to their classmates and you some things about the *Energy Sources* unit that they remember. After a few of them have related their ideas, distribute the writing paper and tell your pupils that they should write a letter with their thoughts to the publisher of the SCIS program (Mr. Andrew McNally) or to one of the collaborating authors (see the bottom of page 3 in the teacher's guide--Mrs. Randle is an elementary school teacher, Dr. Thier and Mr. Webb are science educators, Dr. Karplus and Dr. Berger are university teachers). Encourage the children to express their own ideas about what they especially liked, what gave them difficulty, and how the unit might be improved for the use of other boys and girls. Instead of a letter, some of your pupils may wish to describe the same ideas in the form of a news reporter's story for the local paper.

Review the letters after you collect them to gain suggestions for the teaching of the unit at a later time. Communicate any significant findings to your fellow SCIS teachers, your science consultant, or principal. Then forward the letters to the SCIS Project, Lawrence Hall of Science, University of California, Berkeley, California 94720, for consideration and possible use of the suggestions.

FIGURE 7-13

Example of Intermediate Students' Perceptions of the Classroom Environment (teacher's material page)

Reprinted by permission from *Energy Sources Evaluation Supplement*, written and published by the Science Curriculum Improvement Study. Copyright © 1972 by the Regents of the University of California.

COMMENTARY   Evaluation of your effectiveness as a teacher is an ongoing process that should begin before you start a lesson and continue until you have appropriately assessed your students' growth and change. Assessment should be made with regard to learning new concepts, processes, attitudes and values, and psychomotor and decision-making skills.

The evaluation of your impact as a teacher really continues throughout your professional career. In the next chapter we explore what it means to be a teacher who is responsible for teaching students about science and society. Chapter 8 also provides a self-inventory for teachers as an aid to continuing professional development.

BIBLIOGRAPHY   Bertrand, Arthur, and Cebula, Joseph. *Tests, Measurement, and Evaluation.* Reading, MA: Addison-Wesley, 1980.

Bloom, B. S.; Hastings, J. T.; and Madaus, G. F. *A Taxonomy of Educational Objectives: Handbook 1, The Cognitive Domain.* New York: David McKay, 1980.

Bowyer, Jane; Karplus, Robert; and Randle, Joan. *Models: Electric and Magnetic Interactions Evaluation Supplement,* Card 2. Berkeley: Regents of the University of California, 1972.

Curriculum Criteria Committee on Health. "Health Instruction Framework for California Public Schools, Preschool Through Young Adult Years," adopted by the *California State Board of Education,* 1978.

Ebel, Robert L. *Measuring Educational Achievement.* Englewood Cliffs, NJ: Prentice-Hall. 1965.

Finkelstein, Leonard, and Hammill, Donald. "A Reading Free Science Test." *The Elementary School Journal* (October 1969): 34–37.

Gronlund, Norman. *Stating Behavioral Objectives for Classroom Instruction.* New York: Macmillan, 1970.

Hedges, William D. *Testing and Evaluation in the Sciences.* Belmont, CA: Wadsworth, 1966.

Hull, William. "Things to Think About While Observing." *EES Reader.* Newton, MA: Education Development Center (1970): 153–154.

Krathwohl, David, et al. *Taxonomy of Educational Objectives: Handbook II, Affective Domain.* New York: Longman Inc, 1964.

Mehrens, William A., and Lehmann, Irvin. *Standardized Tests in Education* 3rd ed. New York: Holt, Rinehart, Winston, 1980.

National Science Teachers Association. *Standardized Science Tests.* Washington, D.C.: NSTA, 1973.

Sax, Gilbert. *Principles of Educational Measurement and Evaluation.* Belmont, CA: Wadsworth, 1974.

8

On Being a Teacher and Teaching Science

The purpose of this last chapter in part one is to focus on images of science teaching. You will reflect on the rewards and frustrations of teaching and consider the uniqueness of those who teach. After thinking about your own instructional theories, you will be presented with a self-inventory designed to assist you in discovering your own teaching strengths and, perhaps, a weakness or two as a future teacher. Finally, you will find suggestions for becoming an even better teacher of elementary, middle, or junior high school science.

KEY CONCEPTS AND IMPORTANT IDEAS
☐ effective teaching ☐ personal development ☐ professional development ☐ decision making ☐ direction and flexibility ☐ self-evaluation ☐ scientific knowledge ☐ teaching methods ☐ interpersonal relations ☐ personal enthusiasm

EFFECTIVE TEACHING

Let us begin with some situations that characterize different aspects of teaching. We ask that you respond as you think an effective teacher ought to reply.

What Would You Do?

Here comes Rodney. "Look, I solved the problem you gave me. It was easy after you got me started." Then with gleaming brown eyes and a sheepish smile, he said, "Thanks, I'm glad you're my teacher." What would you do?

1. Ask him if he has solved the other problems.
2. Say nothing, smile, and go on to the next activity.
3. Tell him that it was just part of your job.
4. Encourage him to try the next problem.
5. Give him a hug.
6. Put your arm around him and tell him you're glad he is your student.

You have just sat through one more insufferable faculty meeting. The results of the meeting: your clerical work will increase due to a new grading system; your budget has been reduced by 50 percent (and the price of materials, equipment, and textbooks has gone up 25 percent); your new duty is to supervise the cafeteria; your new class will be a "difficult, but small" group of students; and your salary increase for next year will be 3 percent below the present rate of inflation. What would you do?

1. Quit.
2. Go to the next local teachers association meeting and demand a change in contracts.
3. Ignore the situation because one-fourth of the problems will be changed in the next two weeks; one-fourth of the changes will never be implemented; one-fourth of the problems can

be ignored; and the rest are "part of the territory."
4. Make an appointment with the principal and politely but firmly inform him that he has asked too much of you.
5. Talk over the problems with several colleagues.
6. Go to a psychotherapist to see if it is you or the school system that is sick.

Frustrations and Fulfillments

These two situations summarize two ends of an emotional continuum for teachers. There are fulfillments and frustrations. For some reason more time is often devoted to discussing the frustrating aspects of science teaching than to the fulfilling aspects; yet, without a doubt the latter occur in numerous ways each day. Teaching science is a source of personal fulfillment because it contributes to the development of students and ultimately to the development of society. Occasionally we lose sight of this simple fact due to the effects of daily frustrations and dissatisfactions.

BEING A PERSON, AN EDUCATOR, AND A SCIENCE TEACHER

A review of the research literature on good, ideal, effective, or successful teaching reveals that it is impossible to identify a particular set of variables that constitute "the good teacher." Yet, we all have had good science teachers; and more importantly, we want to be good science teachers. Rather than looking at the research and concluding that we cannot identify the characteristics of the good science teacher, we can take a different view and thus form a different conclusion. Perhaps good science teachers develop their potential, their talents, and their attributes, all of which give them a unique, identifiable style of teaching. It

There are several alternatives teachers may consider to overcome frustration.

is understandable then, that our research efforts cannot find common characteristics: the answer is to be found in uniqueness, not commonality.

As one who would like to do an effective job of teaching science you must consider at least three dimensions of your task. The three dimensions relate to the fact that you are a person, an educator, and one who teaches science perhaps along with other subjects.

The Person

Young children are often surprised to find out their teacher does not live at school. Older students think it is strange to meet their teacher shopping or on a picnic because they do not often perceive their teacher as a person. As teachers, we are all, first and foremost, persons. This means we share qualities with all other persons: we struggle with decisions; we are sad and happy; we succeed and fail; we make mistakes and try to correct our errors; we are frustrated and fulfilled. Students should understand that teachers belong to the community and have hobbies, interests, and ideas that go beyond teaching.

The Educator

As educators, teachers are a part of the helping professions. They identify with people more than with objects and their goal is to help people improve their health, education, and welfare. These are aspects of science teaching that are shared with other helping professionals such as doctors, counselors, social workers, nurses, and psychologists. The shared qualities have to do primarily with **interpersonal relations;** that is, the personal dimensions of educating are those qualities that contribute to effective interactions between people. Effective interactions contribute to personal development.

The Science Teacher

In science, the teacher helps students gain a better understanding of the physical and biological world, learn the methods of gaining that knowledge, and learn the role that science plays in society. Here the qualities of the person, the educator, and the elementary, middle, or junior high school science teacher should unite to achieve the dual goals of furthering the personal development of students and the general development of society through science education.

Several years ago a popular children's book was published entitled *Jonathan Livingston Seagull*. The story is of a seagull who continually developed his potential for flying. He did this in spite of objections from parents, friends, and the "flock." Jonathan "spoke of very simple things—that it is right for a gull to fly, that freedom is the very nature of his being, that whatever stands against that freedom must be set aside, be it ritual or superstition or limitation in any form." Later, in response to cautions that he accomplished his feats only because he was special and gifted, Jonathan mentioned several other gulls who had also developed their potential and asked if they too were special. He answered his own question, "No more than you are, no more than I am. The only difference, the very only one, is that they have begun to understand what they really are and have begun to practice it" (p. 83). Jonathan Livingston worked constantly at being a better seagull; he made this task his goal. He then helped others to develop their potential, and so he became a teacher. This is worthwhile goal: to work constantly to become a better teacher.

DECISION MAKING: A CRUCIAL ASPECT OF SCIENCE TEACHING

During any single class period a science teacher makes numerous decisions about students and the lesson. Often science teachers are unaware of these decisions. Shall I tell John to be quiet? What is Pat doing? Do the students understand density? Should I use a different example to illustrate convection currents? Would it be best for the students to work in groups of two or three on this laboratory? A single decision may not seem important but they all add up to effective instruction and classroom management.

Teaching science is simultaneously directed and flexible. Decision making in science teaching is the process of synthesizing your planned direction with the moment-to-moment, spontaneous events in the classroom.

Direction

Direction in teaching science is provided in two ways: first, through the organization of textbooks, curriculum guides, lesson plans, and objectives; and second, through the science teacher's instructional theory. When situations arise in the classroom, the teacher evaluates the situation, other goals, the curriculum, and the consequences of such choices, and then decides on a course of action. One important variable that is seldom considered in the decision-making process is the direction suggested by the science teacher's own instructional theory. An **instructional theory** provides a frame of reference for decisions, gives consistency to responses, and identifies a general direction teachers can follow in different classroom situations. With the aid of an instructional theory, you are in a better position to make instruction effective and fulfilling. In the final analysis, the individual teacher is the one who must relate possible directions to actual situations when it comes to different students, classrooms, and schools.

Instructional theories evolve from the continual reexamination of learning theories and of one's personal goals and objectives. You might refer to chapters 2 and 3 to help you recall the goals and objectives presented in this text and the theories of learning and development.

In order to help you develop your instructional theory, complete the following exercise:

Developing an Instructional Theory

1. What is the aim of your science teaching? What is the broad goal you hope to achieve through your interaction with students in the science classroom?
   a. What is your primary goal?
   b. What are your secondary aims?

2. Based on the understanding of yourself, students, science, and society, what are your justifications for your goals and aims?
   a. Why are these and not other aims the focus of science education?
   b. In a broad sense, what is to be done or not done to achieve these aims?

3. What are your conclusions about what to do, and how, and when to achieve these aims?
   a. By what specific instructional methods or processes are the aims developed?
   b. What is your science curriculum?
   c. Is there any sequence in your instructional theory?

To sum, you should first: state *what* your aims are; second, justify *why* these aims are important aims; and third, explain *how* you plan to achieve these aims through your science curriculum and instruction.

An instructional theory provides an underlying direction that transcends immediate classroom problems. Going beyond the immediate situation affirms to one's ability as a teacher. Clarifying one's long term goals and intentions helps provide an organizational pattern that slowly becomes a personal style of teaching. The power of an instructional theory is in the direction it provides. There must also be another component; that is, the freedom to deviate from the direct path in response to different classroom situations. This is the spontaneity and flexibility required to teach in any field, including science.

Flexibility

Effective science teachers are able to deviate from the lesson. The degree of flexibility varies with the teacher's goals, students' needs, and environmental contingencies. The problems in the following exercise have no suggested solutions, directions, or ideas; you must resolve the incidents on your own. Based on your sense of direction and flexibility, what would you do as a science teacher in these situations?

☐ You are introducing a biology lesson on predator/prey relationships. The chameleons and crickets you ordered for the children to observe have not arrived. What would you do?

☐ You are teaching a physical science lesson on energy. The class was supposed to read the chapter on energy in the book. You suddenly realize that several of the students cannot read at the level of your textbook. What would you do?

☐ During an earth science lesson on the planets, a student informs you that he has talked with a visitor from a planet called Xerob. He proceeds to describe in detail what the visitor was like. The class is interested and starts asking the student questions. What would you do?

☐ The unit is on health; the lesson on smoking. A student raises her hand and tells about a person she knows who is eighty-five years old and very healthy. And this person has smoked two packages of cigarettes a day for over sixty years. In addition, the person drinks, eats candy, and does not watch his diet. What would you do?

When confronted with problem situations such as these, many teachers respond: "I would have to know more," "I would have to be in the situation," or "it depends, every teacher would probably do something different." This is precisely the point. The individual teacher must decide the course of action in response to the classroom situation. Most teachers, for instance, know intuitively that they are the primary source for effective education. When a teacher wonders how to answer the child's question if teaching by inquiry, there is usually an implicit answer that is often communicated nonverbally or by the tone of voice. That implicit answer is: There is no direct answer except as the individual teacher responds to the situation.

We all enter the classroom with knowledge, techniques, plans, textbooks, and curriculum materials. We combine the resources and build a helping relationship with the students. In the classroom, the helping relationship is characterized by situations requiring us to react spontaneously. We must think critically, diagnose, decide, and respond immediately to conditions present in the educational environment. The creative, insightful, and perceptive teacher does react effectively to the immediate needs and demands of students, classroom, or school.

Developing a consistent direction through instructional theory and flexibility in classroom situations will help to evolve a personal **teaching style** that can help overcome frustrations and contribute to your fulfillment as a science teacher.

Great amounts of time, money, and effort have been expended on improving the elementary, middle, and junior high school science curricula. Most science curricula have carefully structured texts and materials so the science teacher will have maximum opportunity for a good teaching experience. The contribution of new programs and textbooks has been invaluable. But, while curriculum materials are necessary and can account for some success in the classroom, they are not sufficient. They cannot be substituted for effective and successful science teaching. The teacher is still the crucial variable and the one person who must pull all the educational elements together for effective teaching.

TOWARD MORE EFFECTIVE SCIENCE TEACHING

In this section we use self-evaluation to examine some of the immediate concerns of teachers. While concentrating on improving various aspects of science teaching, the self-evaluation feedback may also contribute to your becoming a better educator and a more fulfilled person.

What are your strengths and weaknesses? How do you think you should improve your science teaching? Figure 8–1 is a **self-evaluation inventory** designed to provide answers to these questions. Even though you are not a teacher yet, the inventory may provide insights concerning some of your own characteristics as one who plans to teach science. It is based on items used in a study of teacher effectiveness (Cosgrove 1959) and later modified for the study of science teaching (Bybee 1975). Only selected items are used here.

All five categories—knowledge of science, planning and organization, teaching methods, personal relations, and enthusiasm—are important for effective science teaching. The rating system should help you to identify categories where you are strong and categories in which you would like to improve. Ideally, all categories would have ratings of 5. But really, there are few if any teachers that score that well. Since teachers are continually improving, suggestions for improving teaching within the different categories of this survey seem appropriate.

Remember, you do not have to be a bad science teacher to become a better science teacher. Science teachers want to show evi-

The following statements are descriptions of various facets of science teaching. Read each of the statements carefully. Using the scale given below, indicate how each statement presently characterizes you as a future science teacher.

5–Very characteristic of me. This is a real strength of my future teaching.
4–Frequently characteristic of me. This is a good aspect of my science teaching.
3–Sometimes characteristic of me. I should evaluate this aspect of my science teaching.
2–Seldom characteristic of me. I should improve this aspect of my science teaching.
1–Never characteristic of me. I really need to improve.

As a science teacher I:

_____ 1. Am well read in science.
_____ 2. Have a well-organized science course.
_____ 3. Adjust my teaching to the class situation.
_____ 4. Have a good rapport with my science students.
_____ 5. Enjoy teaching science to students.
_____ 6. Have a thorough knowledge of science.
_____ 7. Always plan and prepare for science class.
_____ 8. Use a variety of techniques in teaching science.
_____ 9. Recognize the unique needs of my science students.
_____10. Am enthusiastic about teaching science.
_____11. Present science concepts that are current and relevant.
_____12. Recognize the need to modify daily and unit plans.
_____13. Facilitate different types of student activities in science.
_____14. Relate well with students on the individual and group levels.

_____15. Get excited when students learn science.
_____16. Am well informed in science-related fields.
_____17. Have thought about the long-range goals of my science class.
_____18. Use different curriculum materials and instructional approaches to teach science.
_____19. Am sincere while helping my science students.
_____20. Make an extra effort to help students learn science.
_____21. Am aware of science-related social issues.
_____22. Have a continuity of course material in science.
_____23. Provide adequate opportunity for active work by science students.
_____24. Listen to student questions and ideas.
_____25. Am excited and energetic when teaching science.

Now go back and add up your responses for the groups of questions listed in the left column. Divide the total number of points by 5. The result should be a number between 1 and 5 for each of the categories listed. Refer back to the 5-point scale for your evaluation.

Questions		Average Score	Category
1, 6, 11, 16, 21	=	_____	Knowledge of science
2, 7, 12, 17, 22	=	_____	Planning and organization
3, 8, 13, 18, 23	=	_____	Teaching methods
4, 9, 14, 19, 24	=	_____	Personal relations
5, 10, 15, 20 + 25	=	_____	Enthusiasm

FIGURE 8–1

Self-evaluation Inventory

SOURCE: Don Cosgrove, "Diagnostic Rating of Teacher Performance," *Journal of Educational Psychology* 50 (1959). Copyright 1959 by the American Psychological Association. Adapted by permission of the publisher and author. Also adapted from Rodger Bybee, "The Ideal Elementary Science Teacher: Perceptions of Children, Pre-Service, and In-Service Elementary Science Teachers." Used by permission.

dence of improvement through their continued involvement in workshops, college courses, attendance at conventions, and so on. The responsibility for improving is yours; likewise, the means of developing as a science teacher is unique to your preferences, problems, and potential. Based on the categories in the self-evaluation inventory, a number of possible suggestions have been outlined below.

Scientific Knowledge

There are a number of relatively easy ways to update and keep abreast of scientific developments.

☐ Read the science content chapters of this book on Environmental Biology, Health Sciences, Earth and Space Sciences, and Physical Sciences.

☐ Read an introductory textbook in the area you feel needs improvement.

☐ Enroll in science courses at a local college or university.

☐ Contact the district or state science supervisor and see if a workshop can be organized.

☐ Subscribe to a magazine such as *Science '84* (title reflects current year)

☐ Read *Scientific American,* or purchase offprints of articles or monographs of accumulated articles on important topics.

☐ Join the National Science Teachers Association (NSTA) and read one of their journals, *Science and Children, The Science Teacher,* or *Journal of College Science Teaching.*

☐ Subscribe to and read *Science News.*

☐ Attend local, state, regional, or national conventions of scientific societies and academies of science, science teachers, and teachers.

☐ Make a point of watching television programs on scientific issues. (NOVA is a science program sponsored by the National Science Foundation and is shown on the Public Broadcasting System.)

☐ Read the science sections of weekly magazines such as *Time* or *Newsweek.*

☐ Go to local museums and planetariums.

Planning and Organization

If this is a concern, part of the problem can be addressed through better lesson planning and classroom organization.

☐ Reread chapter 5, "Planning to Teach," and other chapters on planning in various teaching methods' handbooks.

☐ Find a colleague who seems to be well-organized and ask if the two of you could spend some time planning classes together.

☐ Ask the science supervisor to look over your science program and suggest ways to improve the organization.

☐ Have a colleague observe your teaching and make suggestions concerning your lesson plans and class management.

Teaching Methods

Teachers often find it difficult to break old habits and try new teaching methods, but you can do it. Here are some suggestions.

☐ Reread chapter 4, "Taxonomy of Common Teaching Methods" and try several new methods.

☐ Look over journals such as *Science and Children, The Science Teacher,* and *The American Biology Teacher* for new approaches to science teaching.

☐ Take a professional day and observe several science teachers who use methods that you are interested in adopting.

☐ Read *Models of Teaching* by Bruce Joyce and Marsha Weil. This is an excellent book that presents different teaching methods.

☐ **Team teach** with another science teacher who uses different teaching methods.

☐ Request a visit from the science supervisor for new and different approaches to science teaching.

Interpersonal Relations

This is an area that is essential to effective science teaching, yet often neglected in the education of science teachers. (Sources may be found in the Bibliography at the end of this chapter.)

- ☐ Read *Personalizing Science Teaching* by Rodger Bybee.
- ☐ Contact the Association for Humanistic Psychology (325 North Street, San Francisco, California, 92549) and ask for information on workshops, seminars, and meetings in your region.
- ☐ Read *Reaching Out: Interpersonal Effectiveness and Self-Actualization* by David Johnson.
- ☐ Read *Teacher Effectiveness Training* by Thomas Gordon.
- ☐ Read *The Skills of Teaching: Interpersonal Skills* by Robert Carkhuff et al.
- ☐ Read *Creative Questioning and Sensitive Listening: A Self-Concept Approach* by Robert Sund and Arthur Carin.
- ☐ Read and complete the exercises in "You've Got to Reach Them to Teach Them" by Anita Simon and Claudia Gyram.
- ☐ Read *Human Relations and Your Career: A Guide to Interpersonal Skills* by David Johnson.

Personal Enthusiasm

If you lack enthusiasm for teaching science, the problem is difficult but not impossible to resolve. The reason for the difficulty is that the problem involves a personal dimension of your teaching. We can recommend an introspective route to improvement.

- ☐ Think about your original interest and excitement in science and teaching. What made you choose teaching? Now, think about what you know about yourself and teaching science. What is lacking? Where did the spark of enthusiasm go?
- ☐ List your frustrations about teaching. What are the problems you have encountered as a sci-

ence teacher? What caused you to **"burn out"** or to change your interest in science teaching?

- ☐ Attend meetings of science teachers such as NSTA. This can often inspire enthusiasm through new ideas and new colleagues.
- ☐ Form a group to improve teaching in your school. You will probably find others who share your problem and the discussion and support can certainly assist you to develop new enthusiasm for teaching.
- ☐ Take a professional day and visit other science teachers who are dynamic and enthusiastic.

BECOMING A BETTER SCIENCE TEACHER

Science teachers want to improve and science teachers can improve. Remember that you do not have to be a poor science teacher to become a better one. Often science teachers complain, "I just didn't seem to be effective today; the students seemed confused and frustrated. I didn't get the concept across." Teachers always *want* to improve. They never finish the statement with, "And if you think I was bad today, wait until tomorrow. I'll really be much worse!" Unfortunately, just the desire to improve is not enough; there must be a commitment to change by altering your personal teaching style, and, a plan of action to put the commitment into better teaching.

Educational critics point out the wrongs of education but say little to give it new direction. One result of critical confrontation has been defensiveness by many teachers and administrators. Displacing blame is one manifestation of this defensiveness. The *let's blame somebody or something else* syndrome prompts excuses: "We would if we had money." "I have thirty-five children in my class." "Well, what do you expect? The home has more influence than the short time students are in my classroom." The *let's try*

The intrinsic rewards of teaching increase steadily as your teaching skills increase. (Strix Pix)

something new syndrome is another manifestation of this defensiveness. **Teaching machines, contract performance, accountability, competency-based programs,** and the **open classroom** are all attempts to find new solutions. Here we focus instead on the science teacher developing as a person and as an educator. It might be termed the *let's get with it, people, and perform as professionals* syndrome. You can be in any classroom, with any children, and with any curriculum and still develop your *competency* as a science teacher. These proposals are not carried out easily; they deal, for the most part, with personal improvement and fulfillment. Look at your potential and not your limitations. Becoming aware of the importance of the sug-

gestions and translating them into actual practice can result in your development as a science teacher. The process of becoming a better science teacher is long and it takes courage to overcome the many small barriers to personal and professional growth. It is not something that can occur through purchasing a new set of science materials, or participating in a single workshop, or reading a single book. Becoming a better teacher is a continuous, personal project aimed toward the ideal of being a great educator. And for those who say, "I can't be a great science teacher," we reply, "If not you, then who?" We all have much more ability than we use and that will help you to become a better science teacher, educator, and person.

There is a passage from a children's story written in 1922, *The Velveteen Rabbit,* that certainly captures and exemplifies the process of becoming a better science teacher.

. . . What is REAL?" asked the Rabbit one day . . . "Does it mean having things that buzz inside you and a stick-out handle?"

"Real isn't how you are made," said the Skin Horse. "It's a thing that happens to you. When a child loves you for a long, long time not just to play with, but REALLY loves you, then you become REAL."

"Does it hurt?" asked the Rabbit.

"Sometimes," said the Skin Horse, for he was always truthful. "When you are REAL you don't mind being hurt."

"Does it happen all at once, like being wound up," he asked, "or bit by bit?"

"It doesn't happen all at once," said the Skin Horse. "You become. It takes a long time. That's why it doesn't often happen to people who break easily, or have sharp edges, or who have to be carefully kept. Generally, by the time you are REAL, most of your hair has been loved off, and your eyes drop out and you get loose in the joints and very shabby. But these things don't matter at all, because once you are REAL you can't be ugly except to people who don't understand."

Excerpt from *The Velveteen Rabbit* by Margery Williams. Reprinted by permission of Doubleday & Company, Inc.

COMMENTARY

Becoming a better science teacher is our goal for you. In part one you captured a sense of the excitement about science and the relationship between science, society and teaching. You were challenged to think about your own goals and objectives for teaching science, and you have seen how theories about motivation, learning, and human development are related to teaching about science. You have been given an overview of the rich repertoire of teaching methods and learned the essentials of good planning and classroom management. Part one has closed with suggestions for your own lifelong growth and development as a teacher and a science teacher. As you reflect on part one, recall that it has focused on how to teach science. Part two will focus on what science to teach.

BIBLIOGRAPHY

Bach, Richard. *Jonathan Livingston Seagull.* New York: Macmillan, 1970.

Bybee, Rodger W. *Personalizing Science Teaching.* Washington, D.C.: National Science Teachers Association, 1974.

Bybee, Rodger. "The Teacher I Like Best: Perceptions of Advantaged, Average, and Disadvantaged Science Students." *School Science and Mathematics.* LXIII (May 1973): 384–390.

Bybee, Rodger. "The Ideal Elementary Science Teacher: Perceptions of Children, Pre-Service, and In-Service Elementary Science Teachers." *School Science and Mathematics* LXXV (March 1975): 229–235.

Carkhuff, Robert; Berenson, David; and Pierce, Richard. *The Skills of Teaching: Interpersonal Skills.* Amherst, MA: Human Resources Development Press, 1977.

Combs, Arthur; Avila, Donald; and Purkey, William. *Helping Relationships: Basic Concepts for the Helping Professions.* Boston: Allyn & Bacon, 1979.

Cosgrove, Don. "Diagnostic Rating of Teacher Performance." *Journal of Educational Psychology* 50 (1959): 200–204.

Gordon, Thomas. *Teacher Effectiveness Training.* New York: David McKay, 1974.

Johnson, David. *Human Relations and Your Career: A Guide to Interpersonal Skills.* Englewood Cliffs, NJ: Prentice-Hall, 1978.

Johnson, David. *Reaching Out: Interpersonal Effectiveness and Self-Actualization.* Englewood Cliffs, NJ: Prentice-Hall, 1972.

Jourard, Sidney. *Healthy Personality: An Approach from the Viewpoint of Humanistic Psychology.* New York: Macmillan, 1974.

Joyce, Bruce, and Weil, Marsha. *Models of Teaching.* Englewood Cliffs, NJ: Prentice-Hall, 1972.

Kinget, Marian G. *On Being Human.* New York: Harcourt Brace Jovanovich, 1975.

Simon, Anita, and Gyram, Claudia. *You've Got to Reach 'em to Teach 'em.* Dallas: TA Press, 1977.

Sund, Robert, and Bybee, Rodger. *Becoming a Better Elementary Science Teacher.* Columbus, Ohio: Charles E. Merrill Publishing Company, 1973.

Sund, Robert, and Carin, Arthur. *Creative Questioning and Sensitive Listening: A Self-Concept Approach.* Columbus, Ohio: Charles E. Merrill Publishing Company, 1978.

TWO

Science Content and Activities

Part two of *Science and Society* provides information about science and contains science activities for children and adolescents. It is designed to help enrich your background in science, and at the same time, show how specific science content can be taught in primary, intermediate, and middle, or junior high school classes. Chapters 9, 11, 13, and 15 provide up-to-date, factual knowledge in four areas of science: Environmental Biology, Health and Human Physiology, Earth and Planetary Science, and Physical Science. We have tried to present this information about science in a lively manner for those who feel the need for a stronger science background. Even-numbered chapters offer planning charts and lesson plans which illustrate how specific science concepts are related to the personal experiences of students. Each science lesson plan, preceded by a mini-science lesson for the teacher, provides a plan for integrating science content with teaching objectives and methods. Throughout part two, in both the science content chapters and the science activity chapters, you will find the focus on *Science and Society*.

9

Environmental Biology

The purposes of this chapter are to present some basic topics in biology and the environmental sciences and to provide you with a broader basis for teaching and exploring the special ecological problems facing our world today. This information is offered in the hope that you and your students will then be better able to make decisions about science-related issues that affect the survival of our society. Increasingly, the media educates society about ecological problems and possible solutions. Science teachers can expect many questions from their students about these critical issues.

KEY CONCEPTS AND IMPORTANT IDEAS

☐ environmental biology ☐ diversity of life ☐ classification ☐ kingdoms of living things ☐ growth, development, and change ☐ internal and external mechanisms ☐ biological clocks ☐ the importance of the sun and water ☐ photosynthesis ☐ food chains ☐ ecosystems ☐ limiting factors ☐ habitats ☐ niche ☐ population ☐ natural cycles: water, oxgen-carbon dioxide, nitrogen ☐ community relationships ☐ succession ☐ biosphere ☐ biomes of the world ☐ human impact on the environment ☐ energy resources and reserves ☐ pollution ☐ personal responsibility for decisions about the environment

For thousands of years our ancestors lived in harmony with nature. Life was hard—a constant search for food and a constant struggle for survival—but survival was possible for those who adapted to the environment in which they found themselves. Our ancestors learned, for instance, which parts of nature they could use; they learned which roots and berries could be eaten and where to find water. They learned how to hunt, trap, and fish in order to provide themselves with clothing, food, and bedding. They learned how to predict the changing of the seasons and how those changes would affect their food supplies and needs for shelter. So, from our earliest history we have developed a knowledge of **ecology** and our **environment** in order to survive.

ECOLOGY The study of the relationships between living things and their environment

ENVIRONMENT All the conditions and influences that surround the development of living things

As our knowledge grew, we learned not only to understand the environment but to control some aspects of it. Plants could be grown where we wanted them; animals could be domesticated; and rivers could be dammed. As we began to control the environment, life became easier and the human species became more successful, that is, more populous.

Today, as the world's population grows larger and survival seems more complex, it may appear that we can set ourselves apart from the environment. But that can never happen because we are part of the environment. Everything we do affects the natural environment and it, in turn, affects us. When we forget that the limits of nature apply to us, we get into trouble. Resources become limited. Our waste products pollute the land, water, and air. Our cities and farmlands replace and destroy the **habitats** of plants and animals. If we continue to exploit nature, we may eventually destroy our environment.

HABITAT The place where a plant or animal lives, finds its food and shelter, and raises its young

To avoid this disastrous possibility, some people suggest that we give up our modern technological attempts to control nature and go back to a more natural lifestyle. Others say that seems impractical and probably impossible in today's world. Still others offer a more useful suggestion: we need to take the example of our earliest ancestors—to learn as much as we can about the limits of nature and try to live within nature's limits. Nature is much more complex than our ancestors thought it was. We now know that all living things can affect all other living things in the environment, and that the environment, in turn, affects all living things. The study of the relationships among living things and their relationships with the environment is called ecology. It deals with the dynamic interaction of a variety of the fields in environmental biology, among them **botany, zoology, microbiology,** and **genetics.** Ernst Haeckel (1834–1919), a German zoologist, coined the term *ecology* more than 100 years ago from the Greek words *oikos,* which means home or place to live, and *logos,* which means science. Thus, the emphasis in ecology is on the study of the places in which organisms live—including humans.

BOTANY The study of plants, their life cycles and processes, structure, and classification

ZOOLOGY The study of animals, their life cycles and processes, structure, and classification

MICROBIOLOGY The study of microscopic living things such as bacteria, protozoa, and viruses

GENETICS The study of heredity and variation in all living things

In this chapter, you will learn about (1) the diversity of living organisms; (2) the environment; and (3) the impact that our actions can have on the environment, particularly on the earth's natural resources.

Chemical Waste: How Should We Use the Land?

We first heard about Love Canal in the late 1970s. But the story began in the 1930s when a chemical company loaded its chemical waste into metal drums and buried them in an abandoned hydroelectric canal. In the 1950s the chemical company deeded some of the land to the Niagara, New York School Board. The rest of the land was sold to developers who built and sold modest homes. Love Canal has become a symbol for the question—how should we use the land?

Over the period of forty years water eventually penetrated the landfill at Love Canal and subsequently the chemicals infiltrated the water system. By 1976, residents in the area began noticing a strong odor and oozy slime in the neighborhood. Children with birth defects were born to four families on one block and adults complained of headaches, difficulty breathing, and irritated eyes. Upon investigation a host of poisonous chemicals were found: pesticides capable of damaging human organs, cancer-causing chemicals such as benzene, and a substance named dioxin which has been called the most poisonous substance made by humans. Other chemicals that can cause damage to the brain and central nervous system were also found. In all, over 200 different chemicals had been dumped into Love Canal.

The New York State Health Department investigated the situation at Love Canal. Their worst fears were realized. The area near Love Canal was declared an extremely serious threat to the health of those living near the chemical waste. Approximately $10 million has been spent on relocating families endangered by the chemicals. And $12 million is the estimated cost of cleaning the area.

About the time of Love Canal a series of chemical spills occurred in North Carolina. Upon investigation the substance was the highly toxic, long-lived polychlorinated biphenyl, commonly referred to as PCB.

PCB concentrations in the soil were as high as 7,000 parts per million. For comparison it should be noted that when Congress passed the Toxic Substance Control Act in 1976, it took seriously the detrimental results of PCBs. Congress gave the Environmental Protection Agency the power to ban the manufacture, distribution, and use of this toxic substance. A level of fifty parts per million of PCBs in any substance is due cause for action. The North Carolina PCB spill was the largest in American history.

Later we heard of Jacksonville, Arkansas, the home of a chemical company that produced 2,4,5-T, the herbicide with the deadly poisonous by-product—dioxin. By 1980 the EPA had estimated that there were 30,000 to 50,000 sites in the United States where hazardous wastes had been dumped. About 1,500 posed threats to human health.

For some time many have disposed of chemical waste with an "out of sight–out of mind" approach. Seems simple, but it doesn't work. Suppose a drum of some toxic chemical such as PCB or dioxin is placed in the ground. With time, say fifteen years, the drum finally corrodes, releasing the chemicals into the surrounding soil. With time the chemicals percolate through the soil and enter groundwater and local streams. At this point humans can ingest the chemicals in their water, through direct contact with the hazardous waste or through eating organisms such as fish in which the chemicals have accumulated. One problem with many chemicals that have entered the food chain is that they accumulate, so initial low concentrations become high concentrations and increase their detrimental effects. The accumulation occurs in the fatty tissue of humans. Eventually the chemicals can cause some of the medical problems mentioned earlier.

In many ways we have had better living through chemicals. Pesticides have killed undesirable organisms, herbicides have killed undesirable plants. We have higher crop yields because of chemicals. But we still have the problem of chemical waste disposal in relation to the question of land use.

Questions for Discussion

1. What are the benefits of toxic chemicals to humans?
2. Do the benefits outweigh the risks?
3. Who should pay for the clean up of toxic wastes—the government? The chemical industry?
4. What would you propose as a land use ethic regarding the disposal of wastes?

DIVERSITY OF LIFE

In this section we learn about the classification of living organisms, their modes of growth and change, their external and internal mechanisms of growth and change, and how these mechanisms interact with one another.

Classification

Because there are so many different organisms, scientists have found it necessary to develop systems for classifying them. These systems, called *taxonomies,* reflect the similarities and differences among living things and relationships among groups of living things. Table 9–1 is an example of a system by which any living entity can be classified; this table shows how to describe humans by this classification system.

Kingdoms

Today biologists classify and divide the living world into five kingdoms: *Monera, Protista, Fungi, Plantae* (or plants), and *Animalia* (or animals). (See figure 9–1.)

Monera.   Monerans are one-celled organisms. They are thought to be the oldest, simplest, and most abundant group of living things in our world; they consist of bacteria, viruses, and a blue-green algae.

Blue-green algae are the simplest, and most widespread, oxygen-producing organisms. They are tiny, grow for the most part in fresh water, and cause what appears to be a dark-green scum on the surface of the water. They are classified as Monera because, like bacteria, they contain no nucleus in the cell and because they are single-celled organisms (*mono* means one). Blue-green algae are found in clusters but they have no complex organization, nor do they reproduce sexually. Blue-green algae share a common characteristic with all plants: they contain the green pigment, **chlorophyll.** Chlorophyll is a vital plant constituent that enables the organism to absorb the sun's light energy and change it into chemical energy used for the plant's or algae's growth. This process is called **photosynthesis.**

CHLOROPHYLL The green substance in plants that enables them to use the energy of sunlight to convert carbon dioxide and water into carbohydrates

PHOTOSYNTHESIS The process by which plants convert the sun's energy into chemical energy—or into food for the plant

Protista.   Protistans are somewhat more complex organisms. They can be either single-celled or multi-celled, and they include all other algae, as well as slime molds and protozoa. When you go to the beach and find a piece of seaweed washed up on the sand, you are looking at a species from the Protista kingdom.

Fungi.   Fungi, the third kingdom, include such common organisms as mushrooms, toadstools, and yeast—the latter being that minute organism so important to bread making and the fermentation process that produces beer and wine.

Many plant diseases, such as smut, rust, and wilt, are the result of fungus infections. Fungi also cause diseases in people. However, not all fungi are harmful; some are useful. Fungi help to decompose the dead parts of animals and plants. They free carbon dioxide and other carbon

TABLE 9–1
The biological classification of human beings

Level	Classification	Basis for Classification	Examples
		All living things	Includes bacteria, daisies, earthworms, frogs, cats, monkeys, gorillas, humans
Kingdom	Animalia	Responsive and mobile living things that have a definite adult form and do not undergo photosynthesis	Includes earthworms, frogs, cats, monkeys, gorillas, and humans, but not bacteria or daisies
Phylum	Chordata	Living things that develop a notochord (primitive backbone) at some stage in their lives	Includes frogs, cats, monkeys, gorillas, and humans, but not earthworms
Class	Mammalia	Living things that have improved respiration and circulation, and temperature control mechanisms; upright limbs, highly developed nervous system, and complex behavior paterns: females develop a placenta and mammary glands after puberty	Includes cats, monkeys, gorillas, and humans, but not frogs
Order	Primate	Living things that can coordinate hands and eyes	Includes monkeys, gorillas, and humans, but not cats
Family	Hominidae	Living things with a human-like form	Includes gorillas and humans, but not monkeys
Genus	Homo	Living things with an upright posture	Includes humans, but no other living member of this genus
Species	Homo sapiens	Living things with a larger brain size and higher intelligence level	Includes human beings

ORGANIC Matter composing or derived from living things

HUMUS The organic part of the soil that results from the partial decay of leaves and other living matter

ANTIBIOTICS Chemical substances that inhibit the growth of microorganisms produced by simple organisms

components from the skeletons of leaves, branches, and tree trunks. They are important ingredients in forming the rich **organic humus** needed for plant growth. Yeasts ferment sugar into alcohol. They also produce such **antibiotics** as penicillin and aureomycin as well as vitamins and citric acid.

Plantae. Plants are the fourth kingdom of living things. They are usually divided into two subgroups or divisions: the **bryophytes** and **tracheophytes.** Bryophytes (mosses) are plants that have no true leaves, roots, or stems. Tracheophytes are **vascular** plants. They have distinct

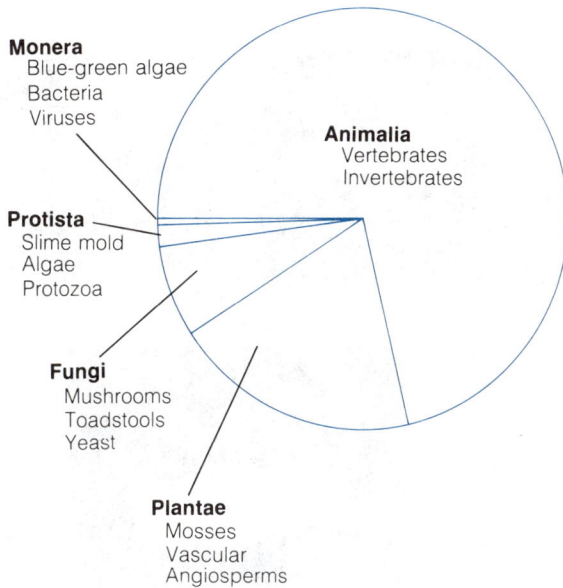

Monera
  Blue-green algae
  Bacteria
  Viruses

Animalia
  Vertebrates
  Invertebrates

Protista
  Slime mold
  Algae
  Protozoa

Fungi
  Mushrooms
  Toadstools
  Yeast

Plantae
  Mosses
  Vascular
  Angiosperms

FIGURE 9–1
The Animalia Kingdom exceeds all other kingdoms in the number of species. The Plantae Kingdom and the Animalia Kingdom both contain a great diversity of species.

systems for conducting food and water throughout the entire plant and they have **leaves, stems,** and **roots** or root-like structures. Vascular plants can be further divided into three groups. The first of these groups includes the fern. The second group, the **gymnosperms,** bear their seeds without cover and include such familiar plants as fir and spruce trees. The third group, the **angiosperms,** are those flowering plants that have covered seeds. The angiosperms are in turn composed of two groups, the **dicots** and the **monocots.**

Vascular plants, which are much more complex than the simpler organisms such as algae, fungi, and lichens, are so named because of their internal circulatory systems. They may also reproduce by **spores** or seeds. Ferns are examples of plants that reproduce by spores.

Seed-bearing plants also range from the very simple to the very complex. The simplest kinds of seed plants are the **conifers,** such as firs, pines, junipers, and cedars. They are a widespread plant on earth, constituting about one-third of all the plants in existing forest areas in the world.

The most complicated form of seed-bearing plant is the flowering plant or the flowering tree. **Fertilization** or **pollination** in this group of plants occurs through the help of insects. A bee moving from blossom to blossom is much more efficient at transferring **pollen** than the wind.

BRYOPHYTES Plants without true leaves, such as mosses and liverworts

TRACHEOPHYTES Vascular plants

VASCULAR Having a system of ducts or tubes for carrying body fluids, such as sap

LEAF The plant part in which food is produced

STEM The plant part that connects the roots and the leaves

ROOT The plant part, usually below the ground, that holds the plant in position, draws water and nutrients from the soil, and stores food

GYMNOSPERM Plants that produce seeds that are not enclosed in a seed case

Kingdom Monera

Kingdom Protista

Kingdom Fungi

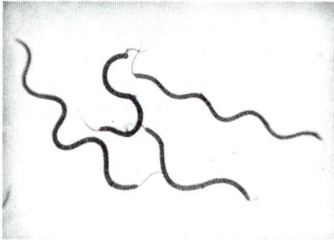
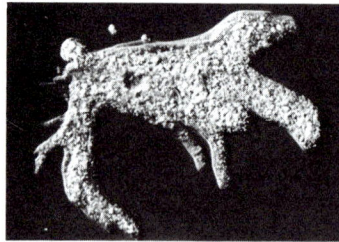

Kingdom Plantae

Kingdom Animalia

From time to time biologists reorganize the classification of living things to incorporate new information and discoveries. Once all living things were divided into two kingdoms: Plants and Animals. Today there are five kingdoms: Monera, Protista, Fungi, Plantae, and Animalia. (Carolina Biological Supply Company)

ANGIOSPERM Plants that produce seeds encased in a covering or ovary

DICOT Plants that produce embryos with two cotyledons (embryonic leaves)

MONOCOT Plants that produce embryos with one cotyledon

SPORES Small reproductive bodies that are able to resist adverse circumstances and can generate a new adult individual

CONIFERS Plants that carry their seeds in cones, such as a pine or fir tree

Thus the amount of pollen produced by flowering plants tends to be greater than that produced by gymnosperms.

Animalia. The fifth kingdom, animals, is enormously varied. Many scarcely look like animals, as one ordinarily thinks of them. Some animals consist of little more than a digestive tract; others pump water rather than blood through their bodies; others lack heads completely. Organisms in this kingdom are mobile during some portion of their lifetime. They respond readily to stimuli, feed on other living things or on dead organic matter, and are multicellular.

There are several major characteristics of animals that are used to classify members of the animal kingdom. The major criterion used for classification is whether the animal has a backbone or not; this divides the animal kingdom into **vertebrates** and **invertebrates.** (figure 9–2) Some of the other characteristics that are used are symmetry, the type of digestive system, the presence or absence of a body cavity, the type of cir-

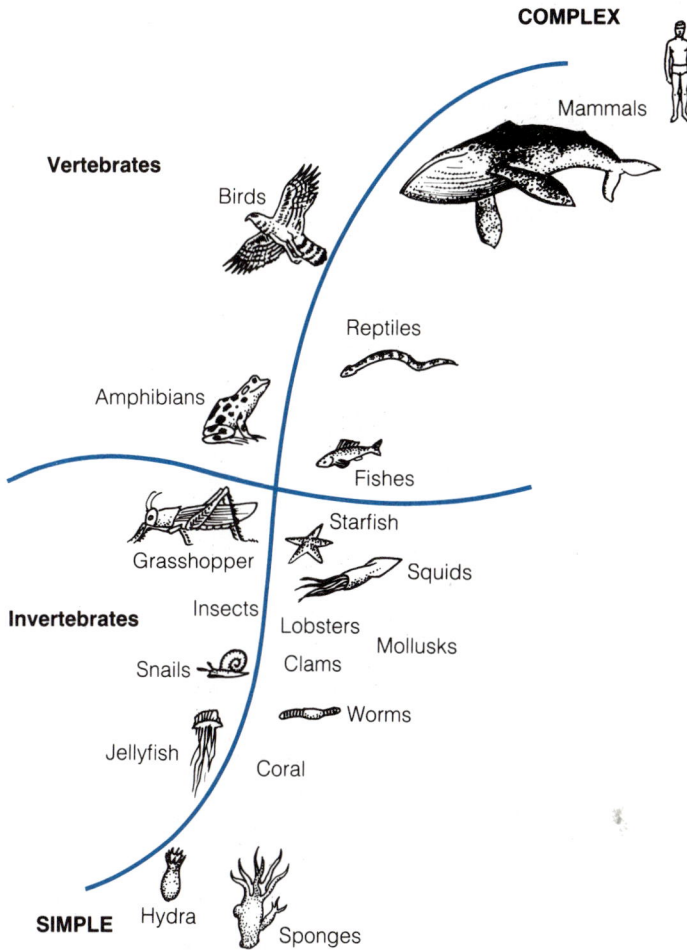

COMPLEX

Mammals

Vertebrates

Birds

Reptiles

Amphibians

Fishes

Starfish

Grasshopper

Squids

Insects

Lobsters

Invertebrates

Mollusks

Snails

Clams

Worms

Jellyfish

Coral

SIMPLE   Hydra

Sponges

FIGURE 9–2
Scientists identify or group animals by specific characteristics. The more complex animals have specialized body parts for specific body functions.

culatory system, and segmentation. Each of these characteristics is discussed briefly.

Symmetry.   **Symmetry** refers to the form of the animal. Some animals have *radial symmetry,* or an equal development of their physical structures around a central or vertical axis. These animals do not move about and gravity seems to play a role in their development. Other animals have *bilateral symmetry,* which is usually associated with organisms that do move around. These animals often have organs, such as eyes and ears, that are paired and placed on either side of a central axis.

FERTILIZATION The combination of female and male sex cells which ultimately produces a new individual

POLLINATION The transfer of pollen from the male part of the plant to the female part

POLLEN The yellow powder-like male sex cells produced by plants

VERTEBRATE Animal possessing a backbone

INVERTEBRATE Animal with no backbone

SYMMETRY Matching form or arrangement in an organism

DIGESTIVE SYSTEM The means by which animals feed, absorb nutrients, and expel waste materials

Digestive System. Some animals have a **digestive system** with only one opening. This is characteristic of stationary animals that feed and expel their waste matter through the same opening. However, other animals have two openings that permit specialization as well as continuous feeding and digesting on an assembly-line basis.

Body Cavity. Some simple organisms have no body cavity. The organs themselves are embedded in an inflexible tissue; this seems to be associated with animals that do not move about very much. However, many animals do have a body cavity in which the organs are either suspended in liquid or are attached by flexible ligaments to help the organs absorb the shock of rapid movement.

CIRCULATORY SYSTEM The means by which animals move nutrients, chemical messages, and waste products through their bodies

SEGMENTATION The ability to move one area of the body independently of the others

Circulatory System. Some animals circulate their nutrients and waste products through their bodies by diffusion; others have a vascular or tube-like system for this purpose. In circulation by *diffusion,* a slow progression of material from cell to cell enables very small animals to live with low energy demands. In more complex animals, a vascular **circulatory system** carries oxygen and nutrients through tubes all over the body regardless of the size of the animal.

Segmentation. A last useful characteristic for classifying animals is **segmentation,** which simply means that there are areas of the body that move separately from one another. Worms, insects, fish, frogs, snakes, and mammals (including humans) are examples of segmented animals. In simple animals, the body is uniform throughout; sponges and flatworms are examples of nonsegmented animals.

Major Groups of Animals

Scientists use twenty-five *phyla,* or groups, to describe animals; these animals are classified according to their similarities and differences.

Sponges (phylum Porifera), for instance, are the most simple and primitive of all multicellular animals. They live in the seas, mostly in shallow waters, and in fresh waters. Sponges come in many shapes and sizes, but all are constructed similarly. They have stiff skeletal structures made of interlocking rods or fibers covered with layers of thin, slimy, flat cells. Numerous holes in the sponge lead into a central cavity lined with special cells. These cells have tiny, whiplike structures that move and create a current that draws water into and through the sponge. The cells ingest food particles such as bacteria and other tiny organisms, and absorb oxygen from the water. The sponge simply sits there, waving its tiny arms, and food and oxygen come to it! Reproduction, too, is simple for sponges. Special cells inside the sponge release eggs that swim about until they find a place where they can attach themselves and grow into mature sponges. Sponges also reproduce by budding, in which a piece of the sponge breaks off and establishes itself as a separate individual. Humans used to be one of the sponge's worst enemies. Sponges were collected by the thousands for many years and sold for use in the bath and kitchen. Now cheap, synthetic sponges are available and there is little call for natural sponges. Except for a few types of fungi, the sponge has few known natural enemies.

More complex creatures, such as *worms* (phylum Annelida), have somewhat more complex anatomies and play important roles in the ecosystem. More than 9,000 types of segmented worms are found around the world in habitats that include the seas, fresh water, and the soil. These animals are constructed as a tube within a tube; the inner tube is a long digestive tract. Worms usually have a head, followed by a series of segments, and an anus. Unlike sponges, worms are important members of **food chains** in almost all ecosystems. Some worms, such as leeches, are parasites and live on or in other animals. Other worms are eaten by insects, fish, birds, and even humans. Worms also contribute to the cycling of essential materials for plant growth. They break up pieces of plant debris in their guts so that these bits of **detritus** are decomposed.

FOOD CHAIN A series of organisms dependent on each other for food

DETRITUS Nonliving particulate and dissolved organic matter in the ecosystem

Higher up in the animal kingdom are the *insects* (phylum Arthropoda). There are more than 1,100,000 species in the animal kingdom. About 900,000 of them are insects, and these account for almost 80 percent of all animal life. Insects have segmented bodies, jointed legs, and well-developed nervous systems. Most, for instance, have elaborate systems of sensory organs and can see, hear, feel, and smell extraordinarily well. Reproduction is sexual, with most insects beginning their lives as fertilized eggs. Some hatch from their eggs as miniature adults, while others (like butterflies and moths) must first go through larval (caterpillar) and pupal (cocoon) stages before emerging as sexually mature, winged adults. We tend to think of insects as pests, crop eaters, and disease carriers (mosquitoes, locusts, and bedbugs), but many are beneficial to humans. They pollinate plants, produce useful substances like honey, control other pest insects, and serve as important links in the food chain. Insects eat plants and waste materials and turn them into protein useful to other animals.

The most complex organisms of the animal kingdom are the *vertebrates*, (phylum Chordata), characterized by their spinal columns. Vertebrates include fish, frogs, snakes, birds, and mammals.

LIVING THINGS GROW, DEVELOP, AND CHANGE

All living things grow and develop. Organisms grow simply by increasing in size. However, living things do more than just get larger; they produce

DEVELOPMENT The process by which living things change in form

GROWTH The process by which living things increase in size

entirely new structures such as horns, hair, flowers, and so forth. These changes are arranged to produce the kind of organism determined by its genetic makeup. Structural change, then, is called **development.** Increased size of the organism is called **growth.**

Internal Mechanisms

In all living things, growth and development are the result of several factors interacting together. Some of these factors originate within the organism (internal) and others occur outside (external). Within organisms, every cell in the body carries coded information, which dictates what new cells will look like and how they will function. This coded information, called DNA (deoxyribonucleic acid), is chemical in nature. DNA, for example, ensures that skin cells will continue to produce other skin cells rather than bone or muscle cells, and that leaf cells will replicate leaf cells rather than stem, root, or flower cells. This genetic factor is one of the most powerful mechanisms affecting growth in plants and animals.

Hormones are a second powerful factor controlling growth and development. Plants and animals produce a variety of hormones which, when released, provide signals for change in organisms. Some hormones trigger the production of hair; others signal the development of eggs; still others promote seed production.

One of the most active areas of scientific research today deals with the manipulation of genetic codes (called genetic engineering) and the production of synthetic hormones. By altering these two powerful internal factors which govern growth and development, scientists hope to produce variations of plants and animals which have more favorable characteristics for survival or are in other ways more valuable to us.

External Mechanisms

Neither genetic codes nor hormones account for all of the growth or development of organisms, however. Environmental factors such as the amount of light, rain, soil nutrients, or radiation can affect plant growth and development just as the amount of available food or other vital factors can affect the growth and development of animals. If genetic information and hormones permit a species of tree to grow to a height of seventy feet, but the environment lacks all of the conditions necessary for an individual tree to reach its inherited potential, it may remain stunted or fail to produce its fruit.

In many plants and animals, growth and development are relatively slow processes. To the observer, a tree, a flower, or a cow may look very much the same from hour to hour. However, changes of another kind can have observable cycles or rhythms.

Biological Clocks

Feeding and migration in animals have been shown to be regulated by certain internal mechanisms that we call biological clocks. Scientists are now studying how these clocks work to alter the way living things adjust to changes in the environment.

Biological clocks, for example, allow fiddler crabs to grow darker and lighter during different times of the day. The fiddler crab has a specific, daily cycle of color changes—dark during the daytime, pale in the evening, and dark again toward daybreak. The darker daytime color helps to conceal the crab from its predators. The greatest darkening is seen to occur about fifty minutes later each day. This delay in darkening corresponds to the low tide schedule where the crabs live. The time of maximum paleness corresponds with high tide. It could be that these crabs use the water level as a cue for changing colors, but scientists have discovered that fiddler crabs kept in the laboratory show the same daily delay in the onset of maximum darkness as do crabs in their natural environment. Even when laboratory crabs are kept in darkness for as long as a month, they seem to persist in their natural rhythm of color changes. This suggests that crabs do not use the sun as a cue for when to change color. The mechanism of biological clocks holds promise for scientists studying the behaviors of many other animals as well.

THE ENVIRONMENT

The environment is all of the external conditions that affect the life of an individual organism. Within this very broad definition, scientists have needed to classify different types of environments. Biologists and ecologists see five levels of organization that are meaningful to our study of life on earth: organisms, populations, communities, ecosystems, and the biosphere. A group of individual organisms of the same species such as rabbits or dandelions is called a **population.** Many populations of different species of plants and animals living in a specific place, such as a field, is called a **community.** Where each individual organism lives is its habitat. The larger environment of the community is called the ecological system, or the **ecosystem.** All of the ecosystems on earth, interacting with one another, form the earth's **biosphere.** All life exists in the air, water, and land.

Our environment has a special relationship with the sun; we will look at some ways that organisms use the sun's energy. We will examine water and food as major environmental components. Then we will turn to the characteristics of ecosystems and to a discussion of the biosphere and the ways that geography affects the earth's biosphere.

POPULATION A group of individual organisms of the same species

COMMUNITY A group of plant and animal populations living and interacting in a given locality

ECOSYSTEM The system of interaction between living organisms and the environment

BIOSPHERE The total of all the various ecosystems on the planet along with their interactions

© 1951 by Edwin Gray

Rachel Carson (1907–1964)

Since the early 1960s we have developed a new awareness of our place in the natural world. One person and one book symbolize this new awareness and the environmental movement. Rachel Carson is the person and *Silent Spring* (1962) is the book.

Rachel Carson was born in Springdale, Pennsylvania on May 27, 1907 and raised in this community and Parnassus, another local town. Following high school, she entered Pennsylvania College for Women at Pittsburgh. Her original intention was to major in communications and become a writer. However, a course in biology caused Rachel Carson to change her major to biology. She received her baccalaureate degree in 1929 and began graduate study at Johns Hopkins University. She received her master's degree in 1932. Her graduate education was extended by work at the Marine Biological Laboratory in Woods Hole, Massachusetts.

In 1936, Carson accepted a job as an aquatic biologist with the United States Bureau of Fisheries, later to become the United States Fish and Wildlife Service. At this point in Rachel Carson's life, her dual interests in biology and writing came together and developed around themes related to the environment. She became editor and chief of the Bureau in 1947 and

The Sun—Our Ultimate Energy Source

The single most important factor in maintaining life on this planet is the energy of the sun. That energy is captured by green plants through the process of photosynthesis. Green plants supply us with oxygen, and they convert carbon dioxide and water into sugar, which in turn ultimately provides food for all living creatures here on earth. In addition, all our fossil fuels, such as coal and oil, are the result of photosynthesis that occurred millions of years ago.

Photosynthesis. The world's green plants produce 150 billion tons of sugar every year. Contrast this with the world's average yearly output of steel or 325 million tons of concrete. This sugar is the sole source of

continued in that position until 1952 when she resigned to devote all her time and talents to writing.

In 1941 her first book, titled *Under the Sea-Wind*, was published. Though the book had good reviews, it was relatively unnoticed by the public. It was late in the 1940s before Carson began her second book. *The Sea Around Us* was published in 1951 and was widely acclaimed. Carson followed this book with *The Edge of the Sea* (1955). All the reviews of Rachel Carson's work support the talent she had for expressing a sensitivity toward nature while presenting scientific information in an interesting manner.

Silent Spring, published in 1962, directed attention to the negative effects of pesticides used to control weeds. Carson warned that the indiscriminate use of chemicals could still the songs of birds, and the leaping of fish, and linger on the soil. Society could soon experience a silent spring. The book caused intense debate. In many cases, Rachel Carson had gone beyond the available evidence and suggested effects for which she had little proof. Predictably, industry attacked the book while defending their products. The government began investigations of DDT and other pesticides. *Silent Spring* and Rachel Carson had the power to gain national attention and make what was her basic conviction ours. During a Congressional hearing she stated: "I deeply believe that we in this generation must come to terms with nature."

The legacy of Rachel Carson continues beyond her early death on April 14, 1964 at the age of fifty-six. For those who supported the environmental movement she became a symbol of sanity and the spokeswoman of their cause. For those who doubted the harmful effects of pesticides, she became a very respected adversary. Battles have been fought and won on both sides of numerous environmental issues. In the process, our awareness and values have been altered in the two decades since Rachel Carson's important book.

energy that organisms have to do the work of adapting, reproducing, and maintaining life. The discovery of photosynthesis goes back two hundred years. At that time people knew that if an animal was kept in a closed jar it would suffocate but they did not understand why. It was not known that air is composed of different gases and that some of these gases are needed for life. In 1775, Joseph Priestly (1733–1804) discovered oxygen and found that animals use up oxygen and replace it with carbon dioxide. He also demonstrated that plants could reverse this process. This happens by the process of photosynthesis (figure 9–3).

Photosynthesis takes place in the **chloroplasts** of plant cells, where chlorophyll absorbs light from the sun. The two other substances that are necessary for photosynthesis are water, which comes up through

CHLOROPLAST The part of the plant cell that contains chlorophyll

FIGURE 9–3
Light energy, along with carbon dioxide and water, is necessary for photosynthesis to take place.

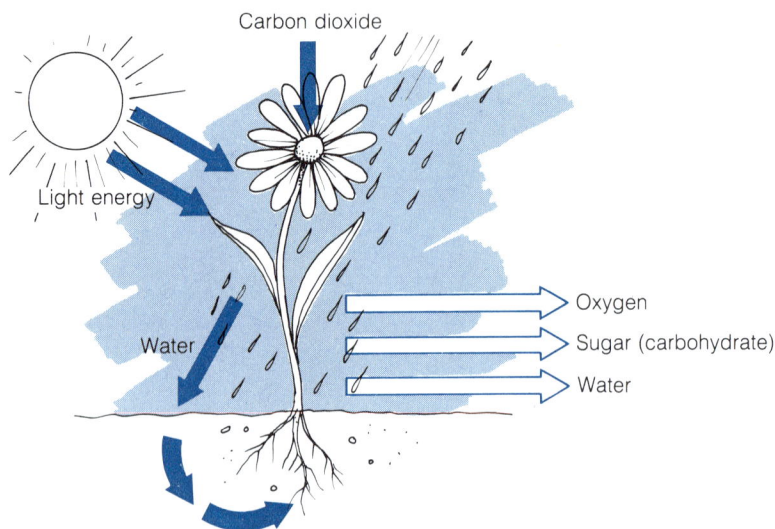

the roots, and carbon dioxide, which comes into the leaf through its many pores. Exactly how photosynthesis works has not been determined yet, but scientists do understand two of its principal activities. First, the direct effect of light is on water, not on carbon dioxide. The light activates the chlorophyll, which breaks down the water molecules. The result is the release of oxygen and an accumulation of energy, which can be stored in the cell. Secondly, carbon dioxide from the air is used, along with the stored energy (in the form of hydrogen ions), to form sugar. This activity has been demonstrated dramatically in experiments using radioactive oxygen. Radioactive elements are used to trace the path of particular kinds of molecules as they go through reactions inside a cell.

As you look at the following equation describing the process of photosynthesis, notice that the molecule containing radioactive oxygen is indicated by an asterisk.

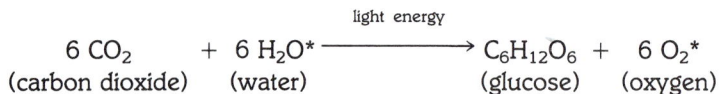

$$6\ CO_2 \quad +\ 6\ H_2O* \xrightarrow{\text{light energy}} C_6H_{12}O_6\ +\ 6\ O_2*$$

$$\text{(carbon dioxide)}\quad \text{(water)} \qquad\qquad \text{(glucose)}\quad \text{(oxygen)}$$

This formula has allowed scientists to prove that the oxygen produced during photosynthesis, and the oxygen used by plants and animals in respiration, comes from water.

Cellular Respiration.   We know how energy is captured from light, meaning that light energy is used for making sugar and oxygen from carbon dioxide and water. But how is the energy that is stored in the sugar released to the cells and used to carry out all the processes of life? Plants and animals reverse this process by **cellular respiration** which breaks the sugar down to carbon dioxide and hydrogen. The hydrogen atoms combine with oxygen to produce water and to release the energy absorbed from the light. This transition from light energy to chemical energy can occur immediately or it can be delayed and stored.

In addition to respiration and photosynthesis, many other biochemical reactions are constantly occurring. **Protoplasm** is synthesized, **amino acids** and other compounds are made, and energy is used. All of these activities are controlled by enzymes. Enzymes are special **proteins** that help stimulate reactions inside cells. Most scientists believe that there are as many kinds of enzymes in a cell as there are substances being made by the cell. All of these compounds come from a relatively few, simple chemical materials that are used over and over again to provide the energy, strength, and substance necessary for growth. The cell is indeed a chemical factory of the highest order.

CELLULAR RESPIRATION The process where hydrogen atoms, obtained from the breakdown of food, combine with oxygen to produce water and a form of energy that cells use to do all work

PROTOPLASM The complex of substances that make up the living cell

AMINO ACID Compound used by cells in the synthesis of proteins

PROTEINS Chains of amino acids linked together in defined ways

Water: An Essential Ingredient of Life

Water, as we have seen, is a necessary ingredient in photosynthesis. But beyond that, water is essential for all living things. It is important in the chemical reactions that take place in cells, and it aids in the movement of life-sustaining substances from one part of a living organism to another part. In brief, nothing goes on in either plants or animals without water.

Plant seeds are a good example of the importance of water. Dry seeds appear to be dead, but in reality their life is only suspended. No chemical reactions are taking place and there are no signs of life. Once that seed is put in water, however, signs of life will soon be seen. When the water content of the seed reaches 8 percent of its total bulk, the process of respiration starts. However, this amount of water is not enough to sustain growth. If the seed does not obtain more water, it will use up its supply of stored food and eventually die. Only when the water content of the seed goes beyond 12 percent will the seed be able to germinate. If the plant is to continue growing, it must replenish the water it uses as well as the water it loses through evaporation. Plants manage to do this in some of the driest and wettest places on the face of the earth.

Parts of the desert in the southwestern United States, for instance, appear to be lifeless most of the year because there is very little rain.

But every once in a while there will be a downpour and the sand will come to life. Long dormant seeds will germinate and produce blossoms and new seeds; these new seeds will then lie dormant until the next downpour. Then the desert will bloom again.

Of course, total precipitation includes both rain and snow; about 10 inches of snow equals 1 inch of rain. At Donner Summit in the Sierra Nevada Mountains of California, for example, 86 percent of the total precipitation falls as snow.

When rainfall is as much as 5 inches a year, the lifeless desert of the southwest becomes dotted with plants such as the Joshua tree, a very water-thrifty plant. Yucca and cacti can also survive in this environment with very little water.

In the colder desert of Nevada and Utah rainfall averages as much as 10 inches per year. In this kind of desert, sagebrush is scattered throughout the landscape. These plants have sturdy roots that grow deep in the ground for water. They also have extensive nets of very shallow roots to absorb the occasional rain that does occur.

Grassland and prairie-like landscapes survive with 25 inches of rainfall a year. These areas have a thick cover of dead blades and roots of grasses to help hold small amounts of moisture. Although prairie grasses may grow as tall as 10 feet, the dry winds that seem to be characteristic of these open areas do not favor the growth of trees.

Once the rainfall increases to an average of 35 inches, as it does along the coast of New Jersey and Florida, forests can develop. However, water is not the only factor that affects the kinds of vegetation that grows in a particular area. For example, because of the mineral composition of the soil along parts of the coastline, only pine trees seem able to survive there.

When rainfall increases to 45 inches per year, as in the Appalachian Mountains from New York to Virginia, different hardwood trees grow well along with the mixture of pines. Finally, in other areas, rainfall is over 100 inches a year and the forest has a dense growth of trees, sometimes reaching a height of 200 feet.

Of course, vegetation is not the only form of living things that needs water. All animals begin life in a watery environment. Eggs are fertilized and embryos develop in a liquid surrounding. In fact, all the known processes of life require water. Plants develop roots to absorb water and animals drink water or get it from their food. It is also important to note that organisms cannot hold water for very long periods. Plants lose their water through small openings in their leaves. The lost water must be replaced in order for photosynthesis and growth to continue.

In a similar way, animals lose large quantities of water through their exhaled breath, through liquid wastes, and through water evaporation from the skin. This water must be replaced for the animal to survive because without water, all plants and animals cease to exist.

Food Chains

The plants and animals in the ecosystem that transfer energy from one to another are members of the food chain. Green plants are the first link in this chain. When an organism eats that plant some of the food energy is passed along and used for growth and for the energy needed by the animal to live and move about. Thus the plant eater, or **herbivore,** is the second link in the food chain. When that animal is eaten by another animal, a transfer of energy is made and another link is added to the food chain. Thus the third link in the food chain consists of the flesh-eating animals called the **carnivores** and **omnivores.** Energy is transferred when an organism eats nonliving matter as well. This eater is called a **detritivore,** a nonphotosynthetic organism that is an important link in the ecosystem food chain.

> **HERBIVORE** A plant-eating animal
>
> **CARNIVORE** Flesh-eating animal
>
> **OMNIVORE** A flesh- and plant-eating animal
>
> **DETRITIVORE** An organism that consumes non-living particulate or dissolved organic matter

It would be very simple if all food chains were the same length. Cow's milk comes from a short, simple chain that has only two links. The grass on which the cow feeds is filled with energy from the sun. The cow in turn uses that energy to produce milk. The grass-to-milk chain is a simple one. Most chains are more complicated because animals eat different kinds of foods. A fox, for example, feeds on several kinds of foods—and takes some of the sun's energy from each of them.

Food chains are usually divided into three groups. The first chain, which we have already discussed, is the *predator chain* (figure 9–4a). Since herbivores, carnivores, and omnivores are predators, they receive energy from the sun second or third hand.

A second kind of chain, the *parasite chain,* occurs when energy goes from a larger animal to a smaller animal (figure 9–4b). A **parasite** is a plant or animal that attaches itself to another, larger organism (plant or animal), called a **host,** that provides the parasite with shelter and a food supply. If the parasite is too greedy, it may kill its host.

> **PARASITE** A plant or animal that lives on another without making any compensation to its host
>
> **HOST** A plant or animal that provides food or shelter for a parasite

In the third chain, the *detritivore chain,* energy from the sun is transferred from nonliving organic matter to animals, fungi, and microorganisms, for example, mushrooms growing on dead wood. *This chain is usually the dominant one in an ecosystem* (figure 9–4c). Even though we can separate food chains in the environment in order to study them, they are woven together in a web-like process. Food chains are an excellent illustration of how all parts of our world depend on one another.

Each time an organism eats some food, only a small part of the energy of the sun is available to it. This results in a food chain being more like a pyramid than a straight line (figure 9–5). If we represent it in that way, the higher a consumer is on the pyramid, the less energy

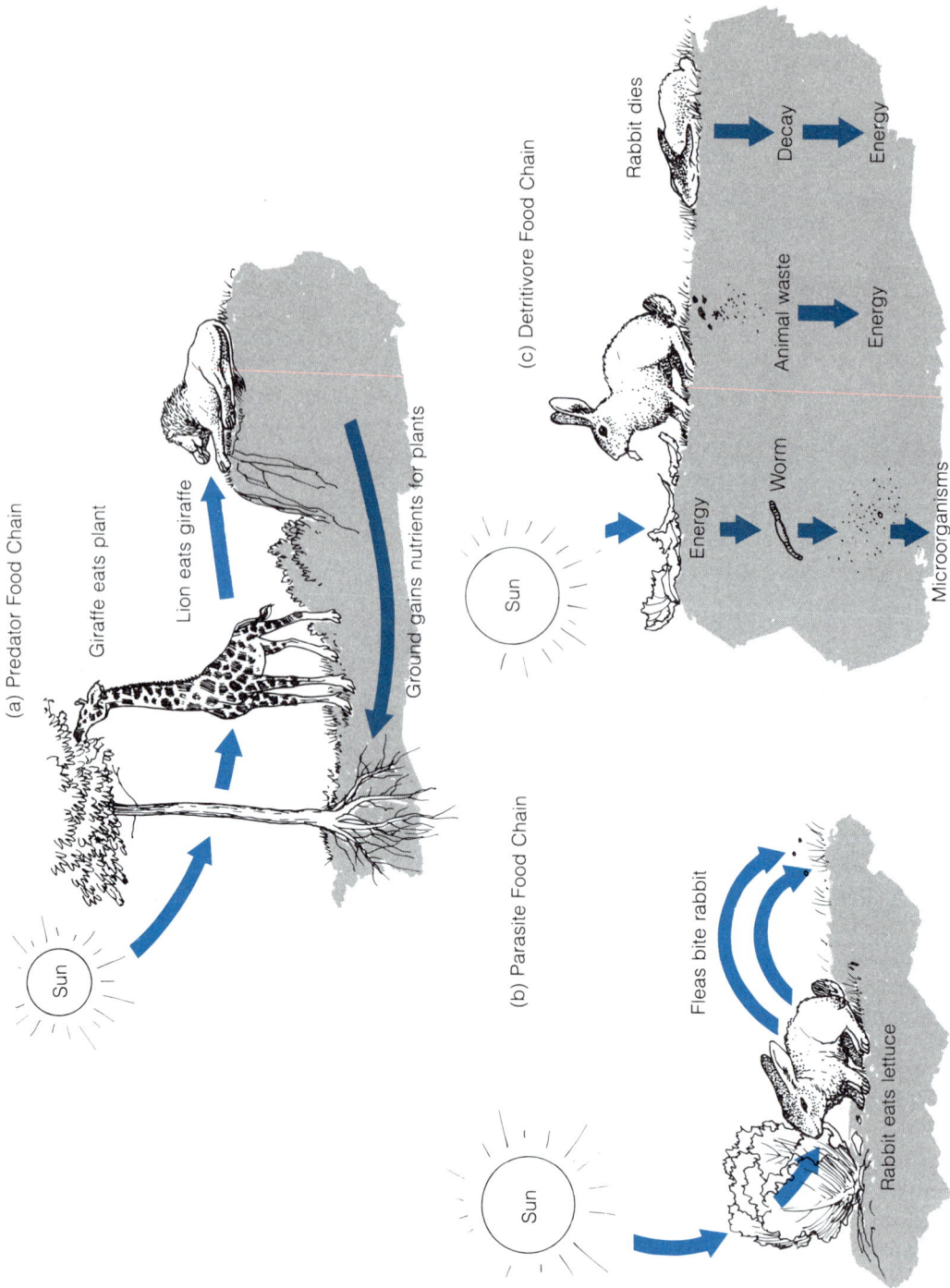

(a) Predator Food Chain

Giraffe eats plant

Lion eats giraffe

Ground gains nutrients for plants

Sun

(b) Parasite Food Chain

Sun

Fleas bite rabbit

Rabbit eats lettuce

(c) Detritivore Food Chain

Rabbit dies

Decay

Energy

Animal waste

Energy

Worm

Energy

Microorganisms

Sun

FIGURE 9–4

(a) In the predator chain, the predator receives the sun's energy second- or thirdhand. (b) The rabbit is the host in this parasite chain. The fleas, which are the parasites, receive the sun's energy thirdhand. (c) The detritivore food chain incorporates the work of microorganisms that return substances to single elements such as carbon, nitrogen, and oxygen.

the consumer receives from the sun. Thus only a small number of consumers can be supported at the top of a pyramid.

Today, there simply is not enough food for all the earth's population to be at the top of the pyramid. One solution to this problem is to recognize that we can use shorter food chains. In China and Southeast Asia, there are too many people and too little land on which to grow food. They use a shorter food chain by eating rice instead of meat; they eat small fish instead of large ones. As a result, less of the sun's energy is lost when it is transferred from one organism to another in the food chain. As the earth becomes more crowded, it will be crucial for us to use our knowledge of the transfer of energy through efficient food chains or else starvation and malnutrition will become common everywhere.

Ecosystems

Organisms that live together in a given area constitute a community. The species populations that are part of the community interact with one another to get shelter, protection, or food. Because of their interactions within a community, both living and nonliving elements are part of an *ecosystem*. The study of ecosystems includes the evaluation of such nonliving elements as heat, light, air, soil, and water. They become essential factors for all forms of life in the community by providing a flow of energy and a cycling of materials.

Carrying Capacity. **Carrying capacity** is a measure of how well an ecosystem can supply a particular species of plant or animal with an adequate supply of food, water, space, and so forth. For example, if too many plants or fish are placed in an aquarium, there simply will not be enough food or space, and some of them will die.

Suppose that the deer population in a particular wooded area increased greatly. If there were not enough shelter, water, or food for them all, what would happen? Some might die of disease or from starvation.

CARRYING CAPACITY The ability of a landscape to support a given plant or animal species

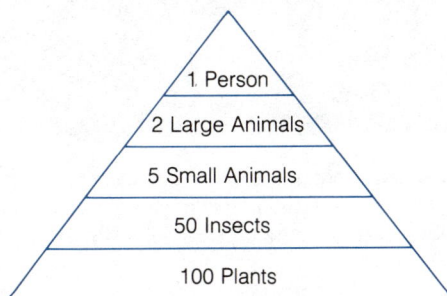

FIGURE 9–5
The plants are the producers in this ecosystem. The primary consumers are the insects and some of the small animals. The large animals and the human are the greatest consumers of energy in the ecosystem.

1 Person
2 Large Animals
5 Small Animals
50 Insects
100 Plants

Others might become prey for predators drawn to an area rich in their food. The lack of shelter might cause others to die from exposure. A few would probably leave to seek a better place to live. The remaining number of deer constitutes the carrying capacity of that wooded area. The equilibrium of that ecosystem, with respect to deer, would be temporarily restored.

Some extreme naturally-occurring events such as fires, earthquakes, or floods may upset the equilibrium of an area temporarily, but usually a new balance evolves as animals and plants repopulate the devastated area.

Limiting Factors.   In an ecosystem, living things must have water, air, energy from the sun, and minerals from the soil. Any one of these requirements may become a **limiting factor.** Each species has its own set of basic requirements; the lack of any single requirement becomes the limiting factor for a population of that species. Without its minimum requirements, individual organisms or species of organisms die.

LIMITING FACTOR An ecological influence that limits or controls the abundance and/or distribution of a species

Each community contains enough natural resources to sustain a population of a certain size. Occasionally a population will increase beyond those natural limits, and destroy their food source in the process. (© Leonard Lee Rue III/Tom Stack & Associates)

In the desert, a limiting factor that often determines the type and number of plants is the amount of available water. In a forest, the amount of available sunlight determines whether certain kinds of plants will grow. For example, sword ferns can grow only where the *range* of light, heat, and moisture are sufficient to allow them to complete their life cycles. Both too little or too much of any one of these factors prevents these ferns from growing in a particular area. If there is too much sunlight, for example, then this is the limiting factor for sword ferns in that ecosystem.

Habitats and Niches.

Within ecosystems, individual organisms live in or are confined to a *habitat*. It is possible to study why living things live where they do. For example, an animal with a very specific diet can live only in a habitat where those particular foods grow. The koala bear of Australia feeds only on eucalyptus leaves and therefore is found only in habitats where these trees grow. The giant pandas of Asia are found only where bamboo grows.

Polar bears are an interesting example of how limiting factors affect the distribution of a species. Polar bears are found naturally in the Arctic. So why are they not found in the cold Antarctic as well? Climate alone is not the factor that limits their habitat. They survive quite well in zoos that are located in much warmer climates. Polar bears live naturally only where three essential conditions exist at the same time: cold water, food (primarily seals), and drift ice. If any one of these factors is absent, so is the polar bear. All three of these conditions exist in the Antarctic, but to get there, polar bears would have to go through the temperate and tropic zones where both the cold water and drift ice are lacking.

When biologists consider where an animal lives and how it relates to its environment, its partners, and its predators, they refer to the animal's **niche** (figure 9–6). To the animal, the local **microclimate** that prevails in its habitat is much more important than the broader climate as viewed by **meteorologists.** A forest rat is barely influenced by the atmospheric conditions several feet above the ground. Much more important to that animal are the temperature, humidity, and light at ground level, underneath or within the shelter.

Even within a niche there are several variations. The smaller the organisms are, the more pronounced these variations become. A rock that is exposed to the sun can influence the life of a plant as significantly as a mountain would influence human activity. That rock can alter local wind and moisture patterns, and it can also retain heat. That is why plants growing in sheltered spots in a northern garden can survive for weeks after killing frosts have destroyed similar plants growing in nearby but unprotected areas of the garden.

NICHE The ecological role of an organism in its ecosystem and its relationship to the total environment

MICROCLIMATE Wind and moisture patterns of a designated environment

METEOROLOGIST One who scientifically studies weather and climate

Polar bears live naturally only where three essential conditions exist at the same time: cold water, food (preferably seals), and drift ice. (Jim Brooks/U.S. Fish and Wildlife Service)

Ecosystem Populations.   Within each community are distinct populations. Because these groups inhabit a particular area, the size of any population can have a dramatic effect on the community. For example, when a particular elephant population in Africa was first protected by law, the population increased so much that the leaf-eating elephants stripped bare all of the trees in their range land. This destruction of the food supply, in turn, eventually had a dramatic effect on the size of the elephant population.

There are other reasons why the size or density of a population can fluctuate widely. Severe weather changes during a single season, changes in the amount of available food, a sudden outbreak of disease, or an increase in the predator population can cause dramatic changes in the size of populations, too. There usually are upper and lower limits to fluctuations in populations. Competition and other forces in the environ-

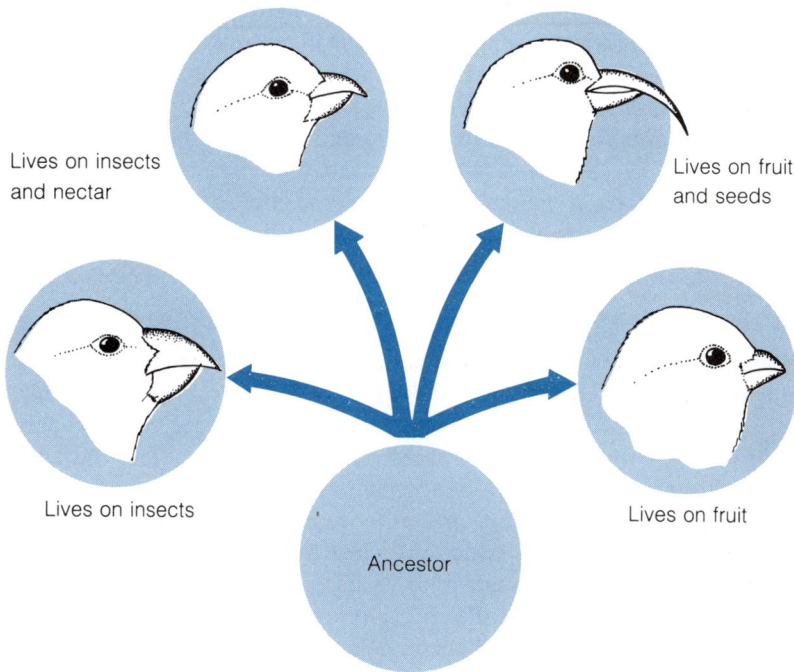

Lives on insects and nectar

Lives on fruit and seeds

Lives on insects

Lives on fruit

Ancestor

FIGURE 9–6
Several species of birds can evolve from one ancestor. Each species has a specialized way to secure food, as evidenced by its beak adaptation. Niches are defined by the organism and its adaptation to its environment.

ment limit the population from increasing beyond a certain maximum. For example, the average density of bobwhite quail in southern Georgia, which has a climate favorable to these birds, ranged between 20 and 100 birds per hundred acres in one study, while in the less favorable climate of Wisconsin, the density averaged from 3 to 10 birds sharing the same amount of land.

Populations in a community can also be influenced markedly by human beings. Hunting is an obvious way, but humans also rearrange or destroy habitats. In some grassland communities, the environment has been modified so much that plant-eating animals can no longer exist. As a result, flesh-eating animals are denied their source of food. The vulnerable grasslands of our country, where vast herds of bison used to roam, have been replaced by checkerboard squares of wheat and corn fields. Although this cropland is needed to feed people, there has been a price to pay. The ecological pyramid of the grasslands has disappeared and many of the animals that formerly inhabited these lands are found only in zoos or wilderness areas.

Cycles. The cycling of materials in the environment is an essential dimension of ecosystems. Materials are taken out of the environment by both plants and animals. Eventually these materials are released back to the environment and are reused by other organisms in bio-geochemical cycles. These cycles are processes by which chemicals from the nonliving environment are made available to living things and vice versa. Animals and plants release carbon dioxide into the air as a result of cellular respiration. That carbon dioxide is reused by the plant to make more sugar, which then is broken down to release more carbon dioxide. Decomposers add minerals to the soil. Plants take up the minerals, and when they die, decomposers return the minerals to the soil. Here are some examples of the most important cycles.

Water Cycle. It has been said that the world's water resources form a single entity. The hydrologic cycle, or water cycle, can be summarized by the statement that annual evaporation equals annual precipitation. Rain that falls on the land and does not evaporate back into the atmosphere descends into the ground water or moves into rivers or streams, from where it eventually goes into the oceans (figure 9–7).

Even though water is not truly lost in this cycle, it does change location, especially with respect to its usefulness to humans. The total amount of water on the earth is estimated at over 1,000 billion cubic kilometers. Much of our fresh water is in the ice of northern Canada and in Siberia, so it is generally inaccessible. Thus it becomes very important that we recognize that maintenance of water quality is an important problem. Lakes, rivers, and groundwater are our only sources of fresh water for drinking, irrigation, and industry. Recycling water so that it is safe and drinkable becomes very costly. By becoming aware of how to conserve water, we will save money and increase the quality of the water supply available to us.

FIGURE 9–7
The water cycle involves the processes of evaporation, condensation, and precipitation.

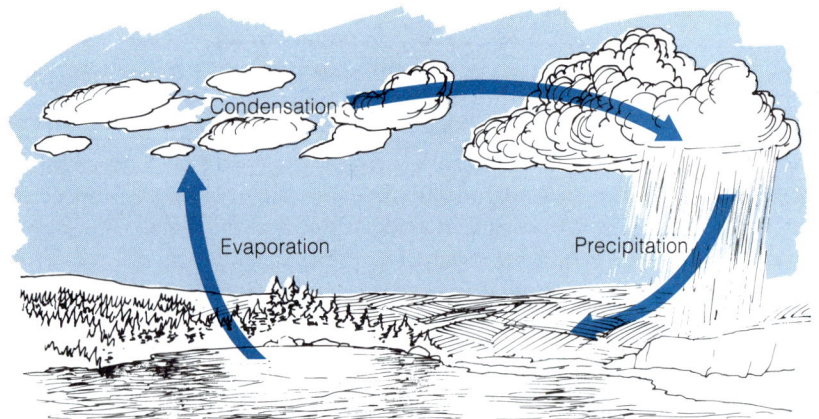

Oxygen-Carbon Dioxide Cycle. Another interaction within the eco-system is the oxygen-carbon dioxide cycle (figure 9–8). During photo-synthesis, plants take in carbon dioxide and give off oxygen. However, plants and animals also give off carbon dioxide during respiration. This carbon dioxide becomes available for plants to use again in photosynthesis. Thus we see that the world of living things forms an interdependent system in which materials such as carbon are used over and over again.

Nitrogen Cycle. Nitrogen is essential to all living things because it is part of all proteins, and is required for growth. Nitrogen, in the form of a gas, makes up almost 80 percent of our atmosphere. However, very few plants and no animals can use nitrogen in that form. It must first be "fixed" into chemical compounds that can be used by plants. This is mainly accomplished by nitrogen-fixing bacteria and algae that are so small that they are only visible under a microscope. Nitrogen fixation can also take place as a physical process in the atmosphere from the effect of cosmic rays and lightning. However, this is a minor source of nitrogen-containing compounds, with respect to the needs of plants at least.

Some bacteria change the gaseous nitrogen into ammonia and sol-uble nitrates, which plants can absorb through their roots. These nitro-gen-containing compounds are then used for growth via incorporation into protoplasm by amino acid and protein synthesis. Once the plants have used the nitrogen for growth, it passes on to other organisms through the food chain. It can be incorporated into proteins of other organisms, or it can return to the soil when animals excrete their waste products, or it is released when they die. Decomposers break down

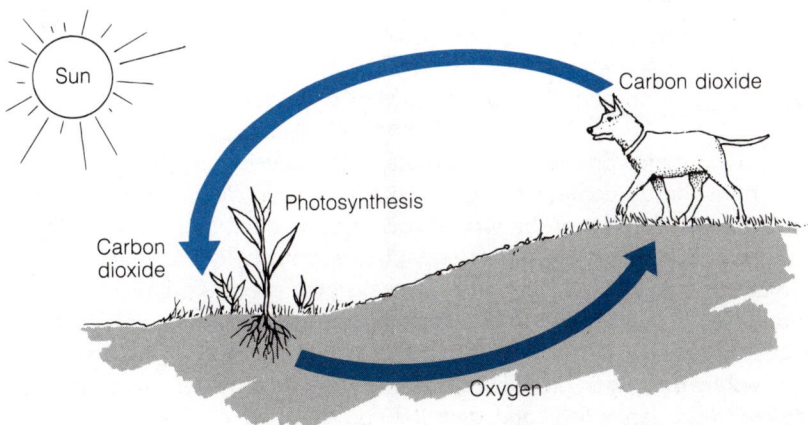

FIGURE 9–8
Through respiration, both plants and animals give off carbon dioxide. In photo-synthesis, these plants use the carbon dioxide and produce oxygen. Animals use the oxygen and give off carbon dioxide.

FIGURE 9–9
The nitrogen from dead organisms is changed by bacteria into a form of nitrogen that plants can absorb.

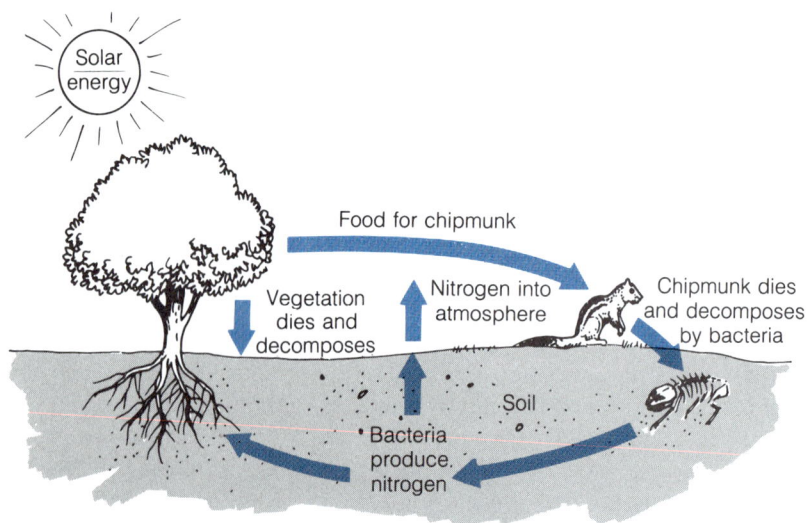

these materials and release nitrogen into the soil as ammonia. As part of this cycle, other kinds of bacteria change ammonia to nitrates and nitrites, and supply plants with these primary nutrients. This cycle then starts over. (See figure 9–9.)

Phosphorus Cycle. Phosphorus also circulates from the nonliving parts of the environment to the living parts and back again (figure 9–10). It is more likely to limit agricultural productivity than any other element. Phosphorus is released from organic matter by cellular metabolism and by specialized bacteria. The phosphorus can then be reused in biological processes or may become insoluble as part of phosphate rocks. Bone and bird dung also tie up a great deal of this element. The greatest reservoir of phosphorus is found in rock, from where it is released slowly into the environment through the action of nitric acid formed by lightning.

One major problem facing many underdeveloped countries is that the phosphorus content of some soils has been decreasing while the phosphorus content of the world's oceans has been increasing. This increase is due to the run-off of phosphorus from land and from sewage discharge into rivers. Fresh water ecosystems are often limited in their productivity by the amount of phosphorus in the water. Discharge of phosphorus and nitrogen into these valuable habitats, a process called **eutrophication,** has brought about massive growths of algae, loss of oxygen and game fish, and deterioration of water quality for industry, agriculture, and domestic supplies. This valuable and essential mineral

EUTROPHICATION Rapid accumulation of phosphates in lakes and reservoirs leading to an increased ratio of plant growth to water volume

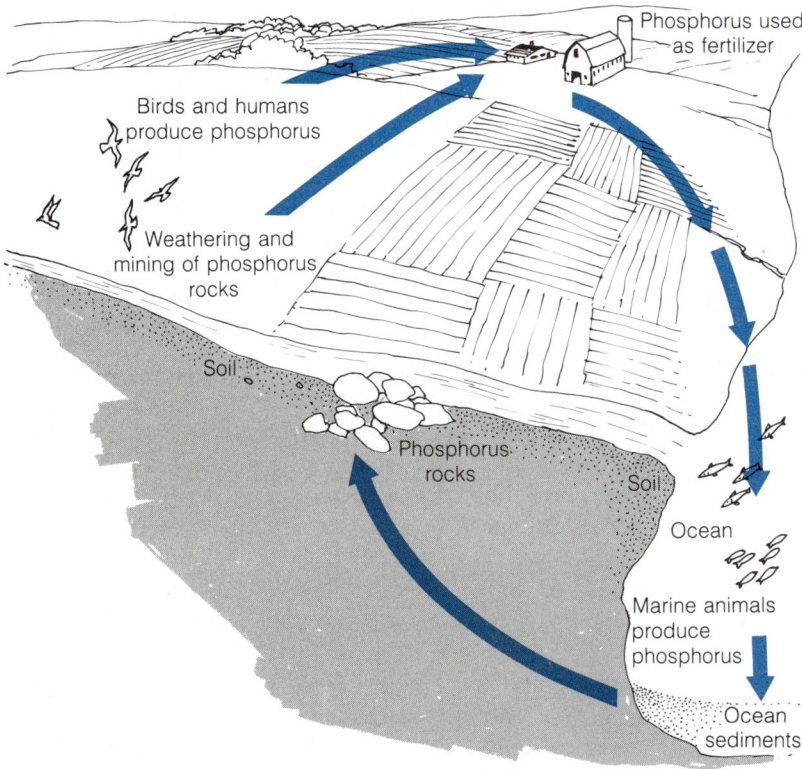

FIGURE 9–10
Phosphorus is an important
fertilizer for soil. Sewage is
a source of phosphorus,
along with plants and animal wastes.

is steadily being lost to aquatic habitats, making it an increasingly rare
and costly fertilizer to obtain and, at the same time, causing the deterioration of our meager fresh water resources. The loss of Lake Erie's
commercial lake trout fishing, for example, was caused by eutrophication from farm run-off, sewage discharge, and the use of phosphorus-
containing detergents.

Community Relationships. The more we look at relationships in
communities, the more we find that they can be placed in several categories. **Mutualism** is a partnership that is beneficial to both a host and
its parasite. For example, flowering plants entice insects and other polli-
nators to visit them by offering **nectar** and pollen, and thus ensure cross-
pollination among these plants. Another example of this kind of relationship is the female wood-boring beetle and the fungus that grows in the
cavity where she lives. The fungus softens the tough wood fibers for the
beetle. When the female beetle lays her eggs, each one is smeared with
some of the fungus spores, ensuring that the new generations of both
fungus and beetle will continue to live in that wood.

MUTUALISM When organ-
isms live together for mutual benefit

NECTAR A sweet liquid
found in flowers and used
by hummingbirds as food

On the seashore you will find the hermit crab, which has a soft shell that offers little protection. However, hermit crabs protect themselves by finding and living in abandoned snail shells. A segmented worm lives inside the shell and helps keep it clean. In repayment for that task, the worm gets fragments of food from the crab. Sometimes hermit crabs are known to obtain a small sea anemone and place it on the back of the shell. The stinging cells in the tentacles of this anemone act as a deterrent to predators that would otherwise destroy the crab.

LEGUME Plants in the pea family noted for their ability to store nitrates

Another example of mutualism is the relationship between **legume** plants, such as beans or peas, and bacteria. Beans and peas have, in their roots, small swellings or nodules that contain soil bacteria. The bacteria gain their food supply from the plant. In return, the bacteria are able to take nitrogen from the air and convert it into nutrients needed by the plant. This beneficial relationship can be disturbed easily if conditions are unfavorable for either the plant or the bacteria. For example, if the bean plant grows in a soil that is deficient in boron, bacteria cannot fix nitrogen and instead will start to feed upon the plant itself; then it becomes a true parasite.

Mistletoe is an example of a partial parasite. It uses its host for both a place in the sun and for its water supply. Aerial plants such as orchids or Spanish moss gain merely a place in the sun from their host. They seem to neither harm nor provide any benefit to their host. A relationship in which one partner benefits without serious harm to the other is called **commensalism.** The host benefits its partner by providing food, transportation, shelter, or simply a place in the environment to live.

COMMENSALISM When a plant or animal lives in or on another but is not parasitic or injurious

Lichens are probably some of the most remarkable groups of simple plants. Grouped together they form a colorful crust on the surface of rocks or the barks of trees. They are found high in the mountains or far into arctic regions. Often they appear as bright orange, yellow, green, or blue-gray crusts. Here where other plants can scarcely grow because of the extreme cold, lichens are the main form of plant life. If you were able to examine a very thin section of a lichen using a microscope you would find that they are actually composed of two separate plants woven together: an algae and a fungus. Because of its chlorophyll, the algae provides the energy from the sunlight; the fungus provides the mineral foods and the shelter. This is a case of **symbiosis.** Symbiosis occurs when two completely different organisms, each unable to survive alone, live together mutually aiding each other.

SYMBIOSIS When two different organisms must live together for survival

Changing Ecosystems.   All of the interactions and characteristics of ecosystems we have discussed continue on a constantly changing stage. The food chain gives each organism a part to play: the predator changes a locality by eating its prey; organisms such as bacteria and yeasts break down nutrients available from dead organisms and thereby bring change to the system. The seasons bring familiar changes such as varying temperatures and precipitation; these in turn, cause concurrent and dramatic changes in the growth and behavior of plants and animals. Animals move in and out of ecosystems with seasonal change, too. Birds migrate southward in fall and winter and northward in spring and early summer. Larger animals climb to graze at higher elevations in warm weather and descend to the valleys in winter for food and shelter. Such changes have obvious impact on all life from the tiny clump of grass to the earth's ecosystem, the biosphere.

Perhaps the most dramatic change occurs as one ecosystem gives way to another. This process, called **succession,** is very slow and often continues for decades. Succession involves gradual but distinct shifts in the environment, called *stages,* where living things slowly change the environment to accommodate new species of plants and animals. When many new species begin to alter their environment, a new stage of succession has begun. For example, on bare rock the first living things to grow are usually lichens. As they live, die, and decay, they form the first elements of soil which in turn creates a suitable environment for small fern-like plants, mosses, and minute animals that can survive in this soil, lodged in crevices of the otherwise bare rock. After several years, accumulated soil from these decomposed simple plants and ani-

SUCCESSION The changes in vegetation and animal life causing one population or community to be replaced by another

Animals such as these Canadian geese move into and out of ecosystems in response to seasonal changes. (© Leonard Lee Rue III/Tom Stack & Associates)

mals allow shrubs to take root. Shade from the shrubs creates a new microhabitat which then permits other small insects and birds to find homes. Eventually seeds from nearby forests may be carried to this area, and seedlings begin to grow into trees. With the presence of trees and shrubs, still other species of plants and animals may join or replace the former inhabitants. A variety of birds and seed-eating mammals such as chipmunks find the new ecosystem to their liking. After a series of such ecological changes, a particular stage of succession remains the same for several decades, and is regarded as having reached a **climax** (implying final) stage of succession.

CLIMAX The stage in succession that gives an ecosystem its final appearance

One of the most interesting discoveries about ecological succession is the predictability of the sequence of ecological stages. Scientists who have studied ecological succession have learned that combinations of environmental conditions allow them to predict how a given environment will change. These conditions include the kind of minerals that are present in the bedrock or soil, the flatness or steepness (slope) of the earth's surface, the elevation of the area above sea level, the amount of rainfall received, and the general range of temperatures that prevail year round. Even the north-south-east-west orientation becomes an important factor in predicting what stages of succession are possible in hilly or mountainous areas, since one side of a hill or mountain will receive different amounts of light and rainfall than another side. By knowing the environmental conditions that exist in a given area of the earth's surface, scientists thus predict what will be the most likely climax stage of vegetation and animal life.

The Biosphere

All life is contained within the biosphere. The thin film that surrounds the earth is less then seven miles thick from the depths of the ocean to the highest portion of the air and yet it is inhabited by more than 1,400,000 different kinds of plants and animals.

On land, the biosphere extends down into the ground as far as the roots of the deepest growing trees—a few meters. In the sea, living things have been detected at a depth of about six miles. However, most marine life is crowded into the upper 500 feet of the ocean surface. Most birds and insects fly relatively near the earth's surface. The biosphere extends only about as high as the tops of the tallest trees, which are perhaps the redwoods of California.

Other than humans, the housefly appears to be the most far-ranging species. It is found almost everywhere that people have gone except the extreme polar regions. It was originally thought to be confined only to the tropical areas of the world, but because of adaptations, it has been able to extend its range. Flies can spend the cooler seasons in a very dormant state and they have adapted to our heated homes. This adapt-

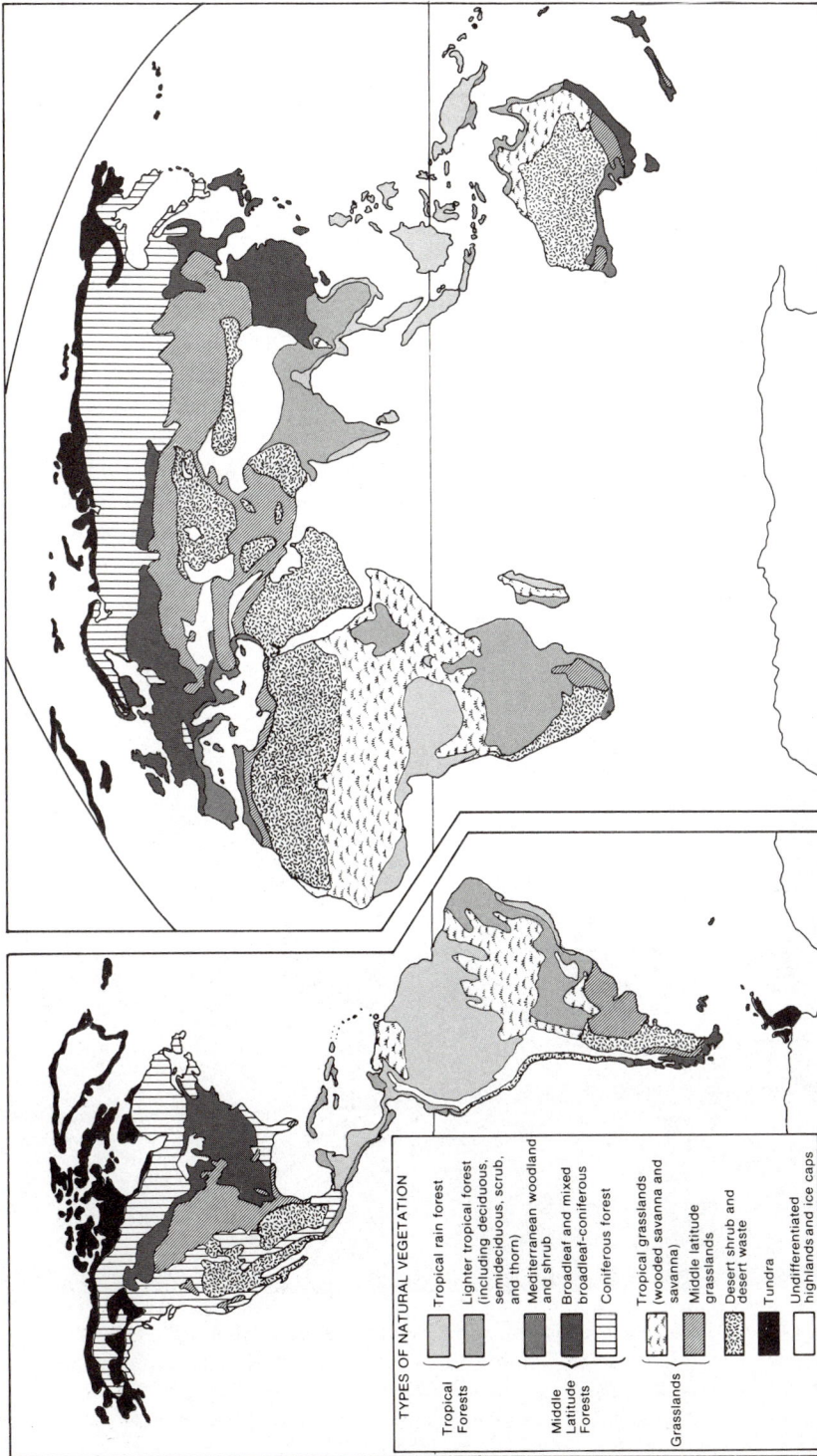

FIGURE 9–11
Communities with the same life forms, geography, and climates comprise the major biomes of the world.

TYPES OF NATURAL VEGETATION

Tropical Forests
- Tropical rain forest
- Lighter tropical forest (including deciduous, semideciduous, scrub, and thorn)

Middle Latitude Forests
- Mediterranean woodland and shrub
- Broadleaf and mixed broadleaf-coniferous
- Coniferous forest

Grasslands
- Tropical grasslands (wooded savanna and savanna)
- Middle latitude grasslands

- Desert shrub and desert waste
- Tundra
- Undifferentiated highlands and ice caps

ability is true for the cockroach, the body louse, the house mouse, and a few other animals that have invaded a wide diversity of environments.

Other than these few exceptions, most plants and animals seem to live only in a very limited range of places in the biosphere. If they are absent from an environment that supports similar organisms, there usually are good reasons for their absence. The amount of heat and cold, dampness or dryness, and the kinds of foods or soils that are available are like invisible fences that serve to keep animals and plants in specific places.

Biomes.   Many of the invisible fences that divide the biosphere are determined by geological features such as mountain ranges, deserts, or islands. So, in addition to looking at the biosphere as a whole, scientists look at large geographic areas that provide specific living conditions because of their climates. These large areas are called **biomes,** and most

BIOME A large geographic area where the community of plants and animals is determined by climate and soil

The coniferous forest biome has temperatures that are usually low all year; the forest often is home for moose, lynx, wolverines, spruce geese, and jays. (From *Plant Biology* by Knut Norstog & Robert W. Long. Copyright © 1976 by W. B. Saunders Company. Reprinted by permission of Holt, Rinehart and Winston, CBS College Publishing.)

of them are named for the plants that dominate them (figure 9 – 11). In adapting to the **climate** of a biome, plants create an environment that is particularly favorable to certain kinds of animal life. So each biome possesses its characteristic combinations of plants and animals.

Coniferous Forest Biome. Broad belts of evergreen forests range from 400 to 800 miles in width and stretch across Canada, Alaska, and Eurasia, covering some higher mountains in other areas as well. This evergreen biome is called the coniferous forest.

In this biome, temperatures are low all year. The length of day and night varies according to the season. The long days of summer are cool and moist, and the shorter days of winter are extremely cold. Snowfall is heavy and may completely cover the trees. Typical **fauna** include the moose, lynx, wolverine, porcupine, great gray owl, spruce goose, and Canadian jay.

The **flora** of a coniferous forest is of great importance to us. For many centuries it has supplied the lumber for building and wood pulp. If we continue to use these trees without adequate preservation of the biome, we could destroy it completely. We must make serious choices and decide what to use and what to save.

Deciduous Forest Biome. Another biome is the temperate **deciduous** forest. In temperate areas where the seasons change from summer to winter and where there is sufficient rainfall, the natural ground cover conditions favor the growth of deciduous trees. As the days grow shorter and the weather turns colder, the trees do not get enough water from the cold ground to replace the moisture they lose from the broad surfaces of their leaves. This loss is speeded up by the strong winter winds. The falling of leaves slows the activity inside trees and decreases their need for water. Trees that thrive in the deciduous biome are beech, maple, hickory, oak, and elm. Bluejays, woodpeckers, chickadees, owls, and hawks build nests, gather food, and breed in this biome. In the spring, different birds come to build their nests and raise their young there. They then **migrate** to warmer climates and other feeding grounds in the winter. Mammals in this biome include the woodchuck, squirrel, raccoon, deer, and black bear.

Tropical Forest Biome. The tropical forest is characterized by rapid plant growth, which is assisted by abundant rains and warm weather. This type of biome is found in South America, Asia, and Africa. The tropical climate is hot and wet year round and there is little change in the length of nights or days. Vegetation appears to be stratified into conspicuous levels or stories, with some forests having as many as five lev-

CLIMATE The average or prevailing weather condition of an area, including air pressure, heat, wind, and moisture

FAUNA The animals in a biome

FLORA The plants in a biome

DECIDUOUS Plants which shed their leaves annually

MIGRATE The practice of some animals to follow a well-established route across one or more biomes — usually on an annual cycle

In the deciduous forest biome are found maple, oak, beech, hickory or elm trees, and among them woodpeckers, chickadees, owls, hawks, squirrels, deer, raccoon, and occasionally bear. (U.S. Dept. of Agriculture)

els. Each level is formed by the crowns of trees or shrubs growing closely together, with many animals living at each level. The trees grow to great heights. Although trees may lose their leaves for short periods of time, the tropical forest is never bare, unlike the trees of deciduous forests during the winter. Animal life is abundant and diverse, including a rich variety of birds, insects, amphibians, reptiles, and mammals.

REEF A chain of rocks, coral, or a ridge of sand near the surface of water

Marine Biome.   Sea **reefs** are marine biomes which flourish with life. Here the clear blue-green or green water is warm and shallow. Australia's Great Barrier Reef is an example. It has a variety of plants and animals, from algae to schools of brightly colored fish. These fish, in turn, attract fish-eating birds and larger fish from deeper ocean waters.

The tropical forest biome has trees that grow to great heights as well as many layers of vegetation at lower levels. The climate is hot and wet year-round. (© Tom Stack & Associates)

Desert Biome. At the other extreme from the watery biome is the desert, which covers about one-fifth of the earth's surface. Desert plants do not grow close to one another. They have long roots that spread out close to the top of the soil in order to gather any rain that falls. This leaves much soil exposed to water and wind. In most deserts, rain can wash away the soil because the plants are not close enough together to hold it in place. A heavy rainfall can result in runoff that is very rapid, causing the flash floods that frequently occur in the deserts. Wind is also an important factor. With little plant cover to act as a windbreak, the soil or sand moves quickly for long distances, scouring plants and rocks. Desert plants are characterized by thick skins and foliage with small surface areas, which help to prevent evaporation and the damage caused by wind and heat.

Large animals are scarce in the desert. However, rodents such as kangaroo rats are abundant. Many live underground during the daytime when temperatures are hottest; they are active only at night, when the desert is cool and the darkness offers them some protection.

The desert biome is subject to hot, dry winds and flash floods. Plants have thick skins and long roots; kangaroo rats and other small rodents live underground during the high daytime temperatures. (© Brian Parker/Tom Stack & Associates)

Tundra Biome. Among the coldest biomes is the tundra. It is found between the northern parts of the coniferous forests and the permanent ice of the polar regions. Tundra also occurs in a modified form in higher mountains. In the tundra, winter lasts for nine, cold months with very little daylight. During the winter the biome seems a frozen and dark waste. The three-month summer has short nights and nearly constant daylight, during which plants seemingly burst from the ground. They quickly mature and develop seeds before the short summer ends. The principal vegetation includes lichens, mosses, and grasses. They make the most of a brief growing season during which the subsoil never completely thaws. They also provide food for the caribou, the largest animal in the Canadian tundra.

Tidal Biome. The seashore provides another type of biome. Life in these tidal biomes is rigorous. The shore is pounded by the waves of the sea and exposed by the tides to both the sun and the air. Yet this is an area where life abounds. Extending from the high to low tide levels,

Above the coniferous forest biome is the tundra biome where summers are short and plants must complete a year's growth in three months of nearly constant daylight. (© Warren Garst/Tom Stack & Associates)

the tidal biome is usually divided into three zones. Sea snails, various crabs, and mollusks are found in the upper zone; brown algae in the middle zone; and kelps, in the lower zone. Each of these organisms reflects a tolerance to varying degrees of exposure to the sun and air.

Grassland Biome. Wide expanses of grass at the foot of mountain areas are typical of the grassland biome. Grasslands are found in the interior of continents where rainfall is insufficient to support trees. Grassland soils are usually rich with humus in the area of taller grasses since the leaves die and decay at the end of each growth season.

Animals that make the grassland their home show adaptations to that area. They burrow into the soft dirt or they run swiftly to escape their predators. For additional protection they also live in colonies or herds. The American bison is perhaps the best example of an important grassland animal in central North America. Ground squirrels, prairie dogs, coyotes, jackrabbits, and prairie chickens also can be found in these grasslands.

Ocean Biome. The largest biome is in the ocean. It may appear to be thinly populated, but it is actually a reservoir of life. It is filled in different regions and at different times with billions of animals and plants so small that they cannot be seen with the unaided eye. These tiny organisms constitute one of the lowest links in the *food chain*. Together, however, all the life in the seas makes up the ocean's biome. Here such limiting

The grassland biome is found in the interior of continents where soils are rich but rainfalls cannot support the growth of trees or shrubs. (USDA—Soil Conservation Service)

SALINITY The quality of being salty

factors as light, water pressure, temperature, **salinity,** and oxygen content are key factors that determine the survival of various species of plants and animals.

THE ENVIRONMENT AND HUMAN IMPACT

We have defined our environment as all the conditions and influences that surround the development of living things. That definition includes a very wide range of environments. We could begin with the tiny, subatomic environment (as discussed in chapter 15), or we could see our environment as including the stars, galaxies, and indeed, the universe (see chapter 13, "Earth and Planetary Sciences"). Some of our human frontiers of knowledge can be characterized by the extent to which we have explored the micro, or very small, subatomic world, and the supermacro, or very large, cosmic environment of space. For the moment, however, our functional environment lies between the subatomic and cosmic extremes. As figure 9–12 shows, human impact has been confined primarily to the environment of the organism, the population, the

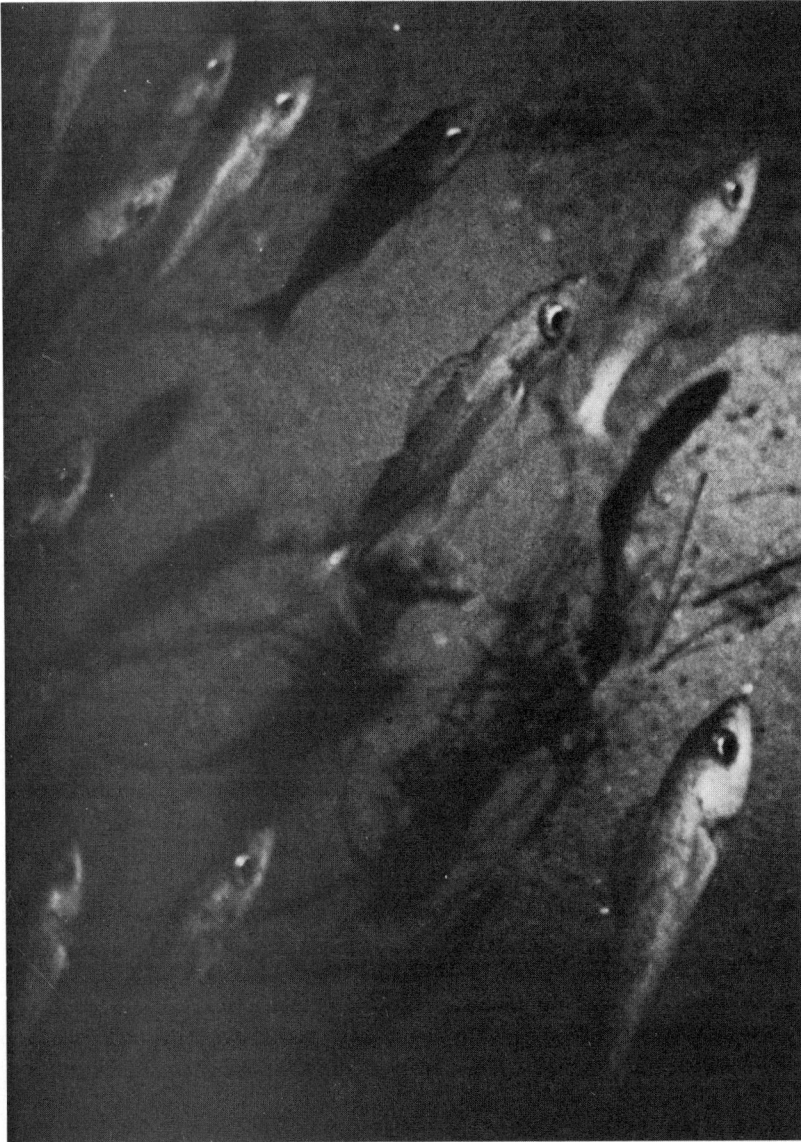

One of the least known but most beautiful of biomes is the ocean biome where variations in water temperature, pressure, and light form invisible boundaries between layers of water and animal life. (Scripps Institution of Oceanography)

community, the ecosystem, and the biosphere. These constitute the realm which ecologists and society as a whole must consider as we face the serious problems that we have created while living in and changing our environment.

We now turn our attention to the relationship between human beings and the natural environment: the human use of natural re-

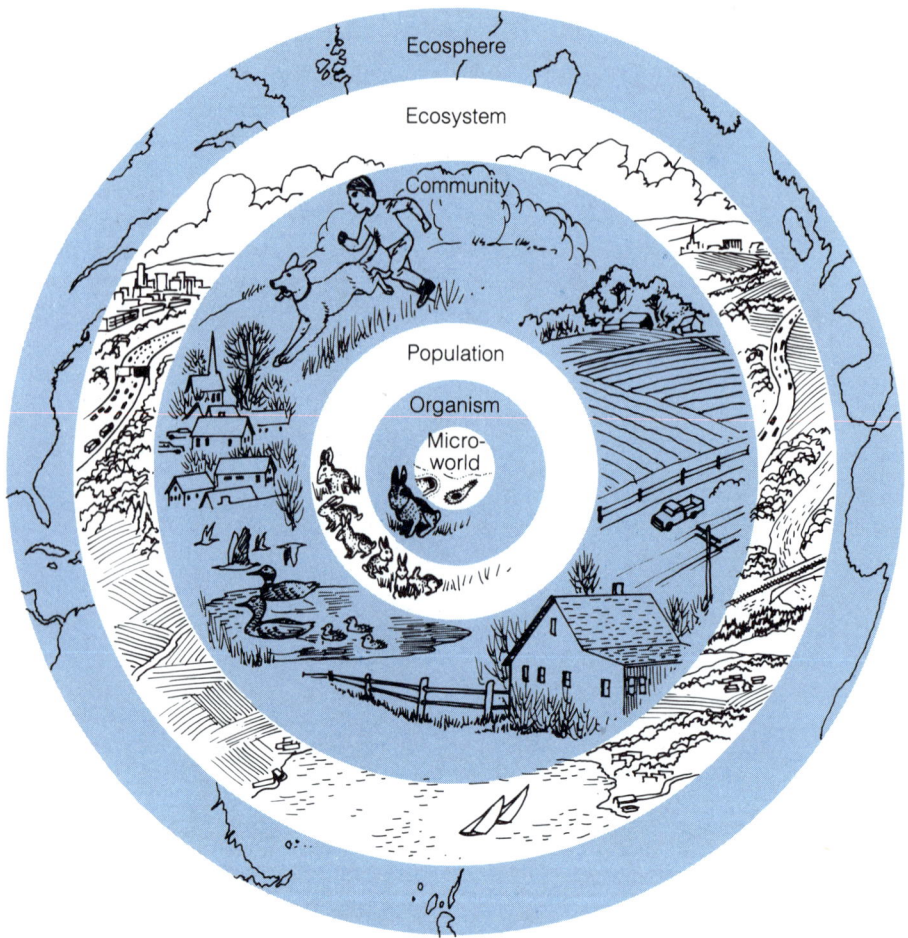

FIGURE 9–12

The ecosphere is the entire world and encompasses each of the environments, including the micro-world, organisms, populations, communities, and ecosystems.

sources, the human impact of that use, and how humans might better use the available resources.

What Are Our Resources?

NATURAL RESOURCES
Energy and matter considered important for human use

A **natural resource** is a naturally occurring organism, substance, or product of an organism or substance (like sunlight) that human beings use or can use in living their lives. Although humans create a great many objects and substances, these are all ultimately made from naturally occurring resources.

Natural resources may be subdivided into **stock-limited** or nonrenewable resources, and **flow-limited** or renewable resources. The terms stock-limited and flow-limited are more useful than renewable and nonrenewable in two respects: first, stock-limited and flow-limited remind us that *all* resources are limited; and second, although the term "renewable" (flow-limited) is not itself misleading, it often seems to be used to mean a boundless supply—we hear "renew" but not "able." In fact, the situation is the reverse of what many have come to imagine is meant by renewable and nonrenewable: We have totally exhausted many renewable resources, notably many species of wildlife; and it is unlikely that we will ever actually exhaust our nonrenewable resources because we will never be able to locate *all* of any material in the vast expanse of the earth's crust.

Stock-limited resources include the planet's atmosphere, land, and water, which can be used over and over again; but the quality of these resources in any given city or region may prevent them from being used. Breathing polluted air can injure people, much of our fresh water is undrinkable, and much of the earth's land will not support human life. Other stock-limited resources are fuels, minerals, and gases.

Flow-limited resources include radiant energy from the sun and all organic life—plants, animals, soil, and bacteria. Each of these resources is renewable (or continuous, as in the case of the sun), but only over a certain amount of time. Again with the exception of the sun, flow-limited resources, even though renew*able,* may be entirely depleted before they have time to renew themselves. Within the last fifty years, human beings have caused 167 species of animals to vanish from the earth. Trees—timber and firewood—are a fast vanishing renewable resource.

We will consider first **energy** resources and then all other resources. Within each of these sections we will continue to distinguish stock-limited from flow-limited resources.

STOCK-LIMITED RESOURCES Minerals, fossil fuels, and other resources, such as water and air, that are present on earth in limited quantities

FLOW-LIMITED RESOURCES Sunlight and other resources that are naturally dispersed and available at limited rates

ENERGY Capacity for producing motion that can be derived from and converted back into mass

Energy Resources

The sun is our most continuous source of energy. Geothermal, tidal, and nuclear power are also energy resources but they are not derived from the sun. Here we take up the ways in which we use energy, and discuss ways in which the severity of our energy crisis might be tempered.

The Sun and Its Products. The sun is a major source of energy, whether we use its energy directly or indirectly. We are able to use the sun's energy directly to dry our clothes; to heat our water, homes, and workplaces; to cook and bake our food; and to make electricity. The sun

Children learn that they can use the sun's energy to cook their hot dogs. (U.S. Dept. of Energy Photo by Jack Schneider)

HYDROELECTRIC POWER (HYDROPOWER) Generation of electricity by free fall of water, creating about 4 percent of the world's electricity

FOSSIL FUELS Organic remains compacted 50 million years ago, now burned for heat in the form of coal, petroleum, and natural gas

also drives the global hydrologic (water) cycle, which allows us to float down rivers and to use the energy of falling water to make electricity by turning **hydroelectric power** generators. And the sun drives the winds that propel sailing and soaring ships, and the windmills that pump water, run machinery, and generate electricity.

By photosynthesis, the sun's energy is stored in food and other organic materials we use for energy: firewood, organic wastes (methane gas and dung, for example), and especially the **fossil fuels** within which the sun's energy was stored by happenstance fifty million years ago.

Fossil Fuels. The fossil fuels are mainly coal, petroleum, and natural gas, and to lesser extents, tar sands and oil shales. All of the energy we receive from the sun is flow-limited except for the fossil fuels, which are

This wind machine on the north shore of Oahu, Hawaii, captures and converts ocean breezes into electricity; it generates enough power for 100 average-size homes. (U.S. Dept. of Energy Photo from NASA)

stock-limited. Yet, because fossil fuels have a higher concentration of energy than flow-limited resources, humans have long preferred coal and oil, for example, to wood or sunlight. Coal provides twice as much heat per pound as does wood, and gasoline gives up half again as much per pound as coal. Such preferences have by now become reflected in the costs of fossil fuel energy: in 1982, firewood cost about 10 cents per pound; gasoline cost about 15 cents per pound but gives more than three times as much heat. Coal, which gives twice the heat of wood, was only 3 cents per pound. For equivalent heat value, natural gas was about 5 cents per pound or about one-third the price of gasoline. So it has actually become cheaper to use energy resources that we will run out of than it is to use energy forms that are continuously available and renewable in our present environment! This has happened as both a cause and consequence of the industrialization that occurred in Europe, the United States, and Canada during the last hundred years or so, and which is now spreading to societies all over the world.

Compared with farms and homes, factories require large amounts of energy in concentrated form. Historically, each factory, if it was not built at a river that ran its mill, had its energy brought to it in the form of wood or coal. With the development of large electric motors and the *power grids* (networks of electric power lines) to run the motors, more energy resources were brought to power plants which in turn converted the resources into electricity. Electricity was then sold to factories and other consumers. For many factories, this was cheaper than making their own electricity. It provided a constant voltage and dependable supply of electricity that was easier to use than coal. Furthermore, as the electrical age came upon us, electrically-operated factories produced electrically-operated goods: lights and radios, then washing machines and hair dryers, and on through TVs, electric waterpicks, and knives. Electricity itself generated the need for electricity.

With the development of internal combustion gasoline and diesel engines, and especially with the development of the automobile, another system—refineries, pipelines, gasoline stations, and so on—was set up for concentrating and distributing the by-products of petroleum, including gasoline, diesel fuel and heating oil, lubricating oil, petrochemicals, propane, butane, and natural gas. With all these systems in place, other resources become more difficult to obtain and use.

Non-solar Energy Resources.   Over 97 percent of the energy humans use comes from the sun. Non-solar sources include tidal power, geothermal power, and nuclear power. Tidal and geothermal power are flow-limited resources; nuclear fuels are stock-limited or nonrenewable.

Tidal Power.   As the moon circles the earth and the earth orbits the sun, the gravitational forces of the moon and the sun pull unequally on the waters of the earth, causing each large body of water to rise and fall twice every day. The energy created by differences between successive high and low tides is called tidal power. However, since this kind of energy can only be harnessed in shallow seas, it is estimated that the world's exploitable tidal power is minimal—about one-third of one percent of the world's exploitable hydropower (the power obtained from running water). Although this means that tidal power is not a global solution to our energy needs, it might be depended upon in appropriate localities. There are tidal-powered electrical generating plants in France and the Soviet Union.

GEOTHERMAL POWER
Energy generated from the hot, molten areas within the earth

TURBINE Engine or motor turned by the pressure of steam, water, air, etc., against the curved vanes of wheels fastened to a drive-shaft

Geothermal Power.   **Geothermal power** is generated by tapping the energy of the hot, molten areas within the earth. There is one operating geothermal power plant in the United States and others in Iceland, Mexico, Italy, Japan, and New Zealand. Using geothermal power currently involves either piping steam from naturally occurring reservoirs of steam deep in the earth, and using that steam to turn **turbines** that generate electricity; or else tapping into very hot water that is sometimes found deep in the earth, and using the steam from that water to run turbines. Now under development is a technique for pumping water down to hot, dry rock, thereby creating steam that is then piped back up to turn turbines. In the near future it appears that geothermal power may be an important source of power but only at particularly favorable locations where reservoirs of steam or hot water are found.

Nuclear Power. The radioactive materials that are used in producing nuclear power—chiefly uranium and thorium, and experimentally, deuterium and lithium—are not solar energy. We have only limited stocks of these materials, with which we create **nuclear reactions.** Such reactions produce heat to boil water into steam. Steam turns the turbines of generators that make electricity. This is called nuclear power.

NUCLEAR REACTION
Combination or splitting of atomic nuclei with release of energy

How We Use Energy. In the United States our total energy uses may be subdivided into four broad categories: residential, commercial, industrial, and transportational. We use nearly 20 percent of the nation's energy in our homes, and half of that is used to heat them. Another 15 percent of residential energy is used to make hot water, and the remaining 35 percent of energy used is divided between refrigeration, air conditioning, lighting, cooking, clothes drying, and other electrical uses. The good news in this budget is that it is quite easy to reduce energy use in the top two categories, heating space and heating water. Keeping heat in a house is as easy as insulating it. Keep the doors and windows shut. Raise window shades or blinds on the sunny side of the house and lower them on the shady side. Hot water heaters can be turned down or off between periods of peak use. Solar panels that heat water "for nothing" are now available at affordable prices and they pay for themselves within a few years.

Another 16 percent of U.S. energy is used commercially—that is, in our shops and offices, while 40 percent is used by our industries. One natural, healthy response to the gasoline and energy crises of the mid-1970s was turning off the lights in electrically lit signs and office buildings that had previously been left on throughout the night.

Transportation takes the last quarter of our nation's energy, and within this category more than half the energy is used for automobiles. Trucks take 21 percent. The really efficient modes of transport—railroads, ships, and buses—account for only 5 percent. Taking into account the number of passengers they normally carry, intercity buses are more efficient with energy than cross-country trains, and both are more efficient than automobiles traveling between cities. Within cities, automobiles are less than half as efficient as they are between cities.

Another critical use of energy resources is to produce usable power. About 45 percent of the world's annual energy needs are supplied with petroleum, 32 percent with coal, 19 percent with natural gas, and 2 percent each with hydropower and **nuclear energy.** This tabulation does not include the energy in food or in the dung, wood, and agricultural wastes that are used for fuel in the many unindustrialized areas of the world. More than 95 percent of the world's energy and U.S. energy is from stock-limited (nonrenewable) fossil fuels—petroleum and coal. The United States and other industrialized nations, which include only one-third of the world's people, use 83 percent of its energy resources. It may seem hard to believe that one hundred years ago we got most of our energy from firewood, or that our energy "needs" per capita have nearly doubled during the last forty years.

Energy Resource Reserves.   How soon will our stock-limited energy resources be depleted? To answer this question we must be clear about what we mean by resources and what we mean by depleted. The

The search for natural resources is really an economic issue. In this strip mine in Wyoming, as in all mining operations, materials are extracted from the reservoir until it costs too much to continue. (U.S. Dept of Energy Photo by Jack Schneider)

Industrial scientists search for ways to dispose of waste without adding to pollution. (U.S. Dept. of Energy Photo by Jack Schneider)

amounts of stock-limited resources we have left—energy and otherwise—will never be known exactly, but as time goes on we think our estimates are getting closer and closer. Natural resources are classified according to how certain we are of their locations and amounts. In descending order of certainty, they are referred to as *proved reserves*, *probable reserves, resources,* and *ultimately recoverable resources.* Proved and probable reserves are often combined and called simply **reserves.**

Depletion is really an economic issue: when resource users decide that it costs too much to extract materials from a proven reserve, we say that the reservoir is depleted.

The U.S. Bureau of Mines and Department of Energy have estimated the reserves of energy resources left in the earth and the amounts that will be left as of the year 2000. Unless we change our usage patterns, they say that the United States will have used up all of its stock-limited energy reserves within ten years and will thereafter have to rely on coal or on imports from other nations to get us through the next

RESERVES Amounts of materials assumed to be currently available for extraction at known locations; the amounts and locations of resources are less well-known

twenty years. At current consumption, the whole world's reserves of petroleum and natural gas are estimated for depletion in the years 2030 and 2059, respectively. Therefore, we will need to use alternative resources in order to conserve fossil fuels and we will need to find ways to use less energy. The next section will give us some ideas about such alternatives.

Alternative Energy Resources and Energy Conservation. Each energy resource is better suited to some tasks than others. Hydropower is best used for making electricity, or, at a local site, turning the wheels of machines directly; it cannot be used for transportation. Windpower is best used in small amounts at many different locations, and may be used to generate electricity or to turn machinery. Submarines and a few large ocean-going ships have been nuclear-powered; otherwise this source of power is used only for making electricity.

Although electric autos are almost pollution-free on the streets, it is only because their pollution was released back at the power plant that made the electricity to charge their batteries. The uses of coal, the only stock-limited source of energy that we can count on in the century ahead, are limited by hazardous wastes that occur when it is burned.

Solar power in the forms of solar windmills, ovens, water heaters, cookstoves, spaceheaters, and panels which generate electricity that can be used and stored are now commercially available. In combination with traditional domestic power, solar power can be depended upon to meet the energy needs of individual households, particularly but not exclusively in sunny areas. The solar technology for heavy industrial use and standby capacity is not available, nor are solar-powered forms of ground and air transportation.

Table 9–2 lists some approaches to energy conservation that will help to alleviate our energy crisis. When facing decisions affecting our environment, students and their families may wish to follow these scientifically-based suggestions.

Non-energy Resources

Not all resources are used for energy, of course. Minerals are one of the most valuable stock-limited (nonrenewable) resources. Air, water, soil, food, and our forests and wilderness areas are considered in this section as flow-limited (renewable) resources.

Minerals. There are seventy-eight vital, nonrenewable resources such as gold, silver, diamond, mercury, sulfur, cobalt, barium, and so on. At current consumption rates, we will be in short supply of 43 percent of these within the next several decades. Since depletion is determined

TABLE 9–2
Approaches to energy conservation

	Short term (5 years)	Middle term (5–15 years)
Leak-plugging	More insulation in buildings More efficient air conditioners Use of heat pumps for heating	More efficient cars More efficient building designs More efficient power plants
Mode-mixing	Switch freight to rail Dress more appropriately for weather Switch containers to glass returnables	Increase durability of goods (more repair, less manufacture) Reliable, convenient urban transit Switch intercity passengers to fast rail carriers
Output-juggling	More sailboats, fewer motorboats More backpacks, fewer trail bikes More books, fewer movies	More communication, less business travel More education, art, music, less motorized recreation More health services, less heavy manufacturing
Belt-tightening	Car pools, lower speed limit, less pleasure driving Houses cooler in winter, warmer in summer	Live closer to work Smaller dwellings

SOURCE: From *Ecoscience: Population, Resources, Environment* by Paul R. Ehrlich, Anne H. Ehrlich, and John P. Holdren. Copyright © 1977 by W. H. Freeman and Company. All rights reserved.

economically, some additional reserves will be found but, again, the cost for doing so will increase. U.S. government sources predict that we will use up all known reserves of asbestos, barium, bismuth, industrial diamond, fluorine, gold, indium, mercury, silver, and sulfur in twenty years—unless we change our usage rates soon.

Air, Water, Soil, Food, Forests, and Wilderness.   The natural resources that industrial societies do not use as energy sources, and which are flow-limited include air, water, soil, food, forests, wilderness, and all life on earth. We can see that, together with the sun, these make up the life-supporting environment of every creature on earth, including industrialized humans. Further, these resources are practically the only life support system for all creatures except humans. Yet the capacity of each of these vital systems to support life on earth is diminishing because of our industrialized way of life.

Air Pollution.   We are taxing the limits of the air we breathe by continuing to release poisons into the atmosphere. As a result, we all now

Paul M. Schrock

United Nations

Acid Rain: Progress or Pollution?

The Industrial Revolution emerged in England in the eighteenth century. By the late nineteenth century, the revolution had spread to America. The Industrial Revolution shifted the production of goods from home to factory. Society, as a whole, experienced a higher standard of living because of industrialization. This was called social progress. Energy for industrialization was provided by fossil fuels, primarily coal. Almost immediately, pollution was recognized as a trade-off for social progress. For over a century the economic advantages of social progress outweighed the ecological inconveniences of environmental pollution. Now many people are questioning the cost of progress and pollution. Acid rain is one of the ecological consequences of social progress. Acid rain is an example of the connection between science and society.

Acid rain refers to atmospheric deposition having an acidity below pH 5.6. Acidity results from the presence of hydrogen ions (H+) in a solution. Acidity, and its opposite, alkalinity, are measures on a pH scale ranging from the most acidic solution, a pH of 0, to the most basic solution, pH 14. A neutral solution has a pH of 7. Each descending number on the pH scale represents a tenfold increase in acidity. A pH of 6 is ten times more acidic than a neutral solution of pH 7. Water with pH of 5 is ten times more acidic—or 100 times more acidic than a neutral solution. One begins to sense there is a tremendous increase in acidity between each number on the pH scale. For example, a pH of 3 is 10,000 times more acidic than neutral water. Rain is not pure or neutral; it commonly has a pH between 5.6 and 7.0. Any wet deposition (rain, snow, sleet, dew, fog, smog, or frost) or dry deposition (dust, fly ash) that has a pH below 5.6 is termed *acid*.

Acid rain results from the presence of sulfuric acid (H_2SO_4) and nitric acid (HNO_3) in the atmosphere. The weak acid rains are formed primarily from plant and industrial emissions of sulfur dioxide (SO_2) and nitrogen oxides (NO_2). This is the link between human activity and the environment.

Millions of tons of sulfur and nitrogen oxides are emitted into the atmosphere each year. These oxides are carried several thousand kilometers from their source. While in the atmosphere, the oxides interact with sunlight, oxidants, and catalysts and are converted to weak sulfuric and nitric acids. Eventually the chemicals return to the earth's surface as acid precipitation.

There are many detrimental consequences of acid rain; we shall examine only one—aquatic environments. Lakes and ponds often serve as an example of an ecosystem. Lakes generally have a fairly high pH, around 8. This is due to natural buffers, or materials, that keep the alkalinity high. But continual acid deposition can gradually lower a lake's pH. As a lake's pH decreases below neutrality, species of salamander eggs fail to hatch. At pH 6.6 snail populations are reduced. Critical changes occur *before* the lake's pH reaches 5.6 (the defined level of acid rain). The number and diversity of plants and animals diminishes. Bacterial decomposers die. The normal reproductive cycle of the fish is disrupted. Toxic metals, such as alu-minum, are leached from soils surrounding the lake and clog the gills of fish. Mercury is converted to its organic form, is absorbed by fish, and thus enters the food chain.

As the pH continues to drop, the vegetation and fish populations continue to change. By the time pH reaches 4.5 there are no fish. Nearly all frogs, insects, and decomposers are gone. Surface insects are abundant since they can escape the water's toxic effects and their natural predators are gone. The lake is crystal clear but dead.

When solutions to the problem of acid rain are sought, the central role of the ethical issues involved becomes apparent. Debates continue about the actual *truth*—causes and effects of acid rain. There are also discussions about what is the greatest good—the economy and employment, or trout streams and a beautiful environment. Inevitably justice becomes central to the resolution of conflicts between industrialists and environmentalists. Society faces a very serious question—are the benefits of progress worth the detrimental effects of pollution?

Questions for Discussion

1. How would you answer the last question: Are the benefits of progress worth the effects of pollution? Is your answer the same if you view this as an individual? From a social perspective?

2. How does acid rain affect the lake ecosystem?

3. What potential effect could acid rain have on humans?

4. How would energy conservation contribute to a reduction of acid rain?

breathe air that contains **pollutants.** Air pollution is particularly dangerous in and near cities and industries, but these wastes are now spread around the globe. Polluted air causes stinging eyes and peeling paint; at times it causes sickness and death.

The air pollutants that are now being monitored (the quantities of which there has been some attempt to control) include the sulfur oxides, which are mainly created in factories; oxides of nitrogen, produced by factories and by autos, trucks, and other transportation; hydrocarbons, which are also produced by trucks, autos, and a variety of industrial processes; carbon monoxide, three-fourths of which is created by autos and trucks; "particulates"—little pieces of matter—that are mainly spewed out by industry; and radioactive wastes from the operation and malfunctioning of nuclear power plants and, occasionally, from nuclear explosions.

Scientists are monitoring the quality of the air and the quantities of sulfur oxides, nitrogen oxides, hydrocarbons, and other pollutants in an effort to establish hazardous limits. (Courtesy of Los Angeles County Air Pollution Control District)

Acid precipitation and smog. The oxides of sulfur and hydrogen are particularly troublesome because they form strong acids (sulfuric acid and nitric acid) in the presence of water vapor and particulate matter, and they become acid rainfall, snowfall, and fog. Acid rain has been known to kill fish, harm vegetation, and deteriorate buildings. **Acid precipitation** can be as strong as pure lemon juice. The hydrocarbons emitted from the incomplete burning of fossil fuels are of interest because some of them react in sunlight to form a number of compounds we call **photochemical smog.** Some of these compounds are known carcinogens; that is, they cause cancer.

Greenhouse effect. Although green plants, animals, and human beings give off carbon dioxide normally, human industry, through burning fossil fuels, has added significantly to the small amount that is already in the air. This addition is important because the carbon dioxide in our atmosphere acts as a one-way filter that holds heat near our earth that would otherwise dissipate into space. The windows in an automobile or greenhouse create this same effect, making it hotter inside than outside, and so the process has been called the greenhouse effect (figure 9–13). There has been concern over the possibility that the temperature of our atmosphere might increase significantly as we continue to add carbon dioxide to it. Were this to happen, the polar icecaps would melt enough to cause serious increases in water levels—flooding—along all the coasts of the earth. However, careful measurements of the temperature of our atmosphere indicate that instead of increasing, the temperature has actually fallen a degree or two during the world's most intense period of industrialization. The explanation for this cooling seems to be that our industrial societies spew out so much waste that they have increased the murkiness of our atmosphere. This reflects the solar radiation (heat) that

ACID PRECIPITATION
Rain, snow, or fog higher in acidity than normal rain

PHOTOCHEMICAL SMOG
Air laden with compounds produced by the reaction of sunlight and hydrocarbons released from incomplete burning of fossil fuels

FIGURE 9–13
The greenhouse effect results when the rays of solar energy are absorbed by objects within a car, house, or greenhouse, and are changed to heat or infrared rays that cannot escape back into the atmosphere. In the atmosphere a similar effect occurs. In addition, the infrared rays are reabsorbed by water and carbon in the atmosphere. When the amounts of water and carbon dioxide change, so does the temperature of the atmosphere.

would otherwise reach the earth and become trapped near us by carbon dioxide. That humans have reduced the amount of sunlight that reaches the earth should probably be no cause for celebration.

Fluorocarbons. Finally, in this connection, our waste **fluorocarbons** may be having an effect on the filtering capacity of the ozone layer (which we learn about in chapter 13) that prevents ultraviolet radiation reaching the earth. Ultraviolet radiation is very harmful in large amounts, and even small amounts can cause skin cancer in humans. Refrigerators contain a fluorocarbon coolant that is released when they are thrown "away," and some aerosol sprays are propelled by fluorocarbons.

FLUOROCARBONS Compounds containing carbon and fluorine, used in aerosols, lubricants, and insulators

Water Pollution. Water becomes polluted because human beings put their wastes into it so that the water will carry it "away." Indeed, we pipe our wastes long distances and into each other's drinking water, causing each little municipality to have to treat its water before it can be used. About one-third of the waste water in this country is from our households; two-thirds comes from industry and commerce.

Contamination of water with human sewage causes a number of water-borne diseases, including cholera, dysentery, infectious hepatitis, and typhoid fever. Nitrate pollution in water, from overuse of inorganic fertilizers, concentrates in crops and also causes disease in farm animals and human infants. The pesticides and herbicides that are used end up in our food chain and thus in our bodies. Sufficient concentrations of these pollutants cause disease in humans and the extinction of certain species of birds. Phosphates from our detergents kill fish. Power plants and factories use water for cooling then discharge it at higher temperatures to flow downstream. Aquatic organisms may be unable to survive at that new temperature, even if the change is only a few degrees. Radioactive substances from commercial and military uses of nuclear materials also end up in our water, as do lead, fluorides, household and industrial chemicals, and sediments from **erosion** caused by agriculture and other human activities.

EROSION Movement of material from one place to another on the earth's surface; the movement of soil to the sea

Soil. The capacity of our earth's soil to grow food and forests is being reduced. Just as important, the absolute amount of soil that covers our land surface is being reduced. Soil is not just dirt. It is an accumulation on the surface of the earth of sand, clay, bacteria, and decayed plant and animal material called humus. Soil supports the growth of vegetation, and as plants grow they incorporate nutrients from the soil, such as nitrates. When plants are taken away after they have grown, as in the case of food and timber, the nutrients are carried away also, and the soil is thereby depleted. These nutrients can only partly be returned by using fertilizers. Humus, which retains water, oxygen, and nitrogen, cannot be

Wind can be a primary cause of soil erosion. (USDA—Soil Conservation Service)

returned by artificial means. Rotating crops that take different nutrients from the soil will, over time, allow soil to restore itself. Not growing too much on a given area within a period of time also allows restoration. It is good practice to allow soil to "lie fallow"—uncultivated—periodically in order for soil to come back.

Erosion is reducing the absolute amount of soil on earth, and thus the absolute amount of our food-growing capacity. Erosion—in this case, the movement of soil from arable land surfaces by wind, rain, and gravity—is caused by overly intensive cultivation of crops and by urbanization, the process by which we pave over land and build structures on it.

Food.   It is estimated that one-sixth of all the human beings on earth do not have enough to eat. This group includes several million Americans, even though the United States is a food-exporting nation.

World hunger has at least five roots: soil depletion; urbanization, whereby farm land is put into non-agricultural use; overpopulation; intensive farming techniques; and the displacement of food crops by more commercially profitable, nonfood, cash crops. The last two causes need explanation. Intensive techniques for raising livestock, especially cattle, involve feeding these animals food grown on land that could otherwise be used to produce food for people. Further, for every pound of beef we produce, we must supply the cow with ten pounds of feed. Ninety percent of U.S. corn is grown for livestock, not for people. Cattle normally eat grass, which humans cannot eat, and then convert it to meat, which humans can eat. The feedlot system of beef cattle production, in effect, puts beef cattle in direct competition with humans for food. Land that would otherwise grow food is used instead for nonfood products like carnations and rubber, and for nonessential foods such as coffee, tea, or cocoa.

Forests and Wilderness.   The forests of the earth are in jeopardy. Firewood and timber, in spite of being flow-limited resources that are *able* to renew themselves, are disappearing because their life-support systems—air, water, and soil—are being poisoned by pollution. They are also disappearing because we are harvesting them in greater quantities than they are renewing themselves. We are failing to note that although wood is renewable, its availability is limited by the *rate* at which it is renewed: That is exactly what we mean by flow-limited.

We need forests; all living things do. They provide air-breathing creatures with oxygen; they allow thousands of species of plants and animals habitats within which to prosper and diversify; they preserve **watersheds** and prevent erosion (and thus allow dams to serve their irrigation purposes longer); they moderate climate; and they give us wood for fuel and for buildings. Sometimes, just their presence seems to make us feel delighted—at one with a world that is meaningful.

WATERSHED Land area drained by a river or a river system

The fact that forests and forest fires go hand in hand is not widely recognized. Yet the fact is this: native Americans are known to have set fire to forests on the slightest of pretexts—to hold a picnic, to rid an area of mosquitoes, or to clear the land for a garden. As a result, and in concert with lightning and other causes of fire, native Americans never really experienced forest fires as we know them today. This is because the understories of forests were always being burned up by small brush fires, and, consequently large forest fires could not develop. It seems by suppressing small forest fires, large ones have become inevitable and devastating.

Wilderness is what the whole surface of the earth was like before humans discovered the technologies of horticulture and agriculture some ten thousand years ago. Humans have spent 99 percent of their history on earth living in wilderness. Our primordial roots are in wilderness,

John Muir (1838–1914)

When he was just eleven years old, John Muir's family moved from Scotland to Wisconsin. There he grew up on a farm surrounded by nature which soon became an integral part of his life and love. He attended the University of Wisconsin and later became a teacher while doing summer work on farms.

By the time he was twenty-nine, he decided to devote his life to nature and the wilderness of his adopted country. Alone and on foot he wandered over America's forests, up its mountains, down through its valleys and into its meadows. He filled his notebooks with sketches and descriptions of the plants and animals he loved. He was excited about his life: "I set forth . . . joyful and free . . . by the wildest, leafied and least trodden way I could find."

And unusual places he did find. From his farm in California, he walked to the towering walls of the Yosemite Valley. Climbing through snow that many times measured up to his waist, he was determined to see the top. He reported that with less than an hour left before sundown, he was almost to the top when he heard a rumble. Above him snow was beginning to break and fall—an avalanche. In a moment it was tossing him toward the valley. In a split-second impulse, he positioned himself like a spread eagle—and rode on top of the snow. On the valley floor he was "on top of the crumpled pile without a bruise or scar."

John Muir convinced the U.S. Congress to establish Yosemite National Park in 1890. He also persuaded then President Theodore Roosevelt to set aside 148 million acres of forest resources, including the Muir Woods in California—a redwood preserve.

Muir loved the wild lands he roamed. His love translated into political action in his efforts to preserve these lands. Today, the Sierra Club which he helped form in 1892 continues his work. John Muir's life demonstrates his belief that "nature has always something rare to show us."

which may be why so many of us experience a special feeling when we are in the wilderness for a while. It is awesome and beautiful.

Wilderness used to be the *habitat* where every living creature *lived* in relationship to every other living creature, regardless of species. Human beings cannot recreate wilderness in zoos, managed forests, or parks. There is no substitute for wilderness. We can only let wilderness be, and enjoy it sparingly, because it is vanishing. Every time a bit of wilderness is taken over by humans, it will not be wilderness any more for a long time. When wilderness is lost, hundreds of species of animals are also lost to the area. Some of them become endangered nationally and some become extinct.

However, increased awareness of the problem has prolonged survival for many species. The Lion Safari zoo near Los Angeles has raised young lions and re-introduced them in protected sanctuaries in Africa. Similar projects have been undertaken for orangutangs in Southeast Asia. Whooping cranes, nearly extinct a few years ago, are now being bred and released to join those that remain in the wild. These humanly assisted birds have now joined their kind and live and migrate with them. And, recently, we just may have saved the great whales.

The recurrent message throughout this chapter can be symbolized by our responses to these animals; we have attempted to restore them to their natural environments and numbers. The future is indeed a bleak one unless we relearn and again act in accordance with the basic limits of nature. Our earliest ancestors survived because they were able to live in harmony with the environment. We have taken precious time to learn that we cannot ignore those limits. We can reverse the frightening course we have taken if we begin to act more responsibly now.

BIBLIOGRAPHY

Bold, H. C.; Alexopoulos, C. J.; and Delevoryas, T. *Morphology of Plants and Fungi.* 4th ed. New York: Harper and Row, 1980.

Bureau of Mines, United States Department of the Interior. *Mineral Facts and Problems.* Bulletins 667 and 671. Washington, D.C.: U.S. Government Printing Office, 1976 and 1981.

Ehrlich, Paul R.; Ehrlich, Anne H.; and Holdren, John P. *Ecoscience: Population, Resources, Environment.* San Franciso: W. H. Freeman, 1977.

Energy Information Administration, U.S. Department of Energy. *1981 Annual Report to Congress.* Vols. 1, 2, and 3. Washington, D.C.: U.S. Printing Office, 1982.

Karmody, E. J. *Concepts of Ecology.* 2nd ed. Englewood Cliffs, NJ: Prentice-Hall, 1976.

Keeton, W. T. *Biological Science.* 4th ed. New York: W. W. Norton, 1980.

Krebs, C. J. *Ecology.* New York: Harper and Row, 1978.

Leet, L. Don; Judson, Sheldon; and Kauffman, Marvin E. *Physical Geology*. 6th ed. Englewood Cliffs, NJ: Prentice-Hall, 1982.

Lehninger, A. *Principles of Biochemistry*. New York: Worth Publishers, 1982.

Miller, G. Tyler, Jr. *Living in the Environment*. Belmont, CA: Wadsworth Publishing Company, 1982.

Nadar, Ralph. Introduction to *Accidents Will Happen: The Case Against Nuclear Power*. Edited by Lee Stephenson and George R. Zacher. New York: Harper and Row, 1979.

Odum, E. P. *Fundamentals of Ecology*. 3rd ed. Philadelphia: W. B. Saunders, 1971.

Simpson, G. G., and Beck, W. S. *Life. An Introduction to Biology*. New York: Harcourt, Brace, and World, 1965.

Turk, J., and Turk, A. *Physical Sciences with Environmental and Other Practical Applications*. Philadelphia: W. B. Saunders, 1977.

10

Teaching Environmental Biology

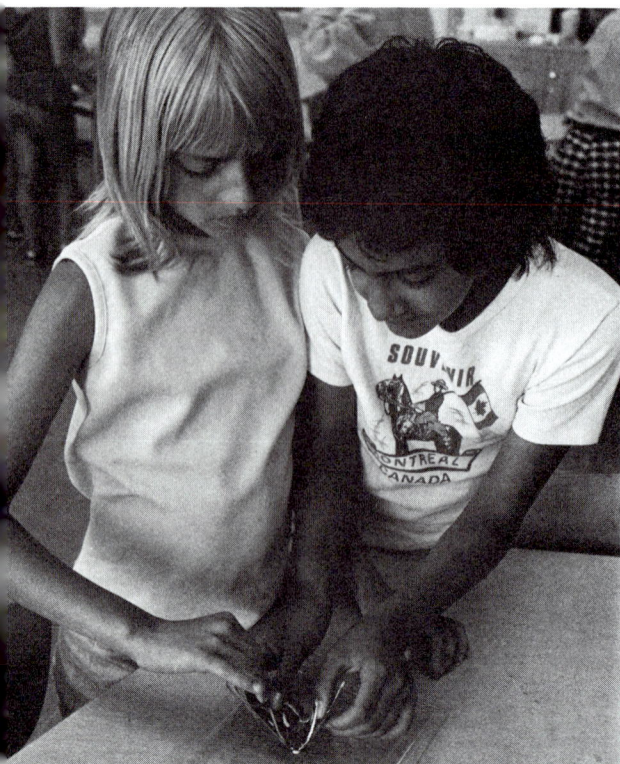

The purpose of chapter 10 is to provide examples of science lessons designed to teach six concepts in **environmental biology** to students from kindergarten through eighth grade. Planning Charts 10–1, 10–2, and 10–3 illustrate how each science lesson allows you to meet several objectives at the same time.

KEY CONCEPTS AND IMPORTANT IDEAS

☐ describing ourselves and others ☐ matching descriptions with people ☐ watching people change ☐ discovering changes outdoors ☐ identifying irreversible changes ☐ starting a nature collection ☐ learning how scientists organize nature ☐ starting gardens in containers ☐ recording changes in plants ☐ recognizing the limits of change ☐ talking to "Old Timers" about environmental changes ☐ field trips to a museum ☐ reading about changes in the environment from old newspapers ☐ visiting an environmental planning office ☐ studying wildlife

CHART 10-1
Planning Chart for Environmental Biology: Kindergarten through Grade 2

Activity	Objectives					Teaching Methods	Related Curriculum
	Conceptual Knowledge	Learning Processes	Psychomotor Skills	Attitudes and Values	Decision-Making Skills		
1. Describing Myself and Others	Living things have identities based on their characteristics.	Observe, measure.	Use and care for simple tools.	Value the uniqueness of one's identity.	Recognize personal decisions that involve self.	Learning center	Art, math
2. Matching Descriptions with People: Guess Who?	Living things have identities based on their characteristics.	Observe, compare.	. . .	Value the uniqueness of one's identity.	. . .	Game evaluation	Language arts
3. Watching People Change	Living things change when they interact with other things (living/nonliving matter) in the environment.	Observe, compare.	. . .	Value the health and welfare of self and others.	Discuss consequences of alternative actions.	Inquiry, discussion	Social studies
4. Discovering Changes Outdoors	Living things change when they interact with other things (living/nonliving matter) in the environment.	Observe, compare, measure.	. . .	Explore the environment and ask questions.	Accept direction in caring for health and welfare of self and others, and wise use of natural environment.	Nature walk	Social studies
5. Identifying Irreversible Changes	Living things can irreversibly change and still keep (conserve) their identities.	Observe, compare, classify.	Use and care for simple tools.	Actively explore environment; combine materials in new ways.	. . .	Discussion, projects, learning center	Language arts, art

CHART 10–2
Planning Chart for Environmental Biology: Grade 3 through Grade 5

Activity	Objectives					Teaching Methods	Related Curriculum
	Conceptual Knowledge	Learning Processes	Psychomotor Skills	Attitudes and Values	Decision-Making Skills		
6. Starting a Nature Collection	Individuals and groups of living things have identities based on their characteristics.	Observe, compare, classify, infer.	Assemble simple projects; use and care for tools and equipment.	Assemble materials in new ways; assume responsibility for the wise use of material resources; value the uniqueness and beauty of objects and organisms in nature.	. . .	Individualized and group projects, exploratory discovery	Art
7. Learning How Scientists Organize Nature	Individuals and groups of living things have identities based on their characteristics.	Observe, compare, measure, gather, and organize information.		. . .	. . .	Lecture, lab	Social studies
8. Starting Gardens in Containers	All groups of living things change; members of the groups interact with each other, with other groups, and with the environment.	Observe, compare, measure.	Use and care for tools and equipment; handle and care for plants; use nontoxic substances; construct simple projects.	Assume responsibility for providing basic needs of plants.	Accept consequences of one's own decisions.	Demonstration, team learning, projects	. . .

9. Recording Changes in Plant Groups	All groups of living things change; members of the groups interact with each other, with other groups, and with the environment.	Observe, compare, gather and organize information, infer.	Use and care for tools and equipment; handle and care for plants; use nontoxic substances; construct simple projects.	Assume responsibility for providing basic needs of plants.	Accept consequences of one's own decisions.	Discussion, lab report, journal keeping	Language arts, math
10. Recognizing the Limits of Change in Plants	When groups of plants are forced to change beyond their limits, groups lose their identity (do not survive) even though matter and energy are conserved.	Gather and organize information, infer.	Handle and care for plants, tools, and equipment.	Explore and ask questions; demonstrate concern for the natural environment and for basic needs of plants.	Describe decisions that involve self and others; describe consequences of various actions or decisions.	Guided discovery, field trip	Social studies

CHART 10–3

Planning Chart for Environmental Biology: Grade 6 through Grade 8

Activity	Objectives					Teaching Methods	Related Curriculum
	Conceptual Knowledge	Learning Processes	Psychomotor Skills	Attitudes and Values	Decision-Making Skills		
11. Talking to "Old Timers"	Environments change as living and nonliving matter interact.	Gather and organize information, infer, describe relation among variables.	Use equipment that requires fine adjustment and discrimination.	Actively explore and ask questions; seek alternative points of view and multiple sources of evidence.	. . .	Student projects, field trips, note-taking, journal-keeping, oral reports	Social studies
12. Field Trip to Museums	Environments change as living and nonliving matter interact.	Gather and organize information, infer, describe relation among variables.	Use equipment that requires fine adjustment and discrimination.	Actively explore and ask questions; seek alternative points of view and multiple sources of evidence.	. . .	Student projects, field trips, note-taking, journal-keeping, oral reports	Social studies
13. Reading Old Newspapers	Environments change as living and nonliving matter interact.	Gather and organize information, infer, describe relation among variables.	Use equipment that requires fine adjustment and discrimination.	Actively explore and ask questions; seek alternative points of view and multiple sources of evidence.	. . .	Student projects, field trips, note-taking, journal-keeping, oral reports	Social studies
14. Visiting a Planning Office	Environments change as living and nonliving matter interact.	Gather and organize information, infer, describe relation among variables.	Use equipment that requires fine adjustment and discrimination.	Actively explore and ask questions; seek alternative points of view and multiple sources of evidence.	. . .	Student projects, field trips, note-taking, journal-keeping, oral reports	. . .
15. Studying Wildlife	Environments change as living and nonliving matter interact.	Gather and organize information, infer, describe relation among variables.	Use equipment that requires fine adjustment and discrimination.	Actively explore and ask questions; seek alternative points of view and multiple sources of evidence.	. . .	Student projects, field trips, note-taking, journal-keeping, oral reports	Social studies

ACTIVITY EB–1

Describing Myself and Others
(Kindergarten–Grade 2)

ACTIVITY OVERVIEW

In this first activity, students are introduced to the concept of *identity* by looking in a mirror and at one of their classmates as they try to find ways to describe the uniqueness of each one's identity. *Estimated time:* Learning Center for one week, plus one class period.

SCIENCE BACKGROUND

Scientists spend a great deal of time studying the identities of living and nonliving matter on earth and in the universe. By studying the structure, organization, and chemistry of living and nonliving matter, scientists assign names to all groups of plants and animals as well as to the parts of their bodies, their life processes, their rhythms or cycles, and so forth. Scientists identify the earth's landforms, the formation of the seas and other bodies of water, the formation of stars or other celestial bodies in the universe, and the gases, liquids, and solids that compose these forms. Naming things is important to scientists because it allows them to communicate with one another about their understanding of the universe and the individuals or matter within it. Thus, identity is essential to scientists.

The identity of human beings has been described by scientists. In our structure, organization, processes, and chemistry, we are vertebrates and mammals and we are composed of cells. That means that our species has backbones and mammary glands that secrete milk. We have an insulating layer of body hair which aids in temperature regu-

lation. We are warm-blooded meaning that we can maintain our body temperature at a high and more or less constant level. As human beings, we are placental mammals: our embryos are nourished within the female's body by a special organ, the placenta, and are not born until development is advanced. And finally, like all living things, we are composed of cells which, in turn, are reproduced by "simple" molecules of **DNA (deoxyribonucleic acid).**

To the children in your classes, these aspects of their identities are not meaningful because they cannot see their backbones, cells, or DNA molecules. But they can see differences between themselves and their classmates. As fellow human beings, their personal or individual identities, (instead of their species identities) are far more meaningful to them.

SOCIAL IMPLICATIONS

People of all ages recognize themselves as individuals who have identities. They value the identities of their family members, friends, pets, and favorite possessions. Identity is a beginning point of reference to the familiar world and thus the beginning point of learning about the identity and the value of less familiar things. In a large and technological society, it is easy to lose sight of the worth of the individual. When decisions are made, they are often made in behalf of minors or for the benefit of the greatest number of individuals in society. When such decisions are contrary to the individual's preference, a part of the individual's self-worth is unexpressed. Therefore the concept of identity reinforces the self-worth of the individual in our society.

But identity is important for a reason that has little to do with recognizing an individual or distinguishing between two individuals. Because you and your students exist and are alive, you all have identities. And when you and your students cease to exist, you will lose your identities. Thinking of identity this way becomes important later when students begin to see how individuals can change and still maintain their identities, or how individuals can change so much that they lose their identities. Thus, students are led to seek ways to preserve the identity of living and nonliving matter in the environment.

MAJOR CONCEPT

Living things have identities based on their characteristics.

OBJECTIVES

- ☐ Describe oneself and another person in such a way that the identity of individuals is recognized by others.
- ☐ Observe and measure.
- ☐ Use tools.
- ☐ Value the uniqueness of the individual.
- ☐ Recognize personal decisions that involve self.

MATERIALS

Mirror, tape measure, bathroom scales, color chart, crayons, paper, and art pencils. If a floor length mirror is available, secure it against the wall temporarily and attach the tape measure to the mirror.

VOCABULARY

color words for hair, skin, eyes, clothing
texture words for clothing
size words for height, weight, and hair length
shape words for articles of clothing

PROCEDURES

A. Set up a learning center with the required materials. Demonstrate how children are to use the learning center, perhaps by using it yourself. Point out the special features that give each person an identity: appearance in height and weight; color, length, and style of hair; color of eyes; clothing; and so forth.

Then invite children to describe themselves, first by looking in the mirror and recording what they see, using crayons and paper, and later by verbally telling someone else. Provide informal assistance to individual children who need to improve their skills of observation and description. Make a temporary display of the completed pictures; they will be used later in Activity EB–2.

EXTENDING THE ACTIVITY

B. On another day, invite the children to choose a partner to describe. Each pair of children use the learning center together. To improve the verbal skills of less articulate children, pair them with partners who have more advanced verbal skills, and ask each pair to work together as a team at writing descriptions of each other. Guide the students in writing their descriptions by saying:

1. Try to describe your partner so that we (members of the class) can guess who it is.

2. Use words to describe your partner's color of hair, eyes, and clothing.
3. Tell how tall your partner is and how much he or she weighs. Use yourself as an example; write a description of yourself on the chalkboard. Then ask someone to read the description aloud with you and ask the class who, in the room, fits the description best.

As students complete their descriptions, have them write at the bottom of the page, the name of the person they have described. Later in the week, children may use these pictures and verbal descriptions to play "Guess Who?" described in Activity EB–2.

C. To assist children in understanding the concept of an individual's identity, have two students who look somewhat alike stand up together, and ask the class how they can tell these two classmates apart. Help students focus on the characteristics that differ between the two children. Repeat this activity several times until children's descriptions are clear in their distinction between pairs of students.

Equip a learning center with simple "tools" that students can use to observe, measure, and record the features that comprise each student's identity.

ACTIVITY EB–2

Matching Descriptions with People: Guess Who?
(Kindergarten–Grade 2)

ACTIVITY OVERVIEW

In this activity, students learn the importance of close observation and careful description as they try to identify individual classmates from pictorial and verbal descriptions. *Estimated time:* one to two class periods.

SCIENCE BACKGROUND

In 1758, a Swedish biologist called Carolus Linnaeus devised a system to catalog or classify every **species** (species is a group that interbreeds) of plants and animals known in the eighteenth century. He called his encyclopedic work *Systema Naturae* and *Species Plantarum*. Linnaeus was an astute observer and believed that a careful description of each species would allow all biologists to recognize, describe, and agree upon the identity of every plant and animal species. Thus, close observation and careful description are essential scientific processes with which to identify new species.

SOCIAL IMPLICATIONS

As the world population grows, the welfare of people is more dependent upon sharing the earth's resources. Thus, it is crucial that people learn to accept and value the uniqueness and differences of their earthly neighbors. Learning to recognize and communicate about the uniqueness of individuals is related to understanding and valuing individual differences. As a teacher, you can help children develop skills to communicate and

recognize the uniqueness of their classmates. Through understanding, they may come to value these differences.

MAJOR CONCEPT

Living things have identities based on their characteristics.

OBJECTIVES

- ☐ Match pictorial and verbal descriptions with real people.
- ☐ Observe and compare.
- ☐ Value the uniqueness of the individual.

MATERIALS

Pictures and verbal descriptions of children in the class from Activity EB–1.

VOCABULARY

Same as in Activity EB–1.

PROCEDURES

A. Display the pictures children have drawn in Activity EB–1. Invite children in the class to guess who the pictures portray.

B. On another day, have children read aloud the verbal descriptions they have written describing their partner, and ask the class to guess the identity of the person being described. Assist children whose descriptions are ambiguous by encouraging them to add details to their descriptions which will allow others to guess the identity of the person they have described.

EXTENDING THE ACTIVITY

C. Have four children stand in front of the class. Ask another student to describe one of the four standing and encourage the other class members to try to guess who is being described.

D. Invite children to give on-the-spot verbal descriptions of plants or animals in the classroom or school ground while classmates try to match the descriptions with the correct objects.

Class members will enjoy guessing who is being described and taking turns describing others to the class.

ACTIVITY EB-3

Watching People Change
(Kindergarten–Grade 2)

ACTIVITY OVERVIEW

In this activity, students' attention is focused on natural changes that occur to them and their classmates during the school year. They talk about, draw pictures of, and pantomime changes in their appearance and behaviors.

This activity can continue throughout the year as children grow, develop, and change. The teacher's role is to focus children's attention on the changes that take place naturally and to help children connect causes with these changes in appearance and behavior. *Estimated time:* 5–30 minutes every few weeks throughout the year.

SCIENCE BACKGROUND

Human conception occurs by the uniting of a male sperm and a female ovum, and the infant is born about 266 days later (or 280 days after the beginning of the female's last, regular menstral period). During that period, the embryo changes rapidly. All of the major organs develop during the first three months while the fetus is less than 8 centimeters long. Fetal movement and heartbeat can be felt in the second three month period and from the sixth to the ninth month, the fetus increases greatly in size; brain cells develop rapidly.

After birth and during the next five to seven years of life, children reach a weight of about 15–25 kilograms (35–45 pounds) and range in height from 100–120 centimeters (40–50 inches), that is, approximately 60–75 percent of their adult height. However, variations in physical growth and development are regulated by quality of diet and genetic inheritance. They learn to walk and play; large muscle coordination develops through running, tumbling, and balancing. Eye-hand coordination begins to develop for written tasks and manipulative activity.

Behavior is regulated by internal forces in part and influenced by external or environmental factors. For example, when children are hungry or tired, their behavior reflects the internal stress they are experiencing. A particularly pleasant or unpleasant companion or a change in weather can bring about a change reflecting the influence of external factors on behavior.

Children learn to talk between their first and second year of life and generally begin to explore the use of written and numerical symbols shortly before they enter school. Reasoning is more or less logical when they focus on a single variable (for example, the length or width of an object). By approximately seven years of age, most children can seriate, that is, place objects in some order, based on a range of sizes, shades of color, and so on.

SOCIAL IMPLICATIONS

Children are extremely interested in the functioning of their own bodies and generally respond with interest to conversations about their own growth, development, skills, and behavior. They are eager to try new tasks and frequently overestimate their ability, coordination, or energy level. Discussions about these topics help students gain important information about their development in relation to those around them. For example, talking about the loss of one's first tooth, or feeling frustrated or angry, or having chicken pox, or getting a new pair of glasses contributes to every child's understanding of how people change.

MAJOR CONCEPT

Living things change because they interact with other living and nonliving things (matter) in the environment.

OBJECTIVES

☐   Describe physical and emotional changes in people.

☐ Observe and compare.

☐ Accept guidance in caring for the welfare of self and others.

☐ Recognize personal decisions that involve self and discuss consequences of various alternatives.

MATERIALS

Art paper, crayons.

VOCABULARY

Words related to growth and development
Words related to emotions and behavior
Words related to changes in clothing

PROCEDURES

A. *Weather* affects the way people feel and behave. It causes them to change what they wear, what they eat, and how hard they play or work. Children will enjoy drawing pictures of the clothes they wear in various kinds of weather. On windy days when children are high-spirited, talk with them about how they feel and behave and ask them to contrast this with the way they feel and behave in other (hot, muggy, calm, or cool) weather.

B. Physiological requirements (food, water, rest, shelter, and safety) affect the way

people look, feel, and behave as well. With proper diet, children grow and change in appearance; they lose teeth, outgrow clothing or need new shoes. Sometimes they become taller and thinner; at other times they stay the same height and fill out instead.

Call children's attention to their restlessness before lunch when they are hungry, to their relaxation or exhaustion after recess, to how energetic they feel on the playground, and how sleepy they feel at night and early morning. Ask children to describe their own behavior in response to these physiological needs. Comment constructively about the causes of these changes in behavior.

EXTENDING THE ACTIVITY

C. *Emotions* are reflected in people's faces and emotions change in response to other people and the environment. Help children understand these changes in their own emotions and the effect their emotions have on other people. Invite children, one at a time, to pantomime specific feelings while the rest of the class describes their feelings in relation to the performer. Have a discussion of responsibility for one's actions, using the decision-making objectives in table 2–5, chapter 2.

ACTIVITY EB–4

Discovering Changes Outdoors
(Kindergarten–Grade 2)

ACTIVITY OVERVIEW

In this activity, students extend their concept of change in living things to include changes that occur in nature outside the classroom. They observe and describe seasonal changes and relate them to changes in classroom plants and animals. *Estimated time:* 30-minute walk once each month.

SCIENCE BACKGROUND

Changes in the plant and animal world that surround you are regulated both by characteristics inherited by the plants and animals themselves and by changes in the **weather.** Weather (the condition of the atmosphere at a given place and time) is regulated by the earth's movement, sun spots, and moving masses of air. It is also a local matter and can be influenced by lakes, forests, mountains, and other natural features. Thus, the changes that occur in the plants and animals in any specific region may differ somewhat from changes that occur in the same species in another region. For example, spring comes to the hills of the San Francisco Bay Area in January and February when winter rains, temperature, and climate cause mustard plants to spring up and flower.

Seasonal changes are also more pronounced in some regions of the country than others, but in general, seasonal changes in weather produce visible changes in plants and animals. In areas where temperatures drop below freezing in the winter, the leaves of deciduous trees change from green to brilliant shades of red, orange, and yellow in the autumn. These changes in color are caused by the disintegration of chlorophyll in the leaves; in some trees, color change is caused by the simultaneous chemical change in colorless sugars, gums, and tannins. The combination of day length, moisture, temperature, and sunlight combine to produce these color changes before the leaves fall in species such as maple, oak, gingko, ailanthus, sycamore, and dogwood.

Near-freezing temperatures also cause many animals to store energy in the form of body fat reserves and to hibernate during the winter. Animals such as deer, ducks, and geese migrate to warmer areas while many insects and other animals are killed off by the winter temperatures.

In the spring, the transition back to warmer temperatures brings new growth and new green leaves on trees and shrubs. Seeds that were left on the ground over winter now sprout and form a new generation of wildflowers, weeds, and grasses. Animals resume their activity and begin new families.

SOCIAL IMPLICATIONS

The lives of people throughout the world and throughout human history have been governed by seasonal changes. The production, harvesting, and preservation of food and the manufacture of clothing and shelter have been linked out of necessity to the seasons, as are recreation, sports, and games. The aesthetic expression of interest and value in seasonal changes is evident in music, art, and literature. Teachers have little difficulty arousing students' interest in seasonal changes since the very nature of change brings anticipation of renewed resources.

MAJOR CONCEPT

Living things change when they interact with other living and nonliving things (matter) in the environment.

OBJECTIVES

☐ Observe, compare, measure.

☐ Explore the environment and ask questions.

☐ Accept direction in caring for the health and welfare of self and others.

☐ Use the natural environment wisely.

MATERIALS

Optional: camera with instant developing film, tape recorder.

VOCABULARY

living things	soil
plants	weather
animals	change
rocks	

PROCEDURES

A. Plan a short (one-half hour) nature walk in the neighborhood near the school. Select a path that takes children through an area with some trees or shrubs and smaller plants (weeds, flowers, or lawns) and some animal life (birds, insects, or small animals such as pill bugs, spiders, and so on). You might consider including the yard of a neighbor near the school who takes pride in his or her yard and would enjoy showing students what changes occur during the year in the yard. Just before the field trip, explain to the children that they will visit this same area often (once a month is about right) to observe changes during the year.

On the first walk, point out a few specific features such as flowers that are in bloom, leaves that are (or are not) on trees, and particular birds or insects that you see. Speculate how these might change by your next visit.

B. If you are in a place where it is permitted, collect a small number of items such as leaves, a cocoon, or an acorn to take back to the class to show what you saw on the first trip there.

Field trips can be to a nearby field, park, or in the school yard. Encourage students to observe the environment carefully with their senses of touch, sight, hearing, and smell.

C. A camera that instantly produces pictures or a tape recorder can be used to record what was seen. These records can be enjoyed much later when children want to compare what they saw during various nature walks or seasons.

D. Especially good times to take this nature walk are:

on a windy day
　to see things blowing as nature is changing and children are restless

after the rain
　to see how plants and animals survived the rain and how wet or soaked things have become

after a change in temperature
　to see how the hot or cold weather has changed trees, shrubs, small plants, and animals

on warm and sunny days
　just to enjoy nature

EXTENDING THE ACTIVITY

E. During or after each walk, talk about the changes that have occurred. Ask children questions like these:

1. Are the leaves, flowers, or bark the same color?
2. Are there as many leaves or flowers; are they the same?
3. Where are the (birds, butterflies, bees, and so forth) that we saw last time? Are they the same? Do they still have food to eat? Where might they go?
4. How does the air feel today? How did it feel the last time we were here?
5. Are things (in nature) as wet or dry as they were on our last visit?

Guide children to recognize through their observations that, in nature, living things change in response to other living and nonliving things.

ACTIVITY EB–5

Identifying Irreversible Changes
(Kindergarten–Grade 2)

ACTIVITY OVERVIEW

In this activity, students identify irreversible changes in order to extend their understanding of the limits of change and the conservation of one's identity. *Estimated time:* Daily for one week.

SCIENCE BACKGROUND

While many changes in living things appear to be temporary or reversible, there is never-

theless within every organism a genetically coded life cycle which results in irreversible changes in the organism. Children become adolescents, then young adults, middle-aged adults and, ultimately, old adults. The life cycle of *Homo sapiens* is never reversed, in spite of the efforts of many people to "return to their youth." The life span for some species is longer than that of humans, a fact that makes it difficult for people to visualize change in these species. For example, some of the Coast Redwood trees in California (*Sequoia sempervirens*) are more than two thousand years old and look like they have always been tall. Regardless of their size (some reach 360 feet), these trees never reverse their height or their life cycle.

SOCIAL IMPLICATIONS

Permanent or irreversible changes in living things have been accepted routinely in the past because changes associated with life cycles were viewed as natural. However, our capacity to control nature has increased; we can, for example, change the weather, the course of rivers, and the variety and rate of plant growth and decay. Scientists have discovered that some technological changes have changed environments in ways that may be harmful to human health or the survival of other species. Insecticides such as DDT and other persistent hydrocarbons used to kill insects on food plants are also passed on to other animals along the food chain, and ultimately affect groups that were to receive the greatest protection or benefit. In the 1950s, no one imagined that the vast oceans or atmosphere could become polluted. But when DDT was found in the bones of penguins in the Antarctic, we were forced to presume that DDT was carried by currents from continents in the northern hemisphere. Another example is the reduction in the volume of lumber harvested in the Sierra Nevada range of California. A reduction in growth of the diameter of ponderosa pine, sugar pine, and incense cedar is presumably due to air pollution. Both examples indicate that human intervention in natural environments and cycles is much more complex than anyone has imagined and might actually and unknowingly cause serious, irreversible changes to the environment.

MAJOR CONCEPT

Living things may change irreversibly and still keep (conserve) their identities.

OBJECTIVES

☐ Describe living things that change irreversibly.
☐ Observe and compare.
☐ Use and care for simple tools.
☐ Assemble simple projects with guidance.
☐ Actively explore the environment.
☐ Combine materials in new ways.

MATERIALS

Old magazines with pictures of people, food, plants or animals; comic sections of newspapers; scissors; tape; long strip of butcher paper (about 5 feet long) posted for a mural.

VOCABULARY

change	different
reversible	irreversible

PROCEDURES

A. Introduce the topic of irreversible change by showing students photographs of yourself taken from childhood to adulthood. Or you may prefer to use pictures of a pet taken during different stages of its development, or a series of pictures showing puppies growing into dogs, kittens growing into cats, calves growing into cows, and so on.

Explain that many changes in nature are permanent. For example, when things grow up, they don't change back to babies again. Cite several other examples.

EXAMPLES OF PERMANENT CHANGES

1. Children grow up and don't become small again.
2. Pencils are sharpened and don't become whole again.
3. Eggs are cooked and don't become raw again.
4. Matches burn and don't become unburned.
5. Food gets eaten and doesn't become uneaten.
6. Plants die and don't come alive again.

Compare these permanent changes with the temporary changes discussed in activities 3 and 4 (changes in the weather or seasons of the year and changes in human behavior).

EXTENDING THE ACTIVITY

B. Invite children to find pictures of irreversible changes. The pictures they find will be cut out and shared with the class. Open a magazine and turn pages, showing children as you turn and saying: Is this picture of something that (has/can/will) change(d) permanently? Find two pictures that will illustrate irreversible change—for example, a puppy and an adult dog. Cut them out of the magazine and tape them onto the mural (butcher paper). Draw an arrow, from the puppy to the dog and explain why this is an example of irreversible change. Place a supply of magazines or comics in the Learning Center and provide guidelines for students using the Learning Center.* As each student chooses cut-out pictures of irreversible change, ask him or her to tell you how the examples illustrate irreversible change. If the student understands the concept, have him or her tape the two pictures on the mural and draw the arrow in the direction of the irreversible change. On the other hand, if the student can't explain how the pictures show irreversible change, help the student find an appropriate picture. In other words, teach the concept on an individual basis.

*For ideas on how to set up a learning center, see the Taxonomy of Common Teaching Methods, chapter 4.

Some changes in nature are irreversible. Students learn to recognize irreversible changes as they put examples on the mural.

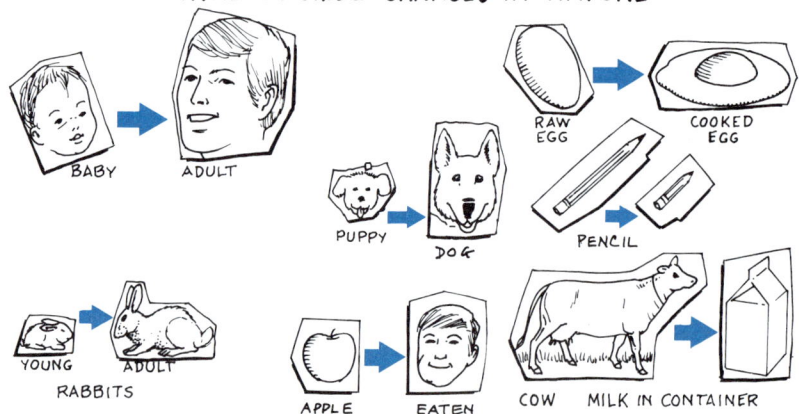

IRREVERSIBLE CHANGES IN NATURE

ACTIVITY EB–6

Starting a Nature Collection
(Grades 3–5)

ACTIVITY OVERVIEW

In this activity, students contribute to a nature collection that will be given to the school as a reference collection. Personal collections may develop as individual projects. *Estimated time:* Ongoing project throughout the year.

SCIENCE BACKGROUND

There are many great nature collections in the world. Like collections of art, many of the most famous nature collections are in museums which have specimens (individual organisms) that have been collected from throughout the world over many decades, identified, and preserved. One purpose of these large collections is to show how widely any given species of plant or animal varies in different environments and to find out if a given population of plants or animals varies noticeably over time.

One of the best known naturalist-collectors was Charles Darwin. As an avid collector of beetles, mollusks, and other small organisms, Darwin was overjoyed at the opportunity to work as a naturalist (without pay) aboard a surveying ship called the H. M. S. Beagle. In 1831, Charles Darwin began the five-year voyage which left the British Isles to explore the coast of South America and the Galapagos Islands. It was a trip that spawned one of the great biological revolutions.

Two nature collections of unusual superiority are found at the Smithsonian Institution in Washington, D.C. and at the British Museum in London. A third is unique in the world: Kew Gardens outside London began in the 1800s as a repository for one of every known plant. Kew now has many acres of living plants. The zoo (zoological garden) is another form of nature collection. Visiting the zoo is an excellent way to learn about the world's wildlife.

SOCIAL IMPLICATIONS

When we find some beautiful, unusual, or curious object in nature, some of us pause and look, others take a photograph, and still others pick up some small sample and take it home to treasure or study. When there were fewer people in the world, fewer endangered species, and fewer local, state, or national parks and wilderness areas, collecting nature was a popular and accepted hobby. Today collecting natural objects is regulated by land ownership, laws, and public attitudes. Concern about the wise use of natural resources and the protection of natural environments has led many people to advise against nature collections altogether. Where nature is protected (wilderness areas, parks, and other such areas), a "hands-off" policy is needed and often crucial to the survival of a habitat, species, or individual organism. However, with an owners' consent on private property, some moderate collecting probably does little harm—unless the organisms are few in number. Autumn leaves, pine cones, beetles, garden snails' shells, and similar things have attracted people for centuries and are not in short supply. Perhaps the most important question to be asked is, "What can be learned by collecting?"

MAJOR CONCEPT

Groups of living things have identities based on their characteristics and behavior.

OBJECTIVES

☐ Describe common characteristics of groups of living and nonliving matter (material).

☐ Observe, compare, classify, and infer.

☐ Assemble simple projects without guidance.

☐ Use and care for tools or equipment that require small adjustments or coordination of senses.

☐ Assemble materials in new ways.

☐ Assume responsibility for the wise use of natural resources.

MATERIALS

Containers students bring from home: clear plastic shoe boxes, wooden or cardboard boxes, 2-pound coffee cans, cigar boxes. Books from the library depicting various aspects of nature including field guides of birds, mammals, insects, fossils, reptiles and amphibians, trees, wildflowers, rocks and minerals, shells, and so on.

VOCABULARY

collection

plant groups

personal property

environment

animal groups

rock and mineral groups

living and nonliving groups

dead

names of various natural objects found, if known

PROCEDURES

A. Invite students to start a nature collection. If some students have begun nature collections, ask them to share their collections with the class. Explain that you are interested in helping the class build a nature collection for the school—one that will be important to them because it will be available for them and others to enjoy for years. Add that some students may want to build their own personal collections as well, but clarify that the nature collection for the school will be one that everyone contributes to and shares. After the collection is completed near the end of the year, it could be presented to the school in a formal assembly while students tell about various contributions to the collection.

Ask students to name the kinds of things that might be included in the nature collection. List their ideas on the chalkboard and add your own ideas to the list as examples. The following list was suggested by one group of students for their nature collection.

OUR NATURE COLLECTION

Bird feathers and broken eggshells

Leaves of trees in the fall and spring

Pictures of mammals or other animals we can't collect

Plaster casts of animal tracks

Butterflies and beetles

As many insects as possible

Seashells that have washed ashore

Flowers that have been pressed in a book

Pine cones from the mountains

Rocks (and minerals)

Fossils of extinct animals

Snakeskins that have been shed

B. Encourage students to bring in various things to add to the nature collection. For simplicity's sake you may want to limit what students bring in to specimens that are found dead: butterflies and other insects, leaves, cones, or blossoms that have fallen. This policy eliminates the need to supervise the making or use of insect killing jars, or to worry about students' personal safety while collecting living things, if you think they might be collecting without adult supervision at home. Most books on insect collecting provide guidelines for making killing jars. If you decide this is an important objective, provide guidelines for their use and assume responsibility for showing students which animals and plants are harmful to them. Of course, you might take students on field trips yourself.

There are three general guidelines that will be helpful for students to follow when they are collecting things.

GUIDELINES FOR COLLECTING

1. Ask permission before collecting in public places, parks, and on personal property. Students should ask parents before collecting in their own yard.
2. Collect only what you need: one of each kind.
3. Leave the rest of the environment as you found it.

Students become familiar with the variety of plants and animals that share the school environment with them. (Strix Pix)

Discuss these and any other guidelines with students in order to avoid criticisms or problems from nature collecting.

C. Preserving and displaying the nature collection will provide students with important skills. In general, most plants can be pressed between sheets of paper and weighted down with a heavy object or a stack of books. Pressed leaves or other parts of plants can then be mounted on stiff paper with tape or glue. A faster procedure is to place plants on sheets of paper or cardboard, cover them with clear plastic laminating film, and then seal the film using a laminating press. Many school district offices use such machines to laminate materials for classroom use. Insects can be mounted on fine (thin) insect pins purchased at hobby stores or university book and supply stores, and can be arranged in boxes that have a cardboard or cellotex-like layer in their bottoms. Rocks, minerals, shells, and fossils are usually placed in boxes and then glued or wired in place.

You might ask some parents to build two or three shallow wooden display cases for the completed collection. These could be designed along the lines of those sold commercially or found in museums and suited for hanging on walls or sitting on tables. Usually such display cases have glass tops or lids to protect the collection from dust and excessive handling.

EXTENDING THE ACTIVITY

D. Consider using photographs from magazines or other printed material for mammals that are not represented in the collection. Interesting arrangements of pictures of animals on chartboard would be an attractive option for students who prefer not to collect natural objects on their own.

One of the major values of a nature collection for the school is its use as a reference. If the collection were on permanent display in some hallway or a corner of the library, students from other classes could benefit from your students'

Use boxes and egg cartons as well as cardboard and wooden boards to display specimens in the nature collection.

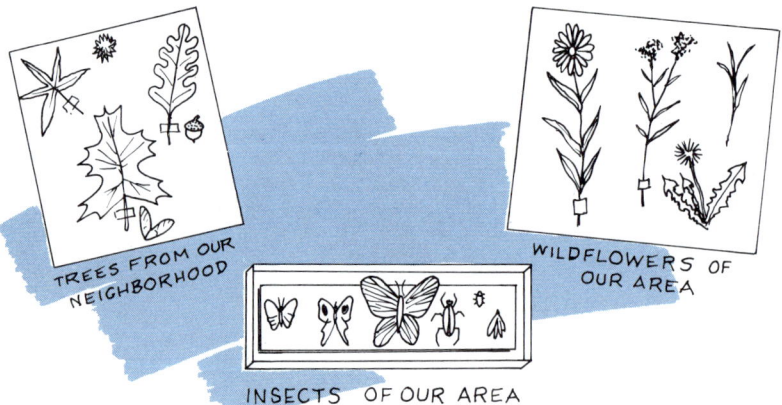

TREES FROM OUR NEIGHBORHOOD

WILDFLOWERS OF OUR AREA

INSECTS OF OUR AREA

learning. As a reference, other students can use the collection when writing reports, drawing pictures of something in nature, or simply satisfying their own curiosity. One permanent collection

started a tradition among students to add to the collection each year. Many of the students also began their own personal collections, thereby sustaining their interest in science.

Students can share learning about local plants, animals, rocks, and shells by making a permanent collection for the school.

ACTIVITY EB–7

Learning How Scientists Organize Nature
(Grades 3–5)

ACTIVITY OVERVIEW

In this activity (a continuation of Activity EB–6), students examine books which have descriptions of plants and animals in them and try to find pictures and descriptions of things in their nature collection. They invite a scientist to visit their class and talk about the way

he or she organizes nature. *Estimated time:* Approximately one week.

SCIENCE BACKGROUND

It is one thing to discover what might be a new **organism** and to describe it; it is quite another thing to place that description in the context of existing knowledge. Yet scientists expect to accomplish both tasks. That expectation has led them to develop a system for organizing bits of information that might otherwise remain unrelated. You can imagine how difficult it would be to keep track of nature if there were no system, especially since there are presently over 1.5 million different species of organisms known in the world. You can see why **systematics** (the study of plant and animal diversity and the relationships among groups of organisms) is so important to science.

The system that scientists use to classify or catalog organisms is called a *taxonomy*. Every time a new species of organism is discovered, its description is added at the appropriate place in the taxonomy. Sometimes, but not often, a new discovery is made that leads scientists to reorganize a part of the taxonomy. That happened in 1969 when R. H. Whitaker [*Science* (1969) 163: 150–160] proposed a new way of classifying organisms into five major groups called kingdoms: Monera, Protista, Fungi, Plantae, and Animalia. Previously, biologists had used only two kingdoms: plants and animals. In this text we focus most of our attention on plants and animals because the other groups are less visible and less available to students of the ages five through fifteen years. (See chapter 9 for additional background.)

Taxonomies have several characteristics:

1. Every organism has a scientific name that is binomial (composed of two names) and always used as a pair. The **genus** (plural **genera**) name is always listed first and followed by the *species* name. For example, the scientific name for all domestic dogs is *Canis familiaris*. Notice that the genus is always capitalized but the species is not. Both parts of the name are underlined or in italics when they

appear in print to indicate that they are the Latin, scientific names of the organism.

2. Scientific names are usually taken from Latin and describe some common or highly visible trait of the organism. For example, *Canis* means dog in Latin, and the fact that domestic dogs are more common in our lives than are coyotes or wolves (which are also *Canis* species) explains the choice of the species designation, *familiaris* for dogs.

3. Taxonomies use a hierarchical or pyramidal system of classification. All organisms are assigned to one of five groups on the initial sorting, and then each of those groups is divided continuously until the species is the smallest, interbreeding group. Taxonomies look somewhat like charts showing family trees.

4. A *key* is used to help readers find information in taxonomies. To use a key, one begins on the first page of the book containing the taxonomy. One decides which of the five kingdoms the organism belongs to: plant, animal, and so on, and then turns to a section of the book containing descriptions of that kingdom. The reader decides among alternatives at each point that a group is divided into smaller groups.

SOCIAL IMPLICATIONS

Through taxonomies, scientists have been able to build upon the knowledge of previous generations of scientists. The system that we now use for classifying organisms was developed in the eighteenth century. Society ultimately benefits from classification systems because they increase knowledge of living things. From this, scientists have been able to develop new food sources and new treatments for disease.

MAJOR CONCEPT

Groups of living things have identities based on their characteristics (structure or appearance) and behavior.

OBJECTIVES

☐ Describe common characteristics of groups of objects in a nature collection.

☐ Observe, compare, and measure.

☐ Gather and organize information.

MATERIALS

Same as in Activity EB–6.

VOCABULARY

Same as in Activity EB–6.

PROCEDURES

A. Introduce students to the assortment of books you have brought in on plants, animals, rocks and minerals, seashells, and so on. Explain why scientists have grouped living and nonliving matter; they give names to the groups (e.g., mammals) and to subgroups (e.g., dogs) in order to be able to communicate with each other about nature and the things they are studying. Ask students to examine the books, sharing them within teams of four to six students. Encourage students to look at the pictures and find examples of things they have seen. If you have been able to obtain a series of field guides that illustrate trees, wildflowers, shrubs, mammals, birds, reptiles and amphibians, insects, seashells, or fossils, students can begin to look for pictures of things in their nature collections. (If the field guides include species from your

With assistance, students can become proficient at using field guides to identify specimens in nature and may begin to add the scientific names of species to their vocabularies.

geographic region, you may be able to identify the species children bring.)

B. Show students how to use field guides when you are able to work with them in small groups. Most field guides have sections of colored illustrations grouped according to similar species. For example, butterflies of all kinds are grouped together but they do not appear in the same category with bees or beetles.

First show students how to search for pictures of the appropriate group (by using the table of contents or by glancing through the book). Then, if they find a picture of a specimen that resembles one in their nature collection, they should hunt for the verbal description of the specimen. Pictures usually list the page numbers of corresponding verbal descriptions. The verbal description describes where the specimen lives or can be found (referred to as its **range**). Both the range and other information in the verbal description may help students to decide if the specimen they have found is in fact the same as the one described in the book. However, field guides generally have some technical terms that may be too difficult for students in the intermediate grades; pictures and the range may be of most help to them.

Once students think they have made a correct identification of their specimen, they can tentatively list its name. Later, if they have the opportunity to talk with someone who has studied the group for a long while (as a hobby or job), they may be able to confirm the identification. Museums also have labeled collections that can be used to compare and confirm identifications.

EXTENDING THE ACTIVITY

C. Invite someone to class who is interested in nature collections. You might consider people from the Department of Fish and Game, Wildlife Management, the Forest Service, Agricultural Extension, 4-H or Scouts, or local Parks and Recreation Departments. If there is a community college or university nearby, someone from the biology department or geology department would be interesting to students as well. Any of these people might be expected to visit your class free of charge, if their schedules permit and if they have an interest in talking with young people. Ask the expert to talk about the way scientists group things in nature and how they decide on the identities of living and nonliving things. Invite the visitor to look at the students' nature collection and comment on students' identifications. You might also ask the visitor to describe his or her job, and at a later time, relate the information to career training.

ACTIVITY EB–8

Starting Gardens in Containers
(Grades 3–5)

ACTIVITY OVERVIEW

In this activity, students plan how and what to plant in container gardens as they bring various kinds of containers, soil, and seeds from home. The activity is psychomotor in nature. *Estimated time:* About 45 minutes.

SCIENCE BACKGROUND

Soil is created when organic matter (living things) decomposes on weathering rocky surfaces. The texture of the soil is determined by the size of the particles from the weathered parent rock. Soils are graded from coarse- to fine-sized particles: *sand, silt, loam,* and *clay.* Water drains most rapidly through sand and most slowly through clay.

The texture and mineral content of the soil determine in part what plants will grow in it; in turn, the plants that grow in a soil contribute to the content of the soil. Beach grass, for example, survives on porous sand dunes along coasts, while lichens cling to rocky crevices in high mountain passes. The sand dunes store minerals left by beach grass that decomposes, and the granite rocks' weathered crevices store minerals left by the lichens. Neither of these soils would support the plants growing in your flower pot or along city streets because the soils are too coarse.

Seeds contain an embryo and stored food and are covered by a seed coat. When a mature seed falls from its parent plant, the embryo becomes dormant until conditions are appropriate for it to germinate. The dormancy period ends when the seed takes in a large amount of water and ruptures the seed coat. Thus, germination can be speeded up by soaking seeds in water overnight before planting them.

SOCIAL IMPLICATIONS

Seed plants have been around for about 230 million years. They have become so diverse that there are now over 250,000 species (groups of interbreeding individuals) known. Throughout the world people depend upon seeds to grow cereal grasses like wheat, rice, and corn, and a variety of fruits and vegetables. Knowledge about seed germination is of great importance for the production of the world's food supply.

MAJOR CONCEPT

All groups of living things change: members of the group interact with each other, with other groups, and with the environment.

OBJECTIVES

☐ Observe; compare; measure.

☐ Use and care for tools and equipment.

☐ Handle and care for plants in the classroom without supervision.

☐ Handle nontoxic substances related to horticulture.

☐ Construct simple projects.

MATERIALS

Rulers (from classroom), soil, gravel, seeds, assortment of small garden tools; carpentry tools are optional. All materials are described in greater detail in a sample sheet to be given to each student.

VOCABULARY

names of plants
  (the names on
  seed packets)
horticulture
germinate
seedling

nutrition require-
  ments
fertilizer
humus
loam
plant shock

PROCEDURES

A. Introduce this activity to the class as a way to observe the way plants change in response to soil, light, water, insects, and other animals or plants. Use the Garden Plan detailed in figure 10–1 as a guide. The plants from these gardens can be transplanted when they are mature, showing how well plants can adapt to a new environment.

B. Once students have brought their garden materials, begin with a demonstration of how to plant seeds. Write on the chalkboard the steps for planting. You may prefer to put it on chart paper or later on

FIGURE 10–1
Garden Plan

GARDEN PLAN

TOOLS: Trowel, hammer, nails, small saw. Carpentry tools are for students who want to make window boxes or planters.

CONTAINERS: Flower pots, cans of various sizes, and saucers or plastic lids to put under pots and cans. All pots and cans should have holes punched in their bottoms for drainage.

Window boxes or planter boxes of various sizes can be constructed from a simple design:

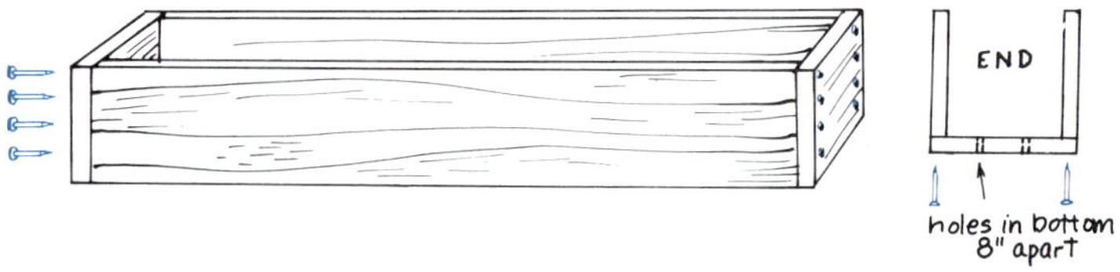

GRAVEL: Into each container, pour a thin (½ inch) layer of gravel to promote water drainage.

SOIL: Choose one kind of soil for each container.
  1. Unaltered soil from a vacant lot, roadside, or alley
  2. Altered soil from a garden or cultivated field of someone you know (soil that has been given fertilizer or other additives)

3. Soil you mix yourself from humus, sand, clay, compost, mulch, or other materials used to mix soil

4. Commercially bagged soil

Fill each container to a level 1 inch from the top

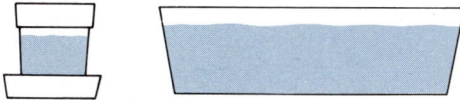

SEEDS: Choose one kind of seed for each container:

1. Seeds you find in nature from trees like oaks, maples, sycamores, pines, and firs and from shrubs and wildflowers

2. Seeds from last year's garden, perhaps from vegetable or flower gardens of people you know

3. Seeds sold commercially in packets or in bulk

MARIGOLDS

NEWSPAPERS: To put on floor, countertop, or outdoor surfaces to make for easy clean up after planting.

NAME _____

TEAM _____

KIND OF GARDEN _____

I WILL BRING _____

_____

a ditto for students to follow. (See figure 10–2, Planting Guide, for the outline). Next, plant your own container garden following the steps on the guide. Then, clean up.

Have students begin to plant their own container gardens. Designate a work place for each team. As teams begin planting, walk around the room and talk with students who are not following in-structions. Find out why they are doing things differently before insisting they follow instructions. Allow enough flexibility for students to satisfy their own curiosity and creativity, but make sure they understand the consequences of departing from the guide.

Planting seeds takes about forty-five minutes. Clean up will be faster if the planting activity is conducted just before recess,

FIGURE 10–2
Planting Guide

PLANTING GUIDE

1. Get materials ready. Look at figure 10–1, the Garden Plan.
2. Put newspapers over the work area to make clean up easier and faster.
3. Pour ½ inch of gravel into the container.
4. Fill the container with soil to a level 1 inch from the top. This leaves space for watering.
5. Prepare the soil if it is completely dry. Example: 1 cup very dry soil soaks up about ½ cup of water. New clay pots also draw water out of soil. Add water slowly. Stop when the soil is damp—not soggy—throughout.
6. Plant the seeds. Follow instructions on the package, or plant seeds to a depth of 1½ times their length. Seeds that are planted too deep often mold.
7. Lightly spray or sprinkle soil with water to settle seeds into place. Recover any that are uncovered. After this initial planting, seeds may be watered every two to three days, twice a week, or once a week. The soil should be kept thoroughly damp (not just on the surface) at all times.
8. Put planted container in its drainage saucer or tray. Place near a window but not in direct sun in areas with high temperatures.
9. Clean up.

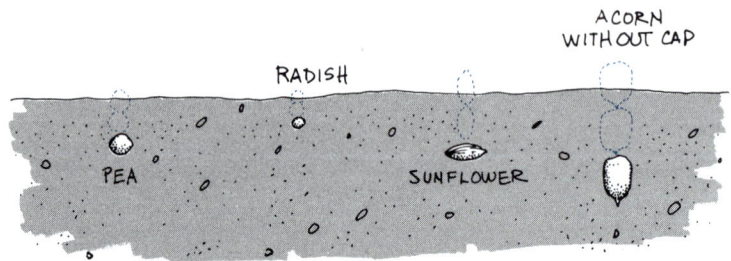

lunch, or time to go home. Before students clean up, have them put labels on their containers. Labels will help to avoid mix-ups of containers and will identify what students have planted. A sample label is shown; recycled paper (clean on one side) can be used for labels that are then taped to containers.

Ask students to place their containers near the window. To prevent water damage to table tops or floors, place folded newspapers under the drainage saucers or trays of containers.

TEAM: (names of students)

KIND OF SOIL:

KINDS OF SEEDS:

DATE PLANTED:

ACTIVITY EB–9

Recording Changes in Plant Groups
(Grades 3–5)

ACTIVITY OVERVIEW

In this activity, students carefully watch as their plants grow in size and develop new parts. They record these changes on a decorated wall chart or in their personal journals. *Estimated time:* 10–15 minutes per day, for two or three days per week, during the time plants are growing.

SCIENCE BACKGROUND

The germination of seeds and the growth of plants are exciting processes to watch. The sun's energy has been converted into living material stored in a seed. When supplied with water, the seed germinates, develops into a seedling, and then grows to a mature individual if it is given adequate support in the form of soil, minerals, water, and sunlight. The life cycles of **herbaceous** plants

(nonwoody plants such as grasses, weeds, wildflowers, and garden plants of most kinds) provide perpetually new sources of energy each year.

What are the basic requirements of plants? Most plants that your students will be concerned with require sunlight, or at least indirect light. Direct sun coming through a classroom window can actually scorch leaves in areas where temperatures are high. Plants also need water—enough so that the soil always feels damp to the touch. They require minerals, too, which enter the roots along with water. In order of their importance to plants, the minerals most needed are nitrogen, potassium, calcium, phosphorus, magnesium, and sulfur. With these basic requirements, plants carry on the process of photosynthesis and feed all other living things in the world. The degree to which plants achieve their maximum growth potential depends upon the acquired amount of each basic need.

The changes described thus far refer to changes in growth and development within individual plants. Entire plant populations also change. Populations may increase or decrease in number of individuals, or they may spread to new locations. Populations can also contract diseases and die out entirely. They may produce unusually large flowers or fewer seeds in a given year or series of years. Any number of changes can affect the entire population of plants growing in a specific place. Perhaps you have seen entire gardens, fields, or forests infected with some insect pests or fungus disease.

SOCIAL IMPLICATIONS

Because we depend so much on plants, any changes that affect large populations of plants are likely to have a social impact. Late rains or early frosts have ruined fruit crops; droughts have caused vineyards to produce exceptional wines; smog has decreased the yield of timber in forests. One of the most tragic and at the same time best documented cases of major plant population change occurred when, in a single week during the summer of 1846, almost the entire potato crop of Ireland was infected by a fungus. Within two years (1845–1847), over a million people or about one-eighth of the population of Ireland died from starvation because they depended on a single crop for their food and economy.

As an agricultural nation, the United States is very careful to check all plant materials entering the country. This preventative action is taken to reduce the likelihood that serious agricultural diseases will be introduced into domestic populations.

MAJOR CONCEPT

Living things change as they interact with each other and the environment.

OBJECTIVES

☐ Observe, compare, measure, gather and organize information, infer.
☐ Use and care for tools and equipment.
☐ Handle and care for plants.
☐ Assume responsibility for basic needs of plants.
☐ Accept consequences of own decisions.

MATERIALS

Same as in Activity EB – 8.

VOCABULARY

germination	watering schedule
seedling	flowering
sprout	fruit
growth rate	transplant

PROCEDURES

A. From the moment they plant their seeds, students will be eager to watch for signs of growth in their container gardens. At this stage, students are often tempted to overwater their gardens, hoping to hurry germination. Explain that overwatering causes seeds to mold and die. One way to preserve students' enthusiasm for helping their seeds to grow is to focus their attention on a wall chart that shows how their gardens are progressing. Create a wall chart of your own design or pattern one after the example in figure 10–3. Invite students to decorate the chart and record their garden's progress.

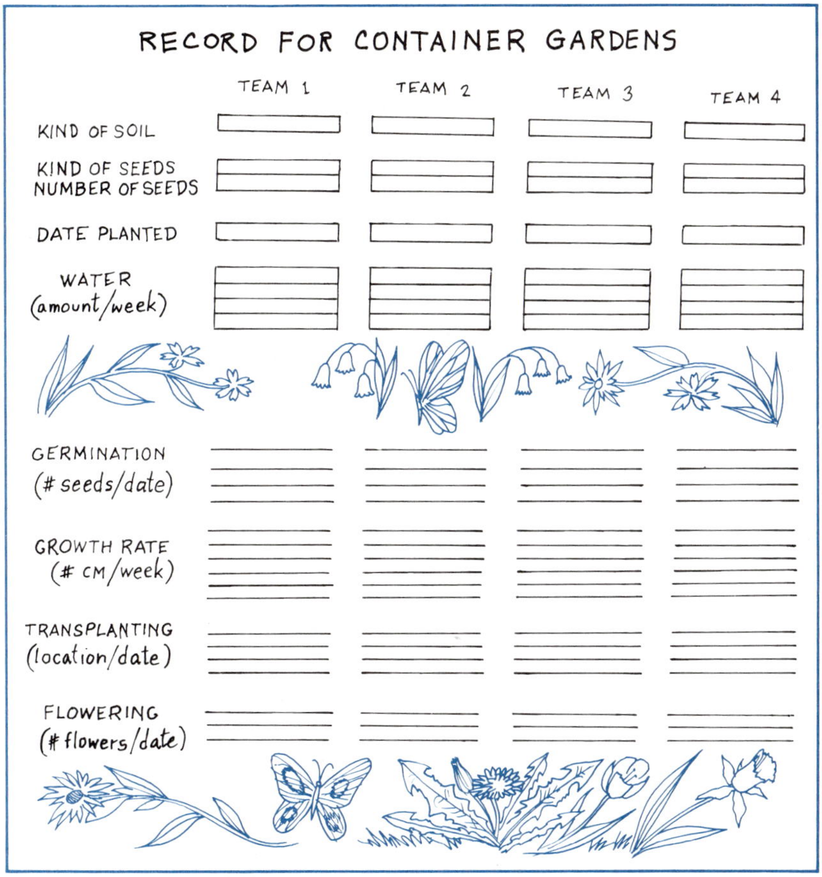

FIGURE 10–3
Wall Chart

RECORD FOR CONTAINER GARDENS

B. Put up the wall chart a day or two after the planting activity. Hold an informal discussion about the chart. Point out the kinds of information to be recorded. Ask students: Why might it be helpful to have such a record? Listen to students' ideas, comment on them, and encourage them to comment on each other's ideas. Finally, help students to understand that the chart will help them to keep track of when they planted their seeds, how much they watered them, and which seeds germinated first in which type of soil. They will then have a record showing what kind of gardens seem to grow best. Following the discussion, help students to record their first information. Recording in pencil is helpful since mistakes can be corrected before the chart is later completed in felt tip pen for the permanent record.

C. About a week after most seeds have germinated, ask students to describe how their plants have changed. Encourage them to be specific in their descriptions and to make observations. While students are looking at their gardens, repeat the questions below as you hear from various students: How have your plants changed? Are some larger than others? How have your plants changed in response to their environment (soil, water light)? This strategy is designed to promote students' careful observation and natural curiosity. Do not attempt to answer all the questions students might ask at this stage. Instead, encourage them to watch carefully to find the answers to their questions.

EXTENDING THE ACTIVITY

D. Introduce students to the idea of keeping their own journals. Tell them that many people use journals to keep their own records of gardening, of field trips, of important things they have seen or done, or of how they spend their money. Scientists keep journals so that they will remember what they have done in their experiment or what they have observed on field trips. Writers keep journals that describe places they have been and what they have seen so that later they can use their journals to help them write books.

Invite students to keep journals about their gardens. In the journal, they might record the same information that is on their label and on the wall chart. In jour-

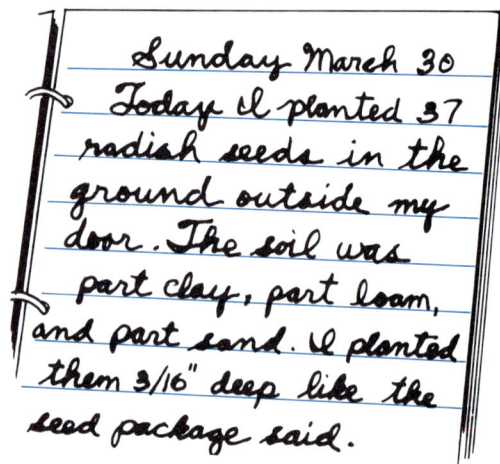

Sunday March 30
Today I planted 37 radish seeds in the ground outside my door. The soil was part clay, part loam, and part sand. I planted them 3/16" deep like the seed package said.

Journals allow students to write and draw their observations and impressions of what is happening around them. Be certain students realize that scientists keep very detailed journals to record observations that are used as a source of future information.

nals, students can describe in greater detail how they started and cared for their garden; what thoughts, ideas, and questions they have about their gardens; and what they discovered while raising their own plants and watching other classmates' gardens. Their journals then become a valuable guide: The next time they plant a garden, they will be able to build on their previous experience.

ACTIVITY EB–10

Recognizing the Limits of Change in Plants
(Grades 3–5)

ACTIVITY OVERVIEW

In this activity, students discover the limits of their plants' capacity to change or adapt to

indoor environments. *Estimated time:* two or three class periods of 30 minutes each, spread over a month.

SCIENCE BACKGROUND

Plants can't move once they put down roots—at least, most do not. Unlike animals, they are "stuck" with their environment. But plants can do many things to protect themselves against harsh environments and have devised ingenious techniques for getting what they need: a home or some place to put down roots, food, a little water, fresh air, sunlight, a few minerals, and at least one surviving offspring.

Arctic plants are a good example of ultimate adaptation. In a frigid land where biting, dry winds race across ice and often carry sand grains and ice crystals, plants either learn to lie low in order to survive or they are pruned by shearing winds. A thin layer of motionless air next to the surface of the black soil offers a slight protection for plants that "think small." Yet even though the dark soil and lichens absorb the sun's energy, and moisture is unlocked from the rock-hard permafrost that helps to hold that heat in, the mercury in a thermometer literally and visibly drops when a cloud passes by. [Material adapted from Richard C. Davids, "When The Icehouse Blooms," *Science* 80 (May/June 1980): pp.56–63.]

Students in this class discover the limits of plants' capacity to change or adapt to indoor environments. (© C. Quinlan)

In this environment plants are often covered with hairs to reduce the drying effects of the wind; some produce only a few leaves on each twig. Flowering plants must respond quickly to the first signs of summer if they are to bloom and produce seeds. Photosynthesis is a twenty-four hour day job during the summer. Flowers like the arctic poppy, twist their "heads" to follow the sun and to collect heat to nurture developing seeds.

It would be nice for plants if they could make all of their changes or adaptations to the environment quickly, but they cannot. Protective adaptations such as hairy leaves or stems that allow flowers to twist and follow the sun occur because species are genetically programmed to produce these characteristics. But having a genetic code that allows a quick response is one thing; changing a genetic code is quite another feat. Changes in genetic codes are called **mutations;** their occurrence is rare and random in the sense that a mutation may occur in any given characteristic (stem height, diameter, or texture) and may either benefit the plant or be lethal to it.

SOCIAL IMPLICATIONS

A Dutch botanist named Hugo de Vries first described mutations in the early 1900s to account for the hereditary changes he observed in the evening primrose. Today, scientists know that X rays, ultraviolet light, and certain chemicals greatly increase the occurrence of mutations. Because mutations are more often detrimental than beneficial to the organism, many scientists are concerned about effects of the increased exposure to radiation that accompany nuclear energy production and the widespread use of X rays. Botanists who have studied the capacity of plants to adapt to various conditions have used their knowledge to develop strains of plants that

produce greater yields or that resist troublesome plant disease.

MAJOR CONCEPT

When plants are forced to change beyond their limits, plant groups lose their identity (do not survive) even though matter and energy are conserved.

OBJECTIVES

☐ Describe what happens when plants fail to adapt to their environment.

☐ Gather and organize information, infer.

☐ Explore and ask questions.

☐ Demonstrate concern for the natural environment and the basic needs of plants.

☐ Describe decisions that involve self and others.

☐ Describe consequences of various actions or decisions.

☐ Handle and care for plants, tools, and equipment.

MATERIALS

(Same as in Activity EB–8).

VOCABULARY

capacity to change	germinate
limits	environment
basic needs	adapt
fruit	

PROCEDURES

A. Whenever you conduct classroom gardening activities, some of the seeds that students plant will not germinate. Some that germinate will never reach the flower or fruit stage. Seeds in nature have many of the same difficulties "growing up" that students have raising them in the classroom. Some seeds get covered too deeply; others receive too much or too little rain or sunlight; still others are overcrowded and die.

The fact that some seeds or plants cannot adapt to their environment (whether that place is in nature or in a classroom container) provides an excellent chance to teach students about the capacity of plants to adapt, to change, or to fail, die, and lose their identities. This activity focuses first on individual plants in the classroom, then on groups of plants in the neighborhood environments, and

finally, focuses on plants in other places in the world.

B. Whenever students' seeds fail to germinate or their plants die, share concern with the students. Help them, on an individual or small group basis, to try to find the cause for the loss of their seeds or plants. Explain that plants, like people and other animals, have limits in their capacity to live in some environments or places. Show students how to try "tracking down" the part of the environment that was too demanding or different for the plants' ability to adapt or change. (Use figure 10–4 as a guideline and a checklist.) Then examine the plants or seeds, and compare them with (1) the checklist, (2) students' Record for Con-

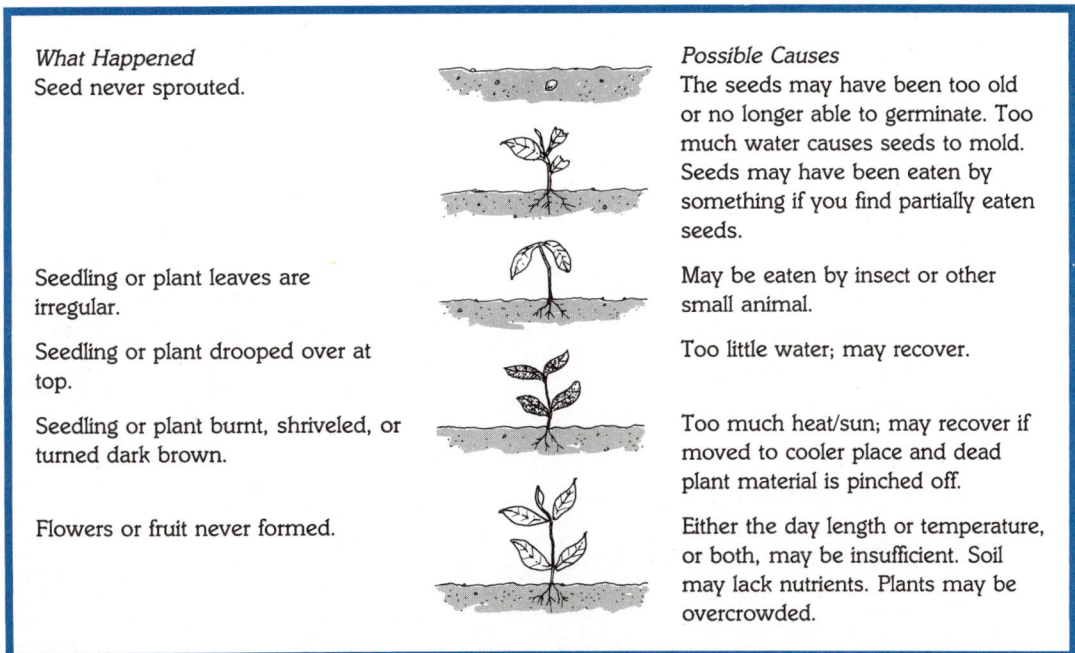

What Happened
Seed never sprouted.

Possible Causes
The seeds may have been too old or no longer able to germinate. Too much water causes seeds to mold. Seeds may have been eaten by something if you find partially eaten seeds.

Seedling or plant leaves are irregular.

May be eaten by insect or other small animal.

Seedling or plant drooped over at top.

Too little water; may recover.

Seedling or plant burnt, shriveled, or turned dark brown.

Too much heat/sun; may recover if moved to cooler place and dead plant material is pinched off.

Flowers or fruit never formed.

Either the day length or temperature, or both, may be insufficient. Soil may lack nutrients. Plants may be overcrowded.

FIGURE 10–4
Checklist for Plant Problems

tainer Gardens (figure 10–3), (3) their journal record, or (4) a verbal description of care given to plants. You cannot always be sure of the cause for the difficulty, but you may be able to help students avoid disappointments the next time they plant seeds.

Encourage the students to replant their containers and to change the conditions that seemed most likely related to their previous problem.

EXTENDING THE ACTIVITY

C. Transplanting young plants before they flower or fruit offers another opportunity to teach students about the limits of plants' capacity to adapt to change in the environment. When plants are about four to six inches high, most are sturdy enough to be transplanted. However, because transplanting can be a shock to plants—just as having a serious accident can be a shock to people—it is helpful to make the change when the plant is small. Transplanting is least harmful when:

1. The soil around the plant is damp but not soggy.
2. All the soil around the roots is transplanted in a block or ball, along with the plant.

3. Transplanting takes place in the evening so that plants are not burnt by bright sun. Newspaper "tents" can be used to protect the plant from bright sun for a few days.
4. Diluted anti-shock liquid (from a nursery) is poured around the plant after it is moved.

When students' plants are large enough to transplant, discuss the advantages and disadvantages of transplanting. If plants survive the transplant, they frequently produce more flowers and fruit than they would have in their former container environment. Yet, not all outdoor environments will be equally good for the kinds of plants students have cultivated. Some may be too hot, others too cool, still others too dark, and so forth. Students should read instructions on seed packages to find the best environment for transplanting the plants from their container gardens. Some students may decide not to transplant their plants at all. They may need to provide additional nutrition for their container gardens in the form of liquid fertilizer or other nursery products. But for those students who do transplant their gardens to the school grounds or at home, follow up their efforts and ask them to report their results to the class.

ACTIVITIES EB–11 to EB–15

Understanding Our Environment:
Today, One Hundred, and Two Hundred Years Ago

(Grades 6–8)

ACTIVITY OVERVIEW

In these activities, students learn about the changes that have taken place in their environment over two centuries. They conduct small research projects on their own or in teams of two to five members. When students share their findings at the end of their studies, they have a better if not a composite picture of ecological change. *Estimated time:* 6-week unit.

SCIENCE BACKGROUND

What is an environment? It is easiest to understand by first visualizing a specific place like a lake, a mountainside, or sand dunes. In each of these places, there is a part that is nonliving which includes the **landform** (the bedrock); and water and minerals found in it; the sunlight that falls on it; and the climate that surrounds it. The combination of these nonliving parts is called the environment.

As you look around though, the environment that you see is not devoid of life. There are plants and animals and other organisms too small to see, for example, bacteria and viruses. These living things and the physical environment interact with each other; that is, they exchange matter and energy and therefore they change one another. Together they form what is called an ecosystem because the enviroment and things living in it work together as a unit or system.

Consider the environment in which you live. If you live in a city, the urban environment consists of landforms which may have been altered somewhat to build transportation systems, housing, and industrial or commercial structures. But the climate and amount of light are, and have been, relatively stable throughout most of the city's history. This environment supports people, their pet animals, some wildlife (perhaps birds and squirrels), insects and smaller organisms, and a mixture of native and cultivated plants.

Environments vary in size. The aquarium environment, containing sand and water, supports a few plants, fish, and snails. Ponds, forests, or lakes are also environments. Some scientists think of that space where life occurs above, below, and on the surface of the earth as a continuous ecosystem called a biosphere because, together, they circulate matter and energy on a global basis.

The surface of the earth is in constant change. New mountains are formed as old ones erode away. The organisms that populate the earth's surface change, too. Species that are no longer suited to an environment become extinct and new species evolve. The weather that surrounds the earth changes its patterns; glaciers form and melt and seas rise and fall. But all of these changes are like the "tick . . . tock" of a large clock, changes on the order of ten to eighty million years in the making.

Nature also has a faster time scale, a "tick-tock" that can be seen in one's lifetime. Changes in environments are on that time scale. Hills erode and again become covered with vegetation. Forest fires destroy many tall trees and shrubs and leave only a few charred stumps. The survivors reseed themselves to begin a new forest. Lakes gradually fill in with silt from nearby hills and become marshes; marshes later turn into meadows that finally become forests. Changes of these kinds are called ecological succession, and refer to the fact that there is an orderly and predictable sequence to the changes that take place in ecosystems or environments.

This sequence has been observed so often that scientists can look at an environment that is in transition and predict what changes might take place next, barring some unforeseen event. The sequence goes something like this: with a starting point of bare rock or cliffs, the first change comes with the development of soil. Small plants like ferns and lichens gain a foothold in tiny crevices where dust has accumulated. When they die and decay, they add to the development of soil. Larger plants which need more soil for their larger root systems move in, and when they die, their decayed material adds to the soil. Minerals from these first two, temporary communities of plants are added to the soil and the rock itself contributes minerals when it gradually weathers. Small animals have also been associated with these two plant communities in succession, adding their remains to the soil. By now the rock or cliff is ready to support shrubs and other small plants which had not been able to survive earlier. Animals are attracted to this new vegetation and become a part of the new ecosystem. But they are not the end of the line. The micro-climate has changed with the addition of shrubs, creating more shade and humidity. This is a condition that allows a few seeds from trees to germinate. As they grow to mature trees, more seedlings sprout to fill in spaces left by other plants that cannot survive with the shade created by the newest ecosystem's trees.

There are limits to the number of changes or stages of succession an ecosystem can accommodate. The final stage, called **climax community,** is determined by the limits of the physical environment. The amount of rainfall, high and low temperatures, and the landform's orientation to light (for example, a northern or southern exposure) limit the kinds of species that populate the area as climax vegetation. Places like deserts and the tundra cannot support forest communities because of their extreme temperatures.

SOCIAL IMPLICATIONS

The recognition of environments and ecosystems helps to focus attention on the interrelatedness of living things and the earth. As we make decisions about changing the environment, we need to see the *ecosystem,* not just the physical environment. Many ecosystems such as swamps, salt marshes, or deserts are thought of as wastelands by some people. These people want such land to be improved for public use. Many of these changes make valuable contributions to the welfare of people; but occasionally some "improvements" backfire and cause damage to the ecosystem. By increasing public awareness of the impact of environmental changes on ecosystems, we may be able to make more informed decisions.

With advancements in engineering and with the availability of heavy machinery, it is now possible to move hillsides, fill in valleys, build dams, and reroute streams and rivers. Because these changes are relatively easy to make (easy in the sense that they can be accomplished in a few years), it is tempting to improve nature where ever it "gets in the way." Yet what can be changed in just a few years by heavy equipment took perhaps hundreds of years to create through natural processes. Some awareness of the nature of ecological succession is of value to people in that it gives them a greater appreciation of the "cost" of creating environments and ecosystems, or restoring them to their natural conditions.

MAJOR CONCEPT

Environments change as living and nonliving matter interact.

OBJECTIVES

- ☐ Gather and organize information; describe relation among variables.
- ☐ Use equipment that requires fine adjustment and discrimination.
- ☐ Actively explore the environment and ask questions; seek alternative points of view and multiple sources of evidence.

MATERIALS

Chalkboard, historical documents (described below), and (optional) portable tape recorder and camera.

VOCABULARY

ecosystem	social history
environment	vegetation
ecosystem	wildlife
natural history	

A guide at a local museum of natural history talks about the plants and animals that were once native and abundant in the surrounding environment. (© C. Quinlan)

PROCEDURES

A. Introduce this six-week unit by showing students pictures from old books which depict people and life in the United States during an earlier period of history. Ask students to describe the clothing, houses, transportation, or other features shown in these old photographs and then invite students to guess when the photographs were taken. Ask them to explain the reasons for their guesses. Inquire if any students' families have old photographs taken of people or the natural environment twenty, fifty, or one hundred years ago; and ask students if their families are willing to share these old photographs with the class. Explain that the class will begin a six-week series of projects studying the changes in their local environment and ecosystem that have taken place over the past two hundred years. Define *environment* and *ecosystem,* using the examples provided in *Science Background.* Then describe the various kinds of activities and projects that will be options for students. The following projects might appeal to individuals or teams. You will want to add your own ideas to the list. After you have read through the projects, turn to Procedure B following the project descriptions for further suggestions.

ACTIVITY EB–11
Talking to "Old Timers"

Talk to "old timers" who have lived in the area all their lives to find out what changes have taken place in the ecosystem and how the environment has influenced people's lives. For example, before the city, suburb, town, or farms were built up as they are today, were there more forests, wildlife, rolling grassy hills or lakes? How did people use the natural resources? When did the environment begin to change, and why?

Students might ask one or several of these long-time residents to visit the class and talk about the changes they have witnessed or students might ask to visit these people in their homes or in homes for the aged. With permission, their conversations could be taped on portable tape recorders to be shared in class and transcribed or replayed by students who need to recall specific details for writing their reports.

Students who choose this project should work together so that they do not interview the same people. Once students have gathered the information, discuss with them how they might most effectively present their findings. Oral reports, perhaps with tapes of actual conversations with "old timers," are very interesting.

ACTIVITY EB–12
Field Trip to Museums

Visit the nearest historical museum or society and find out as much as possible about how the local ecosystem has changed in the past one hundred or two hundred years. Especially ask about photos of people who inhabited the area, of the countryside, of any vegetation and animals, or of natural features such as lakes or rivers. Ask if photocopies can be made of any photographs or printed articles describing the area. Make sketches of any historical objects that show the human impact on the environment or ecosystem, such as axes used for clearing forests, carts used for mining or transportation.

Invite one of the representatives from the local historical society to visit the class and bring any materials which show how the environment or ecosystem has changed over the past one hundred and two hundred years.

ACTIVITY EB–13
Reading Old Newspapers

Go to the library and ask to see the oldest newspapers from the geographic area. Read them to search for information which shows how the climate affected people's lives and the ecosystem in the past, and how people used the environment. Compare these newspaper accounts with recent accounts of the same information. For example, were floods a problem in the past? Are there flood controls or irrigation systems today? Was some area of the city once open space where it is now filled with houses or apartments? Were streets once lined with trees where they are now lined with parking meters? Perhaps before the streets were paved they were made of brick. What happened to the wild animals that once roamed the area? Summarize these newspaper articles and include photocopies of special pictures or drawings.

Visit the city or county planning office in your area to find out what changes have taken place in the environment during the past one hundred years and what changes are planned for the future. Ask the planning engineers (or others who work in the office and have access to the information) how future changes will affect the environment and the wildlife in the area. For example, are new bridges, dams, roads or freeways, housing developments, or other structures planned for areas that are now uncultivated and in their natural state? Look at any maps, blueprints, or other sketches displayed in the office and make sketches of any graphic materials that help illustrate the changes that are planned for the future.

Invite one of the employees from the planning office to visit the class and bring any photographs, charts, or other materials which show how the ecosystem has been or will be changed. Organize a team of classmates to make a large (1 meter by 2 meters) map which illustrates the "before" and "after" character of the area and which depicts man-made changes in the environment.

Locate the nearest museum of natural history to see if there are displays of the native plants and animals that once, or perhaps still, inhabit the area. Look for relief (three-dimensional) maps that show landforms that underlie the area. Find out if the rerouting of rivers or other changes such as floods, erosion, or natural filling in of lakes might have changed the wildlife populations in your area. Museum guides usually know the answers to such questions.

If you can find information about the plants and animals that once lived in your area one hundred and two hundred years ago, draw a **food web** or **food pyramid** to show what plants grew there, what animals ate what plants, and what animals ate other animals as their food source. (See chapter 9 for a description of food chains, food webs, and food pyramids.) Learn how people used these natural food chains to their advantage. For example, did rabbits once eat native plants and did people then eat the rabbits?

Food Web

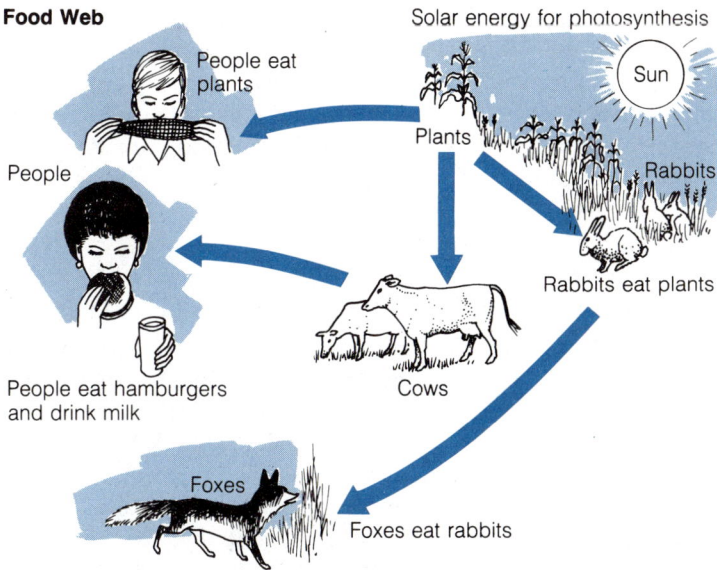

Students can draw arrows to connect pictures of plants and animals that depend on plants or other animals, creating a food web.

Make an insect or plant collection which shows the wildlife of the area today. Ask your teacher or the biology teacher from the closest high school to teach you how to preserve the plants or insects you have selected. Label the collection and display it in class. Someone from a museum of natural history or an agricultural/farm advisor near you may be able to help you identify the plants and insects you have collected.

B. As students select their projects, give assistance by linking them with the resources they need. Remember, the overall purpose of these projects is to show how the local environment or ecosystem has changed over time. If it is easier to find information about changes that have taken place during the past fifty years than over the past one hundred and two hundred years, then adapt this unit or individual projects to fit the resources available. Encourage some students to work in teams, especially those who have less well-developed research skills or those who work best with team support. Permit others who prefer to work alone to do so. As students begin to gather information, help them to organize it and consider various ways to present the information both verbally and graphically. Some students might be able to make three-dimensional maps of papier-maché or plaster while others will do their best work with felt tip pens, chalk, or photocopying. Remember that in addition to the cognitive objectives, this series of research projects is planned to help students develop research skills and learn how to share the information they have gathered.

C. Plan to end this six-week unit with a week-long exhibit, discussion, and sharing. In the discussion, try to bring together the conceptual themes that link these projects:

Identity and Organization
Ecosystems have distinct identities that are based upon the way their physical environments (nonliving) and biological components (living organisms) are organized.

Interaction and Change
Environments change as living and nonliving matter (that is, the wildlife and the environment) interact.

Conservation and Limitations
When an ecosystem's capacity to change is exceeded, the environment and living things within it lose their identities, even though matter and energy are conserved.

EXTENDING THE ACTIVITY

D. Consider inviting other classes, one at a time, to visit your class when students are presenting their final projects. Or plan to display students' work for an Open House at the school. Finally, you might consider displaying some or all of the projects in some public place like a library or community center for the public to enjoy as well.

BIBLIOGRAPHY   ## Teachers' Resources

Brown, Vinson. *Knowing the Outdoors in the Dark.* New York: Macmillan, 1973.

Cornell, Joseph. *Sharing Nature with Children.* Nevada City: Ananda Publications, 1979.

Couchman, J. K. *Examining Your Environment Series.* New York: Holt, Rinehart & Winston.

Denver Public Schools and Colorado State University. *Best of Energy Book.* New York: Cambridge Basic Books, 1980.

Nickelsburg, Janet. *Nature Activities for Early Childhood.* Menlo Park: Addison-Wesley Publishing Co., 1976.

Russell, Helen Ross. *Ten Minute Field Trips: Using the School Grounds for Environmental Studies.* Garden City: Doubleday and Co., 1970.

Scherer, Donald. *Personal Values and Environmental Issues: A Handbook of Strategies Related to Issues of Pollution, Energy, Food, Population, and Land Use.* New York: A & W Publishers (Hart), 1978.

University of Southern California, Department of Mechanical Engineering. *Solar Energy Curriculum for Elementary Schools.* Los Angeles: USC, 1980.

Children's Resources

Animal Homes (film). Los Angeles: Churchill Films, 1980.

Animals and Their Foods (film). Chicago: Coronet, 1978.

Endangered Species Series (free leaflets). Washington, D.C.: U.S. Department of the Interior, 1970s.

Extinction: The Game of Ecology. (board game) Burlington: Carolina Biological Supply Co., 1978.

Johnson, Hannah Lyons. *From Seed to Salad.* New York: Lothrop, Lee & Shepard, 1978.

Life and Death of a Tree (film). Washington, D.C.: National Geographic Society, 1979.

Lowell, Marie (designer). *Predator.* (game). Oakland, CA: Ampersand Press, 1974.

McClung, Robert M. *Mice, Moose, and Men: How Their Populations Rise and Fall.* New York: Morrow, 1973.

Perl, Lila. *The Global Food Shortage: Food Scarcity on Our Planet and What We Can Do About It.* New York: Morrow, 1976.

Ploutz, Paul E. *110 Animals* (card game). Athens: Educational Games, 1977.

Roberts, David. *Animals and Their Babies.* New York: Grosset & Dunlap, 1978.

Schwartz, George L., and Schwartz, Bernice S. *Food Chains and Ecosystems: Ecology for Young Experimenters.* New York: Doubleday Publishing Co., 1974.

Selsam, Millicent E. *How Animals Live Together* (rev. ed.). New York: Morrow, 1979.

Selsam, Millicent E. *How Animals Tell Time.* New York: Morrow, 1967.

Selsam, Millicent E. *How to Be a Nature Detective.* New York: Harper & Row, 1966.

Stevens, Leonard. *How a Law is Made: The Story of a Bill Against Air Pollution.* New York: Thomas Y. Crowell & Co., 1970.

11

Health Sciences and the Human Body

The purposes of this chapter are to present some basic topics and issues in health and human biology and to provide you with a more knowledgeable basis for teaching the health sciences. The reader is encouraged to see current, introductory texts in biology and to use the bibliography at the end of this chapter. Information in this field changes rapidly; keeping up-to-date is therefore an important task for teachers.

KEY CONCEPTS AND IMPORTANT IDEAS
☐ health sciences ☐ structure of cells
☐ mitosis ☐ meiosis ☐ sexual reproduction ☐ development from embryo to fetus ☐ childhood and adolescence ☐ systems of the human body ☐ skeletal and muscular systems ☐ nervous system—the brain ☐ endocrine system ☐ reproductive system ☐ circulatory system ☐ respiratory system ☐ excretory system ☐ immune system ☐ disease ☐ nutrition ☐ mental health ☐ addiction ☐ being healthy

The areas in the science curriculum concerned with the human body are human anatomy and physiology. In elementary and junior high school these subjects are usually taught as health science, and they provide an understanding of how the human body is organized, how it interacts with the environment, and what the limitations are to its development and interactions.

These studies examine the activities that make up human life, and include such questions as how and why we grow; how we obtain oxygen and nutrients from the environment and how they are used in the body; what processes allow us to think, move, communicate, and sense the world around us; where do we get the energy to carry out activities; and finally, what is the process by which we reproduce ourselves?

To reflect on these questions, we look at the simplest component of life—the cell—and describe its many components and functions. Then we examine embryonic development and subsequent events that take place through adulthood. We look at the ways that cells organize into tissues, organs, and organ systems. Finally, we discuss the optimum interaction of these systems and the ways that diseases and nutritional deficiencies can interrupt them.

It should be noted that this chapter can only open the door to many areas in Health Science. The interested reader should consult recent editions of college biology textbooks for more complete information. In reading this and subsequent science resource chapters, it should be kept in mind that many of the explanations presented are the best ones that scientists have at this time for explaining various phenomena. Next year there may be new and better explanations.

THE CELL—THE BASIC UNIT OF LIFE

Every human cell has the same basic structure. Each cell is surrounded by an outer lining or **plasma membrane**, which encircles the **cytoplasm** and the **nucleus** within it (see figure 11–1). The cytoplasm is further divided by intracellular membranes into **organelles**, or units, having many functions. Despite their diversity, each organelle interacts extensively with the others in such a way that each cellular function of the cell, such as secretion, protein synthesis, and so on, is truly the result of the action of all parts of the cell. However, there are a few functions that reside predominately in particular parts of the cell; these are discussed below.

Cell Membrane

The membrane (sometimes called the plasma) of a cell is less than one-millionth of an inch thick. Its function is to hold the cell together and act

PLASMA MEMBRANE The outer cell membrane, which is composed mainly of phosphate-containing fatty acids

CYTOPLASM All the material contained in the cell, except for the plasma membrane and the nucleus

NUCLEUS Cellular organelle surrounded by a specialized membrane, controls many cell functions and contains the organism's hereditary material; found in all human cells except mature red cells and platelets

ORGANELLE A well-defined subcellular structure

FIGURE 11–1
The generalized drawing of
an animal cell contains a
nucleus and cytoplasm.
Within the cytoplasm are
mitochondria, lipids, lyso-
somes, and vacuoles which
collectively perform the
work necessary to keep the
cell alive.

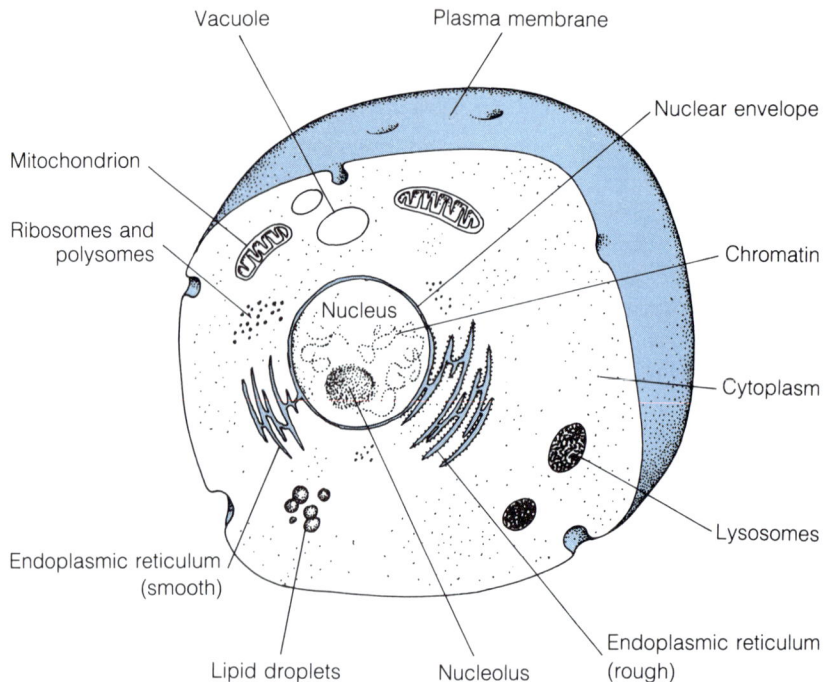

FIGURE 11–1
The generalized drawing of
an animal cell contains a
nucleus and cytoplasm.
Within the cytoplasm are
mitochondria, lipids, lyso-
somes, and vacuoles which
collectively perform the
work necessary to keep the
cell alive.

GLUCOSE The most com-
mon 6-carbon sugar

AMINO ACID A simple nitrogen carboxyl-containing
molecule which is the
building block of protein

PROTEIN Arrangement of
the 20 most commonly oc-
curring amino acids

TISSUE A group of similar
cells that work together to
perform a role in the structuring and functioning of
the body

ORGAN A body part composed of several tissues
grouped together into a
structural and functional
unit

as a selective *gate* for substances (such as **glucose, amino acids,** and
calcium ions) to enter and leave the cell. The importance of this gate-
keeping function of the cell membrane is illustrated by the fact that all
nervous system activity is based on the ability of potassium to pass
through the membrane and the *inability* of sodium ions to pass through
the membrane into the cell. Plasma membranes also contain **proteins**
that allow the cells to aggregate or come together into **tissues** that subsequently form **organs.** Other proteins allow the individual to recognize
its own cells: Should cells become altered, for example, during a viral
infection or after a malignancy, the individual's immune system seeks
them out and destroys them. This action of the immune system is why
transplants are often rejected and why blood types cannot be mixed for
transfusions.

Cytoplasm

The cell's cytoplasm can be thought of as a chemical factory. It contains
many organelles that supply the cell with the chemicals necessary for its
maintenance and growth.

Mitochondria are the units that store and supply chemical energy. Depending on the specific function of the cell, the number of these little power plants ranges from 50 to 5,000 per cell.

Ribosomes are small spherical particles scattered throughout the cytoplasm that are involved in making proteins. Proteins are made from **polymers** of amino acids. Proteins contain 20 different types of amino acids that are strung together one after the other; the order in which these amino acids are strung together determines the protein's function. The human body has hundreds of different proteins. Some of them are used as structural elements; collagen, for example, is found in tendons and cartilage. Some act as carriers of important molecules; hemoglobin, for example, binds oxygen in the lung and carries it throughout the body. Other proteins are capable of contraction; for example, actin and myosin cause muscles to contract. Most importantly, some proteins act as **catalysts**; that is, they speed up chemical changes. Such proteins, called **enzymes,** are by far the largest class of proteins in the body.

Many of the synthetic or chemical reactions that take place within the cell occur very rapidly. This is true even though these same reactions would be slow or nonexistent if they took place in a test tube under the same conditions of temperature and pressure. Enzymes work in living organisms so that reactions can take place rapidly under conditions that do no harm to tissue. Enzymes work in very specific ways. Each enzyme works on a specific chemical or **substrate,** in the cell. Enzymes are responsible for every facet of the body's functions. Pepsin is an example of an enzyme that breaks food proteins down into amino acids; these amino acids, in turn build new proteins that the body needs to maintain itself. There are still other enzymes that break down carbohydrates, fats, and so on. The digestive system then transports nutrients, via the blood, to all parts of the body. The cell membranes have enzymes that transport the amino acids, glucose, and **fatty acids** into the cell where there are other enzymes waiting to transform them into larger compounds that

MITOCHONDRIA Sausage-shaped particles in the cytoplasm that generate, store, and supply energy for metabolism in the cell in the form of high energy bonds

RIBOSOMES Spherical particles in the cytoplasm that are used for protein synthesis

POLYMER A regular arrangement of similar basic subunits produced by one or a few chemical reactions

CATALYST A substance that speeds up a chemical change, but does not change itself

ENZYME A protein that acts as a catalyst

SUBSTRATE A chemical acted upon by an enzyme

FATTY ACIDS Basic subunits of fats

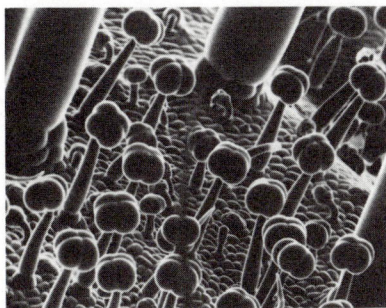

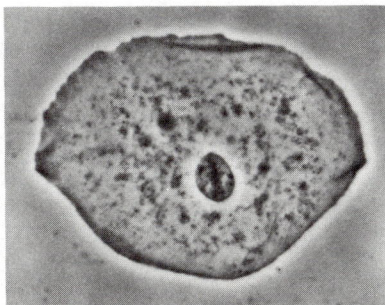

Left, the outer (epidermal) cells of a tomato leaf. *Right,* the human cheek cell shows the relative size of the nucleus to the rest of the cell. (*Left,* courtesy P. Dayanandan; *right,* Carolina Biological Supply Company)

are needed for normal growth, development, and maintenance. Other enzymes break the subunits down further so that they can be used for energy.

COENZYME Substance that bonds with an enzyme, causing a chemical reaction to occur

VITAMIN An organic chemical that usually acts as a coenzyme and is an essential dietary requirement, i.e., the body cannot synthesize it

Many enzymes need what are called **coenzymes,** which are usually compounds related to the various **vitamins.** Although the enzymes can function to a small degree without coenzymes, they function much better with them. This explains why vitamins are so important in our diets. In addition, other enzymes contain various minerals necessary for enzyme functioning, for example, iron, zinc, and calcium to name a few. While there are many more organelles found in the cytoplasm, it is beyond the scope of this book to discuss them further.

The Nucleus

The nucleus is the control center of the cell. It contains the hereditary material that determines the structure and function of each cell. The hereditary material is found in the form of extremely long threads, composed of DNA (**deoxyribonucleic acid**) and proteins, which are called **chromosomes.** Each chromosome, in turn, contains **genes,** the basic unit of inheritance.

DEOXYRIBONUCLEIC ACID (DNA) Part of the chromosone that contains coded information about heredity

CHROMOSOME Thread-like structures containing genetic information

GENE (old) A unit of inheritance; (new) a segment of a chromosome that codes for a functional product

REPLICATION The exact duplication of an entire strand of DNA

TRANSCRIPTION The copying of short stretches of DNA to produce RNA strands

RIBONUCLEIC ACID (RNA) A biological material used to transfer information from DNA

TRANSLATION The process where RNA is used to direct protein synthesis

Among the most significant discoveries of modern science is understanding the structure and the function of DNA. You may have heard of the "double helix," two very long strands of DNA coiled around a common axis (see figure 11–2). Each strand has two very important functions. The first is to act as a template for the synthesis of exact copies of itself, a process known as **replication.** The importance of this process will be seen in the sections on mitosis and meiosis. The second function involves the process of **transcription** where DNA acts as a template for the synthesis of RNA (**ribonucleic acid**) molecules. The ribonucleic acid is used as an informational messenger in the synthesis of proteins. This process is known as **translation.**

Almost all the characteristics that we inherit from our parents are determined by the proteins that our cells produce. Thus, the processes of replication, transcription, and translation determine what we are, both physically and mentally. Our genetic material only provides the framework or, in other words, defines the possible; environment also plays a large role. An individual who has the genetic potential to grow to a height of 6 feet might only grow 5 feet 6 inches if he or she received a very inadequate diet. A young baby who has a genetically determined enzyme deficiency that prevents it from digesting milk sugar would never suffer from that deficiency if fed a formula containing another type of sugar. An individual who has the genetic material to be a fully functional human being will not be one if oxygen deprivation at birth subjected him or her to brain damage.

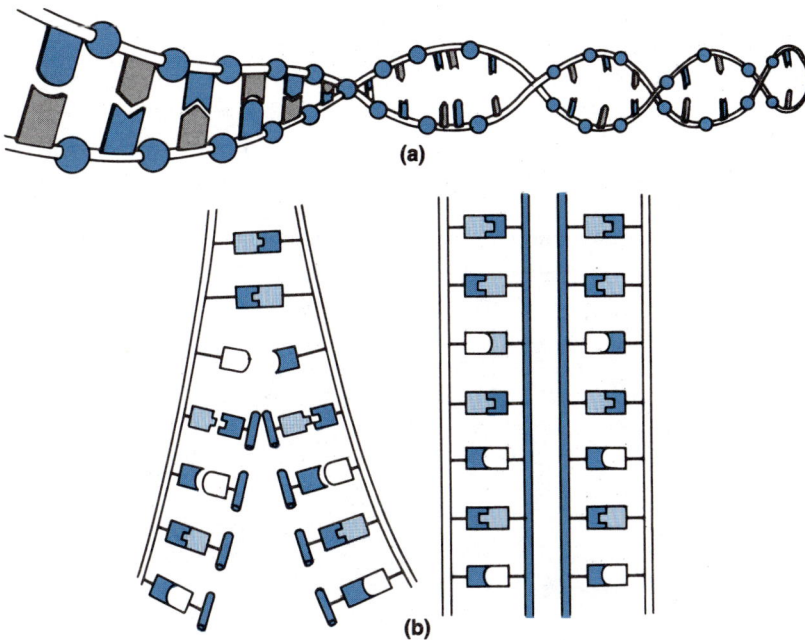

FIGURE 11–2
(a) DNA (Deoxyribonucleic Acid) is the messenger which codes information about hereditary traits through arrangements of nucleotide chains that twist around each other, forming a ladder-like structure called a double helix. *(b)* Replication of DNA involves "unzipping" or splitting the two strands and adding new nucleotides. Specific pairing makes exact replication possible.

Nuclear Control of Cell Function

We are now in a position to understand how the nucleus can control cell functioning. The transcription of RNA takes place within the nucleus. However, this process is influenced by many biological signals, such as hormones, which can stimulate or inhibit the production of particular messenger RNAs, which in turn influence which proteins the cell synthesizes and consequently determines which functions the cell can perform. The nuclear membrane can prevent particular messenger RNAs from entering the cytoplasm and interacting with ribosomes, thus preventing the synthesis of the proteins encoded by those RNAs.

Mitosis

Your body contains about a hundred trillion cells, which all came from a single fertilized egg. In order to produce this enormous number of cells, the individual cells must divide over and over again through a process called **mitosis**. Daughter cells identical with each other—and with the original cell—are created. Although at each division, each daughter cell contains half the cytoplasm of the parent cell, the number and kinds of chromosomes inside the nucleus remain the same. The nuclei of almost all human cells contain 23 pairs of chromosomes (46 in total). One

MITOSIS A process of cell division in which 1 cell produces 2 cells that contain all the inheritance characteristics of the parent cells

member of each pair is of maternal origin and the other is paternal. Thus 50 percent of your genetic material is contributed by your mother and 50 percent by your father. (Later in the section on sexual reproduction you will learn about the process whereby this happens.)

Once the cell receives the signal to divide, and no one is certain of the nature of this signal, the chromosomes replicate, and the extraordinary process of mitosis, as illustrated in figure 11–3, begins. The nuclear membrane disappears and the chromosomes, which have until this time been distributed throughout the nucleus, clump up and appear as distinct threads, known as **chromatids.** The chromatids, which actually contain 2 identical chromosomes, line up across the center of the cell, and the chromosomes separate from each other and move to opposite sides of the cell. A new nuclear membrane surrounds each set of chromosomes and soon a new cell membrane appears across the middle of the old cell.

CHROMATID The 2 daughter strands of a duplicated chromosome that are still joined at their centers

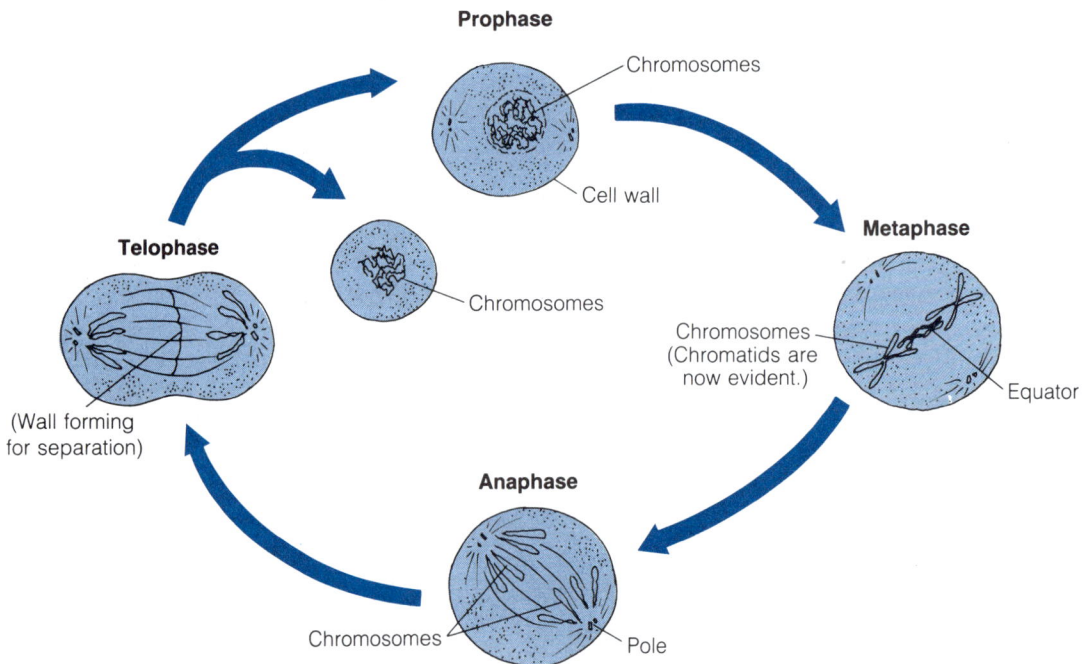

FIGURE 11–3
Through a process called mitosis, cells reproduce themselves with each new cell having exactly the same components as the parent cell. This highly diagrammatic drawing shows only 5 of the 46 chromosomes.

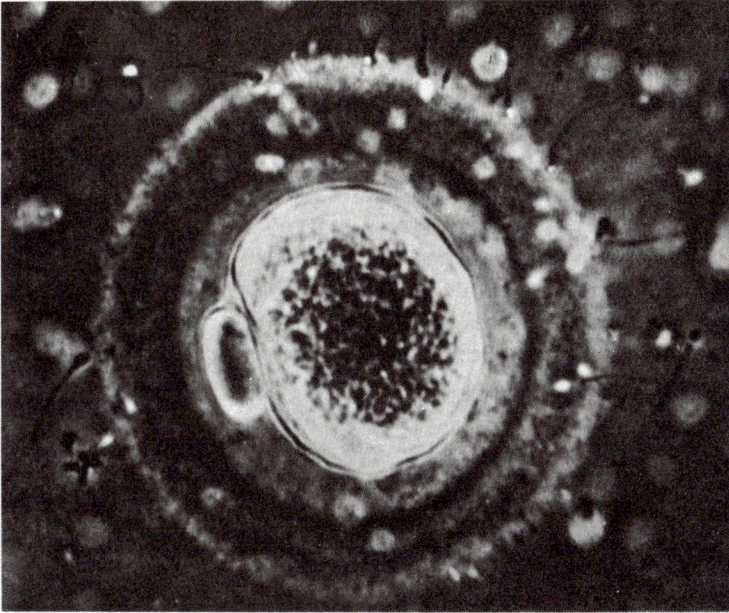

The human egg cell here shows fertilization in progress. Note the sperm entering the corona surrounding the egg cell. (Courtesy Landrum Shuttles)

SEXUAL REPRODUCTION

All advanced organisms create new individuals during **sexual reproduction**. During sexual reproduction 2 cells containing genetic material from each parent unite to form a single cell.

Most physical characteristics of the organism are established at the moment when the 2 **gametes**, the **sperm** and the **egg**, unite. Sex, eye color, hair color and quantity, blood type, and even serious genetic disorders are all determined by the genes of the chromosomes contained in the egg and the sperm. Thus the single cell that represents the first stage of development was itself produced by the union of 2 cells.

As discussed previously, almost all human cells contain 23 different types of chromosomes, giving a total of 46. The gametes are the exception, containing only one representative of each sort of chromosome. Thus once fertilization takes place, the fertilized egg contains the full complement of 46 chromosomes.

Meiosis

When **germ cells** are first produced they contain 46 chromosomes; once they undergo **meiosis** (shown in figure 11–4), the number is reduced to 23 per cell. In the first step, members of each different pair of chromo-

SEXUAL REPRODUCTION Reproduction that depends on the union of 2 gamete cells to form a new individual

GAMETES Cells with one-half the normal amount of DNA that unite at fertilization to produce an organism with a full complement of DNA

SPERM Gamete supplied by the male

EGG Gamete supplied by the female

GERM CELLS Cells that are destined to become gametes even though they contain a normal amount of DNA

MEIOSIS Process whereby germ cells divide to form gametes

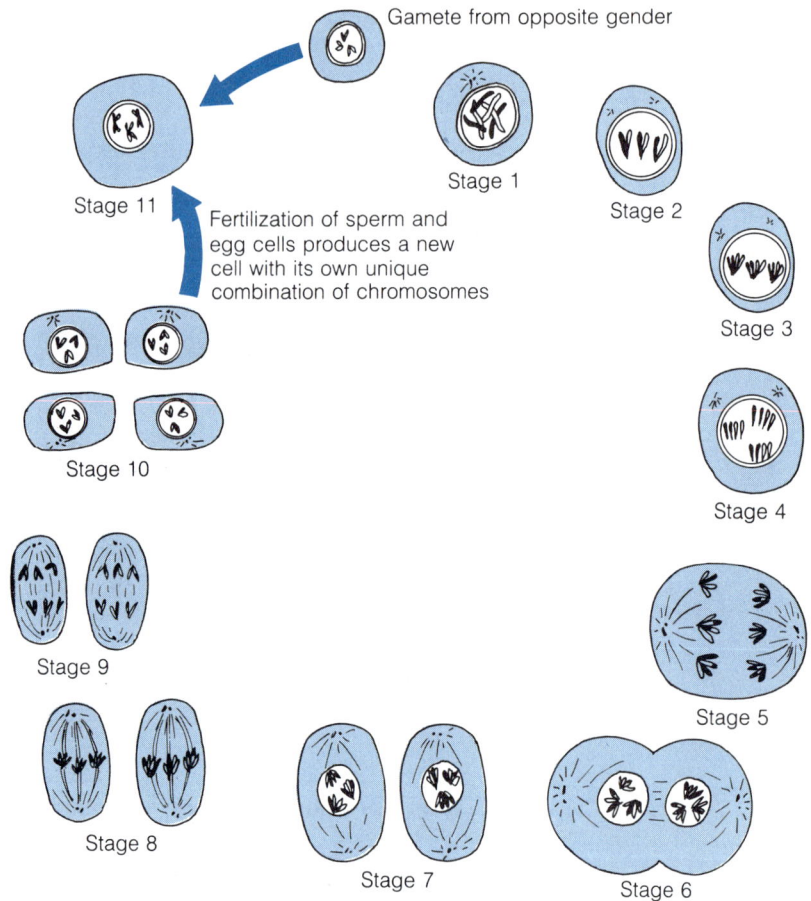

FIGURE 11–4
Meiosis is the process by which cells reproduce and simultaneously reduce their number of chromosomes by one half. Such "halving" is necessary to avoid the doubling effect that would otherwise result when sperm and egg cells unite.

Gamete from opposite gender

Stage 11

Fertilization of sperm and egg cells produces a new cell with its own unique combination of chromosomes

Stage 1

Stage 2

Stage 3

Stage 4

Stage 5

Stage 6

Stage 7

Stage 8

Stage 9

Stage 10

somes move close together and line up across the center of each cell; they then separate so that there are 23 on each side of the cell.

However, since the cell cannot differentiate the source of each chromosome, the side of the cell the chromosomes move to is independent of which parent they came from. Thus the individual germ cell has a 50 percent chance of getting any particular chromosome. Once the chromosomes have assorted, a new cell membrane forms, and the cell divides. If the germ cell being formed is an egg, 3 of the resulting cells degenerate and only 1 is left. If sperm are being formed, all four cells remain viable.

To summarize, during the process of mitosis, 2 cells are formed that contain identical genetic material. During meiosis each cell divides into 2 different pairs of cells that contain half of the genetic material of the original cell arranged in a dissimilar pattern.

The Sperm Cell

Sperm cells are produced in the male by the **testes** in tremendous numbers. The human sperm cell is shaped somewhat like a tadpole. Its small elliptical head, which contains most of the cell, consists of the nucleus and a small amount of cytoplasm. The area behind the head is filled with mitochondria—the power units of the cells. They are particularly abundant in sperm, which need a great deal of energy to power the long thin tails that move them along their journey to the egg. The shape of the sperm and the economy of its structure permits it to move at a relatively rapid rate.

TESTES Pair of egg-shaped organs suspended from the male body, location where sperm cells are produced

The Egg

Compared with the sperm cell, the egg is enormous. As one of the largest cells in the human body, it contains a considerable amount of cytoplasm as well as a small amount of yolk. The yolk is used to nourish the fertilized egg in its early stages. Ovaries contain about 400,000 eggs at the time of birth; however only about 300 are released during a woman's fertile years.

Sex Determination

One of the 23 pairs of chromosomes determines an individual's sex. Females have 2 identical chromosomes, which are called X chromosomes. Males have 1 X chromosome and 1 Y chromosome, which is quite small and contains only 2 or 3 genes. Thus a female is designated XX and a male XY, with respect to their sex chromosomes. In meiosis, when the 2 different members of the chromosome pair separate from each other, all the egg cells will have 1 X chromosome each. On the other hand, 50 percent of the sperm cells have an X chromosome and 50 percent have Y chromosomes. If the egg, which has 1 X, combines with a sperm carrying an X chromosome, the resulting individual will be XX, or female. If the sperm contains a Y chromosome, the resulting individual will be XY, or male.

Some diseases are associated with genes on the X chromosome. Hemophilia, for example, is caused by a deficiency of a blood-clotting protein. If the mother passes the gene on to her daughter, the equivalent gene on the paternal X chromosome is sufficient to produce enough clotting factor. If the offspring is a boy, there is no gene to produce the clotting factor since the Y chromosome does not carry one nor does the defective, maternal X chromosome. This explains why this sex-linked deficiency is found in males.

The Embryo

It is difficult to think of an adult as once having been a tiny, 7-pound individual—and even harder to conceive of him or her as starting as a fertilized egg! However, from birth to adulthood, height increases about 3½ times and weight about 20 times. Nevertheless, all the visible growth and transformations that accompany it are tiny compared with what takes place in the first 2 months of life in the **uterus.**

In 8 brief weeks the **embryo's** length increases about 240 times, its weight increases an incredible millionfold, and a single fertilized egg is transformed into a miniature human.

This tremendous amount of growth and development means that one of the first requirements for the embryo is to secure essential nutrients; the mother's body anticipates this need. Every month her uterus grows a thick, spongy lining that has a rich blood supply. If the egg released during that **menstrual cycle** is not fertilized, the lining is shed and appears as a monthly discharge of blood. However, if conception does occur—meaning the egg and sperm have united—the lining remains and grows thicker in preparation for receiving the developing embryo.

The sperm and egg usually unite in one of the **fallopian tubes**, which connect the **ovaries** to the uterus. The fertilized egg then takes about 3 or 4 days to drift down to the uterus, during which time it continues to divide, forming a cluster of about a dozen cells. About a week after conception, the embryo's cells have divided so many times that they can form a hollow sphere, which implants itself into the lining of the uterus. Almost immediately, the outside portion of the embryo grows very rapidly, and sends out finger-like projections, called villi, that work their way into the uterine wall. The villi allow the growing embryo to anchor itself more firmly in the uterine wall and to obtain an increased amount of nutrients. At this point the embryo is already a complex organism with many differentiated cell types.

By the third week, the beginnings of the **placenta** and **umbilical cord** have formed. These structures ultimately serve to transport both nutrients and oxygen from the maternal circulation to the embryo and to carry away metabolic waste products. However, the maternal-embryonic connection is not a direct one. The placenta functions as a barrier to many harmful substances; unfortunately the placenta is not very efficient at this task, and substances that interfere with embryonic development such as alcohol, caffeine, and narcotics, can still pass through. This is the reason that pregnant women are warned to be very careful about the drugs they take and the foods they eat. The placenta also allows beneficial substances to pass to the embryo, such as the maternal antibodies that provide protection against various diseases.

UTERUS The organ in a female in which unborn young grow and develop

EMBRYO A plant or animal still in an early stage of development

MENSTRUAL CYCLE A monthly cycle in the female during which the egg matures and is released to be made available; hormonal secretions affect the uterus during the cycle so that in the absence of fertilization, a brief period of bleeding occurs

FALLOPIAN TUBES The 2 tubes that serve as a connection between the ovaries and the uterus

OVARIES The organ in the female that contains the gametes

PLACENTA An organ made up of fetal and maternal components that aids in the exchange of materials between the fetus and the mother in mammals

UMBILICAL CORD The stalk containing the blood vessels connecting the placenta and the fetus

First week	Egg is fertilized, divides, and becomes implanted in the uterus	
Second week	The 3 basic layers and the primitive nervous tissue develop	
Third week	The first body segments appear and gradually form into the spine; the brain begins to form	
Fourth week	Heart, blood, circulatory system, and digestive tract form	
Fifth week	Arms and legs begin to take shape; the heart pumps blood	
Sixth week	Eyes and ears form	

TABLE 11–1
Early stages of embryonic development

The many changes that take place during the first 6 weeks of life are summarized in table 11–1. The important point to note here is that the embryo is developing its various organs.

Transition from Embryo to Fetus

By the beginning of the ninth week, the embryo is not yet fully formed, but it is well on the way. Its chances of ultimately being born alive have risen from 90 percent to 95 percent. It is also at this time that the embryo is so well developed that it can now be called a **fetus**. Cartilage is turning into bone, the skeleton is emerging, and the form of vital organs is already outlined. However, much growing remains to be done. Refinement of structures has yet to be accomplished, but the emphasis in development is shifted. From the seventh week on, development of function becomes important. The various organs also begin to function as units. See table 11–2 for a summary of this development.

Parents often wonder whether their baby will be a girl or a boy. It is possible for a physician to determine the infant's sex by sampling the **amniotic fluid,** because it contains some fetal cells. These cells can be analyzed for chromosome number and type (see figure 11–5). This procedure, called **amniocentesis,** is usually performed only when there is a medical need, since it carries a small but real risk to the mother and baby. It is recommended for couples who have had a child with a genetic defect or if the mother is over 35 years old. At this age, it is more likely that she will have a child with **Down syndrome,** a condition that is always associated with the presence of an extra chromosome in the

FETUS An unborn developing animal or human

AMNIOTIC FLUID The fluid surrounding the fetus that protects it from shocks

AMNIOCENTESIS The process by which doctors extract and examine amniotic fluid to determine whether a baby will be affected by Down syndrome; it is also useful in determining whether a baby will be male or female

DOWN SYNDROME A genetic defect that results primarily in mental retardation

TABLE 11–2

Fetal development

Time (in lunar months)	Length (in centimeters)	Developmental Events
3	7	Limbs attain relative length compared with body Teeth are beginning to form External genitalia are developing Liver produces red cells Thyroid becomes differentiated Kidneys are formed Face looks more human Brain is taking final form
4	15	Bone is substituted for cartilage in long bones Uterus is taking definite form Gonads are fully formed Thumb sucking begins Eyes and ears are fully developed Hair follicles develop
5	23	Salivary glands are functional Digestive system begins to function Fingerprints and nails are formed Mother is aware of fetal movements Cerebellum grows rapidly
6 through 10	30–50	Embryo continues to develop more in size By the seventh lunar month, the fetus begins to store fat

NOTE: Lunar months are used to express development time in order to have more regular time periods for studying fetal development.

affected infant's cells. Consequently, amniocentesis has become routine for older pregnant women. It is also the only sure way to tell the sex of the unborn baby.

Birth

Up until the moment of birth, the fetus has all its needs provided by the mother—nutrients, protection, and an environment within which to grow and develop. The fetus does not have to breathe in the womb as oxygen is provided and carbon dioxide is removed via the placenta. However, as soon as the fetus is born, it must breathe on its own almost immediately, or the previous 9 months of development and growth will be lost.

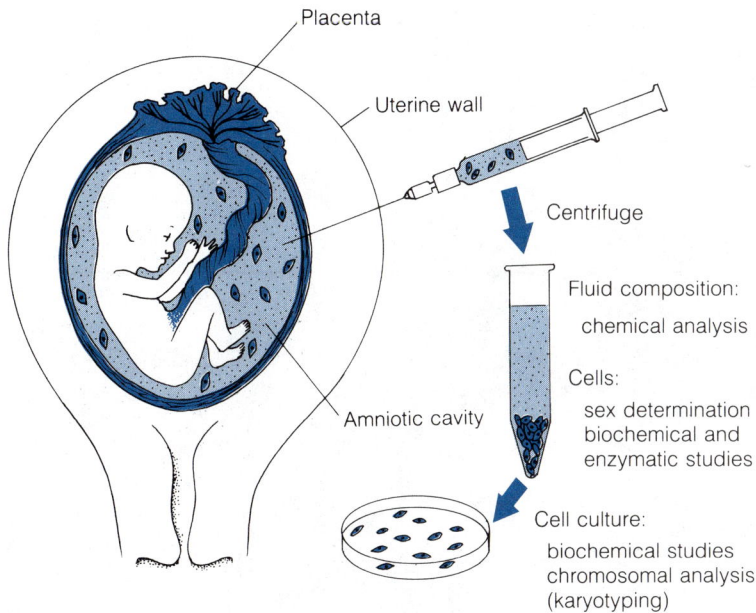

Placenta

Uterine wall

Centrifuge

Fluid composition:
    chemical analysis

Cells:
    sex determination
    biochemical and
    enzymatic studies

Amniotic cavity

Cell culture:
    biochemical studies
    chromosomal analysis
    (karyotyping)

FIGURE 11–5
Amniocentesis makes possible prenatal detection of genetic and chromosomal defects.

From *Human Genetics*, 4th ed., by A. M. Winchester and Thomas R. Mertens. Copyright © 1983 by Bell & Howell Company. Used by permission of Charles E. Merrill Publishing Company.

Many theories have been offered as to what causes the infant to take that important first breath. Until recently it was believed that the shock of birth was the only explanation. A swimmer who plunges into an icy pool or ocean for a swim has to catch his or her breath. It is possible that at birth the fetus has a very similar gasp when it is forced from its warm uterine environment into the relatively cool surroundings. That first gasp ensures survival. Many times the doctor or midwife gives the infant a slap on the buttocks to provide extra stimulation.

There are other factors that may be at work as well. Chemical changes within the body and changes of pressure within the chest cavity may trigger the breathing mechanisms. When the fetus passes through the birth canal, the chest cavity is subjected to a great deal of compression. This forces amniotic fluid and **mucus** in the breathing passages out through the nostrils and mouth. When the restriction is removed immediately after birth, the chest expands, and air rushes in to fill lungs that have until now been deflated and crumpled. Thus, the first sound may not be a cry but a cough that helps force fluid out of the lungs.

No matter which theory is correct, this first independent breath requires a tremendous effort. It takes as much as 10 times the force required of a fully grown adult for the same activity. Thus the first few breaths, which expand the lungs to about 75 percent of their total capacity, put a tremendous strain on the infant.

MUCUS A thick, slippery liquid produced by glands of the mucous membranes

The moment of birth is the single most important event in human life. Here the newborn has just taken a first breath and cries. (© Harvey R. Phillips/Phillips Photo Illustrators)

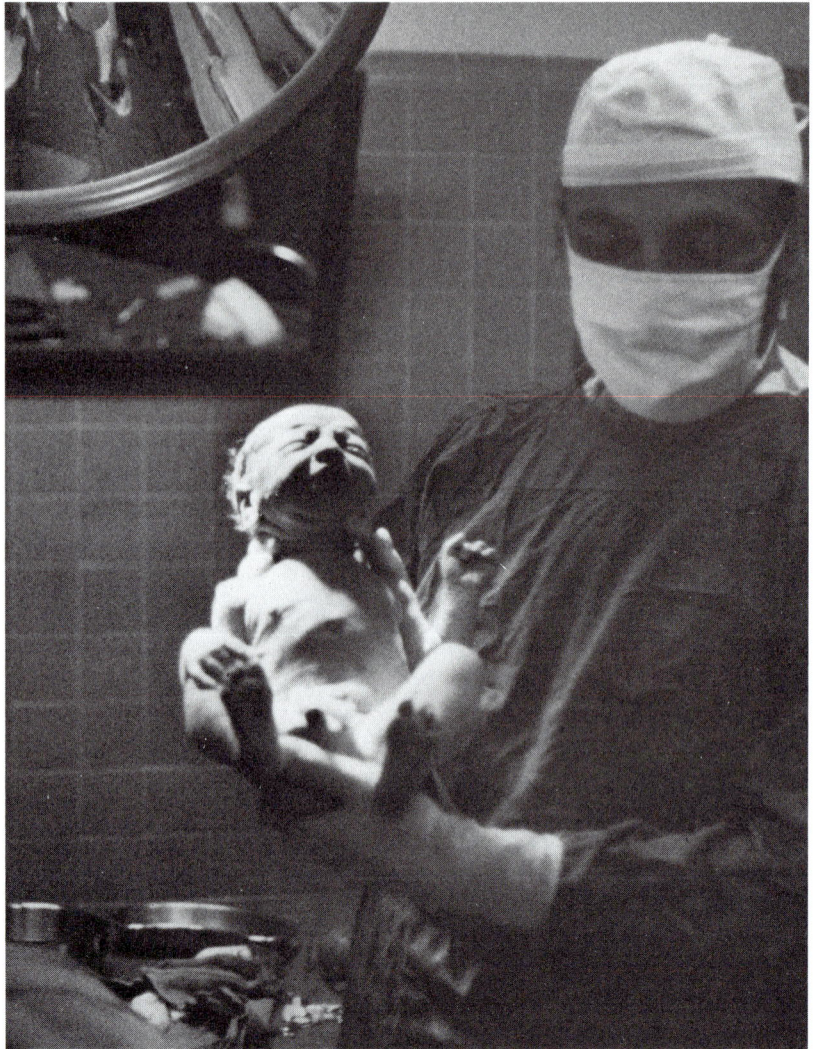

BLOOD The body's circulating fluid that contains many types of cells and carries food, oxygen, waste products, and hormones to and from the cells of the body

Dramatic changes in **blood** circulation take place at birth because fetal blood circulation shunts a great deal of blood away from the heart. However, once the fetus has to breathe on its own and supply its body with oxygen, new patterns of blood circulation are formed.

CHILDHOOD AND ADOLESCENCE

Scientists now know that all children grow in much the same way. Growth is rapid at first, proceeds more slowly, and then another rapid growth spurt is seen at **puberty.** Studies of children's growth patterns show that there is much individual variation. It is also known that even though a child may be ahead or behind his or her contemporaries in physcial or mental growth, it doesn't mean that this position will always be maintained. While the rate of growth depends in part on age, it isn't the only factor; there are internal clocks that regulate growth patterns.

An individual's sex influences growth and development. Girls begin life with certain advantages. Until the age of 5 months, girls lead in their ability to do things such as sitting and crawling. This is true for finer motor coordination skills as well. Girls also learn bladder control earlier than boys. However, there is no clear evidence that girls have more rapid brain development. Chapter 3 gives a more detailed discussion of mental and emotional development during childhood.

PUBERTY The stage of development when it is first possible to reproduce sexually

Working together, the muscles, skeleton, and nervous system allow us to move. (Paul Conklin)

Adolescence

The intensity and duration of the adolescent growth spurt varies widely, but it usually lasts for 2 to 2½ years. The growth that occurs during this short time is enormous and is second only to the growth seen in the uterus. The developmental differences between boys and girls are now most evident. Girls begin their adolescent growth spurt at the average age of 10½ years as compared with 12½ years for boys. In childhood, boys and girls are practically the same height, but once boys reach adolescence, they grow about 8 inches and gain about 45 pounds. At the peak of adolescence, boys grow about 4 inches a year. Girls gain 6½ inches and 35 pounds with a peak average growth of 3 inches per year. During this time the growth of the heart and other vital organs speeds up, the head gets larger, and the front to back measurement of the eyeballs increases. The contours of the face and body show marked changes. The entire profile becomes more angular, and accented with curves, the forehead becomes more prominent, and the chin more pointed. The nose even grows somewhat longer!

Before puberty, girls can hold their own in tug of war games with boys the same age, but this is not usually true after the adolescent growth spurt. From then on, males are usually stronger, because of the increased muscle mass that develops as a result of male **hormones.** Growth slows down and stops after age 19 or 20. On the average, boys achieve 98 percent of their height by the time they are 18—girls by the time they are 16½ years old.

The most striking feature of puberty is the development of sexual maturity. Puberty is triggered by hormones secreted by the endocrine system into the blood stream. The action of these hormones has been reasonably well traced. However why **glands** that have operated in one pattern for several years should gradually shift into a new one is an intriguing puzzle that has occupied the minds of many leading scientists.

The best evidence suggests that the whole process begins in the **hypothalamus,** a part of the brain located at the top of the brain stem. This tiny cluster of cells plays a significant role in controlling blood pressure, body temperature, water balance, thirst, hunger, reproductive behavior, and other functions. The hypothalamus seems to turn on the hormonal process that produces the adolescent growth spurt and sexual maturity. Interestingly enough, at the proper moment it also turns these processes off, too. The hypothalamus seems to function by acting on the body's master control gland—the pituitary gland—which modulates the activity of all the **endocrine glands** in the body.

The primary sexual changes that prepare the body for reproduction are accompanied by a number of secondary ones. In boys, the testes

HORMONE A chemical secreted in one part of the body that exerts specific controls on other parts

GLAND Specialized tissue that secretes specific hormones or proteins

HYPOTHALAMUS The control center of vital functions of the body, located at the top of the brain stem

ENDOCRINE GLAND Gland that produces hormones that are secreted into the blood stream

become larger and capable of manufacturing sperm. The prostate gland develops and begins to secrete the seminal fluid that transports the sperm. The accompanying secondary changes—a deepening voice, coarser facial hair, and pubic hair—announce to the world that the boy is becoming sexually mature. In girls, the growth and development of breasts, uterus, and ovaries come first. Following this, ovulation and menstruation occur.

All of these dramatic physical changes are bound to influence adolescents' feelings about themselves and others. This is a time when the individual is asking such questions as:

"What sort of person am I?"
"Are my thoughts and feelings like others or am I different?"
"Am I better or worse than other people?"
"Would people like me if they knew what I am really like?"
"Am I pretty (or handsome) enough?"
"What sort of people are my parents?"
"How do my parents compare with other kids' parents?"
"What do I think about my friends?"
"What sort of person do I wish to be?"

Adolescents often measure their worth as a person by appearance. Their standard of success is the degree to which they think they meet the standards of attractiveness set by their peers. Because of the comparison phenomenon, late maturers are often most self-critical.

The growing up years in our society offer a real contrast to those in nontechnological societies. There, childhood offers a time to learn to meet the needs of the culture. Sexual and social maturity seem to be reached at the same time. The "in-between" period lasts about 2 to 3 years. In our society, however, the situation is quite different. Our culture demands skills and behavior that cannot be acquired in even a dozen "in-between" years. Thus the time interval between sexual and social maturity can be long and drawn out.

Few people achieve economic independence before their late teens, and those who go on to higher education may not achieve this until their thirties. These intervening years are often difficult. Adolescents who are no longer physically dependent as they were in childhood may be psychologically or economically dependent and therefore have conflicting feelings. They are increasingly held responsible for what they do, and yet they are controlled and supported by others. Many adolescents are ready physically for an adult sexual relationship long before they are ready psychologically and emotionally. These factors often lead to a great deal of frustration for adolescents and young adults.

SYSTEMS OF THE HUMAN BODY

Earlier in this chapter we pointed out that various organs develop as the human embryo becomes a fetus. This section deals with the way cells are organized into organs and organ systems. These systems are responsible for four major functions of the body.

The first of these functions is *support* and *movement,* which involves two organ systems—skeletal and muscular. The second is *control,* which is carried out by the nervous and endocrine systems. The third function is *maintenance,* which is accomplished by the circulatory system, the digestive system, the respiratory system, the excretory system, and the immune system. The fourth major function is *continuity* and has already been discussed briefly. This function is mediated by the reproductive system.

Support and Movement

BONES Hard tissue forming the skeleton of most vertebrates

MUSCLES Tissue composed of contractile cells

LIGAMENTS Connective tissue that links 2 bones at a joint

There are three basic elements in the muscular and skeletal system: (1) **bones,** (2) **muscles,** and (3) **ligaments,** or connective tissue. They provide the support for the body, protection of delicate internal organs, and movement.

Bones. Bones provide the main support for the body. They are capable of supporting and lifting enormous weight and are linked together by ligaments, bands of elastic fiber-like tissue. The linkage points are called joints. A diagram of some major bones is shown in figure 11–6.

Protection of internal organs is provided by bones, joints, and connective tissue. The joints of the bones in the pelvis are rigid since they must support the weight of the upper body and must protect the organs within it. Joints of bones involved in movement, such as hips or shoulders, are more flexible. These joints are lined with **cartilage,** a smooth, glassy tissue that resists friction and is lubricated by a special fluid.

CARTILAGE Dense, fibrous, connective tissue that confers flexibility and strength on joints

Bones come in many sizes and shapes. Skull bones are relatively flat and large, whereas the ear bones are tiny and shaped like their names—the hammer, anvil, and stirrup. The adult has 206 bones. Bone is a living tissue and is renewed completely every 2 years. It is composed of water, various organic materials and minerals, calcium phosphate, and calcium carbonate.

Although the skeleton only makes up 18 percent of the body's weight, it is very strong. The average man exerts 2,000 pounds of pressure per square inch with every step and this increases with such activities as running or jumping. Except in very rare instances the bones take this crushing punishment without a twinge.

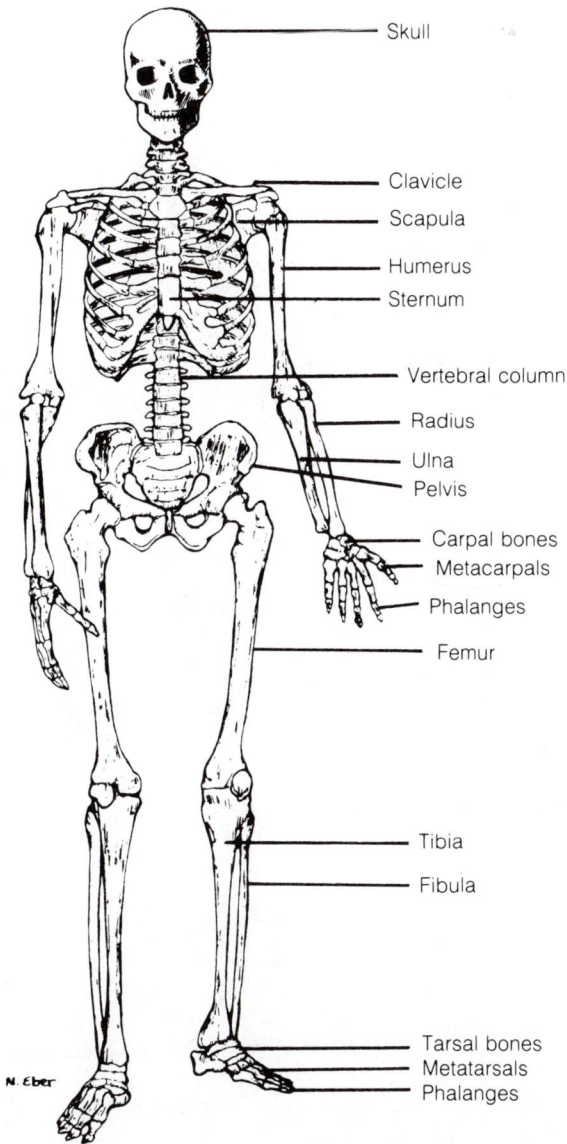

Skull

Clavicle

Scapula

Humerus

Sternum

Vertebral column

Radius

Ulna

Pelvis

Carpal bones

Metacarpals

Phalanges

Femur

Tibia

Fibula

Tarsal bones

Metatarsals

Phalanges

N. Eber

FIGURE 11–6
The human body is supported by a skeleton composed of 260 bones which are connected to each other by elastic fibers called ligaments.

From *Physical Disabilities and Health Impairments: An Introduction,* by John Umbreit (ed.). Copyright © 1983 by Bell & Howell Company. Used by permission of Charles E. Merrill Publishing Company.

Muscles.   Body movement is also a function of the muscular system which is shown in figure 11–7. Skeletal muscles are under voluntary control; that is, they can respond to conscious commands. However, those muscles found in internal organs such as the heart, lungs, stom-

Muscles are tissues that move parts of the body. The muscles shown here are voluntary muscles that move body parts when the person wants to move them. Involuntary muscles move body parts such as the heart and stomach muscles without the person thinking about them.

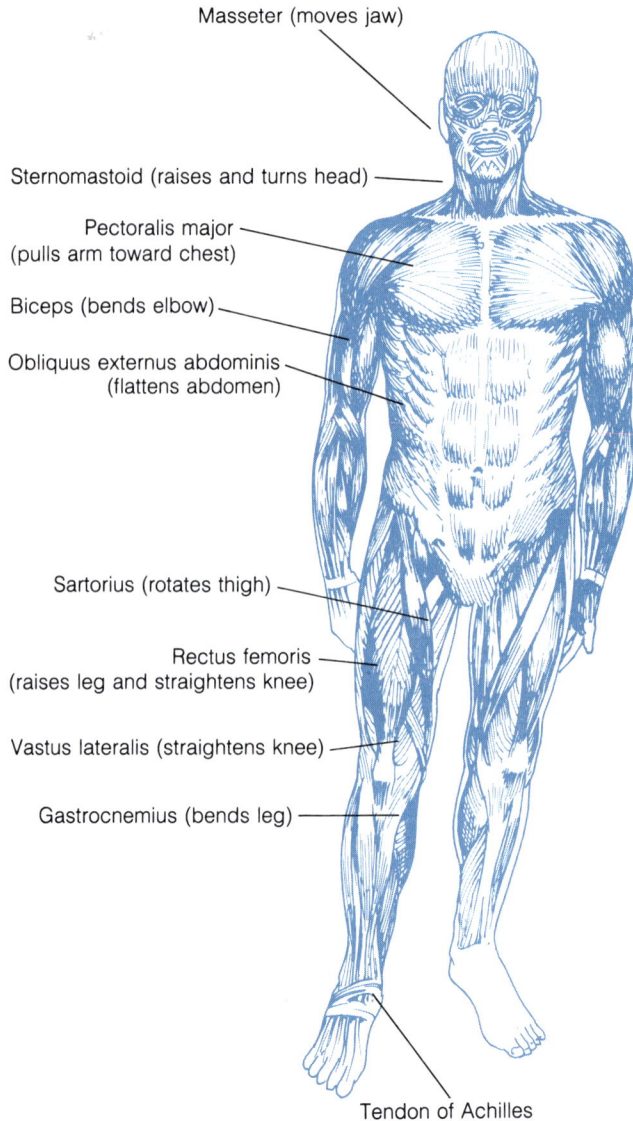

Masseter (moves jaw)

Sternomastoid (raises and turns head)

Pectoralis major (pulls arm toward chest)

Biceps (bends elbow)

Obliquus externus abdominis (flattens abdomen)

Sartorius (rotates thigh)

Rectus femoris (raises leg and straightens knee)

Vastus lateralis (straightens knee)

Gastrocnemius (bends leg)

Tendon of Achilles

ach, and intestines are quite different; they operate without conscious control and are thus called involuntary muscles.

Most skeletal muscles are attached to bones by tendons, which are bands of tissue that are less elastic and thicker than ligaments. Muscles usually function as opposing pairs. While one muscle is relaxed, its partner tugs at the bone much like ropes on a pulley. The movement can

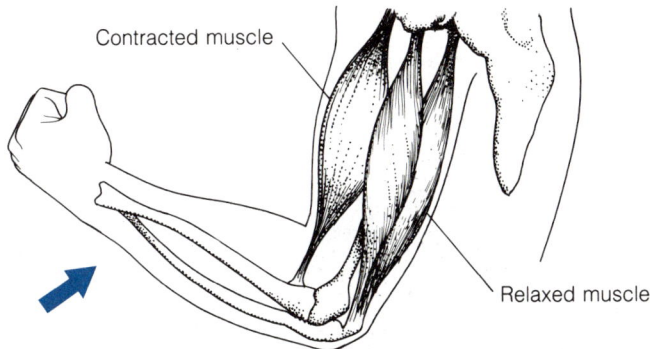

Contracted muscle

Relaxed muscle

FIGURE 11–8
Muscles work in pairs to
move a body part. When
one muscle is contracted,
or shortened and tightened,
the other muscle becomes
relaxed, or longer and
looser.

be reversed by contracting the opposite muscle and allowing the other
to relax. (See figure 11–8.)

Similar to bones, muscles range in size and shape to fit the function they perform. The facial muscles, which permit the eyebrows to
raise and the mouth to frown or smile are less than 3 centimeters long.
The entire floor of the chest cavity is covered by a dome-shaped diaphragm, which is the main muscle involved in breathing, as well as
coughing, sneezing, laughing, and sighing. There are longer, strap-like
muscles in the neck and back that help keep the head upright. The longest muscle, called the sartorius, is found in the thigh and is used in
crossing the legs. The name comes from the Latin for tailor, since ancient tailors sat cross-legged.

All muscles, both skeletal and visceral, cause movement by contraction. This characteristic distinguishes them from other body tissue. In
the case of skeletal muscles, the individual cells are ordinarily long and
thin but when stimulated they become shorter and wider—producing a
tremendous pulling power. Once the stimulus is past, the muscle cells
relax and settle back to their original narrower, longer shape.

Skeletal muscles are stimulated by commands from the motor
nerves in the brain. This signal is received by a special nerve that is
intimately attached to the muscle fiber. The signals from this nerve spark
complex changes in the muscle cell membrane which, in turn, convert
chemical energy within the cells into mechanical energy of motion.

Control Systems

The Nervous System.   The nervous system acts as the body's basic
control center. As shown in figure 11–9, it can be compared to a vast
switchboard of humming, flashing signals constantly sending and receiving messages. Some of these messages are urgent and some are not,

FIGURE 11–9
The nervous system acts
as the control center for the
human body and sends
messages throughout each
24 hours. Every thought,
movement, and bodily
function result from electro-
chemical impulses sent
through the nervous sys-
tem.

From *Physical Disabilities and
Health Impairments: An Intro-
duction,* by John Umbreit (ed.)
Copyright © 1983 by Bell &
Howell Company. Used by per-
mission of Charles E. Merrill
Publishing Company.

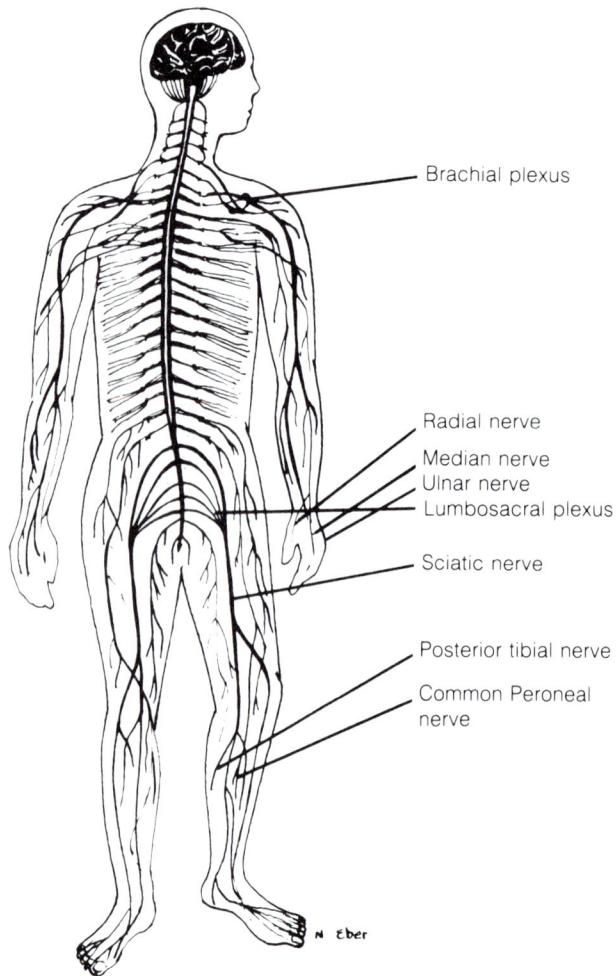

Brachial plexus

Radial nerve
Median nerve
Ulnar nerve
Lumbosacral plexus

Sciatic nerve

Posterior tibial nerve

Common Peroneal
nerve

N Eber

but all are concerned with the effective conduct of the body's affairs.
From birth to death, the nervous system controls every thought, every
emotion, every step taken, and every impression received. Without it,
muscles would not move nor could pleasure or pain be distinguished.

The nervous system operates by electrical-chemical impulses. It
functions as the body's "lookout" and helps gather information about
the outside world as well as the world inside. It is also a clearinghouse
where this information is processed for instant use or stored for future
reference. The nervous system functions as a general communication
center and a headquarters for mapping out strategies and making deci-

sions. In short, the nervous system initiates every bit of muscular activity and regulates all mental and physical functions.

The nervous system is composed of nerve cells, or **neurons,** which are interspersed with other cells. The brain and spinal cord are the main base of operations. From there, the nerve network reaches out to every part of the body.

NEURON A nerve cell

The simplest unit of the nervous system is a neuron. It is different from all other types of cells in the body. From the central part of each neuron (see figure 11–10), fibers extend like delicate tendrils. Each neuron has an axon, which is a very long extension leaving the main body of the neuron; it carries impulses away from the body of the cell. Depending on their location, these axons may be a fraction of an inch long or as long as 3 or 4 feet. The longest one reaches from the base of the brain to the big toe.

It is through these fibers that neurons perform their unique function of transmitting signals from the senses and visceral organs to the brain and of sending directions from the brain to the muscles. When a neuron

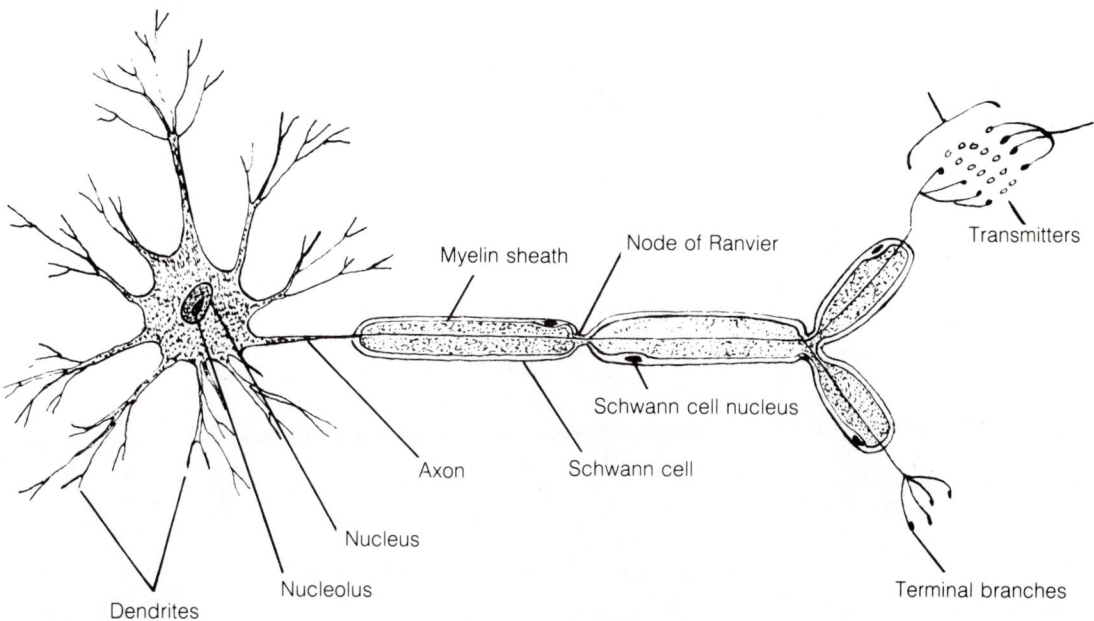

FIGURE 11–10

Neurons (nerve cells) are considered the building blocks of the brain. Each neuron receives information through numerous short branch-like structures called dendrites, and transmits information along a single long myelin-covered fiber called an axon.

From *Physical Disabilities and Health Impairments: An Introduction,* by John Umbreit (ed.) Copyright © 1983 by Bell & Howell Company. Used by permission of Charles E. Merrill Publishing Company.

receives a signal, it relays it across a gap, or **synapse,** to an adjacent
neuron. In this way neurons form a many-linked chain throughout the
body, which functions much like a bucket brigade. Nerve impulses travel
in only one direction: dendrite to cell body to axon to synapse, and so
on to another dendrite.

The Brain.   For all its importance, the brain (shown in figure 11–11)
does not have an impressive appearance. It is white and gray and deeply
furrowed. About the size and shape of an acorn squash, it weighs about
3 pounds. The brain stem is a 4-inch long and 1-inch wide central core
that extends up from the spinal cord. The brain stem controls such functions as heartbeat, respiration, and swallowing; it is considered the most
primitive part of the brain.

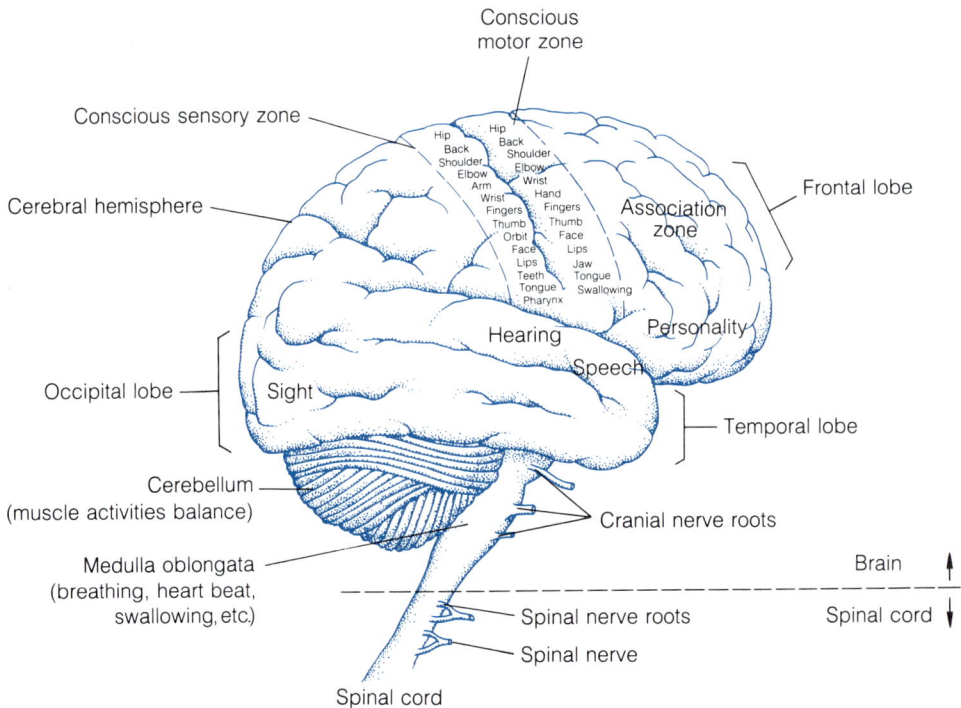

FIGURE 11–11
The human brain weighs between 10,000 and 14,000 grams, slightly less than 3 pounds.
Today neuroscientists continue to map its structure as they attempt to locate and under-
stand the brain's various functions.

From *Physical Disabilities and Health Impairments: An Introduction,* by John Umbreit (ed.) Copyright ©
1983 by Bell & Howell Company. Used by permission of Charles E. Merrill Publishing Company.

On either side of the stem is a mass of tissue called the cerebellum, which is involved in movement and balance. Draped over both the brain stem and the cerebellum is the cerebrum. Its outer layer, or cerebral cortex, would cover more than 2 square feet if it were flattened out. However, it is so folded and convoluted that it fits within the 6 by 8 inch vault of the skull. The cerebrum is divided into 2 hemispheres that are connected by bands that are lighter in color than the exterior tissue. Here—as well as other places in the nervous system—the two colors relate to the way that the neurons are arranged. In most of the larger nerves the axon is covered by a fatty covering, or myelin sheath, that insulates the fiber and enables it to conduct impulses more quickly. In the brain and other parts of the nervous system, the neurons are arranged in such a way that the cell bodies, which are gray in color, are arranged in a layer, and the axons are all positioned in the same direction. The white myelin covering of the axon contrasts with the gray of the cell bodies; this gives a two-tone effect.

The cerebral cortex acts as the general staff of a military unit; it provides a most sophisticated form of control over the incoming signals. It not only records the signals, but acts on them, interprets them, and stores them for future reference. The cerebral cortex is thus responsible for selecting the best course of action to follow at any given time. It is involved in intellectual functioning: thinking, planning, worrying, and problem solving.

Spinal Cord. The spinal cord is about 18 inches long and looks similar to a cable tapered at both ends. At its widest, it is about 1/2 inch across and, like the brain, it is composed of many neurons. The gray matter is found at its center surrounded by the white fibers. The spinal cord is protected by a tunnel of bone formed by the vertebrae. This set of bones is called the spinal column or the backbone. It is not, as is often thought, the spinal cord.

Peripheral Nerves. Networks of nerves are spread throughout the entire body. From the brain, nerve fibers form 12 pairs of cranial nerves that serve the head, eyes, ears, throat, and some organs in the chest and abdomen. From the spinal cord, 31 pairs of nerves pass out through the openings in the vertebrae at various levels. Some of these nerves branch and rebranch from the upper part of the spinal cord to the upper parts of the body—the arms and hands. From the lower part of the spinal cord, nerves branch to form nerve trunks leading to the pelvis, legs, thighs, and feet. Nerves also reach every muscle, blood vessel, and bone in the body from head to toe.

There are three classes of neurons. Sensory neurons carry signals to the spinal cord and brain. Motor neurons carry directions from the brain and spinal cord to the muscles. Connecting neurons shuttle signals back and forth.

An important part of the motor system of neurons is the autonomic nervous system. It operates by causing contraction and relaxation of the visceral or smooth muscles, primarily in the gastrointestinal tract, and controls internal body functions such as respiration and heart rate.

Sensory neurons keep the brain informed with data gained from sensory pick-up units located all through the body. These pick-up units, or **receptors**, which are located in the skin are sensitive to pain or touch or pressure or temperature differences. Receptors in the tongue respond to flavors and odors. The retina of the eye contains special receptors that respond to light and shadows of different intensities.

RECEPTOR A specialized nerve ending that receives stimuli

It is possible to think of a stimulus in terms of only a warning of danger such as a stab of pain or a flash of bright light or a clang of a bell. Actually you receive sensory stimuli all the time—danger or not.

Nerve cells work at a split-second speed 24 hours a day. It may seem that sleep would give your nervous system a rest, but it is quite active even then. Aside from keeping the breathing and heartbeat rates even, the brain uses these periods to search for answers to particularly difficult problems—and to rethink experiences that flowed in too fast to be dealt with during the day.

The nervous system uses energy at an astounding rate. Nerve cells require far more fuel—glucose—than any other kind of cell in the body. They must have a continuous and rich supply of oxygenated blood or they will quickly die. The nerve cells in the brain use as much as 20 percent of the oxygen that is available in the body.

Normally in every 24 hours, some neurons are lost because they wear out and die. Fortunately, great numbers would have to die for you to notice an effect, but small losses do accumulate. Some neurologists believe that in old age, the loss of mental efficiency—or senility—is caused by the day-by-day loss of a few neurons here and there. Whatever the cause, once a neuron is destroyed, it is forever lost. No new nerve cell ever takes its place. If the fiber of a neuron is cut and the cell itself is not killed, a new fiber may eventually grow across the cut, but if the nucleus of the neuron is destroyed, the cell is permanently gone.

The nervous system conducts its complex activities with brilliance. Nerve impulses are transmitted so rapidly that an entire sequence of actions, interpretations, and responses can take place—and we may not even be aware that it has happened.

For example, consider what happens when you cut your arm. First, this may stimulate a kind of protective reaction—a short circuit by which the pain signal from your skin causes the motor nerves in your spinal cord to instantly jerk your hand away from the pain stimulus. This happens before word of the injury ever gets to the brain. The pain stimulus is also picked up by the neurons in the spinal cord. News of the cut is simultaneously echoed to the brain, where a chain of reactions is set off in just a few seconds. The number of individual signals transmitted to or from the brain during this reaction can only be estimated, but it is very large indeed.

The Endocrine System.   The endocrine system is another major control system of the body. The *endocrine* glands produce hormones that act as chemical regulators of body functions; they have an enormous influence over our physical well-being. The power they have is intriguing and not completely understood with respect to internal chemistry.

Glands are usually divided into two kinds: **exocrine glands** and **endocrine glands.** Exocrine glands secrete substances through ducts to surfaces of the body. Examples of these are the salivary, sweat, mucous, and mammary glands. Endocrine glands, on the other hand, produce and secrete their products, almost always hormones, into the blood stream. Table 11–3 summarizes the glands, their hormones, and the

EXOCRINE GLAND A gland that has its secretions carried to body surfaces by ducts

ENDOCRINE GLANDS Glands that produce hormones that are secreted into the blood stream

TABLE 11–3
Endocrine gland functions

Gland	Hormone	Process Affected
Thyroid	Thyroxin	Level of metabolism
Parathyroid	Parathyroid hormone	Calcium balance
Adrenal medulla	Epinephrine	Stimulation of autonomic nervous system
Adrenal cortex	Corticosterone and cortisol	Carbohydrate metabolism
	Aldosterone	Salt balance
Pancreas	Insulin	Glucose metabolism
Ovary	Estrogen	Female sexual development and menstrual cycle
	Progesterone	Control of ovary and uterus during pregnancy
Testis	Testosterone	Male sexual development and activity
Pituitary		
Adenohypophysis	Adrenocorticotropic hormone	Control of adrenal cortex
	Thyrotropic hormone	Control of thyroid
	Gonadotropic hormone	Control of sex glands
	Growth hormone	Growth
Neurohypophysis	Antidiuretic hormone	Kidney action
	Oxytocin	Uterine contractions

SOURCE: G. G. Simpson and W. S. Beck, *Life* (2nd ed). (New York: Harcourt, Brace and World, 1968)

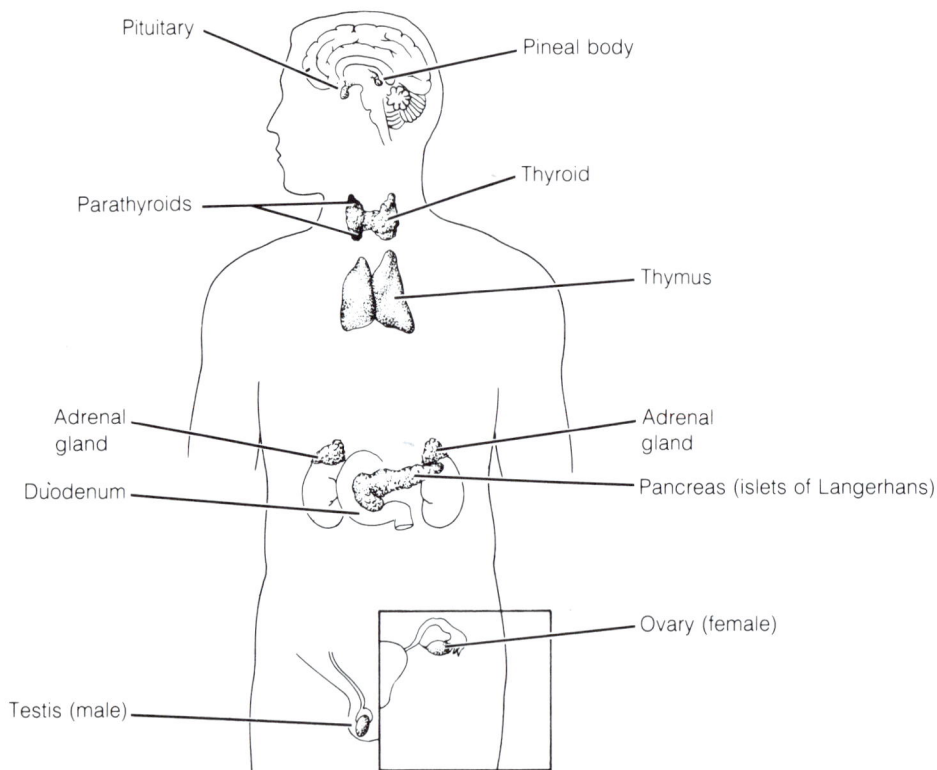

FIGURE 11–12

The human endocrine system is composed of glands which produce hormones that regulate various body functions including metabolism, sexual development, growth, and kidney action.

From *Physical Disabilities and Health Impairments: An Introduction,* by John Umbreit (ed.) Copyright © 1983 by Bell & Howell Company. Used by permission of Charles E. Merrill Publishing Company.

processes affected by them. Figure 11–12 illustrates the location of the most studied endocrine glands.

The hormones only resemble each other in that they are secreted into the blood and act at points distant from the site of secretion. The chemical natures of the hormones are very different from one another: some are proteins, one is an amino acid, and yet others are **steroids.**

A well-known endocrine gland is the thyroid, located at the bottom of the neck just below the Adams apple. Thyroid hormone influences the rate of **metabolism,** or, in other words, the rate at which the body converts food into heat and energy. Small decreases in the amount of thyroid hormone cause chills, drowsiness, and fatigue. Respiration may be slow and the heart rate sluggish. Obesity may result even though

STEROID A large complex molecule of biological origin composed of 4 attached rings of carbon atoms

METABOLISM The sum of chemical reactions within the cell, including both breakdown and synthetic reactions

appetite is below par. If the thyroid is overproducing hormones, the opposite occurs—nervousness, pounding heart, overactivity, and weight loss in the face of increasing appetite. These are all symptoms of the body fires burning out of control.

Another example of an endocrine gland is the islets of Langerhans in the pancreas, which produces insulin. This hormone is an important regulator of how much glucose is taken up by the cells, which in turn affects the production of heat and energy in the body. If the glucose is not taken up by the cells, it passes out of the body by way of the urine. Diabetes is the result of an insulin deficiency and thus testing for it is done by measuring the amount of sugar in the urine.

Two hormones of quite different natures are produced by the adrenal glands. Both these hormones help the body respond to emergency situations. Adrenalin, or epinephrine, which is produced in the central part of the brain or medulla, is a powerful stimulant throughout the body. In times of excitement, adrenalin speeds up respiration, raises blood pressure, sharpens reflexes, and puts the body on guard. Another hormone, cortisone, is produced by the outer part, or cortex. It affects just about every system in the body. It affects the metabolism of all biologically important molecules including proteins and carbohydrates, growth and development in children, salt and water balance, the cardiovascular system, kidneys, skeletal muscle, the immune system, and the nervous system. This group of hormones gives the organism the capacity to resist many types of noxious stimuli and environmental change. In fact, the adrenal cortex is probably the organ that is most responsible for the relative freedom that we and other higher organisms have in a constantly changing environment. It is possible to live without the adrenal cortex, but it would require a highly regulated diet, regular intake of various salts, and a strictly controlled climate.

The most important gland however, is the pituitary. It is called the master gland because it produces hormones that regulate the function of the most important endocrine glands. Like the adrenal, it is divided into two parts—the adenohypophysis and the neurohypophysis. The former produces different hormones that each regulate the thyroid, the adrenals, the gonads, as well as the overall process of growth. The neurohypophysis has a role in regulating uterine contractions, particularly during childbirth. It also produces a hormone that regulates blood pressure and another one that controls the amount of water excreted by the kidney. The pituitary is subject to **feedback** from the glands that it regulates as well as to nervous system control.

From what we now know about the endocrine glands and what has been learned in the last century, there is reason to suspect that we are only now beginning to get a clear picture of their roles in body functioning. Nearly every week there is a new report about the endocrine

FEEDBACK The process by which a control mechanism is regulated through the very effects it brings about system and its hormones. Evidence now seems to point to a direct link between the endocrine system and the brain, and possibly the entire nervous system. In fact, some endocrinologists believe that both systems may actually be different parts of one body-controlling organ system.

The Reproductive System

Earlier in this chapter we described human sexual reproduction in the context of cell division, the fertilization of sex cells, and the development of the embryo and fetus. In this section we complete the story by briefly describing human reproduction from the structural or anatomical point of view.

The male reproductive system consists of several structures all functioning together to accomplish one purpose: to produce male sperm cells and transport them to their ultimate target, female egg cells. Male sperm cells are produced in the testes, a pair of egg-shaped organs suspended from the male body in a sack-like structure called the scrotum. You may be interested to know that scientists think the testes are outside the abdominal cavity rather than inside in order to maintain a lower-than-normal body temperature in the testes. A cooler temperature (3-5° F lower) is thought to be necessary for the production of healthy sperm.

Once males reach the age of puberty, the pituitary gland at the base of the brain stimulates development of the testes, and they in turn begin to produce not only sperm but also a hormone called testosterone which is responsible for the development of secondary sex characteristics. Voice changes and the growth of hair on the face, genital area, and other places on the body are examples of secondary sex characteristics. Within the testes, sperm are produced in small coiled tubes called seminiferous tubules. When sperm leave these small tubes they travel through the epididymis and into the vas deferens where they are mixed with secretions from three glands, the seminal vesicles, the prostate, and the Cowper's gland (see figure 11–13). Together the sperm and secretions from these glands combine to produce a milky fluid called semen which is ejected from the penis and into the female's vagina during intercourse. Between 300-500 million sperm are released in a single act of intercourse.

The female reproductive system is structured to accomplish one central function: to produce female gametes or eggs, receive sperm cells, and provide a safe place for fertilized eggs to develop into new human beings. It is not surprising then to find that all of the structures in the female reproductive system are situated within the protection of the abdominal cavity.

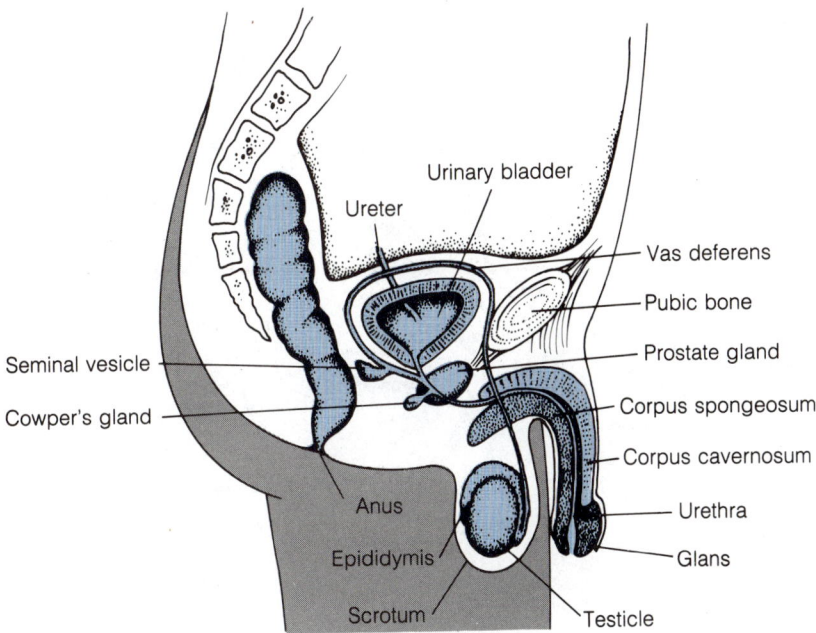

FIGURE 11–13
In the human male repro-
ductive system, sperm are
carried out of the body
through the urethra, a duct
which also expels urine
from the body.

Urinary bladder

Ureter

Vas deferens

Pubic bone

Prostate gland

Seminal vesicle

Corpus spongeosum

Cowper's gland

Corpus cavernosum

Anus

Urethra

Epididymis

Glans

Scrotum

Testicle

At puberty (generally between the ages of 10 and 15 years), the female reproductive system begins to produce hormones that trigger the development of secondary sex characteristics. In the female, these include the development of breasts, and hair in the genital area and other places on the body. Inside the body, nearly 500,000 immature egg cells that have been dormant in the ovaries since birth begin to mature, one per month. Located one on each side of the abdomen, the ovaries alternate, each releasing one mature egg every other month. Upon release, the mature egg cell, or ovum, passes through a tube called the fallopian tube, or oviduct, and into the pear-shaped uterus (see figure 11–14).

If the egg is fertilized by a sperm while en route to the uterus, it embeds itself (called implantation) in the wall of the uterus which thickens and becomes enriched with blood vessels to begin the 9-month developmental process which results in the birth of a new individual. Eggs which are not fertilized are transported out of the female body through the vagina. The process by which the human embryo becomes a fully developed individual capable of living outside the uterus, or womb, was described earlier (see p. 364, "The Embryo").

Maintenance Systems

The third group of body systems is concerned with the maintenance of the body. These systems include the digestive system, the circulatory

FIGURE 11–14
The human female repro-
ductive system has a separate passageway, the va-
gina, to carry eggs from the
body, while urine is ex-
pelled through the urethra.

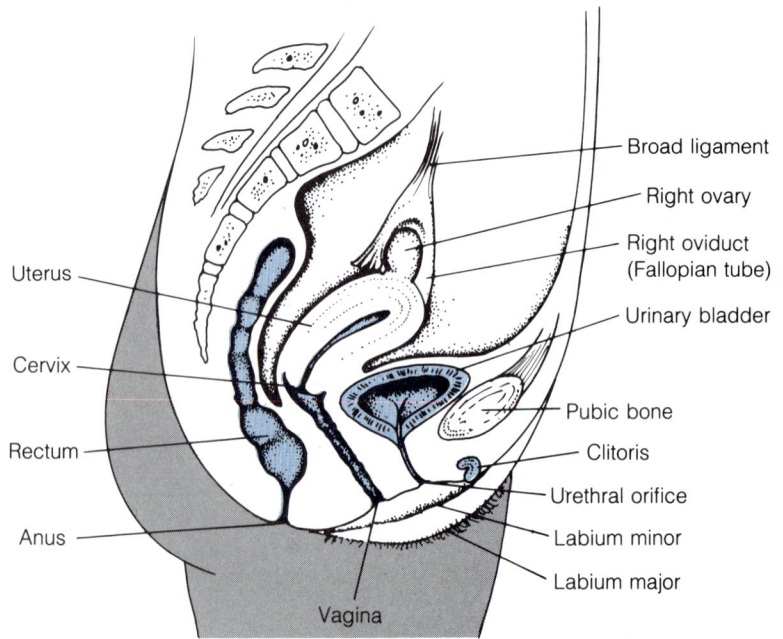

FIGURE 11–14
The human female reproductive system has a separate passageway, the vagina, to carry eggs from the body, while urine is expelled through the urethra.

system, the respiratory system, the excretory system, and the immune
system. Together these systems are responsible for securing food for the
body's cells, circulating it, using it, and getting rid of the waste products
of metabolism.

In addition to providing the cells with nutrients, these systems provide other means of communications among parts of the body, protection against disease, and regulation of salt and water balances.

Digestive System.   The body's digestive system, as seen in figure 11–
15, is basically a twisted and convoluted tube. If its coils and turns were
straightened out, it would be about 30 feet long. Within this 30 feet, the
injested food, which is chemically quite complex, is broken down, modified, and changed into the simpler substances that the body needs for
energy and maintenance.

When food contacts the tissues in the mouth, they send a message
to the brain that results in an "order" to the salivary glands to secrete
saliva. As the teeth grind the food, the saliva lubricates it, which helps it
on its way through the digestive system. Saliva also contains enzymes
that aid in the breakdown of starch in the food.

As the teeth finish their job, the tongue helps move the food on its
way. As the food is swallowed, it passes beyond voluntary control. Mus-

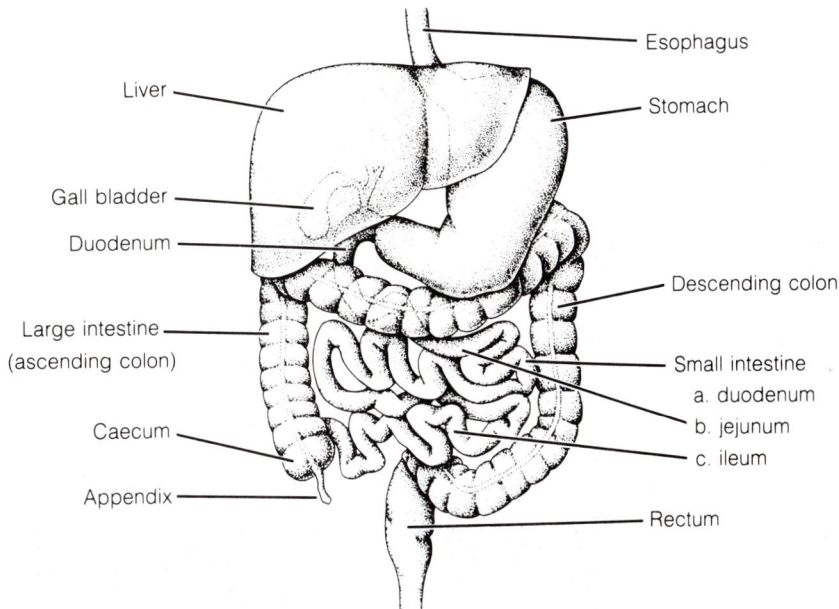

FIGURE 11–15

The human digestive system is very much like an elongated tube which expands and contracts in rhythmic movements as food is digested.

From *Physical Disabilities and Health Impairments: An Introduction,* by John Umbreit (ed.). Copyright © 1983 by Bell and Howell Company. Used by permission of Charles E. Merrill Publishing Company.

cle contractions triggered by the swallowing action automatically select the correct path for the food. To prevent the food from going down the windpipe and possibly causing suffocation, a small "trap door," called the epiglottis, closes off the windpipe after each swallow. The food now travels down the **esophagus** into the stomach. At this point it needs no help from the earth's gravitational pull to make its way down to the stomach. The astronauts' digestive systems worked quite well under conditions of zero gravity, because the power to move food comes from the smooth muscles that line the full length of the digestive tract.

The nerve endings in the esophagus cause waves to be sent up and down the tube that help force food toward the stomach. This wave action is called **peristalsis,** which is also seen in a number of other places in the digestive system.

In the meantime, special messages have been sent to the stomach to alert it to begin releasing gastric juices in preparation for the arriving food. Hydrochloric **acid** and the enzyme pepsin are used to break down protein material into simpler amino acids. Although hydrochloric acid

ESOPHAGUS The passageway from the mouth to the stomach

PERISTALSIS Rhythmic wave-like' contractions and dilations of the walls of the digestive tract

ACID A substance that dissolves in water and increases the hydrogen ion content of the solution

and pepsin would, on their own, be strong enough to eat a hole through the stomach's muscular wall, the stomach has a protective lining that prevents such action.

The stomach is on the left side of the body directly under the diaphragm and is protected by the rib cage. It is a pouch about 10 inches long. When it is full, it can stretch to hold as much as 2 quarts of food; when it is empty, it collapses and looks much like a deflated balloon.

Once food enters the stomach, it is worked on both mechanically and chemically. The movement of the stomach walls mashes, kneads, and mixes in the digestive juices. The stomach also serves as a storage tank holding the food until the next part of the system, the small intestine, is ready to receive it.

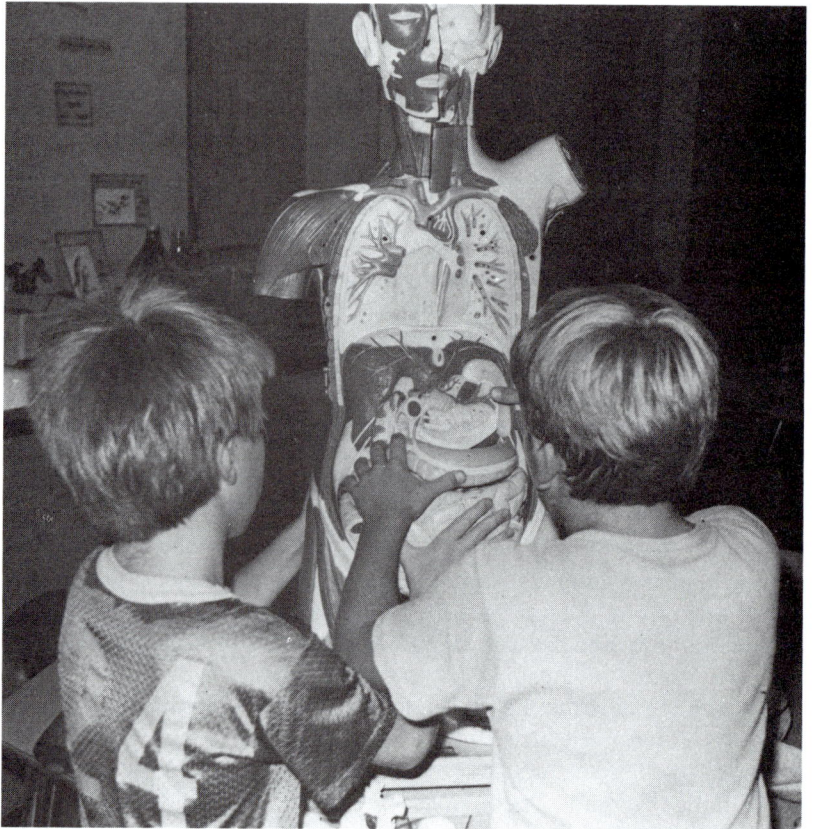

Once food enters the stomach, it is worked on both mechanically and chemically.

In the duodenum, the first 10 inches of the small intestine, the action switches. The acid that was added in the stomach stimulates intestinal hormones, which signal the liver and pancreas to release bile and pancreatic juice, respectively, to the small intestine. These digestive juices and those secreted by the intestine are **alkaline**—that is they **neutralize** acid and allow the enzymes of the small intestine to break the food down even further. Cellulose and some fats resist this change, but the rest is almost ready to yield the nutrients that the body needs.

Food goes into the small intestine in very small quantities. Here the partially digested food undergoes constant bombardment and any bit that happens to escape assault in one place is likely to be hit further on. This is where the most powerful of the digestive enzymes work. It is for this reason that the duodenum is an essential component of the digestive system for securing nutrients in the entire body.

Some of the most fascinating structures in the whole body are to be found in the small intestine. The villi are tiny soft, hair-like projections that protrude in countless millions along the wall of the small intestine. They resemble the nap in a bath towel. Although they are minute, villi perform an outstanding function: they separate the molecules of protein, sugar, and fat from the undigestable material. These little fingers then send the nutrients on their way to the cells in the rest of the body by way of the hepatic portal vein to the liver. All nutrients go to the liver before they go out to the individual body cells by means of the circulatory system.

The material that remains undigested passes into the large intestine. It spends about 10 to 12 hours there, during which time large quantities of water and a few nutrients are absorbed. The bacteria *Escherichia* is a constant inhabitant of the human intestinal tract and also plays a role in breaking down unabsorbed material into simple compounds that can be used for energy and synthesis of large molecules.

The Liver.   The liver is the largest, and probably the most versatile organ in the entire body. In fact, if it were removed you would die within 24 hours. On the other hand, the liver is so efficient at its biochemical job that you could get along if it were working at 1/4 its capacity. The liver has the unusual ability to replace its missing or damaged parts.

In addition to its role in digestion, the liver filters out old blood cells, removes harmful chemicals and drugs, and manufactures many other materials needed by the body, such as blood proteins and cholesterol.

A unique thing about the liver is that it can build new sugars called glycogen, which is a convenient storage form of glucose. The liver can also convert glycogen into glucose and release it into the blood as the

ALKALI Any chemical that dissolves in water and decreases the hydrogen ion content of the solution

NEUTRAL The state when the concentration of hydrogen ions in solution equals that of pure water

body needs it. Thus one of the liver's most important functions is to maintain the necessary glucose level in the blood.

The digestive system works without any conscious intervention. It is an enormous and elaborate mechanism that is coordinated and controlled by the nervous sytem. There is a network of nerve cells that relays messages back and forth between the digestive system and the brain. As far as we know, one cannot willfully interfere with this relay system, but there is a relationship between emotional upset and digestive tract problems.

Through experimentation, doctors know that strong emotional reactions can cause too much acid to be produced by the stomach. When some of this escapes from the stomach to the esophagus as part of a gas bubble, or goes into the duodenum with food, these sections of the digestive tract suffer because they lack a protective lining. This hyperacidity can cause minor irritation—heartburn—or, if it is serious enough, an ulcer or a break in the digestive system wall.

Circulatory System.

The circulatory system works in cooperation with the digestive system. Nutrients that are needed by the cells must get from the small intestine to the point of use. Day and night, the body's trillions of cells are loading and unloading. They are constantly taking in food and oxygen and exchanging waste products. The blood also carries various chemical messengers, hormones, and enzymes that are necessary for integrated body functioning. Although these processes slow down a bit during sleep, they never stop. In fact, if they did, death would be almost instantaneous.

The circulatory system is like a system of freeways with underpasses, cloverleafs, small roads, and quiet streets (figure 11–16). In terms of the body systems, these are arteries, arterioles, capillaries, venules, and veins. The organ that pumps the blood through this freeway system, of course, is the heart. The heart's walls are composed of thick muscle that divides the heart into four chambers: a right and left **atrium** and a right and left **ventricle**. Carrying blood into and out of these chambers are major **arteries** and **veins**.

ATRIUM The upper chambers of the heart

VENTRICLES The lower chambers of the heart

VEIN A vessel that brings blood to the heart

ARTERY A vessel that carries blood away from the heart

AORTA The largest artery in the body

In one day, the heart pumps the 10 pints of blood in the body through more than 1,000 complete circuits. This is as if the heart actually pumped 5,000 to 6,000 quarts of blood every day. Even though the heart is so strong that it is capable of such incredible feats, it is not much larger than a fist and weighs less than a pound. It points downward in the chest cavity at about mid-center of the body line.

The **aorta,** which is the largest artery in the body, is about 1 inch wide. It curves in a large arch up from the heart and down along the backbone into the abdomen. From it, other arteries branch into the head, digestive organs, and other places in the body. Branching from the

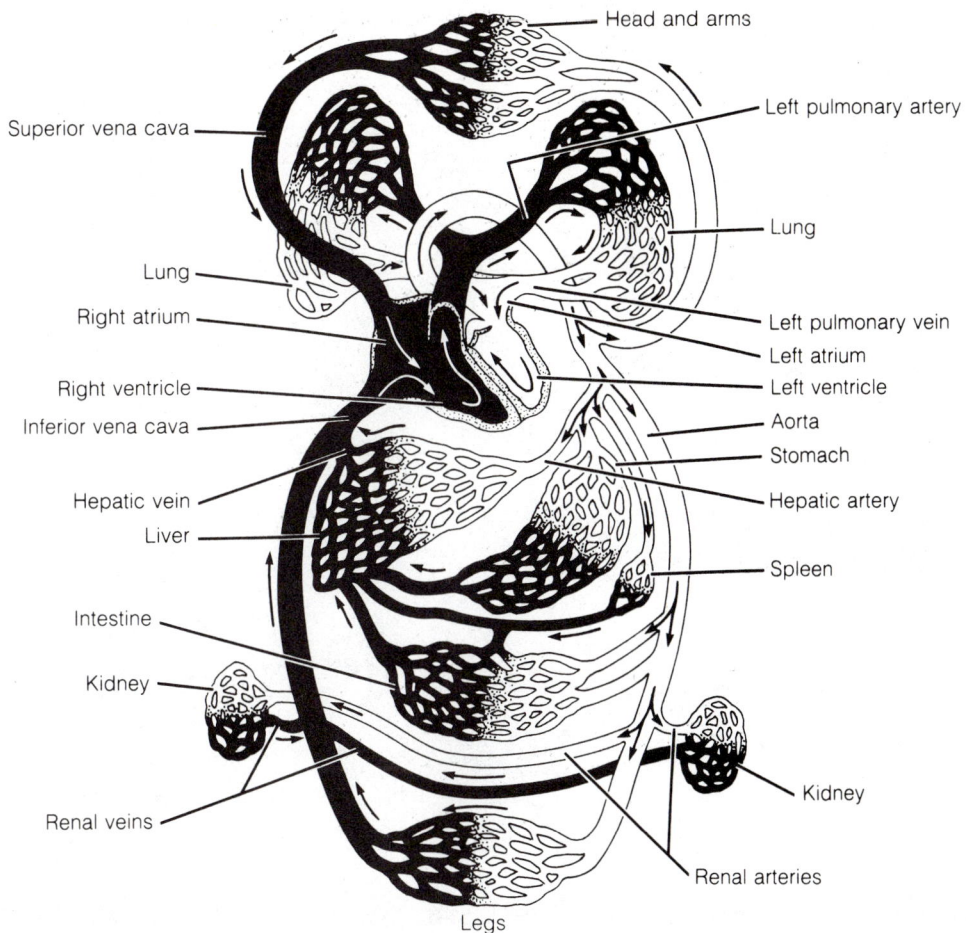

Head and arms

Left pulmonary artery

Superior vena cava

Lung

Lung

Left pulmonary vein

Right atrium

Left atrium

Right ventricle

Left ventricle

Inferior vena cava

Aorta

Stomach

Hepatic vein

Hepatic artery

Liver

Spleen

Intestine

Kidney

Kidney

Renal veins

Renal arteries

Legs

FIGURE 11–16

The heart is the pump for the circulatory system and functions to move the blood throughout the body, carrying nutrients and oxygen to all parts of the body and carrying away the body's waste products.

From *Physical Disabilities and Health Impairments: An Introduction,* by John Umbreit (ed.). Copyright © 1983 by Bell and Howell Company. Used by permission of Charles E. Merrill Publishing Company.

arteries, we can see arterioles, and from these arterioles, capillaries branch off, forming networks all over the body.

The larger arteries have tougher muscular walls than their branches. Healthy arterial walls stretch and rebound with each heartbeat. You feel this movement as your pulse. As fatty materials are deposited in the artery walls, they harden and become less flexible. The arteries become clogged and constricted. This causes a disease commonly seen

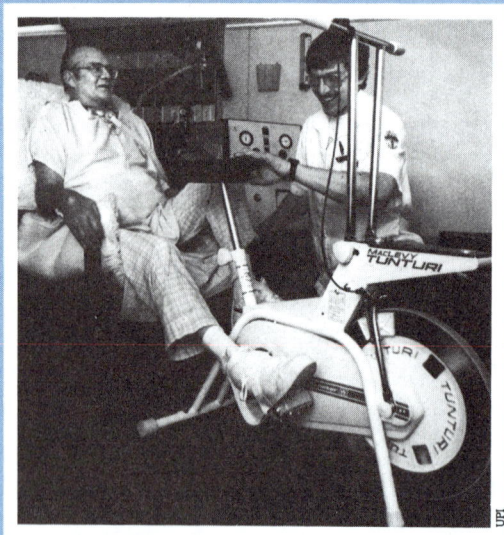

The Artificial Heart: Is It Worth It?

We commonly hear that advances in science and technology will improve the quality of life and living. Is this true? The case of Dr. Barney Clark, a 61 year-old dentist, may clarify some issues related to science, technology, and health.

In late November 1982, Barney Clark was in the final stages of cardiomyopathy, a disease that results in the progressive deterioration of the heart muscle. Clark's skin was blue from lack of oxygen as his heart could only pump about 1 liter of blood per minute, 1/7 the normal amount. By December 1, doctors estimated that Clark had only hours to live. Doctors at the University of Utah Medical Center decided that Clark would be the first recipient of an artificial heart. A delicate 7½-hour operation that required the removal of 2/3 of Clark's heart and the attachment of a mechanical polyurethane heart was completed. After the operation, Clark's heart required compressed air and a long tube attaching the heart to an electric pump.

The operation was a remarkable example of technological achievement. After 25 years of research, physicians were able to replace a human heart with a plastic heart made by humans.

Life for Barney Clark after the surgery was painful and precarious. Clark had a siege of convulsions, severe nose bleeds,

OXYGEN A colorless, tasteless, and odorless gas that is essential to all living things in converting food to energy

CARBON DIOXIDE A colorless, tasteless, and odorless gas that is released by living things when food is converted to energy

VENA CAVA The largest vein in the body

in older people, called arteriosclerosis, that results in high blood pressure and heart attacks.

By the time blood reaches the capillaries, it is moving at a relatively slow pace. As it goes through these smaller vessels, it discharges its food and **oxygen**. Some fluid leaks through the thin capillary walls and surrounds the adjacent cells. The blood also removes waste products such as **carbon dioxide,** urea, and uric acid from the cells.

When the blood gives up oxygen and takes on carbon dioxide, it changes from a bright red to a color that is duller. It is now on its way back to the heart—going from the capillaries to the venules, then to the small veins, into larger veins, and then into the largest vein—the **vena cava** which is just above and below the heart. Coming from the vena

and three trips back to surgery to correct various problems. For periods he confused time, place, and persons. One hundred and twelve days after receiving the artificial heart, Barney Clark died. The heart, after 13,000,000 beats, was still going, but the rest of his vital organs simply stopped.

You might ask the obvious question— was it worth it? The obvious counter question is—for whom? Society? Science? Technology? Barney Clark?

The artificial heart raises social questions. The actual heart and operation are still in the early stages of research even though millions of dollars and a quarter century of work have already been invested. Greater public financing for research and development will be required for continued progress.

The cost of the heart was over $9,000 and additional equipment was another $7,000. Clark's hospital bill was an estimated $200,000. The Office of Technology Assessment estimates that 66,000 Americans a year might qualify for an artificial heart. Very few individuals could afford an artificial heart. Should the U.S. government pay? If so, there would be an estimated additional tax burden of over $5 billion. Such a bill represents a significant drain on the nation's health care system.

Barney Clark's answer to the question was given without hesitation: "It is worth it." He advised others to have the operation "if the alternative is they either die or have it done." He concluded with the thought, "All in all, it has been a pleasure to be able to help people."

Questions for Discussion

1. Should more money be spent on prevention of heart disease thus reducing the need for artificial hearts?
2. Do you think *your* quality of life would be better or worse if *you* had an artificial heart?
3. What are some of the important ethical problems concerning the use of artificial hearts?
4. Do you think the operation was worth it? For society? Science? Technology? Barney Clark? How would you justify your answer?

cava, the blood enters the right collecting chamber or atrium of the heart. It is then pumped to the right ventricle and then via the pulmonary artery to the lungs, where the blood is supplied with fresh oxygen. From the lungs the blood proceeds via the pulmonary vein to the left atrium, then the left ventricle and out to the body again via the aorta.

The amazing fact is that the entire process from the first entry to the return to the same chamber takes about 20 seconds, about as long as it takes to read this paragraph!

Today scientists have studied some of the most basic questions about how the circulatory system functions. But the causes of the very common ailments such as high blood pressure, hardening of the arteries, and heart attacks are still unknown.

FIGURE 11–17
The respiratory system purifies the air we breathe. Small hairs in the nose and mucus that line the wind-pipe (trachea) help in this process.

From *Physical Disabilities and Health Impairments: An Intro-duction,* by John Umbreit (ed.). Copyright © 1983 by Bell and Howell Company. Used by per-mission of Charles E. Merrill Publishing Company.

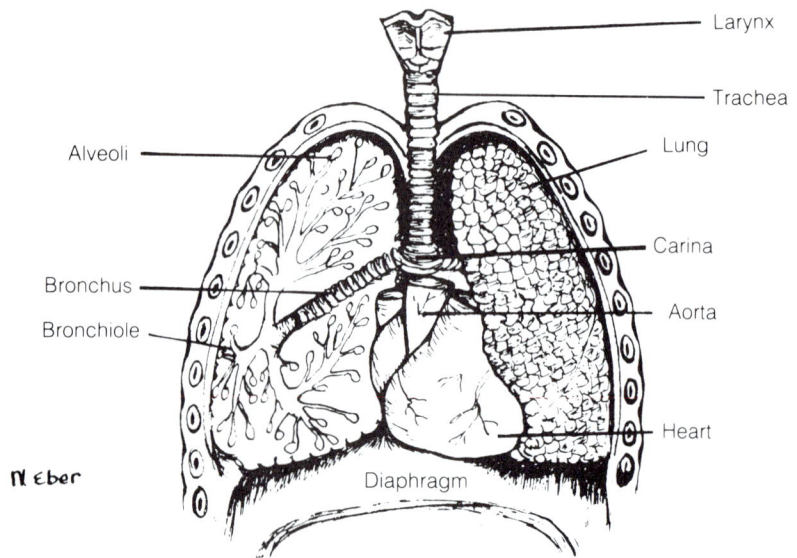

The respiratory system purifies the air we breathe. Small hairs in the nose and mucus that line the windpipe (trachea) help in this process.

The Respiratory System.   The respiratory system, shown in figure 11–17, consists of the lungs and a number of associated structures whose primary functions are to provide the organism with oxygen from the air and to remove excess carbon dioxide from the blood stream.

Air travels from the environment through the nose, the pharynx, the trachea or windpipe, the bronchi (which are two large tubes that are formed when the windpipe divides to go to each lung), and then finally to the main bronchioles, which are like branches or twigs of a tree.

As air moves through these channels, it is cleansed, warmed, and humidified. In the nose, some dust particles and bacteria drop off simply because they are blocked when going through the nasal passage. Other particles are trapped by mucus and tiny hairs in the nose.

In the windpipe, most of the remaining bacteria are stopped by the mucous lining. When these linings are sufficiently irritated, the accumu-lated impurities can cause sneezing or coughing. Once beyond the nose and windpipe, air gets its toughest screening. In the bronchi and bron-chioles, the air is conducted directly to the lungs. This air must be cleansed as much as possible. It is here that environmental factors play an important role. If you could compare a baby's lungs with those of an adult smoker, you would find bright pink lungs as contrasted with the dull gray and spotted black lungs of the adult.

In order for the body's trillions of cells to obtain enough oxygen, 20 times the surface area of the skin is needed. The lungs provide this area because they are filled with tiny pockets formed of membranes so thin that a sheet of the finest paper would seem thick by comparison. In order to understand how oxygen gets transferred into the blood, a description of these little sacs, called **alveoli,** is in order.

The alveoli, which make up the largest portion of the lung, are located at the end points of the bronchioles. When the lungs expand, 300 million of these little sacs open up and provide about 600 square feet of surface area; this is equal to the floor space of a 20-foot by 30-foot room.

ALVEOLI Microscopic air sacs that are located at the end of the airways in the lungs

Even with sufficient surface area, the oxygen cannot be transferred across the walls of the aveoli unless it is dissolved in a fluid. The moisture on the surface of the walls and in the inspired air serve this function.

Blood carried by the pulmonary artery is to be reoxygenated in its passage through the lung. In the lung, this artery divides and redivides into smaller and smaller blood vessels that become parallel to the air sac vessels. Here the layers of capillaries and alveoli lie in direct contact. A double membrane is formed so that oxygen can move from one side to the other as the blood moves by on the other side. Here the oxygen molecules are "snatched" by the **hemoglobin** in the red cells. The hemoglobin "locks" the oxygen in a chemical exchange, so the oxygen cannot escape back to the air sac but is instead carried back to the heart and on to the cells.

HEMOGLOBIN The major protein of the red cells, which carries oxygen from the lungs to the body tissues and carbon dioxide from the tissues to the lungs

The lungs are not finished when they have delivered oxygen to the blood. They help remove the carbon dioxide that is produced as a result of energy production in the rest of the body. Carbon dioxide seeps out from the blood by a mechanism similar to the one which "snatches" oxygen in. Although both of these gases—oxygen and carbon dioxide—go through the same membranes, they seem to have nothing to do with one another. They are like two total strangers checking in and out of a hotel at the same time.

From the lungs, carbon dioxide leaves the body by the same route as the air came in. An excess of carbon dioxide would be poisonous to the body. However, a small amount is left in the blood because carbon dioxide is an important regulator of body chemistry.

Under normal conditions, the respiratory system works so smoothly that we are not even aware of it. This system can be interfered with in a number of ways. A piece of food in the windpipe or inhalation of chemical fumes can cause breathing to stop. There are also many illnesses that affect this system. Common colds cause mucus to collect in the nasal passages and block the flow of incoming air. Bronchitis is caused by a swelling of the airways that obstructs airflow to the lungs. Asthma, caused by allergies, anxiety, or tension, makes the muscles in the bronchi contract and causes wheezing and gasping for breath. Tuberculosis, pneumonia, and emphysema are all conditions in which the aveoli are damaged or blocked so that oxygen and carbon dioxide cannot be exchanged. These diseases can be extremely difficult to treat.

Excretory System. In addition to carbon dioxide, cells give off many substances that are poisonous in high concentrations; nitrogen compounds, sulfates, and phosphates are a few. These substances are like the ashes left after the fire has burned to completion. As waste products, they have to be removed if the cells are to continue functioning. Figure 11–18 shows the excretory system.

The removal of these wastes is the task of the kidneys, the dark red, bean-shaped organs located just behind the stomach and liver. Ridding the body of wastes is only one of the kidneys' jobs. They also regulate the blood chemicals and play an important role in maintaining the correct balance between salt and water in the body. The importance of both the lungs and the kidneys is shown by the fact that they are paired. Most people can survive the damage or actual removal of one lung or one kidney; The other member of the pair can fill in and do the work of both.

The method by which the kidneys remove wastes from the blood is fascinating. Blood goes into the kidney and is immediately channeled into microscopic clusters of capillaries. The capillaries are tightly enclosed in a double membrane that leads to a small tube. Together, this cluster of capillaries, the double membrane, and the tube make up a nephron.

FIGURE 11–18

The kidneys rid the body of wastes, regulate blood chemicals, and help maintain the balance between salt and water in the body.

From *Physical Disabilities and Health Impairments: An Introduction,* by John Umbreit (ed.). Copyright © 1983 by Bell and Howell Company. Used by permission of Charles E. Merrill Publishing Company.

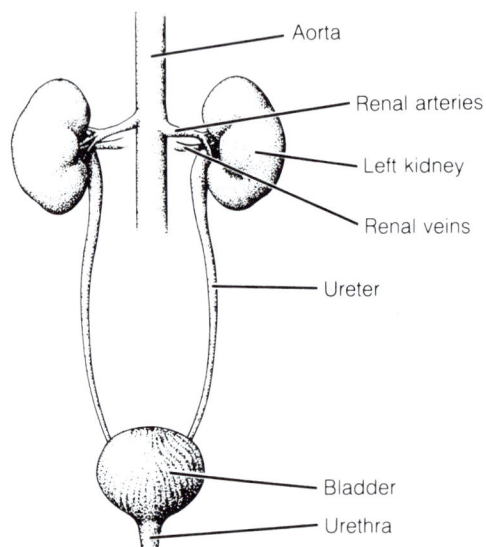

There are about 2.5 million nephrons in each kidney. If they and all their tubes were straightened out, they would stretch about 50 miles.

As blood goes through the nephrons, the waste materials and water move out through the membrane into the tubes in much the same way as carbon dioxide moves out of the lungs. The tubes separate out any sugars, salts, or other useful material that may have been carried over and send them back to the blood. The waste products are neatly trapped inside the tube along with the water. The body can't afford to lose all that water, which is equivalent to about 42 gallons a day. Therefore, about 99 percent of this fluid is reabsorbed back into the blood. The remaining 1 percent along with the waste product are converted into urine as they flow through the tubes. The urine goes into the ureters, which connect the kidneys to the bladder; here the urine awaits elimination.

The Immunity System.   All vertebrates possess **immunity,** a surveillance mechanism called the immune system that protects them from disease-causing organisms, including **bacteria** and **viruses,** and from cells that have turned cancerous.

There are two different sorts of immune responses. Although both sorts are seen during a response to any kind of infection, the cellular response is primarily effective against parasites, cancer cells, foreign cells, and certain viral infections. The other kind is mediated by antibodies and is known as the humoral response; it is most active against bacterial and viral infections. It is also involved in many allergic responses, such as hay fever, bee stings, and food allergies.

Organs of the Immune System.   The immune system is composed of several organs: the spleen, thymus, bone marrow, adenoids, tonsils, Peyers patches in the small intestine, and the appendix. The system includes the lymph nodes, which are found near the body's entrance ways, and under the jaw, the armpits, and the groin, among other places, and which act as way stations for circulating **lymphocytes**. Lymphocytes are white blood cells that actually carry out the surveillance. Except for the bone marrow and thymus, these organs act as traps for foreign organisms and cells, which are then presented to the lymphocytes. These act to stun and kill the invaders. The total number of lymphocytes in the body is about a million million. If all these cells were gathered in one place they would form a ball about the size of the liver. Although these cells are free to roam all over the body, they can be considered as a single organ since they have a single function.

Keys to understanding the most essential concept of the immune system are the questions "How do cells of the immune system distin-

IMMUNITY The state of being able to resist disease

BACTERIA Small, single-celled organisms that do not contain a nuclear membrane; may be harmful (causing disease) or beneficial

VIRUS A protein-coated, minute organism containing either DNA or RNA that is a parasite dependent on nutrients in cells in order to live

LYMPHOCYTES White blood cells involved in destroying disease-causing organisms

PATHOGEN Disease-causing organism

ANTIGEN Substance that provokes an immune response

guish self from non-self?" or, "How does the immune system recognize an object as foreign?" Apparently all lymphocytes have on their surfaces protein molecules whose shapes are complementary to those found on the surface of many **pathogenic** organisms. Any shape that can be recognized by lymphocyte surface proteins will be termed an **antigen** for our purposes. But each cell, and this is the key point, can recognize *only one* antigen. All the daughter cells of that cell recognize only that same antigen. It has been estimated that lymphocytes can recognize approximately 1 million shapes. It is this incredible feat, which usually takes place unbeknownst to us, that makes our lives possible in the face of the constant onslaught of pathogenic organisms that we are exposed to everyday. It is a very flexible and adapting system in that, among these millions of shapes, there are proteins complementary to antigens of new strains of bacteria or viruses that did not even exist when the individual was developing.

Occasionally the immune systems go awry, and what was once self is recognized as nonself and thus becomes the proper subject for an immune response. Diseases such as arthritis, multiple sclerosis, and rheumatic fever are all considered examples of this problem and we call them auto-immune diseases. In the case of rheumatic fever, the heart becomes infected by a streptococcus bacteria. This infection actually does little damage on its own. The lymphocytes respond by making antibodies to the bacterial antigens. Unfortunately, these are very similar to certain proteins in the heart valve. The antibodies bind to the heart valve tissue as well and cause considerable damage.

T CELLS Lymphocytes that mature in the thymus gland

Cellular Immunity.   The cellular immune response is directed primarily against cells that are either infected with virus or have become cancerous. The cells involved in this sort of response develop in the thymus and are known as **T cells**. If a foreign cell is encountered by a lymphocyte, the foreign cell, after a short lag period, is either killed by a lymphocyte or by another cell recruited by a lymphocyte. This response is extremely specific. Certain cells, and only those cells, have proteins that specifically recognize the foreign cell or organism in question; these cells divide many, many times and provide a strong defense against the foreign cell. These cells can be removed from an infected animal and injected into another one. That second animal can respond immediately; that is, there is no lag period when the animal is exposed to the same, specific organism. If the animal is exposed to a different organism, however, a lag period is seen.

B CELLS Lymphocytes that can mature into antibody-secreting cells

Humoral Immunity.   This type of immunity is mediated by **B cells**. These cells were originally found to develop in an organ of the chicken called the Bursa of Fabrisius. This organ does not exist in humans but the name of the immunity remains. On their surfaces, B cells have protein molecules known as antibodies. Each antibody molecule is capable of recognizing only one molecular shape: its antigen. Once a B cell encounters its complementary antigen, the cell divides many times. Some of the daughter cells become plasma cells that secrete, into the circulation, large amounts of the same antibody found on the cell's surface.

When antibodies bind to their antigens, the invading organism or toxin is neutralized and eventually destroyed. If the individual is exposed to the same antigen later on in life, the lymphocytes will respond much more quickly, producing an antibody and either preventing the multiplication of the organism or neutralizing the toxin. This is what is meant by *building up immunity.*

Vaccinations. The specificity of antibodies allows them to be transferred, by vaccination, to individuals who have been exposed to a particular disease-causing organism for the first time. These are individuals who have not been able to build up any immunity.

The case of tetanus is a good example. An individual who has received a puncture wound (the type of wound in which tetanus organisms grow best) will receive a vaccination. This is an injection of a serum taken from a horse that has received a previous injection of tetanus toxin. The horse, therefore, has made antibodies to the toxin. This vaccination will protect the injured individual for only about 20 to 30 days afterward. This is known as conveying *passive immunity*; that is, the immunity is not a result of the individual's own immune response.

In order to make the immunity more long-lasting, the individual must actively make antibodies; in this example, antibodies to the tetanus-causing organism. To accomplish this, the individual receives an injection of killed bacteria. The immune system responds to this injection by making antibodies and thereby giving protection against the disease. If the individual is exposed to the same organism again, the immune system remembers and increases the numbers of antibodies with which to defend against the challenging organism. Today in the United States, children receive vaccinations against rubella (German measles), diptheria, whooping cough, tetanus, measles, polio, and smallpox. In other countries, a vaccine against tuberculosis (BCG) is also given routinely.

HEALTH AND DISEASE

For the greater part of our lives, most of us enjoy good health and think relatively little about disease. The pursuit of good health has become something of a national goal. Anti-smoking campaigns, weight-controlling clubs, and exercise-oriented programs abound. Food and drink that are low in calories and cholesterol, packed with vitamins, and packaged without preservatives compete for advertising time on television or space on the grocer's shelf. It is not surprising to walk down a main street of any American town or city and find the majority of those encountered appearing to enjoy fine health.

Because we are generally a healthy nation, we forget that we live in a world that is filled with disease-causing organisms called pathogens. The air that we breathe, the food that we eat, and the things that we touch are literally teeming with bacteria, viruses, and fungi which have the potential to cause disease. Given their abundance and variety, it seems somewhat incredible that we are not sick more often than we are. But the human species has gradually acquired the capacity to withstand many illnesses in spite of the continuous exposure to billions of microorganisms called germs. Our bodies have developed an immunity to many common diseases.

Through medical research, scientists have added greatly to our understanding of disease and have found many ways to control, cure, or prevent illnesses that were once feared or fatal. In fact, many scientists

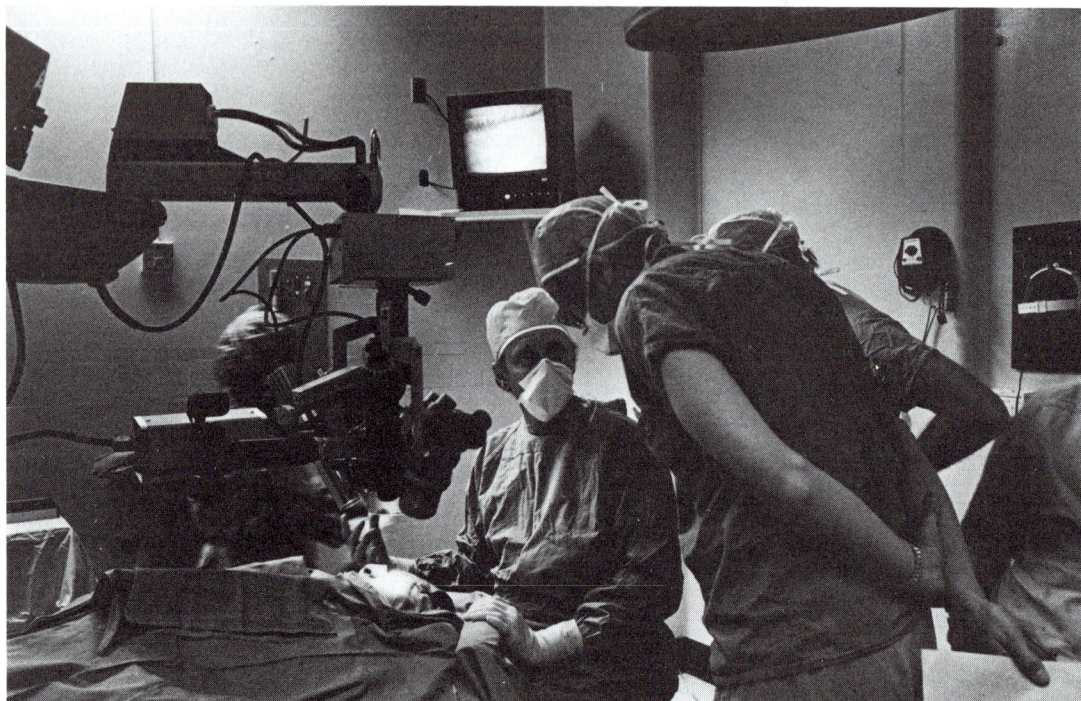

Dramatic technological advances in medical research have lengthened lives and improved their quality. (© 1981 Laimute E. Druskis/Art Resources, Inc.)

consider the biggest triumph of medicine today to be the control and sometimes complete eradication of infectious disease through the discovery of antibiotics like penicillin and the development of vaccines for diseases like smallpox and polio. Without the identification of bacteria and viruses, none of these discoveries would have been possible.

More spectacular to the general public, however, are the more dramatic technological advances in medical research. Fiberoptics, nuclear imaging, and sensitive tests for biochemical analysis of tissues and fluids have become invaluable aids to the correct diagnosis of disease and disorders. The CAT-Scan (computerized axial tomography) and PET-Scan (positron emission tomography) now make possible the diagnosis of lesions on the brain, often at an early enough stage to prevent irreversible damage. Heart-lung machines have enabled surgeons to carry out organ transplants and open heart surgery. Kidney dialysis machines have prolonged lives when human kidneys cease to function normally, and intensive care units in hospitals have allowed doctors and nurses to save many lives today that previously would have been lost without highly specialized monitoring machines.

In this chapter, we describe some of the most common diseases and disorders and the scientific efforts to prevent, control, or cure them. You will discover that there are many ways to protect yourself against disease and to ensure good health.

Disease

Growing up in the twentieth century makes it difficult for us to imagine the fear of disease that haunted the world's populations in the past. For centuries, contagious diseases such as bubonic plague (called Black Death) spread uncontrolled through towns and villages throughout Europe, sometimes killing 75 percent of the population. It was not until the late 1800s that germs, small organisms invisible to the naked eye, were found to be the cause of many diseases. Before that time, evil spirits and witchcraft were thought by many to govern life.

Today, modern science categorizes diseases according to their causative agents. We will describe first two general classes of diseases: those caused by bacteria and those caused by virus. Then we consider parasitic, or vector-born, diseases and genetically inherited diseases. Cancer and heart disease are handled as special cases because of their current importance to American families.

Bacterial Diseases.   There are hundreds of different bacteria which cause illness and can even be fatal unless they are treated. Many familiar bacterial diseases no longer threaten your life because scientists have identified the microbe which causes each disease, and have either de-

Courtesy of Burndy Library

Louis Pasteur (1822–1895)

Pasteur was the first to clearly show that all living things come from other living things—a revolutionary concept for those scientists who believed that life was generated from the nonliving.

Pasteur was the son of a French tanner. In school he was known to be a not-very-bright student who was careful and deliberate in his work, especially in the area of his best skills—art. By the time he was twenty-six, however, he had become much better known for his work in chemistry on the structure of crystals.

His discoveries related to bacteria were his best known, however. He believed that wine turned bitter because of microbes that contaminated it while it was being made. He developed the process by which these could be killed or controlled by using heat. This process, pasteurization, is credited for saving the French wine industry and is today still used in milk processing.

Like others, Pasteur believed that similar microbes were responsible for many diseases. His unique discovery was that if these microbes were weakened in the laboratory, they could then be injected into the body and help the body develop defenses, or immunity, against the germs. He proved his theory in developing vaccinations for diseases of sheep and chickens, among others. His best known experiment involved using the same process to treat rabies in a boy. The parents of the small boy who had been bitten by a rabid dog asked Pasteur for help. In spite of the tremendous risk, he finally agreed. The vaccination proved adequate and the boy recovered!

veloped an antibiotic to treat the disease or a vaccination to prevent the disease from attacking you. Several of the most widely known bacterial diseases are tuberculosis, typhoid fever, pneumonia, scarlet fever, syphilis, rheumatic fever, bubonic plague, cholera, and food poisoning.

The old adage, an ounce of prevention is worth a pound of cure, is valuable advice to keep in mind as you think about protecting yourself from bacterial diseases. Highly infectious diseases can often be prevented by knowing how they are transmitted. For example, some diseases are transmitted through *direct contact* with the sick person. That is why individuals with diseases such as scarlet fever are quarantined; yet other contagious diseases such as syphilis and gonorrhea, which also are spread through direct contact, are not quarantined. They can only be prevented by avoiding contact with the diseased individual. *Breathing bacteria through the respiratory tract* is another way that disease is transmitted. Some protection is afforded if people cover their mouths or noses when they cough or sneeze, but bacteria can still travel on dust particles in the air. That is one reason why doctors and nurses wear gauze masks during surgery. Still other *bacteria can be carried on the skin* without the carrier getting the disease. Restaurants require those who prepare or serve food to wash their hands before handling food for just this reason. *Insects* such as flies and fleas *can also be carriers* of bacterial disease. Bubonic plague, for example, is transmitted by fleas that are found on rats and then bite people.

The simple health rules which often are taught in elementary school help to increase protection from many bacterial diseases. These rules include (1) avoiding contact with persons who are known to be diseased or appear ill; (2) washing our hands before eating and following trips to the restrooms; (3) covering our nose and mouth when sneezing or coughing; (4) staying home when we appear to be ill; (5) preventing insects from contacting our food or biting our skin; (6) keeping foods refrigerated to prevent contamination; and (7) following the advice of physicians, especially with regard to the treatments they prescribe for illnesses.

Viral Diseases.   If you imagine bacteria as being so small that they require an optical microscope to be seen, think of virus as being too small to be seen in an optical microscope. These tiny microorganisms which can be seen only through electron microscopes cause many of the most common diseases, including influenza, the common cold, measles, poliomyelitis, yellow fever, and rabies. Fortunately vaccinations exist for some of the most serious of these diseases (polio, measles, and yellow fever, for example). But because virus have the capacity to reproduce so rapidly, they also form mutations more frequently. This fact makes the development of vaccinations challenging. Perhaps you have heard of vaccinations being developed for one strain of flu (influenza) only to become ineffective as a vaccination the following year because a new strain of flu has developed.

Parasitic or Vector Born Diseases.   Earlier we mentioned that insects can be carriers of disease. These parasitic agents cause many of the most deadly ailments. They have life cycles that may be quite complicated. The protozoa that cause malaria must live in humans during

Some diseases such as malaria are called vector born because they are carried from one host to another by some organism (frequently an insect). (Courtesy of the Center for Disease Control, Atlanta)

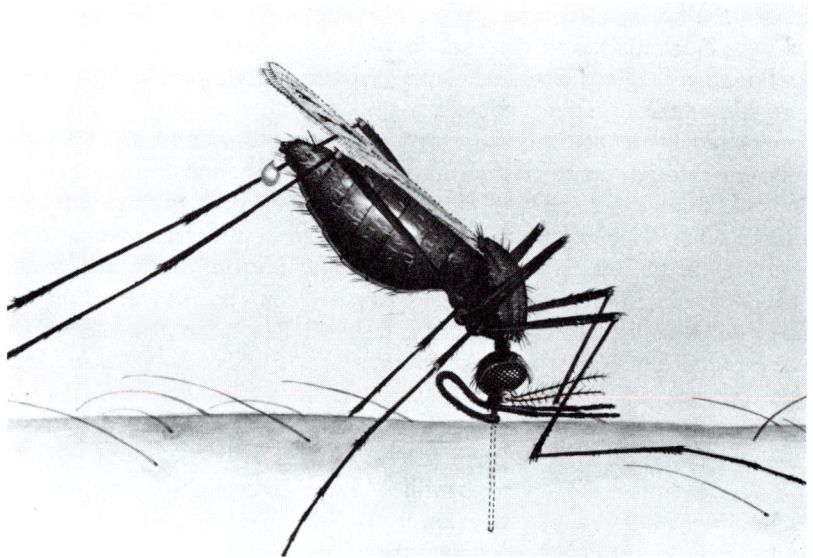

one part of their cycle and in the body of a specific mosquito during the other part. Diseases of this type are called vector born because they are carried from one host to another by some organism—often an insect. Without a human to serve as a host, some diseases would cease to exist. In others, humans serve only as an infrequent and maybe accidental storing place.

In either instance, the key to controlling vector-born disease lies in understanding the complicated life cycle of the organism that causes the disease. This kind of control requires research and planning. It may involve attacking the vector or the animal host as well as treating the disease in humans.

Genetic Diseases.   The interruption of healthy functioning of the body's trillions of cells may come from disease carriers, microbes, nutritional problems, or environmental poisons. However, many illnesses are inherited from the unique set of genes with which we are born.

Exactly how many of these disorders exist is not known but both heredity and environment are involved in the extent to which many diseases are seen.

Today's study of genetics enables us to know that success in fighting off an infection and other threats to health depends in part on our genetic endowment. Some rare diseases are known definitely to be inherited.

The role of genes and the body's adaptation to the environment is illustrated by the ways in which different people react to sunlight. On a day at the beach, people with fair skin must be careful not to get badly burned. Darker-skinned people have a greater capacity for exposure to sunlight and have much less likelihood of skin damage. As one consequence, fair-skinned people are more likely to develop skin cancers than are darker-skinned people.

Thus, although today we do not believe that cancer itself is an inherited disease, some people may be more vulnerable to it because of their genes. The genes and the environment are so intertwined that it makes the geneticists' task a most challenging one. Even when a disease has a clear genetic origin, it is often difficult to assess which parent may be the carrier.

New medical knowledge does help to provide cures and abolish unknowledgeable attitudes about disease. At one time, muscular dystrophy, cystic fibrosis, hemophilia, diabetes, and epilepsy were considered incurable diseases. Research is helping to cure some of these diseases by uncovering their origins. Drugs have helped to control diseases such as diabetes and epilepsy. Effective treatment is being developed for hemophilia and cystic fibrosis, enabling some people to enjoy a new climate of hope.

Health

Health is a function of many things, among them nutrition, attitudes, the absence of physically detrimental habits or addictions, and the ability to function effectively within the environment (see figure 11–19). We look briefly into each of these areas.

Nutrition.   Good nutrition can make a significant contribution toward healthy functioning. Poverty and hunger are unchanging facts of life for

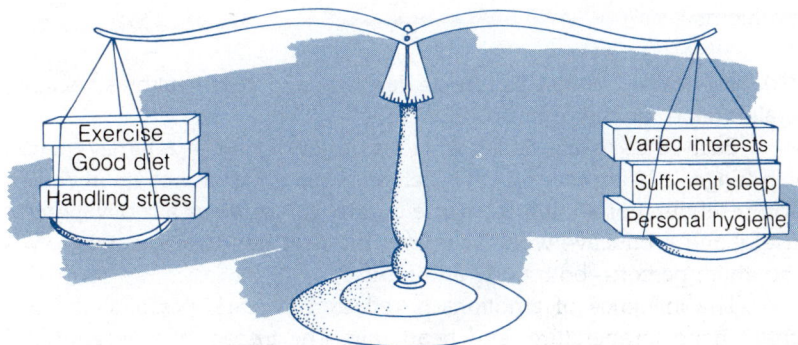

Exercise
Good diet
Handling stress

Varied interests
Sufficient sleep
Personal hygiene

FIGURE 11–19
Healthful living involves many areas of a person's life.

many of the world's people. Many of them have a great deal of difficulty getting any food at all. In some parts of the world, meat and dairy products that contain the most complete sources of protein are nonexistent. In these regions, animals cannot be raised for food since they return only a fraction of the calories or energy they consume. This is an important consideration in countries where a poor rice crop translates into the death of millions. Here the polite greeting between friends is "Have you eaten?"

While malnutrition resulting from a lack of food is a worldwide problem, it presents us with a cruel paradox. We have knowledge to produce enough food to feed the entire world's population an adequate diet. This technology could be available. It really is a question of priorities; that is, deciding whether money and resources should be spent for food.

Many nutritional diseases, such as scurvy and pellegra, have been diagnosed and cured. However, overeating can also be a deadly affliction. With an abundance of food, the nutritional problems of the past have evolved into a new set of problems for us. Obesity is a common malady in the United States today. Obesity is caused by too much food plus too much fat and refined foods. It is estimated that 1 out of 5 American men and 1 out of 4 American women are more than 10 percent overweight. Obesity is also linked to a shorter life expectancy; it is clearly linked to arteriosclerosis, high blood pressure, diabetes, and many other diseases. An older person living a sedentary life in an urban area needs far fewer calories than a young construction worker or farmer. Although we do less physical labor to burn these calories than did our ancestors, we continue to eat as much—or even more.

During recent years, five diseases have emerged as major health problems in our society and are the primary killers and cripplers in our day: arteriosclerosis, cancer, arthritis, bronchitis, and emphysema.

Our rich diet, stress, smoking, and the polluted atmosphere in which we live are all key contributors to these largely human-made problems.

Mental Health. Stress can be as dangerous to your health as bacteria, malnutrition, or chemical pollution.

Each of us meets challenge in his or her own way. A family quarrel can trigger a heart attack in one person while it may make another person resentful. For a third, the same quarrel can serve as a motivation for useful and productive work. Whatever the response to stress, it involves the entire person—both body and mind.

The influence of emotions is reflected in tears, perspiration, face color, hand temperature, and heart rate. The impact of a resentment-provoking situation may be found in the stomach, urinary tract, posture, and facial expression. We react to our environment when it appears threatening. These reactions are both physical and psychological.

Addiction to Smoking, Alcohol, and Other Drugs. In our society today, many people attempt to cope with stress by smoking or by using alcohol or other drugs. Often, people underestimate the harmful effects of these substances, and only later do they find that they are harmful or even lethal. The dangers of smoking today are well known. A cigarette package makes some of these dangers explicit.

In the early 1900s lung cancer was a rarity. Today it is a leading cause of death among American men. The cure rate for lung cancer is equally sobering: Of every 100 patients who have lung cancer, 90 die within 5 years.

Emphysema and chronic bronchitis are two other presistent problems associated with smoking. In emphysema, the alveoli in the lungs become much less flexible and rupture. In bronchitis, the lining of the bronchi become inflamed and the smoke causes the bronchial cells to manufacture excess mucus. In both these diseases, the heart must increase blood circulation because the blood itself has not been able to pick up an adequate oxygen supply from the lungs. Death from both respiratory and heart failure occur twice as often among smokers as nonsmokers. The evidence shows clearly that smokers die at a much earlier age than nonsmokers.

Smoking is correlated with non-fatal sickness as well. We know that American workers who smoke have 1/3 more absenteeism due to illness than their counterparts who do not smoke.

Is smoking an environmental hazard to others? A person aged 35–65 who daily smokes a pack or more of cigarettes is an environmental hazard equal to all other hazards to your life combined!

While many adult smokers quit smoking, there are still about 1 million young people who take up the habit each year. In a number of studies about smoking among children and teenagers, almost all conclude that there are few smokers below the age of 10 or 12. Smoking increases rapidly in junior high school with regular smoking beginning by the eighth or ninth grade. During high school the proportion of smokers increases. Today it is estimated that by the age of 18, out of every 100 boys, 40 smoke regularly. The same statistic for girls is 22 out of every 100. Today smoking by those who are older than 18 years of age has increased from about 600 cigarettes per person in the 1920s to more than 4,000 cigarettes per person in the 1960s. This is a good news/bad news story. The good news is that these numbers are slowly decreasing; the bad news is the senseless destruction of health that these statistics indicate. In recent years, the trend has been toward a slight decrease in smoking among boys and a steady increase among girls.

Studies have shown that persons are much more likely to smoke if parents or older brothers and sisters smoke or friends smoke. Smoking begins at earlier ages among children who have lower goals, less ability, and who have demonstrated less achievement. Smoking is known as a danger to everyone's environmental health.

Marijuana abuse and alcohol abuse are related issues. The introduction to the instructional program *Well Being* aptly describes the impact of alcohol and other drugs as environmental health hazards:

> The most important effect of alcohol is on the brain. After a certain amount of alcohol reaches the human brain, judgment becomes impaired. As more alcohol reaches the brain, one becomes unable to organize the movements of different muscles in teamwork. This muscular disharmony becomes greater as still more alcohol reaches the brain. Finally, with still additional alcohol, individuals lose consciousness. And if, under certain conditions, enough added alcohol should reach the brain after one is unconscious, death can occur.
>
> In considering the problem of alcohol input, the following points are important.
> Over the centuries humans have found both advantages and disadvantages in drinking alcohol and have tried many ways to prevent its abuse. Regardless of the kind of wine, beer, or liquor, drinks get their "kick" from the ethyl alcohol they contain.
> Ethyl alcohol is an anesthetic (a depressant, not a stimulant) that numbs behavioral controls and judgment, and affects vision, coordination, and speech.
> It takes a lot longer for the body to get rid of the effects of alcohol than it takes to drink it—time during which it is not safe to do things like drive a car.
> The custom of drinking for some people becomes a habit, and this habit can become an addiction, just like the addiction to narcotics, barbiturates, or other drugs.
> There is help available in most communities for those who have become addicted and, by their own unaided efforts, cannot get along without alcohol.
> A drug is a substance used as a medicine or in making medicine. Properly prescribed for legitimate medical purposes, all drugs have served some need of mankind. Their danger lies not in their use but in their abuse or misuse. Sometimes they become poisons, producing unwanted reactions, stupor, addiction, and even death. In many cases they can cause physical and psychological dependence and physical tolerance.
> Defining a term like *addiction* is difficult. It is made more so by the various ways in which it is used. One definition of addiction might be the compulsive use of a drug, so that the user finds it difficult to lead a normal

life. The addict is "hooked"—he or she takes the drug because he or she has to. In this respect, any chemical input can to some degree become addictive if it interferes with the usual, constructive pursuits in life.

Physical tolerance is caused by the adaptation of the body to some chemical input that is consumed. The body tolerates larger and larger doses and, in fact, it requires these larger doses for the drug to have a desired effect. The degree of tolerance depends on the drug and the dose. For opiates, tolerance develops very readily. With alcohol, tolerance can be induced only if the dose is greater than a certain minimum. But in any case, tolerance results in physical dependence.

When the body has become so accustomed to the presence of a drug that unpleasant physical symptoms result when the dosage is stopped, the individual has become physically dependent on the drug. The key to its identification—the operational definition of physical dependency—is always the presence of withdrawal symptoms. Sometimes these can be extremely harmful to the individual.

The addict may or may not be physically dependent on a drug, but may be psychologically dependent on it, as evidenced by a controlling desire, with or without the physical need, to continue taking the drug and to obtain it by any means.

The abused drugs generally fall into two main groups—depressants and stimulants. Properly used, the first classification can relieve pain, lessen nervousness, and produce sleep. The second group are often used to reduce fatigue, overcome sleepiness, and treat overweight patients. But improperly used, both can cause addiction, serious physical reactions, and in some tragic instances death.

One of the difficulties with the depressant-stimulant classification system is that there are several drugs that can have either effect, depending on the mood of the user. Among these are marijuana, which is legally classified as a narcotic, and the so-called hallucinogens such as LSD, mescaline, and psilocybin. Not only do their effects vary with mood, but they differ from person to person. And to complicate matters further, when these drugs are used in conjunction with other drugs, such as alcohol, the results are even more unpredictable and sometimes violent.

Narcotics can also act either as stimulants or as depressants. The principal stimulant narcotic is cocaine, which is a drug extracted from coca leaves. It often causes unprovoked, violent behavior, and terrifying hallucinations. The opium derivatives (heroin, morphine, codeine, paragoric, etc.) are sedatives. The synthetic opiate equivalents produce essentially the same reactions of a dreamlike trance and withdrawal from reality. Incidentally, heroin is the most addictive of any known drug. (ISCS, *Well-Being: Probing the Natural World/3* [Morristown, NJ: Silver Burdett General Learning Corporation, 1972], pp. T-8-9).

In efforts to combat disease, hundreds of chemical substances that act powerfully on the entire body have been developed. Each year, some of these substances used, and sometimes abused, have been shown to be toxic in ways we never suspected. Our society, in creating a drug-oriented health system, has helped to create disease-causing drug abuse as well.

Being Healthy. Clearly medicine has a long way to go in helping us toward good health. Through science and technology, we will gain further understanding of infectious, immunological, and malignant organic diseases which, in turn, will give us the confidence to look forward to better health in the future.

What does it mean to be healthy?

Good health means quite different things to an astronaut, a fashion model, a lumberjack, or a Wall Street stockbroker. Each way of life requires different levels of physical activity, different food needs, and each lifestyle involves different kinds of stresses.

An adequate measure of health includes but is not solely the absence of disease. Health is the ability to function effectively within the environment. Since each individual's relationship to the environment continually changes, health is the ability to adapt to the many microbes, irritants, pressures, and problems that are daily challenges to us all.

BIBLIOGRAPHY

Bier, O. G.; da Silva, W. Dias; Götze D.; and Mota, I. *Fundamentals of Immunology.* New York: Springer-Verlag, 1981.

Bloom, W., and Fawcett, D. W. *A Textbook of Histology.* 10th ed. Philadelphia: W. B. Saunders, 1975.

Boston Women's Health Collective. *Our Bodies, Ourselves.* New York: Simon and Schuster, 1976.

Braude, A. *Microbiology.* Philadelphia: W. B. Saunders, 1982.

Green, J. H., and Silver, P. H. S. *An Introduction to Human Anatomy.* London: Oxford University Press, 1981.

Karp, G. *Cell Biology.* New York: McGraw-Hill Book Company, 1979.

Langman, J. *Medical Embryology.* Baltimore: Williams and Wilkins, 1969.

Nilsson, L. *Behold Man.* Boston: Little, Brown & Co., 1973.

Watson, J. D. *Molecular Biology of the Gene.* 3rd ed. Menlo Park, CA: W. A. Benjamin, 1976.

12

Teaching the Health Sciences

The purpose of chapter 12 is to provide examples of science lessons designed to teach health science to students from kindergarten through eighth grades. Planning Charts 12–1, 12–2, and 12–3 illustrate how each science lesson will allow you to meet several objectives at the same time.

KEY CONCEPTS AND IMPORTANT IDEAS

☐ using your senses ☐ immunizations and community health ☐ you and your vision ☐ basic first aid in emergency situations ☐ participating in a physical fitness program ☐ food, population, and world health

CHART 12–1

Planning Chart for Health Sciences: Kindergarten through Grade 2

| Activity | Objectives | | | | | Teaching Methods | Related Curriculum |
	Conceptual Knowledge	Learning Processes	Psychomotor Skills	Attitudes and Values	Decision-Making Skills		
1. Using Your Senses	The use of the five senses increases one's ability to function in the environment.	Compare, describe.	Operate a tape recorder.	Appreciate how the senses contribute to enjoyment of life.	Become aware of the sensory properties of objects.	Learning centers	Writing, art
2. Learning about Immunizations and Community Health	Immunizations can help control infectious diseases in the community.	Construct and interpret graphs.		Primary responsibility for improving and maintaining the health of the community is a function of its members.	Describe consequences of individual actions on community health.	Guided discovery	Mathematics

CHART 12–2
Planning Chart for Health Sciences: Grades 3 through 5

| Activity | Objectives | | | | | Teaching Methods | Related Curriculum |
	Conceptual Knowledge	Learning Processes	Psychomotor Skills	Attitudes and Values	Decision-Making Skills		
3. You and Your Vision	The different parts of the eyes have different functions.	Describe, predict, measure.	Handle experimental equipment.	Appreciate that the eye is a delicate, yet well protected organ: assume responsibility for personal practices that contribute to visual health.	Become aware of information needed to make visual health decisions.	Lecture/demonstration	Art
4. Applying Basic First Aid in Emergency Situation	Body temperature and other signals indicate the kind of care required by an individual in an injury situation.	Observe, measure, gather information.	Handle first-aid materials.	Understand that preparedness in dealing with emergency situations contributes to community wellness.	Evaluate an injury situation for the purpose of determining appropriate care.	Teacher demonstration/competency-based instruction	Reading

CHART 12–3

Planning Chart for Health Sciences: Grades 6 through 8

Activity	Objectives						Teaching Methods	Related Curriculum
	Conceptual Knowledge	Learning Processes	Psychomotor Skills	Attitudes and Values	Decision-Making Skills			
5. Participating in a Physical Fitness Program	There is a cause and effect relationship between physical activity and cardiopulmonary endurance.	Measure, predict, form hypotheses, and gather and organize information.	Participate in physical activities such as running and jumping.	Value physical activity as an important factor in achieving life-long health.	Become aware of cause and effect relationships.		Teacher-directed discovery	Physical education
6. Food, Population, and World Health	The relationship between population growth, and food supply and distribution affects the health of individuals in a society.	Interpret data.		Appreciate the importance of food as a factor in maintaining the health of people throughout the world.	Understand consequenses of population density and food supply.		Simulation and individual research	Geography

ACTIVITY HS–1

Using Your Senses
(Kindergarten–Grade 2)

ACTIVITY OVERVIEW

In this activity, the students conduct independent explorations using the senses of touch, taste, smell, hearing, and sight. The students make tape recordings or written records of their observations for later discussion. *Estimated time:* one to five class periods in the Learning Center.

SCIENCE BACKGROUND

The seventeenth century English philosopher Thomas Hobbes stated: "There is no conception in a man's mind which hath not at first, totally, or by parts, been begotten upon the organs of sense." *All* organisms, not just humans, gain information about environmental conditions through the use of their sense organs. Any fairly rapid physical or chemical change occurring in the environment acts as a **sensory stimulus** if the organism possesses appropriate nervous structures to receive the information.

Sensory stimuli can be divided into three categories: mechanical, electromagnetic (light), and chemical. Organs of touch and hearing depend on the registration of mechanical changes to parts of the body. Sensitive sites are situated on the surface of the organism or in deeper tissues and special organs. In the sense of sight, the eye perceives the visible part of the **electromagnetic spectrum**. In the senses of smell and taste, the sites most sensitive to chemical stimuli are the nose and tongue. The sensory stimuli are translated

into a coded message in the form of an electrical pulse. This **coded message** is then conveyed to the central nervous system, especially the brain, where it is translated.

SOCIAL IMPLICATIONS

Developing the use of the five senses increases our ability to learn, communicate, work, and recreate in the environment. All humans, from birth through old age, learn by means of selecting and processing sense data. Although all of the senses provide rich input, research indicates that young children learn better from visual and kinesthetic (touching) cues than from auditory cues. Children's ability to clearly communicate observations gained through the senses is often limited because their descriptive vocabularies are weak, making it difficult for them to describe fine distinctions. The work and play environment is filled with sounds, smells, sights, and tastes that delight children and adults. Experiences in discriminating finely among these various stimuli can help children develop a sensitivity that contributes to the enjoyment of daily living.

MAJOR CONCEPT

Living and nonliving things (matter) have identities and are organized based on their characteristics.

Use a variety of foods and other natural objects as stimuli for student's use of smell and touch.

OBJECTIVES

☐ Observe and compare textures, odors, sounds, tastes, and colors.

☐ Control and handle nontoxic substances.

☐ Actively explore the environment.

☐ Recognize personal decisions that involve self.

MATERIALS

Texture/touch cards; texture objects, such as broken eggshells, noodles, beans, carpet pieces, silk, cotton balls, hairbrush, sticky tape; substances with odor, such as perfumed cotton, spices, coffee, onion, garlic, and orange peel; opaque containers for "smell jars"; substances for testing, such as vinegar, salt water, quinine water, and sugar water; cassette tape recorder; objects for visual classification, such as blocks, classroom materials, or objects from nature.

VOCABULARY

bitter	sight
color	slippery
furry	smooth
hard	soft
pointed	sound
prickly	sour
rough	sticky
salty	sweet
scratchy	taste buds
shape	texture
sharp	

PROCEDURES

A. Set up a Touch and Smell Station where children can explore a variety of texture cards and then record their descriptions on a tape recorder or on paper.

1. Prepare a set of texture cards by attaching to each card an object or objects for the children to touch.
2. Invite the children to go to the station in pairs. One child sits behind the other and hands one card at a time to his or her partner. The other child, reaching back, feels but does not see the touch card and verbally describes the texture, using as many descriptive words as possible. (Make sure the students record their descriptions.) The partners then switch roles.
3. When all students have completed the task, share their written and recorded observations. List vocabulary on the chalkboard, and extend the list when possible. Have the students try to match descriptions with touch cards.
4. Set up stations in a similar manner at which children can observe and describe odors from "Smell Jars," sounds from tape recordings, and the tastes of sourness, sweetness, saltiness, and bitterness.

B. Prepare a Vision Station where children can classify a set of objects. Ask the students to describe the visual properties used to group the objects.

Use a variety of fabrics and materials on poster board or index cards to make texture cards.

Stock the Vision Station with objects of various sizes, shapes, and observable textures. Encourage students to group the objects in several different ways.

EXTENDING THE ACTIVITY

C. Invite the children to locate the specific taste buds that are sensitive to bitter, sour, sweet, and salty tastes. Using cotton swabs, dab quinine water, vinegar, sugar water, and salt water on one area of the tongue at a time. Have the children draw maps of the tongues and indicate the locations of the four kinds of taste buds.

D. Invite the students to taste and identify foods such as onions, potatoes, and apples with and without the use of their noses. Ask them if the ability to taste is affected by the sense of smell.

E. Give each of the children a small piece of bread, and ask them to suggest ways to make it taste sweet, sour, bitter, and salty. Try their suggestions, such as adding sugar.

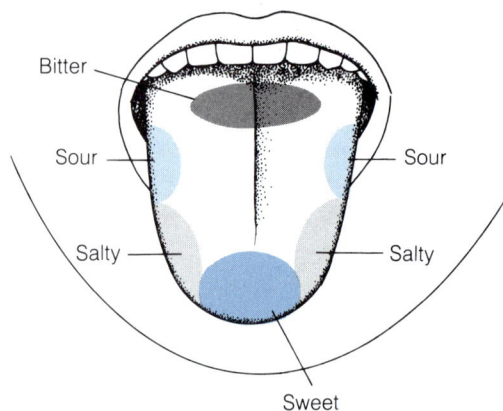

Specific areas of the tongue provide part of the ability to taste certain substances.

ACTIVITY HS–2

Immunizations and Community Health
(Kindergarten–Grade 2)

ACTIVITY OVERVIEW

In this activity, the children discuss personal immunization data. *Estimated time:* one to two class periods.

SCIENCE BACKGROUND

Historically, infectious disease has been a serious threat to human life. Disease is presently controlled by good nutrition, sanitation, housing, and the more recent developments of immunizations and antibiotics. **Immunization** is the most significant contribution of science to the control of disease.

Immunization works with the body's own defense system to resist the invasion of disease-causing organisms. The body develops immunity to disease naturally by making antibodies in response to the invasion of a disease-causing organism. These antibodies remain and prevent reinfection when or if the organism attacks again. Immunization occurs when a small dosage of an infecting agent is given to promote an immune response without causing the disease to develop. This agent can be a small dose of the disease itself, a similar disease, or dead bacteria or viruses of the disease.

SOCIAL IMPLICATIONS

The impact of immunizations on public health has been clear. In the past decade, smallpox has virtually been erased from the face of the earth. Polio, which crippled 20,000 a year during the early 1950s, now afflicts only a few children annually. Unfortunately, the infrequent occurrences of the diseases that immunizations have helped to eliminate has led to public complacency about immunizations in this country. The Center for Disease Control in Atlanta, Georgia recently released figures indicating that nearly twenty million children under 16 had not been immunized against one or more serious, communicable diseases.

The U.S. Public Health Service recommends that children, by the age of 6, should be immunized against polio, diphtheria, pertussis (whooping cough), tetanus, and measles. In fact, many school districts require proof of immunizations prior to school entrance. However, disease prevention does not stop at 6 years of age. Some immunizations require periodic boosters for continued effectiveness. Immunizations against rubella and mumps are recommended under certain conditions for some young adults. Additional immunizations are recommended or required for travel in some foreign countries.

While primary-age children may have had little control over their early immunization record, they can contribute to the control of infectious disease by practicing good personal health habits. Appreciating how immunization helps control infectious disease can motivate children to take an interest in the continuation and renewal of their immunizations. The primary responsibility for maintaining the health of the community is a function of its members. Even the smallest members can participate.

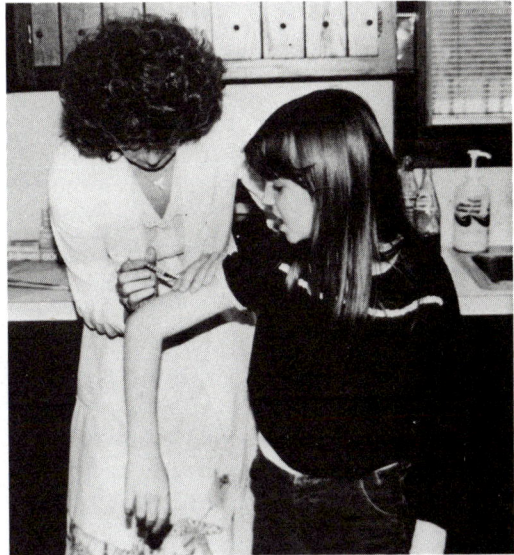

If students understand how immunization helps control infectious diseases, they will be more motivated to take an interest in the continuation and renewal of their own immunizations. (Courtesy of B. Young and Amy Ross)

MAJOR CONCEPTS

All living and nonliving things (matter) change when they interact with each other. The human body makes antibodies in response to disease-causing organisms.

OBJECTIVES

☐ Recalling observations.

☐ Using tools (crayons) to draw.

☐ Accepting direction in caring for health and welfare of self and others in class.

MATERIALS

Paper, crayons, personal immunization records, and (optional) **syringe** used for giving injections and a dab of cotton. Plastic syringes may be obtained from the school nurse, a hospital, doctor, or the county health office.

VOCABULARY

community	epidemic
contagious	immunization
disease	vaccination

PROCEDURES

A. Begin with a class discussion in which the children share some of their own disease experiences. Discuss how diseases are contracted: from being too close to people who cough, sneeze, or spit; from eating food that other people have eaten; from dirty hands; from towels and dishes used by others; from flies and mosquitoes; from sick animals; from putting foreign objects in the mouth.

B. Ask the children to suggest ways of protecting themselves and others from contagious diseases. Some protecting practices include: always washing hands *before* eating and *after* using the toilet, covering sneezes and coughs, staying home from school when ill, getting enough rest, and eating nutritious food.

C. Introduce the role of immunization in disease control and ask the children to recall any immunizations they might have received. Ask them why they think they received those immunizations even though they were healthy at the time. Suggested questions might be:

1. Has anyone received the same immunization as someone else?
2. Why do you think so many of you have received the same immunization?
3. Why do you think you were given these immunizations even though you were healthy at the time?
4. Has anyone ever had a reaction (stiff or sore arm) to an injection? If not, do students recall how they felt when they were sick? (Their reactions will provide examples of the body changing in response to bacteria or virus.) Has anyone ever received an injection when he or she was sick? (Penicillin and sulfa are examples of antibiotics interacting with disease-causing organisms.

D. If you have obtained a syringe used to give injections, ask two or three students to role play a doctor giving a shot to a patient. Ask students why doctors use cotton saturated with alcohol to wipe the skin before giving injections. (Alcohol acts as an antiseptic to clean the skin and to prevent bacteria on the skin from entering the body with the needle.)

EXTENDING THE ACTIVITY

E. Raise the question of what could happen if some of the children in the school community were not immunized against a particular disease—measles, for example.

F. Invite the children to write (or dictate) and illustrate a composite story in response to the question in point E above.

ACTIVITY HS–3

You and Your Vision
(Grades 3–5)

ACTIVITY OVERVIEW

In this activity, the students conduct three simple experiments to learn the functions of the various parts of the human eye. *Estimated time:* three class periods.

SCIENCE BACKGROUND

The human eye is a globe-shaped, delicate organ that fits into a depression in the skull. The major parts of the eye include the cornea, lens, retina, iris, pupil, optic nerve, and muscles. The sensitive inner eye structure is housed within a protective white cover, the eyeball itself.

Vision is a function of the nervous system. Sight results from messages that are sent to the brain through a series of nerves within the eye. Light rays first enter the eye through the **cornea,** a transparent and windowlike covering over the eyeball which protects the iris and the pupil. The direction of entering light rays is altered by a clear, crystalline refracting device known as the **lens,** the shape of which is controlled by tiny muscles within the eye. The light rays are altered so that they fall in proper focus on the light sensitive **retina,** a layer on the inside of the eye rather like a coat of paint. The retina consists of modified nerve cells which connect to larger nerve cells, which in turn, connect to the brain via the optic nerve. The image created by the light rays is conducted by the **optic nerve** to the optical centers of the brain located at the back of the head. Here, the **impulses** are interpreted and we "see."

The colored part of the eye is called the **iris.** In its center is an opening known as the *pupil.* The size of the pupil changes as the iris expands and contracts to control the amount of light that enters the eye. The iris expands as light decreases and contracts as light increases. Controlled by tiny muscles, this process is important because excessive light or brightness can injure the eye. The eye has two internal chambers, one in front of the lens and another behind the lens. A water fluid known as **aqueous humor** fills the front chamber; the second chamber is filled with a jellylike substance, the *vitreous.* Three pairs of muscles attached to the eyeball govern the movement of the entire organ, allowing one to look up and down or from side to side, and to roll the eye around on its axis.

In addition to the internal parts of the eye, several accessory parts contribute to its healthy condition. The bones of the skull form the *eye socket* which surrounds the eyeball and protects it from injury. The *eyebrows, eyelids,* and *eyelashes* protect the eye from outside forces and keep foreign objects from entering it. A reflex action causes rapid closure of the eyelids whenever an object suddenly confronts the eyes. The *tear ducts* are small glands located at the innermost corners of the eyes. They produce tears that keep eye sockets and eyeballs moist. Tears also serve to wash away dust or other matter that may enter the eye. Professional eye care, when indicated, can be obtained from an *ophthalmologist,* an M.D. specializing in eye care. *Optometrists* are nonmedical practitioners licensed to correct **refractive errors.** Their practice is limited to prescribing and fitting glasses.

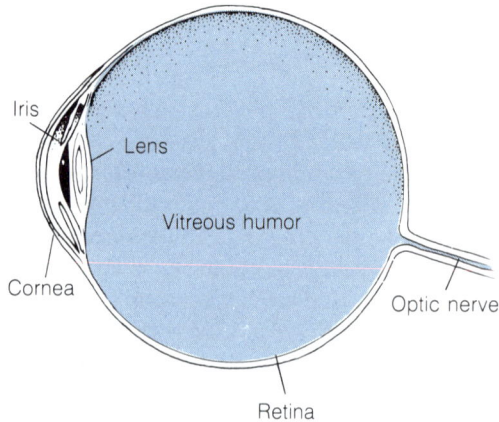

The human eye is a very valuable sense organ. The lens focuses the image, the iris regulates the amount of light, and the retina serves to transmit nerve messages to the brain.

SOCIAL IMPLICATIONS

The eye is a very important organ of the body. Sight enables us to learn. As young children, we learned better from visual and kinesthetic (touch) stimuli than we did from other types of stimuli. Certain personal habits not uncommon to children may endanger the eye. Putting pencils or sharp objects near the eyes; throwing rocks, sand, or other things at playmates; and reading with improper light are some of the practices that can prove harmful to healthy vision.

Knowledge of the basic structure and function of the human eye and an appreciation of its delicate nature despite its natural protections, can encourage personal practices that promote and protect visual health for oneself and others.

Reading with proper light is an important practice that you can encourage as your students develop personal habits that promote and protect visual health. (Paul Conklin)

MAJOR CONCEPT

Living and nonliving matter (including organs such as the human eye) are organized and have specific functions.

OBJECTIVES

- ☐ Compare parts of the eye and the differences between peripheral and frontal vision.
- ☐ Use tools and materials that require coordination of senses.
- ☐ Assume responsibility for the health and welfare of self and others.
- ☐ Describe consequences of various alternatives (regarding vision) to self and others.

MATERIALS

Model of the human eye, a wall chart of the human eye or hand-drawn diagram, and a ditto diagram of eye; water; pencil; onions; knife; poster paper; and marking pens.

VOCABULARY

aqueous humor	ophthalmologist
cornea	optic nerve
eyebrows	optometrist
eyelashes	pupil
eyelids	retina
eye sockets	tear ducts
iris	vitreous
lens	

PROCEDURES

A. Using a model of the eye, a wall chart, or a hand-drawn diagram, have the students locate the parts of the eye. Describe the function of each part. Ask the students to list all of the parts of the eye that have a protective function.

B. Invite an ophthalmologist to the class to discuss vision disorders of near- and far-sightedness and corrective glasses. Or you may wish to encourage interested students to conduct independent research studies on these common disorders.

C. Demonstrate methods of flushing the eye with water to remove foreign objects or liquids from the eye. Cup a hand under running water and splash water into the eye, or allow water from a drinking fountain to run into the eye. The immediate use of water is important to dilute, if not remove, any foreign liquid or matter. Stress that a physician should be contacted in such circumstances.

D. Invite a group of interested students to make a class poster listing practices that promote healthy vision. This list could include practices for TV viewing: sitting at least ten feet from the screen with eyes at about the same level as the screen, never viewing TV in a completely darkened room, and resting the eyes from time to time by looking around the room.

E. Invite the students to arrange a table display of items designed to protect the eyes that might include sunglasses, safety goggles, and eye shields.

F. Invite the students to explore the use and importance of peripheral vision. Allow the students to make and wear, for a given period of time, simple tunnel-vision masks that limit peripheral vision. Have selected students test their fields of vision by looking straight ahead (not moving their heads) and determining the points to their sides at which they can identify shapes, colors, printed material, and so on.

G. Invite the students to imagine what their lives would be like without vision. Allow individual students to move around the classroom blindfolded. Ask them to describe what other senses and external cues are most helpful when eyesight is removed temporarily.

EXTENDING THE ACTIVITY

H. Invite the students to use a glass of water and a pencil to assist them in understanding how light rays bend as they pass through the cornea and lens of the eye. Ask them to imagine the water as representing the cornea and the pencil as a ray of light. Immerse the pencil in the water. The pencil will appear to bend at the surface of the water.

I. Demonstrate the function of tears by asking students in groups of three to chop an onion. The fumes from the onion contain a substance irritating to eyes. The eyes automatically water to flush out the irritant.

Adapted, with permission, from *My Body Is Mine, My Emotions Are Mine, My Environment Is Mine: A Health Education Curriculum.* Albany, N.Y.: Albany Unified School District, 1978.

ACTIVITY HS–4

Basic First Aid in Emergency Situations
(Grades 3–5)

ACTIVITY OVERVIEW

In this activity, the students practice and demonstrate skill in administering basic first-aid treatment of four serious or emergency conditions: shock, bleeding, loss of breathing, and broken bones. Skill in administering emergency first-aid care (while waiting for medical aid) can be learned by the intermediate student. Detailed first-aid instruction is available through the American National Red Cross. *Estimated time:* one or more class periods.

Shock

Symptoms of *shock* include: pale, cold, clammy skin that is mottled in color; apathy; nausea; and weak, irregular breathing. Because shock is not always easily determined, a victim of any serious injury must be treated for shock. Early treatment *can* prevent an injured person from going into shock. Treatment for shock includes the following four points:

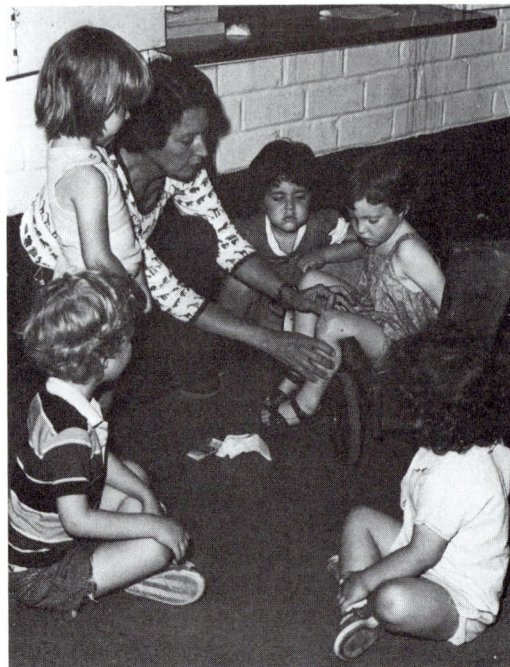

Every injury provides the opportunity for teaching good practices in first aid. (Ben Chandler)

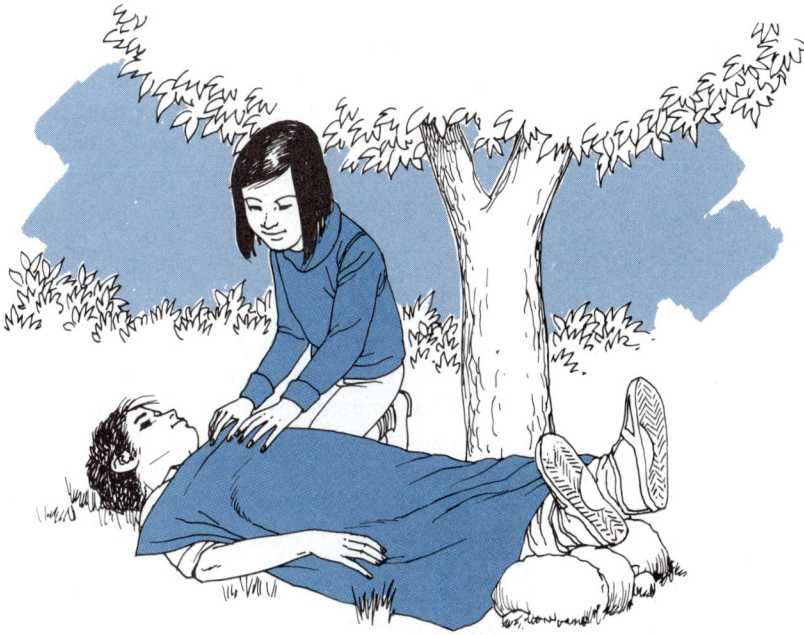

Be sure the victim is warm in treating for shock.

1. To minimize any work for the body, have the victim lie down.
2. If the victim can breathe sufficiently, elevate the legs about eight to twelve inches, unless that causes pain. If the head is injured or the victim has trouble breathing, elevate the neck and shoulders.
3. Cover the victim lightly with a blanket to keep him or her warm. If the victim is cold or damp, place a blanket under the body as well. Be sure not to move the patient if a back or neck injury is suspected.
4. If the weather is warm, provide the victim with shade and as much comfort as possible.

Bleeding

To control *bleeding*, immediate *direct pressure* and elevation of the injury are important. Instructions are as follows:

1. Using a gauze pad, clean cloth, clothing, or your bare hand if necessary, press down with the heel of the hand directly over the wound.
2. Elevate injured limbs higher than the heart unless you suspect a fracture.
3. Continue to press until bleeding stops. Apply additional cloth pads over the initial pad if necessary.
4. When bleeding is controlled, bandage or tie the pads on the wound firmly—but not too tightly.
5. Check the pulse below the wound. If you cannot feel a pulse, loosen the bandage until a pulse returns.
6. Treat victims of severe cuts for shock.

Loss of Breathing

For someone who has *stopped breathing*, *mouth-to-mouth breathing* is the best way to get air into the lungs and to initiate breathing. *Preparation instructions* include the following three points:

1. Turn the victim's head to the side, and use two fingers to remove from the mouth anything that might obstruct air from passing through.
2. Placing one hand on the victim's forehead and the other hand under the neck, tip the head far back and keep it tipped. This prevents the tongue from blocking the airway. In this position, victims sometimes resume breathing.
3. Listen for breathing. If breathing does not resume, pinch the nose closed to prevent air leakage.

Breathing instructions include:

1. Taking a deep breath, open your mouth wide and place it over the victim's mouth, making a tight seal.
2. Blow hard to fill the victim's lungs and watch the chest rise.
3. While watching the chest fall, listen for air to come out.
4. Repeat the blow-and-listen steps once every five seconds, twelve times per minute, until the victim breathes spontaneously or medical aid arrives.
5. If air cannot be blown into the lungs, repeat preparation procedures and try again.

Sometimes it may be impossible to make a seal over the victim's mouth due to injury or size. In this case, *mouth-to-nose breathing* is necessary. Breathing instructions remain the same, except that the mouth is now placed over the victim's nose, and his mouth is held closed when breaths are given. The victim's mouth is then held open to expel air.

Broken Bones

The recommended treatment for dealing with *broken bones* is to keep the victim immobile. The victim of a suspected fracture should lie still if a doctor is on the way. If the victim must be moved to get to a doctor, the broken bones must be kept stationary by immobilizing joints on both sides of the break. A *splint* can be made from (1) wood with cloth padding when possible; (2) rolled magazines or newspapers; or (3) a tongue depressor or the like for broken fingers. The splint should be tied with cloth strips on either side of the break and at intervals along the splint. When it cannot be determined if a bone is broken, treat the victim if it is, and immobilize the injured area as described.

SOCIAL IMPLICATIONS

Preparedness in dealing with emergency situations contributes to community wellness. Even the younger members of the community can contribute to the health and safety of an injured person by evaluating the injury, contacting medical aid, and administering appropriate basic, first-aid care.

MAJOR CONCEPTS

Individuals and groups change when they interact with each other. Primary responsibility for improving and maintaining the health of the community is a function of its members.

OBJECTIVES

☐ Observe, compare, and evaluate an injury for the purpose of determining appropriate care.

☐ Use equipment that requires coordination of senses.

☐ Assume responsibility for health and welfare of others.

☐ Consider consequences of various actions and select one, accepting responsibility for the consequences of that choice.

MATERIALS

Blankets (two–three), clean gauze and clean cloths, wooden splints and boards of various lengths, large dolls for mouth-to-mouth breathing, tongue depressors, magazines, newspapers, and a telephone directory.

VOCABULARY

direct pressure
elevate
emergency
immobilize
injury
mouth-to-mouth
  breathing

mouth-to-nose
  breathing
shock
splint
wound

PROCEDURES

A. Prepare a First-Aid Learning Station. Include:
   1. Charts with printed instructions on first-aid treatment for shock, bleeding, loss of breathing, and broken bones.

2. First-aid materials (i.e., blankets, cloths, and so forth as listed in the Materials section above.
3. Large dolls for practicing. (You may want to ask the students to bring these.)
4. A local telephone directory.

B. Invite the class or groups of students to join you at the First-Aid Station for an introduction. Define emergency injuries, their evidence, and symptoms. The students may enjoy sharing personal experiences of their own injuries and broken bones. Discuss community resources for emergency help and locate emergency phone numbers in the telephone directory. Demonstrate basic steps in first-aid care, using the dolls and volunteer students. Invite the children to try the procedures with your help and supervision.

C. During free time or designated time in the following weeks, invite pairs of students

Supply the First-Aid Learning Station with a variety of materials for practicing simple first-aid techniques.

to practice the given first-aid procedures, using dolls or each other. You may want to assign a single procedure, such as controlling bleeding, to be practiced and mastered during a given week. Also ask the students to make a list of emergency telephone numbers for their use at home.

D. During a demonstration or practice session, you might ask the students to consider the following questions:

1. Suppose you witness a friend being thrown from his bicycle after being hit by a car. Your friend immediately gets to his feet, though trembling, and insists that he is all right. What would you advise your friend to do? What actions would you take?

2. Suppose you cut your feet badly on a broken bottle at the seashore. Your towel is up on the beach with your friends. Should you try to walk up from the shore to get your towel to use in stopping the bleeding? If not, what should you do first?

3. Which joints would you need to immobilize in case of a broken forearm? How do you suppose you could immobilize a suspected broken foot? (Remove or cut away shoe. Immobi-

lize with a padded splint, using a blanket, pillow, or other material. Tie it snugly but not too tightly.)

EXTENDING THE ACTIVITY

E. Check community libraries for films on standard first aid.

F. Invite the students to put on one-act plays in which they demonstrate to other classes the basic first-aid treatments they have learned. Encourage the students to use props and costumes to create real-life emergency situations. Some sample themes might include the following scenarios.

☐ A woman cuts her hand badly with a knife while dining at a restaurant. Someone at another table knows how to stop the bleeding and instructs the waiter on how to get help.

☐ A boy injures his leg seriously during an after-school football game. Team members treat him for his broken leg and for shock.

G. Contact your local Red Cross for student materials or for a guest speaker.

ACTIVITY HS–5

Participating in a Physical Fitness Program
(Grades 6–8)

ACTIVITY OVERVIEW

In this activity, the students plan and participate in personal exercise programs that utilize the **overload principle** for the purpose of improving their **cardiopulmonary endurance**. *Estimated time:* introductory discussion and

trial run should take one 55-minute class period. Planning session for personal activity programs can be one class period. Daily activity periods to conduct exercise and record pulse include 15-minute daily sessions for four weeks. One class period each for mid-

way retest run and discussion, evaluation retest run, and concluding discussion.

SCIENCE BACKGROUND

Cardiopulmonary endurance (heart and lung
working capacity) is an important indication
of physical fitness. The heart's capacity for
work can be determined by measuring the
speed with which an elevated heartbeat rate
recovers or returns to its normal rate following strenuous exercise.

Although heart (beat) rates vary from person
to person, our resting heart rate, indicated by
our pulse rate, is rather stable. As we exercise, the amount of energy used by the cells
increases each minute. The heartbeat rate el-
evates, replenishes energy, and removes
waste products from the cells. Only when
waste levels in the blood have reverted back
to levels of the resting state does the heartbeat rate return to normal. A rapid recovery
rate usually implies a heart and a circulatory
system that is efficient in removing waste
products from body muscles and other cells.
A person whose pulse recovers rapidly can
usually work for longer periods of time, can
feel prepared to exercise again sooner, and
will appear less winded after exercise than
do individuals with longer heartbeat recovery
times.

The potential exists for improving the
strength and endurance of the heart. The
strength of a muscle can be increased
through the stress of being made to work
harder than it normally works. This is commonly called the *overload principle.* Overloading a muscle or muscles can be accomplished by several means: increasing the
frequency of exercise; increasing the duration
of exercise; and/or increasing the intensity of
exercise—for example, running a little farther
or faster each time you run.

When applying the overload principle, however, it becomes important not to push too
hard or too fast. We each respond uniquely
to exercise. Overload programs should be
adjusted individually. Increases in work and
activity should be gradual so the heart, lungs,
and muscles can adjust to new work requirements. Only through experimentation can we
discover our own consistent, comfortable overload. Monitoring our pulse rates is a convenient way to measure how our bodies are
adjusting to stress. Excessive stress is characterized by weakness, shakiness, faintness,
light-headedness, throbbing in the head, or
the feeling that the heart is pounding. Excessive stress should be avoided.

Caution: Because students will vary widely in
their capacity for exercise and stress, teachers must be careful that students do not ov-
ertax themselves or feel the need to keep up
with others in the class. Therefore we recommend that you ask a local physician to suggest safe limits for students at your particular
grade level.

SOCIAL IMPLICATIONS

Personal responsibility for developing healthy
life habits is enhanced by understanding the
benefits of exercise on the cardiopulmonary
system. These benefits include a lower probability of heart disease; weight control; higher
energy levels; and improvement in sleep patterns, digestion, elimination, and self-image.
Through demonstrable participation, the students learn to value physical activity as an
important factor in achieving lifelong health.

MAJOR CONCEPTS

All living things change when they interact
with the environment. There are limits to
one's capacity to change.

Understanding the importance of daily exercise is an important step in building personal responsibility for healthy life habits. (© Joanne Meldrum)

OBJECTIVES

☐ Measure.

☐ Gather and organize information.

☐ Construct and interpret graphs.

☐ Use equipment that requires coordination of senses.

☐ Assume responsibility for health of self.

☐ Describe decisions that involve self.

MATERIALS

One 4″ × 6″ index card for each student, graph paper, wristwatch with second hand, and (optional) whistle.

VOCABULARY

cardiopulmonary
  system
endurance
overload principle

pulse rate
physical fitness
stress
recovery rate

PROCEDURES

A. Introduction. Introduce the words **physical fitness** and invite the students to comment on their meanings. What are some of the characteristics of a person who is physically fit?

  1. Define *cardiopulmonary endurance,* and explain the implications that **pulse rate** and pulse **recovery rate** have on heart and lung working efficiency. Ask the pupils to consider ways of improving the working capacity of the heart and lungs.

  2. Explain the *overload principle* to the students. Invite them to participate in a four-week activity program for the purpose of increasing personal cardiopulmonary endurance. Define and discuss **stress**.

  3. Give each student an index card. These cards can be folded later and carried during trial runs for recording pulse rates. The students may title the cards "Personal Data Card" and, using both sides, may set up the cards as illustrated on page 431.

B. First Trial Run. Invite the students to count their resting pulse rates. Next, invite the students to predict how their pulse rates might change after running. How long will it take the pulse to return to its normal resting rate? Will the pulse rate slow down at the same speed that it accelerates? Then invite the students to conduct their first trial run on a track or in a gymnasium that has been divided into 100-meter or 11-yard segments.

1. With data cards in hand, the students run for two minutes or until they are breathing sufficiently hard. (Before the students run they should set their pencils down.)

2. At the end of two minutes, blow a

(Front)

Personal Data Card

Name: Date:

Resting Pulse

	Pulses in 15 seconds	x 4 =	Pulses in 1 minute
Test One			
Test Two			
Test Three			

Average Resting Pulse Per Minute _____

(Back)

Pulse and Exercise

	Trial 1 Date: Distance:		Trial 2 Date: Distance:		Trial 3 Date: Distance:	
	Pulses in 15 seconds	Pulses in 1 minute	Pulses in 15 seconds	Pulses in 1 minute	Pulses in 15 seconds	Pulses in 1 minute
1						
2						
3						
4						
5						
6						
7						
8						

whistle or call the students to a designated area and ask them to record their pulse rates immediately.

3. Within fifteen seconds, begin pulse counting procedures identical to those conducted prior to running (students counting for fifteen seconds).

4. Repeat pulse-counting procedure once a minute for eight consecutive minutes.

5. On their personal data cards, the students note their pulse rates for fifteen-second counts, multiply by four, and record their pulse beats for each minute.

6. The students also note the number of lengths (running area lengths) or meters completed during the run (distance).

C. Individual Programs. Display and explain a list of activities that the students can use in planning their own exercise programs. The students may want to suggest additional activities such as swimming (if available), soccer, or basketball. Possible activities include:

1. *Running.* Begin with 100 meters, increasing the distance each time you run. Time yourself at each distance. Then try to increase your speed each time you run.

2. *Jumping Rope.* Begin by jumping for one minute, increasing your jumping time each time you jump rope. When you reach a plateau, count the number of times you can jump during that period. Try to build up the number of times you jump during subsequent exercise periods.

3. *Relay Races.* Design a series of relay races that include jumping, running, skipping, hopping, and so on. Each relay race should incorporate about five minutes of movement which includes running.

4. *Obstacle Course.* Put together an obstacle course that provides at least five minutes of movement. Activities can include boxes to jump over, hoops to hop in, ropes to jump between, markers to run around, and a good, hard run to the finish line.

For a few days, add activities to the relay races and obstacle course. Then try to increase the number of times you run the race or course, or work to complete them faster each time you exercise.

D. The Schedule. Invite the students to write out an exercise schedule for two weeks and remind the students to consider the overload principle in scheduling activities.

1. Allow one or two fifteen-minute periods each day for students to carry out their exercise programs. Request that the students keep a record of pulse rates immediately before and after exercising.

2. At the end of two weeks, conduct the second test run, the midrun test, after which the students record pulse rates on personal data cards. Discuss the results and invite the students to modify their activity schedules for the following two weeks.

EXTENDING THE ACTIVITY

E. Using their own personal data, the students can record results of test runs on a simple line graph. Simply draw the axis of a graph and mark pulse- and minute-scales as shown. Resting pulse can be indicated with a dotted horizontal line. Resulting pulse rates from test runs can be illustrated with lines of differing colors. The students can compare curves of trial

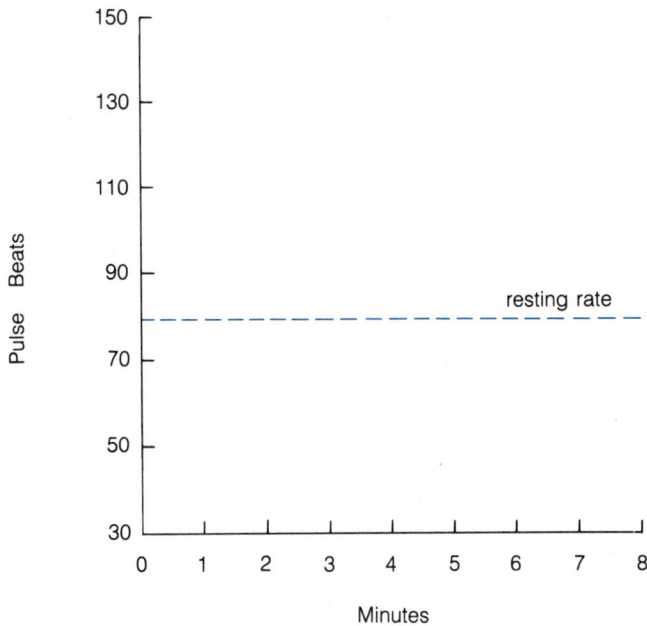

Students can record their pulse rates on this simple graph.

runs. Invite them to speculate on what the curve of a marathon runner might look like.

F. Invite the students to raise questions that can be answered through experimentation:

1. How long will it take for your body to lose the benefits it gained during your activity program?

2. How does your diet, the amount you eat, or the time you eat before exercise affect how you feel after exercising?

3. Does the time of day you exercise have any effect on how you feel after exercising?

G. Invite the students to keep a journal of their physical and/or emotional feelings following exercise periods during the four-week period.

From *Health Activities Project*, Trial Edition. Berkeley, CA: University of California, Lawrence Hall of Science, 1976, 1977, 1978.

ACTIVITY HS–6

Food, Population, and World Health
(Grades 6–8)

ACTIVITY OVERVIEW

In this activity, the students simulate the population, land area, and food production conditions of the five continents. The students then select particular nations and conduct individual research reports, with emphases on population, food production and supply, and health. *Estimated time:* simulation and discussion may occupy one class period. Research projects may take one or two weeks. Sharing of reports may take one or more class periods.

SCIENCE BACKGROUND

Two-thirds of the world's people live in *diet deficient* regions where the average amounts of food eaten do not contain sufficient amounts of calories, protein, and fats. **Undernutrition**, another name for calorie deficiency, results from a lack of quantity rather than from poor quality of food. Prolonged undernutrition can cause edema and irremedial mental and physical disability. **Malnutrition** occurs when food quality lacks vital content, particularly protein, even though quantities may be sufficient. Prolonged malnutrition can lead to high infant mortality, increased susceptibility to disease, shortened life span, and weakened mental ability.

Insufficient food production in developing nations results from numerous factors, including unpredictable climatic changes, population growth, agricultural resource abuse, and limited access to modern agricultural technology. Floods and droughts cause regional scarcities occurring as a result of limited local production. In spite of economic insecurity, the birth rate is highest in developing countries. Low-income nations need people to provide labor, but inadequate medical care and malnutrition result in high death rates, particularly among children. This, in turn, necessitates a higher birth rate. In an unstable economy, children are needed to assure the welfare of the aged, who lack social security and pensions. An understanding of information and techniques for planned parenthood is also lacking in such economies.

Traditional farming and grazing methods practiced in developing nations assume an endless **regeneration** capacity of the earth. However, population density and poor climatic conditions in many areas fail to allow the earth sufficient time to regenerate. Developing nations lacking modern agricultural technology, equipment, irrigation, and fertilizers are less able to produce efficiently or to react to climatic and social conditions. Modern technology is also energy dependent, creating an additional burden for those areas in which energy is limited.

SOCIAL IMPLICATIONS

By even the most conservative estimates, at least 460 million people are threatened with starvation today. In nations unable to produce enough food for their own populations, what prevents food from crossing borders and reaching the people? Many factors contribute to world hunger. The physical distances that food aid must travel not only leads to high transportation costs for nations that may lack exports to pay for incoming food, but also results in much food spoilage due to poor storage and distribution within the recipient nations. Sadly, conflicting political ideologies play a part in world hunger. Cultural differences, too, present a problem. Certain foods, though plentiful, are not culturally acceptable to the customs or habits of some needy nations.

We find that the interrelationship between population growth and **food production,** supply, and **food distribution** affects the health of individuals in any society. Finding solutions to problems that plague the world community begins with cognizance of and sensitivity to the pertinent facts. As students or adults, appreciating the importance of food as a basic factor in maintaining the health of people throughout the world invites self-examination of our personal roles as world citizens; it can be a first step in solving this global problem.

MAJOR CONCEPTS

There are limits to the capacity of the environment to provide for populations. The relationship between population growth and food supply and distribution affects the health of individuals in a society.

OBJECTIVES

☐ Interpret the relationship between population density and food supply.

☐ Construct complex projects.

☐ Demonstrate to others the need and means to use natural resources wisely.

☐ Describe immediate- and long-range consequences of various alternatives to self and others in the environment.

MATERIALS

Chairs, 1 box of crackers.

VOCABULARY

food distribution   malnutrition
food production   undernutrition

PROCEDURES

A. Use figures for population, land area, and food production given in chart 12–4 to arouse interest and inquiry regarding population growth and food supply.

   1. Divide the classroom chairs proportionately to the land area of the five continents.

   2. Next, divide the students proportionately to the population. There will not be enough chairs for all students in some continents and a surplus will exist in others.

   3. Using chart 12–4 figures for food production, distribute crackers proportionately to students. Ask the students to note whether there are enough crackers for people in each continent. You might ask if there would be enough crackers for all class members if no continental boundaries existed.

   4. Have the children act out scenarios on how to obtain enough chairs and crackers for their continent.

B. Invite each student to select a nation of interest for a written report and display of related materials. Nations in the continents of Africa, Asia, and Latin America should be included particularly. At least one student should also select the United States for the purpose of comparing various existing conditions. Research on each country should concentrate on responding to the following issues:

   1. Describe the common diet within the country.

   2. What kinds of foods are grown and produced? Describe climatic or other conditions that affect food production.

   3. Are the supplies produced within the country sufficient for the needs of the population?

   4. If not, from which areas of the world does the country obtain additional food?

   5. Are there indications that the country receives aid from the United Nations, CARE, the United States, or some other country?

   6. Is any progress being made to overcome any problems related to food production?

	Africa	Europe	Asia	North America	South America
Land area in millions of miles	12	4	16	10	6
Population in millions	400	600	2,400 (2 billion, 400 million)	400	200
Food production in millions of metric tons	100	400	400	400	100

	Africa	Europe	Asia	North America	South America
Land area in millions of miles	ℎℎℎ ℎℎℎ	ℎℎ	ℎℎℎℎ ℎℎℎℎ	ℎℎℎ ℎℎ	ℎℎℎ
Population in millions	⚡⚡	⚡⚡⚡	⚡⚡⚡⚡ ⚡⚡⚡⚡ ⚡⚡⚡⚡	⚡⚡	⚡
Food production in millions of metric tons	)(	)( )( )( )(	)( )( )( )(	)( )( )( )(	)(

KEY: ℎ = 2 million square miles

⚡ = 200 million people

)(= 100 million metric tons food production

CHART 12–4
Land Area, Population, and Food Production of the Five Major Continents: 1971

EXTENDING THE ACTIVITY

C. Follow up the reports with a discussion. Encourage the students to draw some conclusions. Questions such as the following might be written on the chalkboard:

1. Does food production seem to be a global problem or is there a problem with food distribution?
2. What roles do undernutrition and malnutrition play in physical disability and disease in the world?

3. How do, and how can, the masses contribute to the problems and solutions of food production, food distribution, undernutrition, malnutrition, and disease?

4. How would you define **overpopulation**?

Adapted, with permission, from *My Body Is Mine, My Emotions Are Mine, My Environmment Is Mine: A Health Education Curriculum*. Albany, N.Y.: Albany Unified School District, 1978.

BIBLIOGRAPHY

Teachers' Resources

Ash, Joan, and Stevenson, Michael. *Health: A Multimedia Source Guide*. New York: R. R. Bowker Co., 1976.

Bershad, D., and Bernick, D. *From the Inside Out*. Boston: Learning for Life/ MSH, 1979.

Jenkins, Dorothy M. *Human Sexuality: Curriculum Guide for a Course in Human Sexuality*. Los Angeles: Los Angeles Regional Family Planning Council, 1976.

Kellogg Company. *Fitness Focus*. St. Paul: Kellogg Company Nutrition Unit, 1980.

Children's Resources

After the Ouch (film). Los Angeles: Churchill Films, 1979.

Allison, Linda. *Blood and Guts: A Working Guide to Your Own Insides*. Boston: Little, Brown & Co., 1976.

Burns, Sheila H. *Allergies and You*. New York: Julian Messner, 1980.

Madison, Arnold. *Smoking and You*. New York: Julian Messner, 1975.

A Physical Examination (film). Santa Monica, CA: Ramsgate Films, 1979.

Showers, Paul. *No Measles, No Mumps for Me*. New York: Thomas Y. Crowell & Co., 1980.

Staying Healthy (filmstrips). Santa Monica, CA: BFA Educational Media, 1978.

13

Earth and Planetary Sciences

The purposes of this chapter are to present some basic topics and issues in geology, meteorology, and astronomy, and to provide you with a more knowledgeable basis for teaching earth and planetary sciences.

These discussions are necessarily brief ones. The reader is encouraged to see introductory texts in the earth sciences and to use the bibliography at the end of this chapter. Information in these fields changes rapidly; keeping up-to-date is therefore an important task for the science teacher.

KEY CONCEPTS AND IMPORTANT IDEAS

☐ historical geology ☐ geologic time scale ☐ radiometric dating ☐ age of the earth ☐ physical geology ☐ common minerals ☐ igneous, sedimentary, and metamorphic rocks ☐ the rock cycle ☐ structure of the earth ☐ plate tectonics ☐ continental drift ☐ earth's magnetic field ☐ earthquakes and volcanoes ☐ weathering and erosion ☐ meteorology ☐ structure of the earth's atmosphere ☐ hydrologic cycle ☐ astronomy ☐ the solar system ☐ constellations, stars, and galaxies ☐ the universe ☐ measuring distances in space

In order for us to understand the complex ecological crisis facing our society, we must learn more about the earth and the universe. The study of the earth and the universe is called earth and planetary science. To understand the subject, we must use an interdisciplinary approach that includes the fields of **astronomy, meteorology, oceanography, geology,** and **geophysics**. Geography and engineering are closely allied fields. And, while prerequisites to serious study in these areas are mathematics, physics, chemistry, and biology, a prerequisite for *everyone* in our society is a basic knowledge of the major areas and perspectives in earth and planetary science. We must all take responsibility for ourselves and the world in which we live. The areas of study in the earth and planetary sciences have grown to include many hitherto separate sciences, but geology was the first discipline to focus on the earth as its main subject of study.

ASTRONOMY The study of celestial bodies

METEOROLOGY The study of weather and climate

OCEANOGRAPHY The study of the oceans

GEOLOGY The study of the earth, its physical properties, and the processes that have changed the earth over time

GEOPHYSICS The study of the physical properties of the earth; the applied physical measurement of geological events

GEOLOGY

Geology is an organized body of knowledge that includes historical and physical geology. *Historical geology* is the branch that deals with the history of the earth, how various life forms have evolved throughout time, and how the earth has changed over billions of years. *Physical geology* describes the properties of matter composing the earth, the processes by which matter is formed and interacts, and the nature and development of landforms as we know them today.

Modern geology as a field dates from 1785 when James Hutton (1726–1797) set forth the *doctrine of uniformitarianism*. The doctrine states, simply, that the forces that operate to change the earth's surface now have worked in the same ways in the past. Today, this allows scientists in all the earth sciences—geologists, **geochemists, paleontologists**, meteorologists, astronomers, **astrophysicists**, and many others—to study *current* geological events and to make inferences about *past* geologic events. For example, the eruption of Mount St. Helens in May 1980 allowed scientists to study volcanic eruptions and earth movements, and to make statements about how all volcanoes work and affect the lands surrounding them.

GEOCHEMIST One who studies the chemical composition of the earth's crust and the chemical changes that occur there

PALEONTOLOGIST One who studies the fossil remains of animals and plants

ASTROPHYSICIST One who studies the physical properties of heavenly bodies

Because the earth is composed largely of rock, part of our study of the earth is the study of the composition and varieties of rocks, and processes they undergo. We often cannot see the important geological changes taking place in our environment because rock processes can take thousands, millions, and billions of years to accomplish. It only takes an instant for a particle of sand or soil to slip into a gully or stream. Yet that particle, once carried through creeks, rivers, and finally to the sea, can lie covered by other sediment for centuries and then become part of a new mountain range. How long does it take to build a mountain? To make coal? To create deserts or lakes?

Historical Geology

GEOLOGIC TIME SCALE A history and relative calendar of the earth

The **geologic time scale** and the geologic column in figure 13–1 represent a history and relative time scale of the earth. The column shows that geologists place the earliest existence of primitive marine plants and invertebrates and one-celled animals on the earth at over 570 million years ago. It shows the existence of coal-forming swamps in the Pennsylvanian period some 300 million years ago. It also dates the beginning of both the Appalachian Mountain and Rocky Mountain chains at 470 million years ago and about 65 million years ago, respectively. The geologic column puts in perspective about how long, relative to one another, the major geologic time periods or divisions have lasted during the history of the earth.

Time. Time, then, is important to geologists; they have constructed both geologic and relative time scales. *Relative time* indicates which events took place before or after other events, but it does not say exactly when the events took place in absolute time. *Absolute time* is a recorded, historical date, let's say the year you began college or the actual day and year of your birth. Although geologists would like to be able to pinpoint time in terms as specific as the date given on your birth certificate, geologic dating will probably never be that precise. Yet, as we shall see presently, by dating fossils we can say that dinosaurs lived and that they preceded amphibia and fish. By dating rocks, we can date the formation of major geologic landforms and by using relative information, we can trace the origin, history, and major geologic processes of the earth.

Radiometric Dating. The oldest dated rocks on earth were formed about 3.8 billion years ago; the origin of the earth is now estimated at a time 4.5 billion years ago. Our knowledge of these dates is based on a process that determines the amount of radioactive material that remains in rocks or **fossils. Radiometric dating** methods are the most effective means of measuring absolute time and absolute age.

FOSSIL Any trace or imprint of a plant or animal that has been preserved in the earth's crust

RADIOMETRIC DATING The measurement of geologic time by the isotopic abundance and known disintegration rates of radioactive isotopes

Rocks and objects of geologic interest contain unstable radioactive elements that decay and yield more stable nuclear elements at a constant rate.

For example, the radioactive element Uranium decays into helium and lead at such a rate that half of the uranium is left after 4.5 billion years. By measuring the proportions of uranium and lead in an object we may therefore calculate when it was formed and trapped the ura-

Marie Curie (1867–1934)

French Embassy Press and Information Division

Through her work in the late 1800s, Marie Curie helped build a bridge between chemistry and physics. Her work was so outstanding that Marie Curie is the only person to have won the Nobel Prize twice: once for research done with her husband and once for her own research. Marie (or Marja as was her name in Poland) was known as an outstanding student in high school. She had a dream of going to college but in Poland in 1890 women were not permitted to go.

Marie moved to Paris and in two years completed her master's degree in physics and mathematics plus scoring the highest in the class. One of her teachers, Pierre Curie—a confirmed bachelor of 35—believed that women could not understand the real need a scientist had to devote his life to his study. But in Marie he found a girl who asked questions about his work with crystals, magnetic fields, and electricity—and who understood his answers!

Within two years of their marriage, they had a daughter Irene and a devotion to the laboratory. Based on the work of Becquerel who in 1896 discovered that uranium gave off radiations that could be captured in photographs, Marie's husband invented a way to measure these radiations. Then Marie discovered that more elements than uranium gave off these radiations. As they continued their work, they discovered two new elements for the periodic table in addition to their discovery of radioactivity, for which they received the Nobel Prize in chemistry in 1904. After Pierre's death, Marie won her second Nobel Prize in chemistry in 1911 for her discovery of uses for radioactive elements.

In later years, Marie's second daughter, Eve, wrote about what she called her mother's "terrible patience" which characterized her during her long radium experiments. She called her mother "a physicist, a chemist, a specialized worker, an engineer, and a laboratory man all at once."

Her daughter Irene and Irene's husband Frederic also received honors for their work on radioactivity. But there was a high cost to this talented family. Marie, Irene, and Frederic all died from excessive exposure to the radioactivity they studied.

Era[a]	Period	Epoch	Millions of yr. Duration	Before present	Major physical events
Cenozoic	Quaternary	Holocene	0.01		**Continental glaciations**
Mesozoic		Pleistocene	1.5–2		
	Tertiary	Pliocene	5–5.5		
Paleozoic		Miocene	19		
		Oligocene	11–12		**Yellowstone volcanics begin**
		Eocene	15–17		
		Paleocene	11–12	65	
					Beginning of Rocky Mountains
	Cretaceous		71		
					North America separates from Eurasia
	Jurassic		54–59		
	Triassic		30–35	225	**North America begins to separate from Africa** **Atlantic basin originates**
	Permian		55		**Climax of Appalachian mountain building**
	Pennsylvanian[d]		45		
Precambrian	Mississippian[d]		20		
	Devonian		50		
	Silurian		35–45		
	Ordovician		60–70		**Beginning of Appalachian Mountains**
	Cambrian		70		
	Precambrian[e]			570	**Oldest dated rocks (±3.8 billion yr ago)** **Origin of earth (±4.5 billion yr ago)**

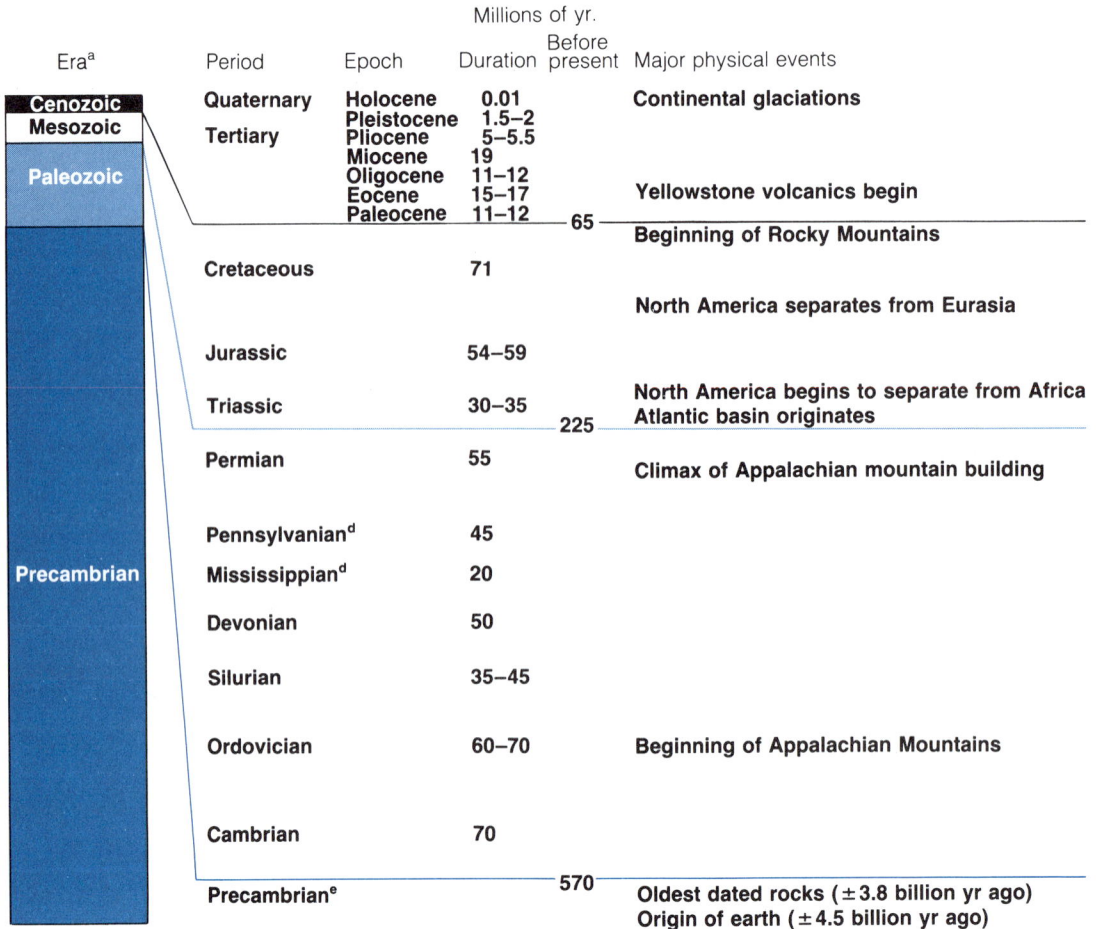

[a]Duration on approximately uniform time scale.
[b]This column does not give the complete time range of the forms listed. For example, fish are known from pre-Silurian rocks and obviously exist today; but when the Silurian and Devonian rocks were being formed, fish represented the most advanced form of animal life.

FIGURE 13–1

The Geologic Time Scale and The Geologic Column

From L.D. Leet, S. Judson, and M.E. Kauffman, *Physical Geology*, 5th ed., © 1978, end paper. Reprinted by permission of Prentice-Hall, Inc., Englewood Cliffs, N.J. Data in first five columns of figure from W. B. Harland, A. Gilbert Smith, and B. Wilcock (eds.), "The Phanerozoic Time-Scale," *Quart. J. Geol. Soc. London*, vol. 120S (suppl.), 1964.

Aspects of the life record

Major events	Dominant forms[b]		Systems
	Homo		**Quaternary**[c]
			Tertiary[c]
Grasses become abundant	**Mammals**	**Flowering plants**	
Horses first appear			
Extinction of dinosaurs			**Cretaceous**
Birds first appear			**Jurassic**
Dinosaurs first appear	**Reptiles**	**Conifer and cycad plants**	**Triassic**
			Permian
	Amphibia		**Pennsylvanian**
Coal-forming swamps			**Mississippian**
		Spore-bearing land plants	**Devonian**
	Fish		**Silurian**
Vertebrates first appear (fish)			**Ordovician**
	Marine invertebrates		
		Marine plants	
First abundant fossil record (marine invertebrates)			**Cambrian**
	Primitive marine plants and invertebrates **One-celled organisms**		**Precambrian**

[c]Some geologists prefer to use the term Cenozoic for the Quaternary and the Tertiary.
[d]In most European and some American literature Pennsylvanian and Mississippian are combined in a period called the Carboniferous.
[e]Subdivisions not firmly established.

nium. Uranium 238 is used to date events from a few million years ago all the way back to the time of the origin of the earth. Carbon 14, as well as other radioactive **isotopes,** is used to determine the ages of objects believed to be less than 50,000 years old.

Age of the Earth. Radiometric measures have been used to try to determine the age of the earth. The oldest rocks yet found on earth have been dated at well over 3 billion years old but there is reason to believe

ISOTOPE Alternative form of an element produced by variations in numbers of neutrons in the nucleus

METEORITE Stony or metallic body fallen to earth from outer space

that the earth is older than the rocks we can date today. Radiometric dating of **meteorite** fragments from space yields ages of up to 4.5 billion years. Rocks returned to earth from the moon by the Apollo Mission have been dated at about 4.6 billion years old. Since it is generally assumed that all of the solar system came into being at the same time, most geologists are prepared to accept this range, 4.5 to 4.6 billion years, as the age of the earth. However, as in any discussion of origins, we must remain tentative in our conclusions. Science teachers should emphasize that the scientific method is a continuous process whereby scientific "truth" is always in question.

EARTH'S CRUST Outermost layer or shell of the earth

Fossils. A fossil is evidence of past plant or animal life, such as a dinosaur bone, an impression of a leaf, or a footprint from a previous geological period, preserved in a rock formation in the **earth's crust.** Fossil remains tell us the history of animal life and also the history of the earth's crust. If fossil remains of animals that lived in the sea millions of years ago are found at the top of a mountain, we know that the mountain top was once under or very near a sea. Most fossils are found in sedimentary rocks; they are most abundant in mudstone, shale, and limestone.

The remains of plants and animals are destroyed if left exposed at the earth's surface. The exceptions become fossils: sea creatures that sink to the bottom and then are covered and preserved by layer after layer of sediment, and land animals and amphibians that died in now-extinct swamps and lakes and were covered by sand or mud before they could decompose. Some have been preserved in ice.

PALEONTOLOGY The study of fossil remains of animals and plants

ALGAE One-celled plants

Paleontology is the study of life of the past. Living things today are a continuation of unbroken chains of ancestors and descendants that date back at least 3 billion years. (See figure 13–1 for a geologic calendar of time periods.) The earliest fossils yet found are of bacteria and **algae** that lived about 3.1 billion years ago in the Precambrian period. Since then, plants of various kinds have thrived in the waters of the earth. Land plants appeared 400 million years ago, and the modern plant forms that we see today appeared 100 million years ago.

Animals without backbones—invertebrates—evolved during the late Precambrian period and became abundant about 600 million years ago and still exist today. Animals with vertebrae—vertebrates—include fish, amphibians, reptiles, birds, and mammals. Many early vertebrate groups are still with us, although individual species have become extinct.

EXTINCTION No longer in existence

There may have been as many as 500 million different plant and animal species in the earth's past, however only one million survive today. This suggests that **extinction** is as much a part of life as is the creation of new life forms.

From fossils of shelled invertebrates, scientists are able to identify areas of the world which were once under water, but now are above water. (E. B. Hardin, U.S. Geological Survey)

The earliest reptiles, about 250 million years ago, were 1 to 8 feet in length. They grew, very significantly, in size, number, and variety, during the next 160 million years, and then they suffered massive extinction. Unlike mammals, reptiles grow in size until they die, the most spectacular examples being the dinosaurs.

The mammals appeared when the reptiles were at the height of their evolutionary success, about 110 million years ago. The earliest were marsupials, like kangaroos and opossums, who carried their offspring to term in "pouches" outside their bodies. They were small, rodent-sized animals. Later the placental mammals took over. In contrast with the marsupials, these animals carried their offspring within them until the babies were able to survive in the earth's natural environment.

From amongst the mammals, the primates began to appear about 50 million years ago. Primates have flexible hands, fingers, feet, and toes. The first primates were like tree lemurs and shrews. The higher primates, monkeys and apes, appeared about 14 million years ago; and from amongst the higher primates, about one million years ago, most scientists believe that modern humans emerged.

Plant fossils tell scientists about the climatic and other conditions that existed millions years ago during previous periods of earth's history. (G. E. Becraft, U.S. Geological Survey)

ROCKS Aggregates of one or more minerals or a body of undifferentiated mineral matter

MINERAL Naturally occurring, inorganic element or compound having an orderly internal structure

LANDFORMS Features of the earth's surface attributable to natural causes

LANDSCAPE The combination of features seen on the earth's surface

ELEMENT Unique combination of protons, neutrons, and electrons that cannot be broken down by ordinary chemical methods; one of the more than 100 "building blocks" of which all matter is composed

COMPOUND Substances that can be reduced into simpler, pure substances, each with a definite proportion by weight

Fossil remains have allowed us to trace and date the plant and animal life of the past as well as rocks and other inorganic material. Just as the age of a rock in which a fossil is found will tell the age of the fossil, the fossil can date the rock, and also tell us where the rock was when the fossil was formed.

Physical Geology

Estimating the age of the earth is a more speculative task than is understanding the earth itself. Beyond the asphalt, concrete, and buildings of our cities, suburbs, and rural areas, we see soil and its parent material, rock. **Rocks** are so common that we rarely think of them. Yet, except for the atmosphere, the earth is made of rock materials. Therefore, an understanding of the world around us requires us to become more familiar with them.

Rocks and Minerals.   Rocks are aggregates of **minerals** of one or more kinds in varying proportions, and they comprise the **landforms** and **landscapes** and the interior formations of the earth. It seems obvious from this definition, then, that we need to go back one step and discover more about minerals.

Minerals are naturally occurring solid **elements** or **compounds**. They are inorganic, or nonliving, and are crystalline in nature. Importantly for the elementary and junior high school science teacher, minerals are large enough to be observed and identified by the unaided eye.

Identifying Minerals.   Minerals have observable features that we can identify and classify. We identify minerals according to their physical characteristics: crystal form, hardness, specific gravity, cleavage, stria-

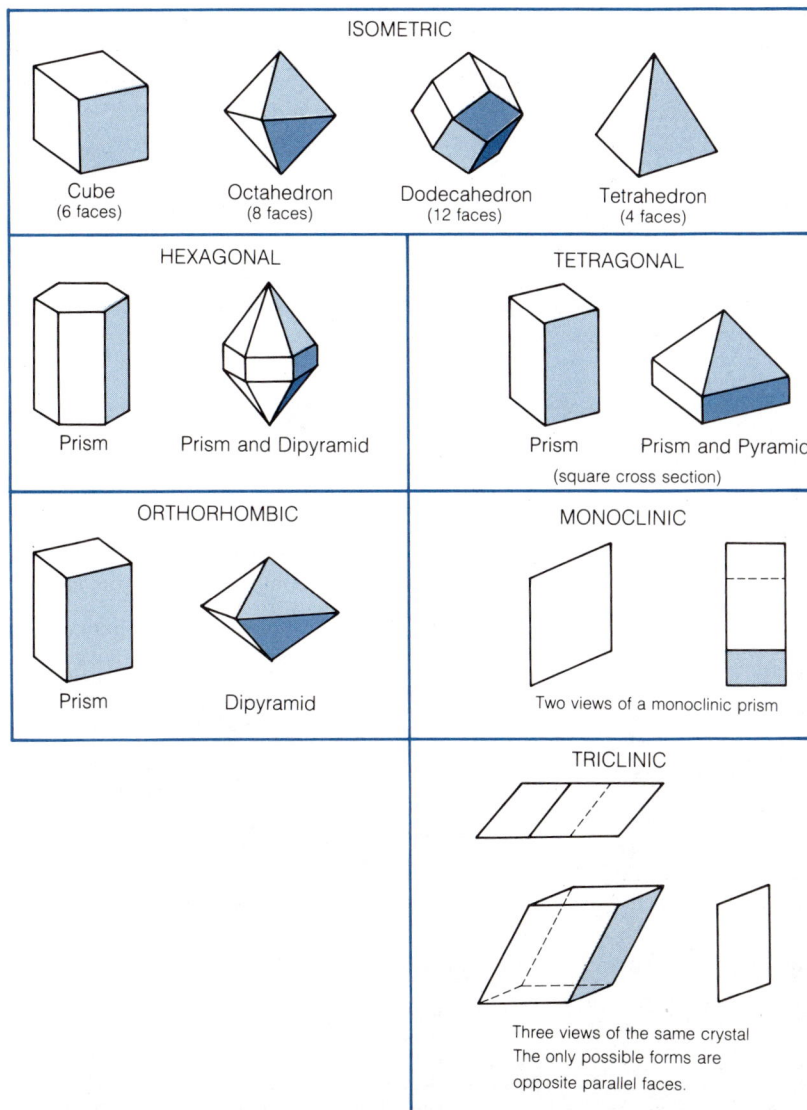

FIGURE 13–2

Examples of some of the more common crystal forms, shown and named for reference only.

From *General Geology,* 4th ed., by Robert J. Foster. Copyright © 1983 by Bell & Howell Company. Used by permission of Charles E. Merrill Publishing Company.

tions, and color streak. The *crystalline form* of minerals usually appears as a structure of surfaces that intersect at symmetrical angles. Figure 13–2 shows some basic crystalline structures. *Cleavage* is the tendency of a mineral to break in specific directions on its smooth surfaces but not on other surfaces. We can observe cleavage by seeing how a mineral breaks when we hit it on a smooth surface. Some minerals have *striations*—parallel, thin lines or bands that appear across their crystal faces. *Hardness* is a physical property of minerals that is easy to demonstrate; it refers to the mineral's "scratchability." If the mineral can be scratched by your fingernail, it is classified in the lowest range of hardness. Table 13–1 shows the categories of mineral scratchability. The hardest mineral known is diamond because it cannot be scratched by any other substance.

SPECIFIC GRAVITY Ratio between weight of a given volume of material and weight of an equal volume of water

Specific gravity refers to the mineral's weight per unit volume. It is usually compared to the weight of the same volume of water. *Color* is not a reliable physical property by which to classify most minerals. *Streak* refers to the color left on a piece of porcelain after a mineral has been rubbed across it and has left a fine, powdery residue. The streak is

TABLE 13–1
Mohs scale of hardness

Scale	Mineral	Test
1	Talc	[Softest]
2	Gypsum	
2.5		Fingernail
3	Calcite	Copper coin
4	Fluorite	
5	Apatite	
5.5–6		Knife blade or plate glass
6	Orthoclase	
6.5–7		Steel file
7	Quartz	
8	Topaz	
9	Corundum	
10	Diamond	[Hardest]

SOURCE: L. D. Leet, S. Judson, and M. E. Kauffman, *Physical Geology*, 5th ed., © 1978, p. 436. Reprinted by permission of Prentice-Hall, Inc., Englewood Cliffs, N. J.

not always the color of the mineral itself. In fact, the same mineral may be found in different colors, but the fine powder of their streaks will be the same color.

The Most Common Minerals.   There are about 2000 minerals found in the earth's crust but only one family, the *silicates*, comprises 90 percent of the minerals of the earth. Silicates are rock-forming compounds that contain the elements silicon and oxygen plus one or more metals. Seventy-seven of the known elements are called metals. They have in common the properties that they conduct electricity, fuse (unite with other materials upon melting), are opaque (impervious to visible light), and have a metallic luster (they are relatively shiny). For a discussion of what elements are, see chapter 15.

There are two main classifications of silicates: metals (ferro-magnesians) and non-metals (nonferromagnesians). Metals contain iron and magnesium and they are all very dark or black. Nonferromagne-sians, or non-metals, are so named because they do not contain iron or magnesium; they include three important minerals: muscovite, feldspar, and quartz. Muscovite is another name for the familiar sparkling mica. Feldspar is white, gray, pinkish, or opalescent, and it comprises nearly 54 percent of all the minerals in the earth's crust. Because of its crystal-line structure, quartz is a relatively hard mineral that appears smokey to clear in color. In less common varieties, it is purple or violet amethyst, rose-red or pink rose, smokey yellow to brown, or milky in appearance. Other familiar mineral families include oxides (of which ice is an example!) carbonates (the garden additive, dolomite), and halides (halite or common rock salt).

The rock you can find lying unheeded outside your doorstep, then, is a mixture of many minerals and is inorganic or nonliving. It is proba-bly a silicate and is probably gray, white, or pinkish in color because it has feldspar in it. It is also probably granite (or perhaps limestone). Ge-ologists estimate that the rock at your doorstep is probably at least 60 million years old! And it could be even older. If you lived in West Green-land where the oldest dated rocks have been found, the rock outside your door might be about 3.8 billion years old.

Rock Familes.   All rocks are classified by geologists into three families: **igneous rocks**, **sedimentary rocks**, and **metamorphic rocks**. They base these classes on the ways in which each kind of rock is formed.

Igneous rock.   Igneous rocks are those that were once part of a hot, molten mass under the surface of the earth called *magma*. Magma cools under the earth's surface and gradually surfaces, usually weathering at the earth's surface. Magma is called lava when it erupts from volcanoes,

IGNEOUS ROCK A rock or mineral that solidified from molten or partly-molten material

SEDIMENTARY ROCK A rock resulting from the consolidation of loose sedi-ments that have accumu-lated in layers

METAMORPHIC ROCK Any rock derived from pre-existing rocks in response to marked changes in tem-perature, pressure, stress, and chemical environment

cools on the earth's surface, and crystallizes into rock. Magma forms many types of rocks, among them **granite**, the glassy **obsidian**, and **basalt**. These rocks form major components of the earth's structure and are involved in the processes of structural change. The light grey or pink color rock so common to us is granite, an igneous rock that forms the continental crusts of the earth.

GRANITE Coarse-grained, igneous rock dominated by light-colored minerals

OBSIDIAN A glassy equivalent to granite

BASALT A fine-grained igneous rock dominated by dark-colored minerals

LITHIFICATION Process by which unconsolidated, rock-forming materials are converted into the consolidated, rock state

Sedimentary rock.   Sedimentary rock is formed by **lithification**, a process by which particles from the breakdown of preexisting rocks are consolidated into a new kind of rock. Weathering breaks down the rocks and transports them to places where they become layered deposits, usually of different types of rock material: sandstone, mudstone, shale, limestone, dolomite, chalk, and coal. Note that coal and chalk are organic rocks not found in other rock families.

Metamorphic rock.   Metamorphic rock is the third largest family of rocks. They are formed under the earth's surface by subjecting igneous and sedimentary rocks to heat, pressure, and chemical activity. Major types of metamorphic rock are slate, phyllite, schist, amphibolite, gneiss, marble, and quartzite.

Until now our study of geology has focused on the basic building blocks of the earth. Next we examine how the earth continuously changes using these building blocks. Some of the processes of change are very dramatic and highly visible while others are quite subtle.

Processes of Change.   Glaciers, earthquakes, and volcanoes alter the earth's surface dramatically. But studies have shown that daily changes occur when a drop of warm rainwater hits a rock or when a wave reaches the seashore. Land moves particle by particle to the sea; plates move only inches or centimeters each year; mountains build just as slowly. Such small, subtle processes are the major forces for change of the earth's surface. We begin this section by looking at the way rocks themselves change.

ROCK CYCLE A sequence of events involving the formation, alteration, destruction, and re-formation of rocks

The Rock Cycle.   The **rock cycle** shows how the three rock families interact with one another in a never-ending cycle of change. The cycle is pictured in figure 13–3. Because the cycle is basic to geology, and includes some important ideas and processes, such as weathering, let us look at some examples. Magma cools into igneous rock. If it cools below the earth's surface, it is gradually uncovered by weathering and faulting and joins igneous rock that cooled initially at the earth's surface. Here weathering continues to erode the rock until it becomes layered sedimentary rock. Sedimentary rock is pushed under the surface and changed by that movement to create metamorphic rock. As metamorphic rock, it undergoes continual pressure and heat until it begins to melt into magma again. Note in the illustration that the cycle may be interrupted and that, for example, igneous rock may become metamorphic, or that metamorphic rock may forgo the magma and igneous states and

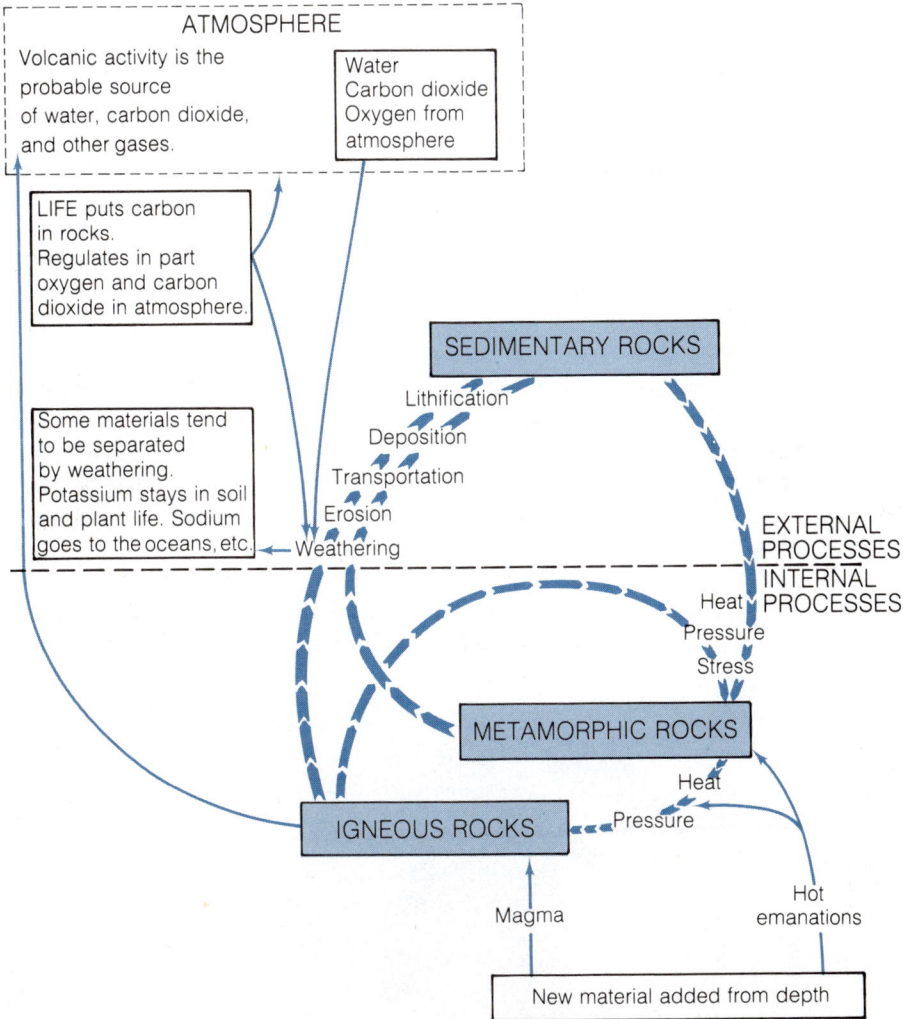

FIGURE 13–3

The middle of this diagram, showing the relationships among igneous, sedimentary, and metamorphic rocks, is what is generally considered the rock cycle. The upper and lower parts of the figure show how material is added to and subtracted from the rock cycle.

From *Physical Geology: The Structure and Processes of the Earth* by B. Clark Birchfiel et al. Copyright © 1982 by Bell & Howell Company. Used by permission of Charles E. Merrill Publishing Company.

begin to weather into sedimentary rock instead. Note also the processes by which these changes occur: melting, weathering, transportation, erosion, and lithification. The concept of the rock cycle has been attributed to James Hutton, the father of geology, in the late eighteenth century. It is a theory that has withstood more than a century of scientific observations.

Structure of the Earth. There are three main parts, or zones, of the earth. From the center to the surface, they are: the **core,** the **mantle,** and the **crust**.

Core. Imagine for a moment that we could find a way to travel down toward the center of the earth. We need to imagine this journey, of course, because no one has ever really penetrated the earth more than about 4 miles (6.5 km) and the core begins about 1,800 miles (2,900 km) below the surface. The earth is about 7,926 miles (12,783 km) in diameter; its center is thus approximately 3,963 miles (6,392 km) from the surface of its crust.

Everything we know about the core comes from **seismic data** and by knowledge of matter believed to be similar to core material, especially meteorites that have landed here from space. Figure 13–4 illustrates the main structures of the earth. The core is divided into two concentric layers, the inner and outer cores. Both layers are probably made primarily of iron.

The core is white hot, with temperatures reaching from 4,000 to 9,000° (2200° C to 5,000° C), and the pressures are 2 to 3 million times greater than pressures at the surface. These pressures have compressed

CORE Innermost zone of earth, surrounded by the mantle

MANTLE Intermediate zone of the earth; surrounded by crust resting on the core

CRUST The outer zone of the earth consists of the continental crust and the oceanic crust

SEISMIC DATA Information gained from a seismograph or instrument for recording the earth's vibrations during earthquakes or volcanic eruptions

FIGURE 13–4
The three major subdivisions of the earth—core, mantle, and crust—are shown in the cross section.

From *Essentials of Oceanography* by Harold V. Thurman. Copyright © 1983 by Bell & Howell Company. Used by permission of Charles E. Merrill Publishing Company.

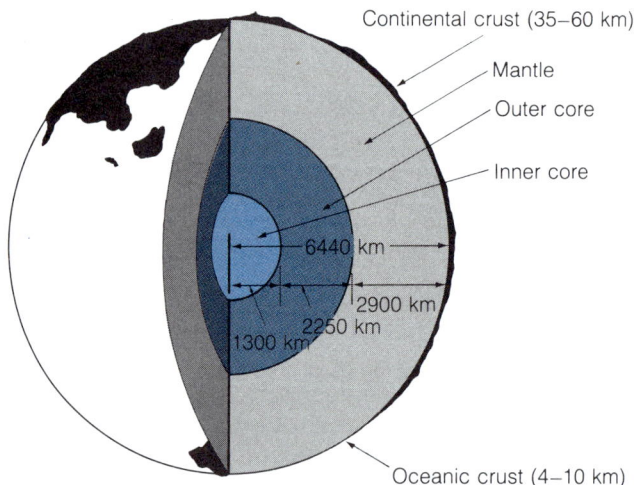

Continental crust (35–60 km)
Mantle
Outer core
Inner core
6440 km
2900 km
2250 km
1300 km
Oceanic crust (4–10 km)

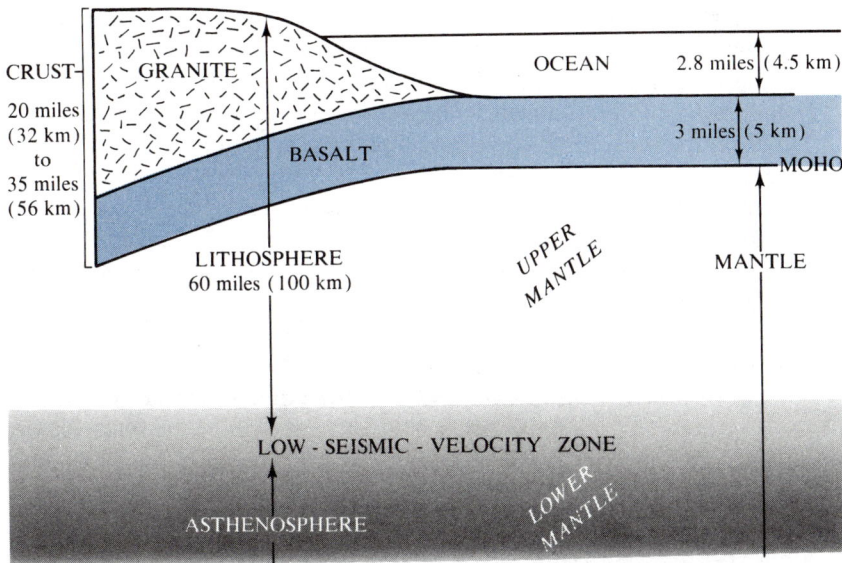

FIGURE 13-5

The crust and upper mantle form the lithosphere, in which the rocks behave as solids. In the asthenosphere, the rocks flow.

From *General Geology*, 4th ed., by Robert J. Foster. Copyright © 1983 by Bell & Howell Company. Used by permission of Charles E. Merrill Publishing Company.

the inner core into a solid mass that is about 1,600 miles (2,581 km) in diameter. The outer core is liquid, according to seismic data, and is about 1,400 miles (2,250 km) thick.

The mantle.   The mantle extends from 5 to 20 miles (8 to 32 km) below the earth's surface, down about 1,800 miles (2,903 km). The crust and mantle are shown in figure 13-5. The upper mantle, along with the crust, is referred to as the **lithosphere**, and the lower mantle is termed the **asthenosphere**. The asthenosphere is the warmer, lower section of the mantle. At a depth of about 43.5 miles (70 km), the temperature of the earth (rising at about 15° C each kilometer of depth) has reached about 1,000 degrees Celsius, a point at which rock will flow slowly if there is sufficient pressure. This *slow flow* creates what seems to be a soft zone upon which the lithosphere appears to float.

The lithosphere, the upper portion of the mantle, is a more rigid shell that is cooler than the asthenosphere. Because the lithosphere rests and moves along with the asthenosphere, it essentially carries the crust (with the oceans and continents) on its top in a piggyback fashion.

LITHOSPHERE The outer, solid portion of the earth, which includes the crust and the upper mantle, active component in plate tectonics

ASTHENOSPHERE The weak layer or shell of the earth below the lithosphere where adjustments in crustal, vertical movement take place

The Crust.   Much of what we know of the crust, the earth's outermost and thinnest layer, comes from studying measurements of earthquakes. A Yugoslavian seismologist, A. Mohorovičić (1857–1936) recorded the waves produced from an earthquake in Croatia in 1909, and noticed that the speed of his seismic wave readings increased abruptly at a depth of about 30 miles (48 km) below the surface. He concluded that the increase indicated a change in the material through which the waves passed. The finding became known as the *Mohorovičić discontinuity,* or simply the *Moho,* and we now accept this wave change as indicating the location of the base of the earth's crust all over the globe. Figure 13–5 shows the granite and basalt layers of crust lying above the mantle—the material that caused the change in Mohorovičić's recordings.

It became clear, however, that the depth of the Moho discontinuity varied depending upon the kinds of landforms on it or the existence of oceans above it. The Moho is deeper under the continents; continental crust is composed of the igneous rock, granite, which is underlaid by a basalt rock base.

The crust under the oceans is much thinner; it is about 2½ to 6 miles (4 to 10 km) deep and is composed of the igneous rock, basalt. Figure 13–5 shows the differing widths of the continental and oceanic crusts.

Oceans.   The oceans cover more than 70 percent of the earth's crust or surface. Most of the waters of the oceans lie in an ocean basin that is located between the continents in the southern hemisphere and that extends northward to form the Pacific, Atlantic, and Indian Oceans. The Arctic Ocean connects with the Pacific Ocean through the narrow Bering Strait (between the United States and Russia) and with the Atlantic Ocean through the Davis Strait (between Canada and Greenland) and through several small seas (between Greenland and Norway). Oceans, and in fact all the waters of the earth, are connected. The waters in oceans are completely mixed about once every 2,000 years.

Oceans are salty because they contain dissolved minerals, and these minerals are carried back and forth, between land and sea, as part of the *rock cycle.* Most of the sedimentary rocks that now cover three-fourths of the earth's land surface were formed on ocean floors.

The Earth as a Heat Engine.   A heat engine is a device for getting useful mechanical work out of heat. It has been helpful to some people to see the earth as a heat engine, too. This is because both the internal heat of the mantle and core and the external heat energy from our sun provide the energy that keeps the lithospheric plates and crust of the earth in continual motion. The moving parts of the heat engine are the crust, the lithosphere, the plastic or slow-flowing asthenosphere, the air or atmosphere, and the oceans. How these solid elements of the "heat engine, earth" work together is described by the theories of plate tectonics, continental drift, and sea floor spreading.

Plate tectonics.   During the last twenty years, there has been a revolution in the thinking of earth scientists. Earlier, geologists believed that the motion of the crust was mainly vertical; that is, major geological features such as mountains were formed by large earth masses being pushed upward or downward by shifts in and below the 5- to 20-mile (8 km to 32 km) crust of the earth. They assumed that the crust was a continuous covering much like a hard, relatively thin shell. Now, however, scientists believe that the crust is not a continuous shell, but rather is composed of massive and separate crustal fragments called **plates**. The vertical movement that creates mountains and other major landforms is, by comparison, a small by-product of the *horizontal movements* of these world-girding plates. Plates form both continents and ocean floors. They push against, and slip over one another at their boundaries in regions of the world where there are concentrations of earthquakes. This newer theory is called **plate tectonics**. Figure 13–6 shows the boundaries of what are believed to be the earth's major plates.

Continental drift.   By studying changes in the sea floors, scientists are now substantiating a theory of **continental drift** first offered by Alfred Wegener, a German meteorologist, in 1912. He conjectured that the continents were once one large land mass that he named Pangaea (Greek for "all earth"). Over millions of years, they separated and moved around the globe. Figure 13–7 shows how the continents may have looked before they began to drift apart toward their present positions around the world.

Certainly, if you observe in this figure how easily the east coast of South America seems to fit with the west coast of Africa, you can see why this theory appeared reasonable. However, in the absence of supporting research, Wegener's work remained unaccepted. Recent studies involving measurements of the vertical movement of landforms of the crust and of the earth's magnetism lent credence to the theory. New technologies have enabled oceanographers and geologists to gather certain kinds of data from the ocean floors. Results of this research yielded further and persuasive evidence that moved most earth scientists to accept new ideas of how the earth's crust works and changes.

Accurate measurements of the vertical movements of different landforms show that the crust lowers or becomes depressed under a sudden, new load such as the filling in of a large lake behind a new dam. Analyses of rock strata show that the crust was lowered under massive glaciers, and further, rose again and is still rising as the remnants of these glaciers melt today. Layers of what we know were shal-

PLATES Segments of the earth's crust, varying in thickness and including part of the upper mantle above the asthenosphere

PLATE TECTONICS Theory of worldwide dynamics of the many rigid plates of the earth's crust; a branch of geology dealing with the broad architecture of the outer part of the earth

CONTINENTAL DRIFT Slow, horizontal movement of the continents; involves plates as vehicles of transport in the mantle

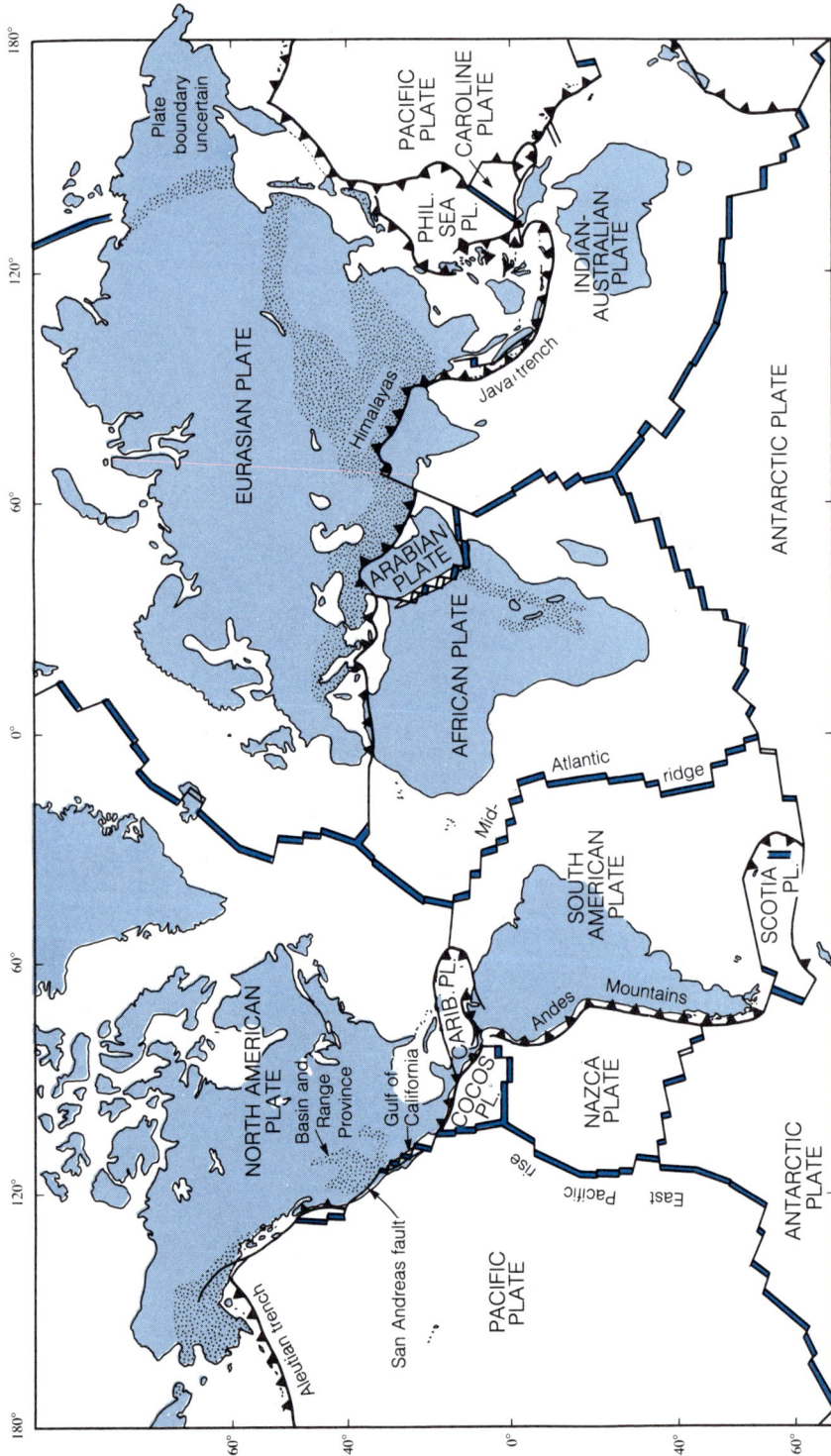

FIGURE 13-6

Simplified map of the lithospheric plates of the world. Plate boundaries are divided into spreading axes (double lines), transform faults (single lines), and convergent or subduction zones (barbed lines on overriding plate). Stippled areas within continents are regions of active deformation away from the plate boundary. (Modified slightly from W. Hamilton, 1977)

From *Physical Geology: The Structure and Processes of the Earth* by B. Clark Birchfiel et al. Copyright © 1982 by Bell & Howell Company. Used by permission of Charles E. Merrill Publishing Company.

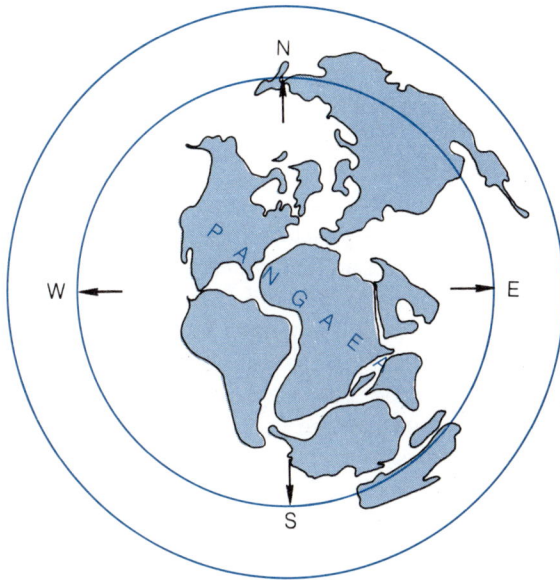

FIGURE 13–7
Continental reconstruction for about 200 million years ago showing one major continent, called Pangaea.

From *Physical Geology: The Structure and Processes of the Earth* by B. Clark Birchfiel et al. Copyright © 1982 by Bell & Howell Company. Used by permission of Charles E. Merrill Publishing Company.

low-water sediments are found deep in the crust; this is possible only if shallow basins did, in fact, sink into and below the crust. It seemed as though the stiff, crustal layers were floating in a soft, plastic base. And, indeed, geologists now think that the continental and oceanic crustal plates do float in the slow-flowing asthenosphere. The principle by which this occurs is called **isostasy**.

    Isostasy is a condition of balance and of floating equilibrium. Note the blocks of wood floating in water in figure 13–8. Each block sinks to a different level but the amount of each block above the water line is *proportionately* the same for each one. Each block shows about one-third of its length above water and this amount is balanced by the two-thirds of each block floating below surface. If another block of wood were added, the whole would submerge a little bit more. The earth's

ISOSTASY Ideal condition of balance attained by earth materials of differing densities

FIGURE 13–8
The floating blocks of wood illustrate how material of a single density (wood) maintains a constant level or buoyancy in the water. Segments of the earth's lithosphere vary in density and yet are thought to float in the asthenosphere in a balanced condition called isostasy.

From L.D. Leet, S. Judson, and M.E. Kauffman, *Physical Geology*, 5th ed., © 1978, p. 10. Reprinted by permission of Prentice-Hall, Inc., Englewood Cliffs, N. J.

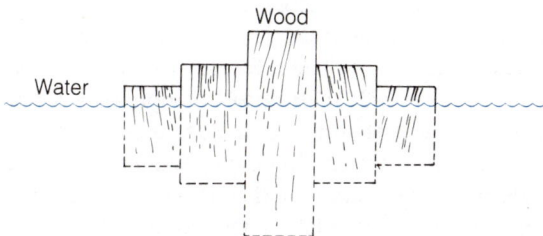

plates behave much like the wooden blocks; they rise and fall in response to increased or decreased loads. The dynamics of isostasy are also used to explain why mountains are not quickly worn down to flatlands by weathering or do not sink into the slow-flow layers of the mantle.

Sea floor spreading and the earth's magnetism. Although we still do not fully understand the magnetism of the earth, we know that a **magnetic field** exists within and around the earth because the earth behaves like a magnet. It has a magnetic field that allows compass dials to "point north." We also know that the magnetic field is not located at the north pole but that it moves around the position of the pole of rotation and varies in strength over time. The magnetic field forms around the earth's axis, but there is no particular reason for the magnetic pole to be at the northern end of the axis as it is today. Geophysicists have now determined that the poles actually have been reversed in the geologic past. Let us consider how this discovery is made.

MAGNETIC FIELD A region in which invisible forces cause some objects (those having positive or negative charges) to move toward or away from other objects

Rocks, or more properly the minerals in rocks, ordinarily become magnetized as they cool and crystallize. With recent strides in the study of rock magnetism, scientists have measured the polarity (the north or south magnetic orientation) of the magnetic field that is preserved in newly formed rocks. Through the use of improved methods for dating the time at which rocks became magnetized, scientists have demonstrated the geologic and relative time periods when the earth's magnetic poles have been reversed.

Applying the results of rock magnetism studies to examinations of rocks in the sea floor, geophysicists have determined that the sea floor continually yields new rock and that the crust is thereby always renewing itself. Magma emerges from the asthenosphere through ridges in the sea floor; it cools into rock and then spreads outward from the ridges and toward the continental plates and landmasses. This process poses a problem: if there is new crust being formed all the time, older crust must disappear somewhere else. Research findings indicate that the sea floor spreads between 2 to 3.5 inches (5 cm to 9 cm) per year. From year to year, these amounts do not seem very great; however after three or four billion years have elapsed, it is indeed a very sizeable distance to consider in our understanding of the earth.

The expanding sea floor gets rid of the excess crust by slipping under and against adjoining plates. Remember that the plates are separate rock masses lying with their edges or boundaries under or over a neighboring plate. Here at the boundaries, pressure is created by one plate slipping over and past another plate. This action produces heat that begins to melt the rock even before it slips into the more molten or slow-flowing mantle below. The plate edge melts, becomes metamorphic rock, and finally magma, and disappears. It is much like the way butter softens and melts when heated. Further, the process follows the rock cycle—a theory originally put forth in the late eighteenth century.

Sometimes the floor slips into a deep ocean trench that is really the point where plate edges rest on one another. Other times an ocean plate meets and slips under a continental plate.

Plate movement.   Sea floor spreading is only one of several ways that the plates may move and interact. Figure 13–9 suggests three of the ways that plates may interact with one another.

Earthquakes.   **Earthquakes** are ground vibrations of the earth. During an earthquake rock and soil are deformed in ways that tell geologists that earthquakes are an ancient geologic event. Because people have suffered from disasters resulting from earthquakes and their side effects, it is important to understand them. Although earthquake prediction in our country is not now possible, there has been a dramatic increase in the number of scientists who are studying earthquakes and earthquake prediction. In addition, during the last decade the Chinese have made great strides in this area. **Seismology** is the study of earthquakes and **seismologists** are the scientists who study them around the world.

Noticeable earthquake activities have varying effects. Most earthquakes are not felt by people but, in the case of stronger quakes, nearly everyone feels them. At such times, people are often frightened. On rare occasions, as during severe earthquakes, panic is widespread. Some quakes can be felt only indoors; other quakes can be felt indoors and outdoors. In mild quakes, a few dishes may rattle but not break. In

EARTHQUAKE A sudden motion or trembling in the earth caused by the abrupt release of slowly accumulated strain

SEISMOLOGY Scientific study of earthquakes and other earth vibrations

SEISMOLOGIST One who studies earthquakes and other earth vibrations

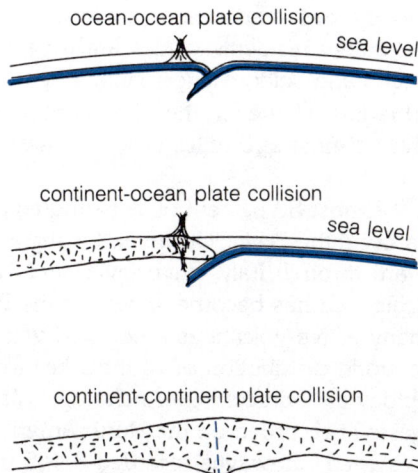

ocean-ocean plate collision

continent-ocean plate collision

continent-continent plate collision

FIGURE 13–9
Illustrated are three of the possible ways plates can interact with each other.

From *Physical Geology: The Structure and Processes of the Earth* by B. Clark Birchfiel et al. Copyright © 1982 by Bell & Howell Company. Used by permission of Charles E. Merrill Publishing Company.

stronger quakes, trees may shake, glass breaks, wall plaster falls, and chimneys are damaged. In severe quakes, one can see waves in the ground.

One such severe earthquake shook Yokohama and Tokyo, Japan on September 1, 1923. There were only 5 minutes of fast and short vibrations followed by slower, more violent motions. But both cities entered a 30-minute period of destruction and fire after which 99,333 persons were reported killed; 43,476 were missing; 103,733 were injured; and 576,262 houses were destroyed.

Side effects of earthquakes.   Although an earthquake can be frightening, often the side effects are more damaging. Fires, **seismic sea waves** and landslides bring the greatest destruction.

SEISMIC SEA WAVE
Large ocean wave generated by vibrations of an earthquake (sometimes mistakenly called a tidal wave)

Scientists estimate that up to 95 percent of the damage done by earthquakes is due to fires. Broken gas lines fuel the fires, and broken water pipes cause a loss of water with which to fight fires. Most fires rage out of control. Firestorms can mount that kill thousands. Damage to homes and structures is less when structures rest on bedrock rather than loose soil. Steel-frame buildings seem to ride through earthquakes when other types of structures around them are destroyed completely.

Seismic sea waves, often called by their Japanese name *tsunamis* (tsoo-NAH-meez), are another destructive effect of earthquakes. They are great ocean waves set up by the shock waves of an earthquake. At shorelines, some have been reported as high as 100 feet (33 m). Successive waves travel thousands of miles across oceans devastating islands in their path and bringing damage to distant continental coastlines. Fortunately, only a few of the thousands of sizeable earthquakes each year set tsunamis in motion and much effort has gone into seismic-sea-wave warning systems, especially in lands bordering the Pacific Ocean.

In mountainous areas, landslides or underwater slides are another threat to earthquake area residents. Slides are usually confined to soils sensitive to vibrations and seldom affect lands beyond a 25- to 30-mile (40 km to 48 km) radius. However, the 1964 earthquake in Alaska produced slides in areas almost 100 miles (160 km) away.

Earthquake belts.   Almost 80 percent of all earthquakes occur in known earthquake belts around the Pacific Ocean and along a zone that extends from Spain eastward through Italy, Yugoslavia, and Turkey into Southeast Asia. The Pacific area has become known as the Pacific Ring of Fire because of the many active volcanoes associated with the zone. Figure 13–10 shows the world distribution of earthquakes for a 9-year period. The 1964 Good Friday earthquake near Anchorage, Alaska (in the Pacific ring) is the strongest yet recorded on the North American continent. Killing 131 people, it started sea waves recorded 8,445 miles (13,621 km)

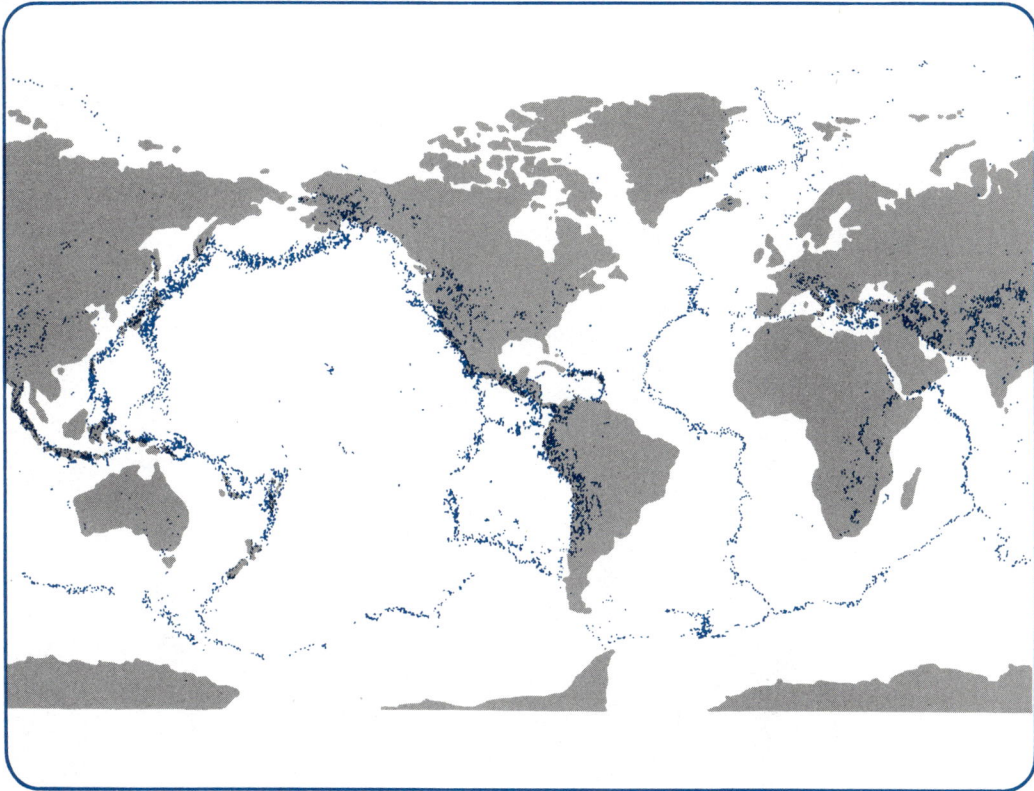

FIGURE 13–10

Earthquakes are distributed in patterns around the world. The heavy concentration of earthquakes around the rim of the Pacific Ocean is known as the Ring of Fire.

From *Earth Science*, 3rd ed., by Edward J. Tarbuck and Frederick K. Lutgens. Copyright © 1982 by Bell & Howell Company. Used by permission of Charles E. Merrill Publishing Company.

away in Antarctica. The ocean floor rose 50 feet (15 m), the Mississippi River rose and fell 5 feet (1.5 m), ships were wrenched from their moorings 3,200 miles (5,161 km) away along the Gulf of Mexico, and a strip of land 4,000 feet (1,219 m) by 400 feet (121 m) fell into the sea.

Faults.   Earthquakes occur along major fault zones. A **fault** is a rupture or large crack in the earth's crust. Pressures build up around the area of the fault or fault line and the already cracked crust slips, each side of the crack moving in a different direction. The slippage sets off an earthquake. The San Andreas fault, which extends along part of the California coastline, caused the disastrous San Francisco earthquake of 1906.

FAULT A rupture or crack in the earth's crust along which there has been differential stress

Flying over California, it is possible to see places where the San Andreas fault is clearly defined. Scientists now think this fault is evidence that the Pacific plate (left) is moving in a northwesterly direction past the North American plate (right). (R. E. Wallace, U.S. Geological Survey)

SEISMOGRAPH An instrument that detects, magnifies, and records vibrations of the earth; the resulting record is a seismograph

Measuring earthquakes. A **seismograph**, the instrument that seismologists and geophysicists use to study earthquakes, is really two instruments in one: a seismometer which responds to ground motions and earthquakes, and a seismogram which records ground motions over time. Scientists use different kinds of seismographs that will record either vertical or horizontal earth movements in identifiable wave patterns. The principle of the seismograph is shown in figure 13–11.

From news sources we have all heard of the Richter scale. It is a numerical identification of the strength or magnitude of an earthquake. It is based on the calculated total energy generated and is reported on a 10-point scale of increasing magnitude. Each point on the scale represents a tenfold change; a recording of 2 is 10 times larger than a recording of 1; a recording of 3 is 100 times larger than a recording of 1. The most intense earthquakes register about 8.6 on this scale. Surprisingly, there are more than 1,000 shocks or earthquakes in the world each day registering magnitudes of 2 or more on the Richter scale.

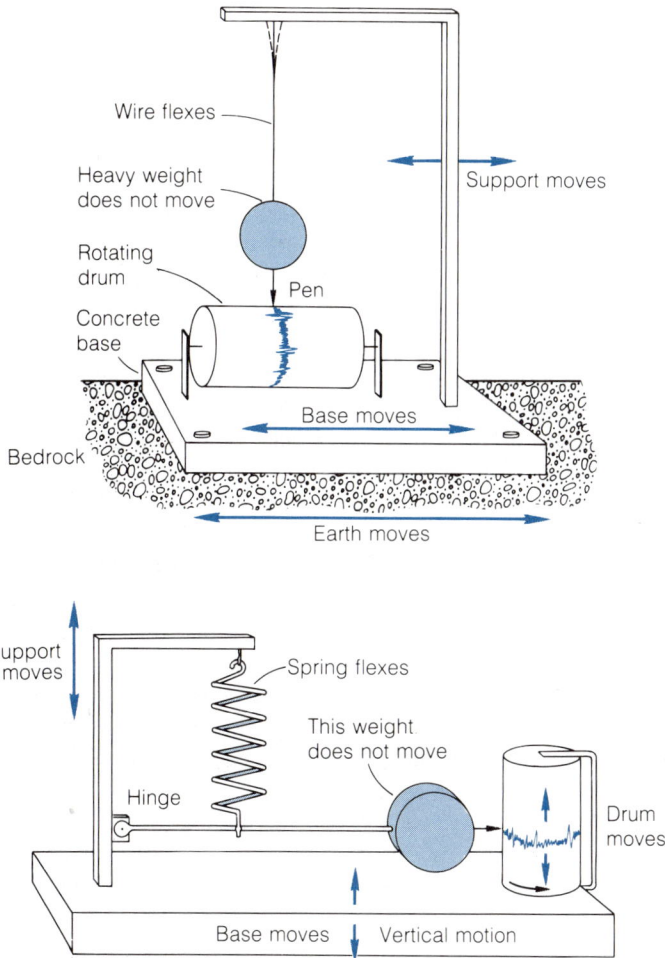

FIGURE 13–11
The seismograph on the left shows how horizontal motion is measured; the one on the right measures vertical motion.

Left, *Earth Science*, 3rd ed., by Edward J. Tarbuck and Frederick K. Lutgens. Copyright © 1982 by Bell & Howell Company. Used by permission of Charles E. Merrill Publishing Company. Right, *The Earth Sciences* by Arthur N. Strahler. Copyright © 1963 by Arthur N. Strahler. Used by permission of Harper & Row, Publishers.

Volcanoes. Earthquakes and volcanoes are found in similar zones or belts. In the light of plate tectonic theory, scientists see that both occurrences may have the same cause: deeper movements of the convergent boundaries of ocean and continental plates. Certainly the Pacific Ring of Fire, the earthquake belt so named because of the many volcanoes associated with it, tends to substantiate this interpretation.

Volcanoes are normally cone-shaped landforms formed by the accumulation of outpoured lava (lava, remember, is magma that has surfaced). Lava builds up around a central vent that extends down into the hot asthenosphere. A volcano "blows" or erupts because shifting plates

VOLCANO Landform that is built by the accumulation of lava around a central vent or fissure

and sea floor spreading create horizontal pressures on the soft, slow-flowing asthenosphere. Molten material is forced upward through the volcanic vent to relieve the building pressures below. Volcanoes are classified as either shield volcanoes, composite volcanoes, or combinations of these two types depending upon their formation. *Shield* volcanoes are built by lava pouring out of the central vent or from lateral vents or fissures on the volcano's slopes. *Composite volcanoes* are developed by lava outpourings as well but these volcanoes also yield *ejecta* such as ash, cinders, and larger bits of igneous lava called lapilli, blocks, bombs, and pumice.

Mount St. Helens.   Mount St. Helens is a composite volcano in a chain of young volcanic mountains that extends from Lassen Peak in northern California, north to Mt. Garibaldi in British Columbia, Canada. Mount St. Helens was known to have erupted in 1900 B.C. and in 1500 A.D. By 1980, it rose 9,677 feet (2,950 m) above Spirit Lake and the surrounding countryside in southwestern Washington State. This and other volcanoes in the Cascade Range were probably formed by the relatively small, converging plate called Juan de Fuca, moving beneath the American plate off the western coast some 50 miles (80 km) away. The pressure caused by the two plates sliding past one another and the lowering of the Juan de Fuca boundary into the lithospheric magma created pockets of melted rock or magma. At 8:32 A.M. on May 18, 1980, pressures were sufficient to force the molten material up 60 miles (97 km) through Mount St. Helens' central vent to erupt in a spectacular lateral and vertical blast. The eruption blew the top off the mountain, leaving a crater 1.2 miles wide by 2.4 miles long (1.9 km by 3.9 km) and devastating more than 150 miles of forest and streams, including 26 lakes.

Mountains.   Mountains sit atop thick layers of dense granite and basalt rock, all of which float in the slow-flow mantle, according to the principle of isostasy. Like an iceberg or the wooden blocks in figure 13–8, each mountain has a root that may represent most of the mountain's mass. There are three main types of mountains: volcanic mountains, folded mountains, and block mountains (see figure 13–12). They are classed according to the ways in which they are formed.

Volcanic mountains.   Volcanic mountains form by the deposition of lava flowing from the central vents of active volcanoes. Volcanic mountains may be active, dormant (sleeping), or extinct, depending on the length of time since volcanic activity has been recorded—and those records maintained for later generations.

A.

B.

C.

D.

This sequence of four photographs tells the story of Mt. St. Helens, the volcano that made history in the state of Washington. A series of mild earthquakes first warned geologists in mid-March 1980 (A) that the snow-capped mountain was restless. By late April a bulge developed on the north side and on May 18, 1981 at about 8:30 A.M. the volcano erupted (B) spilling lava and mud flows for miles around (C) and killing much vegetation. During the next three months, several additional eruptions occurred and clouds continued to hover (D) over the crater. (U.S. Geological Survey, A, B, and D by Austin Post)

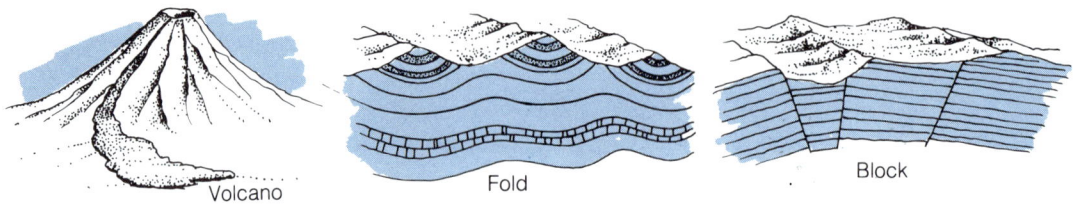

FIGURE 13–12

Illustrated are the three main types of mountains.

From *General Geology*, 4th ed., by Robert J. Foster. Copyright © 1983 by Bell & Howell Company. Used by permission of Charles E. Merrill Publishing Company.

Folded mountains.   Folded mountains are mountains that consist primarily of elevated, folded sedimentary rocks. When expanses of crust are exposed to extreme horizontal pressures, the crust buckles into rounded peaks that, from a distance, look like ocean waves. It is hard for us to imagine that the crust is elastic enough to form folded mountains but, over millions of years, this is exactly what happens. The Appalachian Mountains are a good example of folded mountains.

Block mountains.   Block mountains are layered expanses of crust that, when under horizontal pressures, break along the fault lines that are already present in their structures. Some peaks rise upward; adjoining, faulted sections fall or tilt in more irregular, rectangular shapes. The Rocky Mountains are block mountains.

Weathering and Erosion.   Rocks formed within the earth change when they are exposed to water, air, and living matter on the earth's surface. They change by **weathering** and **erosion**. In general, these processes break down what preceding processes have built up and changed. If there were no weathering and erosion, we would have no soil in which to grow food or to support vegetation and life on the earth. We would have no fossil fuels with which to power our civilization. Further, weathering and erosion are the processes that transform the land we see everywhere. They shape the **topography** of the world.

WEATHERING The process by which landforms and rock materials are decomposed

EROSION The general process or group of processes whereby the materials of the earth's crust are loosened, dissolved, or worn away, and simultaneously, moved from one place to another

TOPOGRAPHY The general shape and feature of land

Weathering.   Rocks weather in two ways: chemical weathering and physical or mechanical weathering. *Chemical weathering* is the *decomposition* of rock into new materials by chemical processes. The mineral feldspar, for example, decomposes into clay; and rocks containing iron oxide simply rust. *Physical weathering* is the *disintegration* of rock. Rock is broken down in many ways. Some of these are exposure to frost, heat, and organic activity. Water expands as it forms ice or frost; when water seeps into cracks in rocks and freezes, the cracks are enlarged and the rocks shatter. Heat expands rock; when rock expands during the day's heat and then contracts at night, layers of rock are loosened. Living organisms such as tree roots, burrowing animals, and earthworms break up loose rock and soil as well.

Erosion.   Erosion is the movement of material from one place to another. The agents or vehicles for this movement are water, ice, wind, and gravity. The transportation of material happens in four ways: picking it up is called *entrainment* or local erosion; *transportation* is just moving it along. *Deposition* is the depositing of materials in a new place. The familiar type of erosion that involves downward movement of slope material is called *mass movement* or *mass wasting*.

Soil.   When rock has been broken down by weathering and moved by erosion into layers on the surface, it becomes **soil**. Here it is modified by physical, chemical, and biological processes so that it will support plant growth. Although weathered materials are basic to the process, soil formation involves other factors as well: climate, the kinds of organic or living components in it, the type of landform on which it rests, the types of rocks from which it came (parent material), and the length of time involved in soil formation. Groundwater makes an important contribution to soil formation.

SOIL Material that forms on the earth's surface as a result of organic and inorganic processes

Groundwater.   One erosional process, **leaching**, goes on by the action of groundwater. Water in the ground comes from rain, ponds, streams, and the like. It sinks in and seeps down into the earth, seeking the path of least resistance, until it reaches the bedrock. Here it saturates the bedrock. The upper surface of the **zone of saturation** is called the water table (see figure 13–13). The ground above the water table and, generally, that land found up to the surface, is called the **zone of aeration**. It is here that groundwater leaches or breaks down the rock particles, playing a critical role in soil formation. This is also the process by which caves are formed, especially in limestone deposits.

LEACHING The filtering of water through soil that causes weathering and furthers soil formation

ZONE OF SATURATION Underground region within which all openings fill with water

ZONE OF AERATION Zone immediately below the earth's surface in which openings are filled with both air and water; area of intense weathering and soil formation

Running water.   Running water, like groundwater, is an erosional and weathering force. It creates the preponderance of landforms including streams, valleys and canyons, floodplains, and lakes. These forms result from the erosion processes of entrainment or local erosion, transportation, and deposition.

Running water erodes rock material by picking up soil and other particles along the sides and bottoms of river beds. Turbulent waters lift particles into the flow; these moving particles then help the stream to dislodge other particles by abrasion. Water transports these rock materi-

Gully erosion in a newly plowed field. (USDA—Soil Conservation Service)

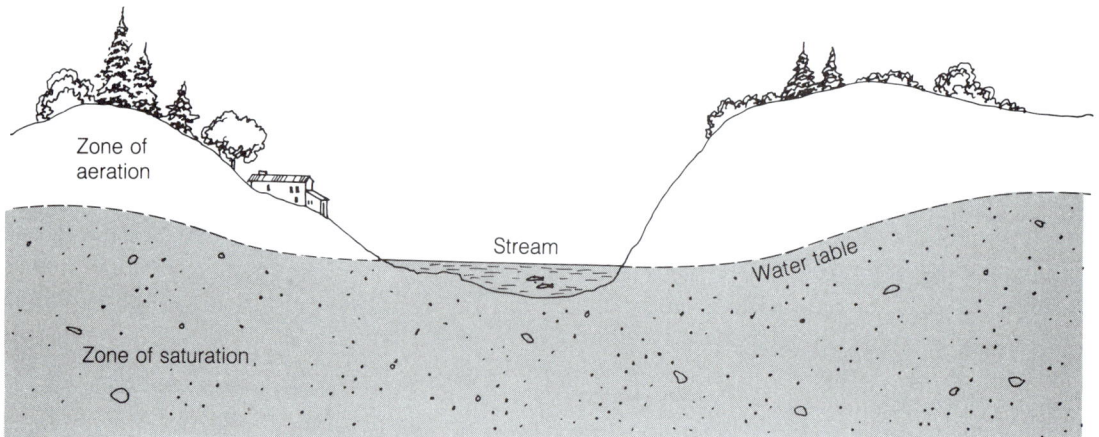

FIGURE 13–13

The water in the zone of saturation is called groundwater; the upper limit of this zone is known as the water table.

als by dissolving them, by carrying them along, and by pushing or rolling heavier bits along the bottom of riverbeds.

In this way, small streams grow to become faster-flowing rivers. Erosion continues as the river forms channels; channel walls erode (by mass movement and mass wasting) and valleys are formed. Rivers overflow eroding valley walls and (by deposition) create floodplains. In over ten million years of erosion, one river, the Colorado, carved the monumental Grand Canyon in Arizona.

Running water deposits or redistributes nutrient-rich materials beneficial to crops and other plant growth. Such deposition creates landforms such as **river deltas** and **oxbow lakes**. And deposition brings rock particles to the sea floors where sediments will become layered sedimentary rock.

RIVER DELTA An area, often triangular-shaped, where a river stream flows into a larger body of water and deposits sediment at its mouth

Glaciers. Glaciers are great rivers of ice that move very slowly. They advance and recede in cycles lasting tens of thousands of years. Yet they are a most significant agent of weathering and erosion. Landforms of over half of the North American continent have been shaped by glaciers.

OXBOW LAKE A lake formed by the meander or curving of a river which then gets cut off from the main flow

A glacier is a large mass of a form of ice called **glacier ice**. Glacier ice is classified as a metamorphic rock because it is formed by a variety of crystals of frozen water vapor—first into granular ice that is like the big snowbanks which outlast the rest of winter as spring approaches, and then into glacier ice, a true solid composed of interlocking crystals. Once formed, glaciers last for a very long time.

GLACIER ICE A unique form of ice made by compression and recrystallization of snow

Glaciers move an average of an inch or so a day; and their movement is often accompanied by creaking, crackling noises as fissures form in areas of greatest stress. Each year thousands of tourists visit the Columbia Glacier in Alaska to watch this movement. They are spellbound at the earshattering sounds when "calves" break off the glacier and plunge into the ocean, causing ocean waves of such height that the touring boats remain over a mile away to avoid damage.

Glaciers now cover about 10 percent of the land surface of the earth and are found on all the major continents: North and South America, Europe and Asia, Africa, and Antarctica. They are not confined to polar areas: Mount Kenya in East Africa, about 70 miles (113 km) from the equator (but 16,700 feet or 5,100 meters above sea level) supports 10 or more glaciers.

Glaciers transform the surface of the earth by erosion and by deposition—that is, by carrying material from one place to another, and by leaving deposits along their paths. During the peak of the last glacial epoch, some 20,000 years ago, about one-fourth of the earth's crust was covered with glaciers, which, in North America, extended southward to where New Jersey, St. Louis, and Arizona are located now. Their comings and goings sculpted and modified many of the major features

This photo of a glacier in the Olympic Peninsula of Washington State, with its rows of crevasses, shows the far-from-smooth nature of the glacier's surface. It is easy to see how glaciers gouge out valleys and scour rock surfaces beneath them. (U.S. Geological Survey)

of the earth we know now, including the Great Lakes and all but a few of the lesser lakes in the midwest, and all the surrounding landforms.

So far, our study of the earth has focused on the composition, structure, and processes of the earth. The rock cycle (figure 13–3) has been said to represent a capsule view of geologic identities and processes; it shows how they work continually together to form and change the earth.

Still, the world we live in is one of land and air. We experience warmth, cold, and different kinds of weather. Air, or the atmosphere, also has a composition and structure and it, too, interacts with the earth in cycles. Through the study of meteorology, the next section, we understand that part of our world.

And beyond the atmosphere, we can see our sun, moon, and other celestial bodies. Astronomy and space studies have shown that the moon, other planets, our sun, and the stars are all composed of matter found or understood on earth. We are learning about the earth in the larger context of space. Therefore, a section on astronomy follows the one on meteorology, for it too is a part of the earth and planetary sciences.

METEOROLOGY AND THE EARTH'S ATMOSPHERE

Beyond the crust of the earth there is another layer that is significant to our study of earth and planetary sciences. It is the atmospheric envelope that surrounds the earth. The **atmosphere** dominates our lives and that of all creatures on the land just as water dominates the lives of fish. Without the atmosphere's oxygen we would die. Without the carbon dioxide in the atmosphere, most plants could not produce carbohydrates, our primary food. Further, the carbon dioxide, moisture, and particles keep most of the sun's heat in our atmosphere. Without the thin layer of ozone, a rare form of oxygen that is produced by ultraviolet rays 25 to 30 miles (40 to 48 km) above us, the ultraviolet rays would reach the earth's surface and destroy life on earth.

Weather as we know it occurs in the atmosphere. Without *weathering* and weather's contribution to erosion, there would be no soil for plants.

ATMOSPHERE A mixture of invisible gases, dust particles, and water vapor; generally seen as consisting of five layers: troposphere, stratosphere, mesosphere, thermosphere, and exosphere, according to how temperatures change with altitude

WEATHER The general, day-to-day condition of the atmosphere in a particular place and time, involving temperature, moisture, clouds, wind, and so forth

Composition of the Atmosphere

We live at the bottom of an ocean of air called the atmosphere. This ocean extends from the surface of the earth out thousands of miles until it thins into outer space.

Air is a mixture of invisible gases, moisture, and tiny particles of matter. The gases are mainly: nitrogen (78 percent), a generally inert element; oxygen (21 percent), a generally reactive element that is capable of combining with most other elements; argon (1 percent), another inert element; carbon dioxide (0.03 percent), a gaseous compound that is important for plants and for the maintenance of the heat balance on earth; and traces of several other gaseous elements and compounds.

Layers of the Atmosphere

Moving outward from the earth's surface, the earth's atmosphere is now usually subdivided into five layers: the troposphere, stratosphere, mesosphere, thermosphere, and exosphere. (See figure 13–14.)

The Troposphere.   The first region, or *troposphere*, includes about three-quarters of the bulk of the whole envelope of atmosphere that surrounds the earth. It includes the haze, smog, dust, clouds, and the storms. At its outermost limit there are hot and cold air exchanges at

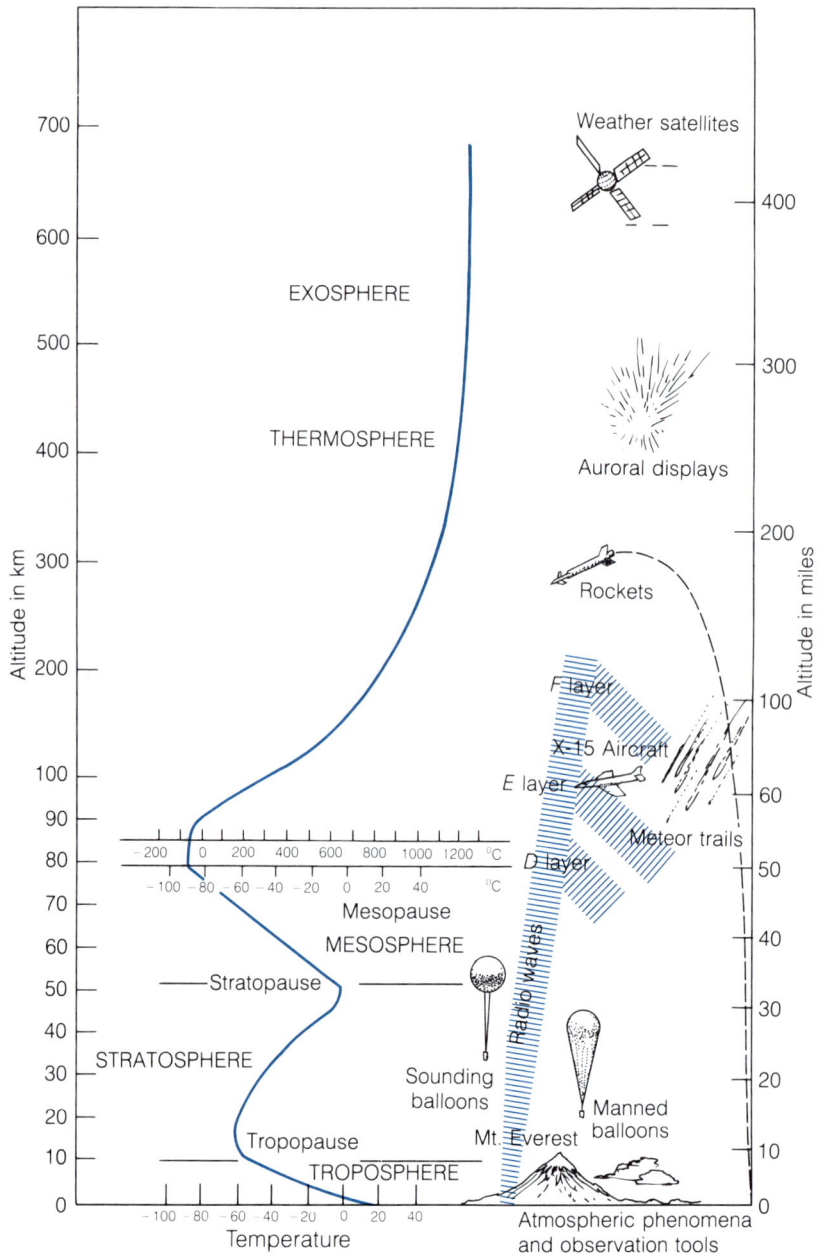

FIGURE 13–14
The earth's atmosphere consists of five layers, so divided to reflect temperature changes with increased altitude.

From *Meteorology*, 4th ed., by Albert Miller and Richard A. Anthes. Copyright © 1980 by Bell and Howell Company. Used by permission of Charles E. Merrill Publishing Company.

altitudes as low as 5 miles above the surface near the poles or as high as 10 miles above the surface near the equator. It is in this zone that the air temperature drops at the rate of about 3.5° F at every 1,000 feet of elevation.

The Stratosphere.

Beyond the 5- to 10-mile (8- to 16-km) layer of troposphere is a layer another 20 miles (32 km) thick called the *stratosphere*. It includes almost all the remainder of the mass of the atmosphere—only 1 percent lies beyond it. Before space travel, it was the outer limit of human travel. The **ozone layer** is here; it is a band of rare gas that at ground-level pressure would be less than one-eighth of an inch thick and it screens out most of the **ultraviolet rays** that would otherwise destroy us. As of the early 1980s, evidence indicated that fluorocarbons used in most spray cans and refrigerators might well be breaking down this protective layer; this would significantly increase skin cancer. Some corrective measures have been taken.

The Mesosphere.

The layer beyond the stratosphere is called the *mesosphere* (middle sphere), and extends another 20 miles out. Until recently it was customary to subdivide the atmosphere into three layers: the troposphere, the stratosphere, and the ionosphere. The mesosphere would be the lower portion of what we used to call the ionosphere. The atmosphere is now subdivided according to how temperatures change with increased altitude: Temperature *decreases* with altitude in the troposphere, closest to the earth. But in the stratosphere, temperature *increases* with altitude (as does the speed of sound, oddly enough), from about 70° F *below* zero at the top of the troposphere to about freezing at the top of the stratosphere. As we rise in the mesosphere, however, the temperature falls again, this time to a very chilly 130° F below zero. Fortunately there is no wind chill factor because there is no wind.

Beyond the mesosphere there is again some increase in temperature. It rises again but overshoots our comfort zone, varying from 1,100° F to 3,600° F, some 350 miles (565 km) above sea level (and 300 miles or 484 km beyond the mesosphere). It is within this layer that some of the more beautiful celestial events occur: white hot meteors burn up as "shooting stars" when they enter our atmosphere; the **aurora borealis**, or *northern lights* and *southern lights,* originate here. We used to believe these spectacular displays were reflections of the sun from the earth's polar icecaps. Now we understand they are caused by the glowing of **molecules** and **atoms** in the atmosphere as they are bombarded by electrically-charged particles from the sun. These particles make up what is called the "**solar wind**."

OZONE LAYER Thin layer of a rare form of oxygen in the stratosphere that is formed by solar ultraviolet radiation; decreases the amount of ultraviolet rays that reach the earth

ULTRAVIOLET (UV) RAYS Radiation having wavelengths between visible light and x-rays; in excessive amounts, UV rays disrupt terrestrial life

AURORA BOREALIS Luminous streamers, arches, and other patterns, consisting of glowing molecules and atoms in the mesosphere

MOLECULES The smallest unit of a pure substance (an element or compound) that can exist and still retain the physical and chemical properties of the substance. Examples include oxygen (O_2) and water (H_2O)

ATOM A building block of matter; a combination of protons, neutrons, and electrons

SOLAR WIND The continuous ejection of **plasma** from the sun's surface into and through interplanetary space

PLASMA Charged particles containing about equal numbers of positive ions and electrons; like a gas, but is a good conductor of electricity and is affected by a magnetic field, including the earth's magnetic field

The Thermosphere.   The *thermosphere* is the upper portion of what used to be called the *ionosphere;* it is the electrically-charged sphere that reflects radio waves back to the earth. It makes possible short-wave radio communication between stations on earth that would otherwise be too distant to receive each other's communications. (Short wave radio waves are restricted by the curvature of the earth; longer radio waves are not.) But when the sun's surface activity increases the solar wind into a solar gale of electrically-charged particles, short wave radio communication is often disrupted.

The Exosphere.   Three hundred-fifty (565 km) miles beyond the top of the thermosphere we reach what is now called the *exosphere* (outer sphere), assumed to extend some 3,000 to 10,000 miles (4,839 to 16,129 km) before it tapers off into the emptiness of space.

The Atmosphere as a Heat Engine

The earth's atmosphere is never at rest. It is driven constantly, mainly by (1) the radiant energy of the sun—a capriciously varying energy source itself; (2) the temperatures and emissions of the earth; and (3) the rotation and texture of the earth. So we can think of the atmosphere as a sun-powered mass, with a minor input from the hot core of the earth and a mechanical assist from the spinning of our planet. The atmosphere moves in *generally* predictable patterns about our globe.

Weather

Meteorology is the study of weather. As such it is the science of our atmosphere and the phenomena in it: **clouds,** rain, snow, tornadoes, hurricanes, temperatures, pressure, hail, and that elusive quality we call climate. Meteorologists also try to predict the weather and have developed some techniques for altering patterns of **precipitation.**

All the weather most of us have any interest in occurs in the troposphere, the 5 to 10 mile (8- to 16-km) layer of atmosphere next to the earth. This layer is driven by the heat of the sun and the temperatures, gyrations, and emissions of the earth. The atmosphere's weather can be understood to be the result of *masses* of warm and cold air, moving in this layer over the surface of the earth, primarily from west to east in the temperate latitudes. Any distinguishable air mass is called a **front**. Relatively warm masses are called *warm fronts;* cooler ones are called *cold fronts.* When indistinguishable fronts of equal mass meet, we call the result a stationary front.

Winds.   The earth is not uniformly warmed everywhere by the sun's rays. Sunlight arrives on the earth nearly vertically in the equatorial zone

CLOUD Visible mass of particles of water or ice in the atmosphere

PRECIPITATION Falling rain, snow, sleet, and other moisture

FRONT Boundary between air masses that differ in temperature, pressure, or moisture

but at an angle in most areas of the globe. On June 21 at noon, the sun's rays come in at a vertical angle only as far north as about Key West, Florida. Thus, the air near the equator is much warmer than the air near the poles. It is this difference in temperature that makes our atmosphere act like a heat engine and causes winds to blow.

If it were not for the earth's rotation, surface winds would thus blow directly north or south—but it is not quite that simple. The earth's rotation causes a force that deflects a moving body to the right of the predicted path in the Northern Hemisphere and to the left of the predicted path in the Southern Hemisphere. This deflection has been called the **Coriolis effect** after the nineteenth century French mathematician,

CORIOLIS EFFECT The tendency of any moving object, on or starting from the surface of earth, to continue in the direction in which the earth's rotation propels it

FIGURE 13–15
The earth's rotation causes a deflection called the Coriolis effect.

From *Earth Science*, 3rd ed., by Edward J. Tarbuck and Frederick K. Lutgens. Copyright © 1982 by Bell & Howell Company. Used by permission of Charles E. Merrill Publishing Company.

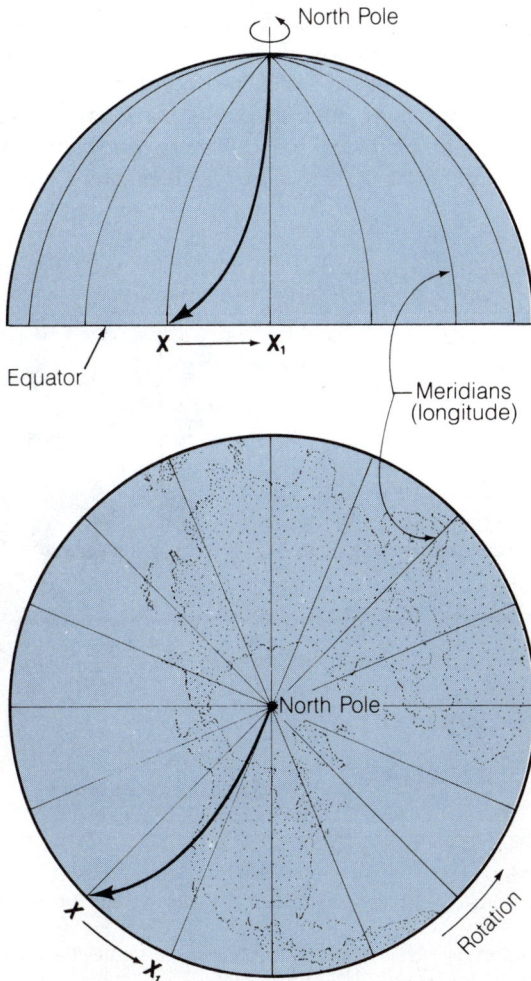

Gaspard G. de Coriolis, who, in 1835, was the first person to study it experimentally. This effect is responsible for the counter-clockwise motion in the northern hemisphere (and the clockwise motion in the southern hemisphere) of water going down a drain. Coriolis demonstrated that this deflection is produced by the acceleration of any body (including the atmosphere) moving at a constant speed above the earth with respect to the surface of the rotating earth (see figure 13–15).

Clouds. Clouds are forms of moisture in the air. There are two types of clouds: those formed by upward currents of air that rise until the moisture in the air reaches its condensation point and is seen as lumpy or billowy cumulus clouds; and those that develop horizontally and lie in sheets or formless layers like fog. These last are called stratus clouds.

Clouds near the earth are called either cumulus or stratus unless they are producing precipitation. If they are producing rain, snow, or other precipitation, they are called some combined form of the word *nimbo*, (which means rain), such as cumulonimbus or nimbostratus. If clouds are ragged and broken, we call them *fracto*, meaning broken. And if they are high, we call them *alto*, which means high. Similarly, we

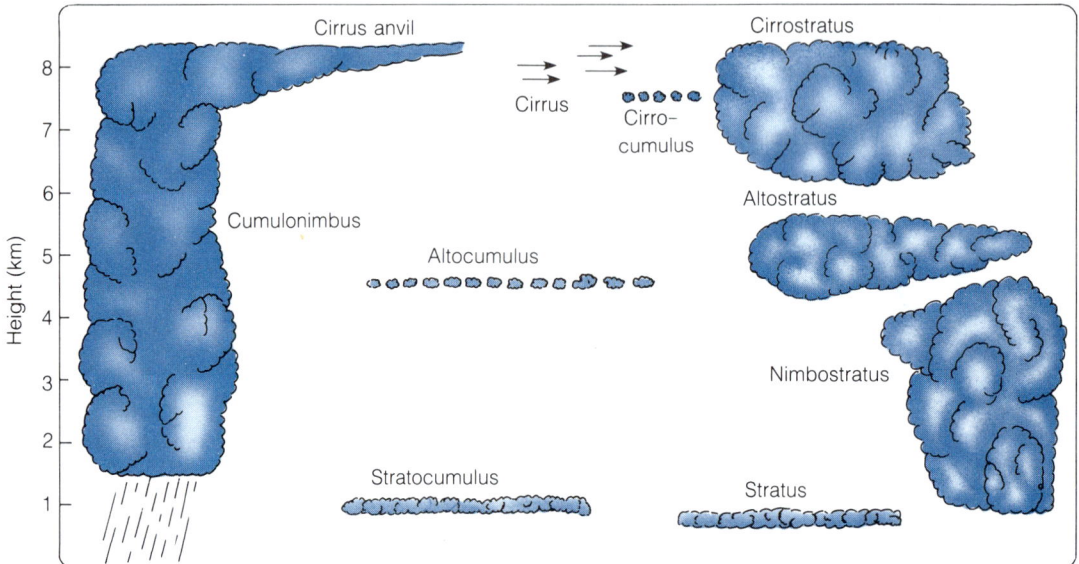

FIGURE 13–16
Clouds are given names which describe their shape, altitude, and whether or not they are producing precipitation.

The Atmosphere, 3rd ed., by Richard A. Anthes et al. Copyright © 1981 by Bell and Howell Company. Used by permission of Charles E. Merrill Publishing Company.

call curly-looking clouds *cirro,* which means curly, as in cirrocumulus or cirrostratus. Finally, there is at high altitudes in the troposphere a fibrous type of cloud that appears as curly wisps that we call *cirrus.* Figure 13–16 illustrates cloud types and typical elevations.

Fronts. Our atmosphere moves over us from west to east, at about 600 miles (968 km) per day. Cold air masses move somewhat more rapidly than warm air masses. Different air masses, having had different experiences as they travel over the surface of the earth, have different temperatures. When two different air masses meet, they do not ordinarily mix, unless their temperatures, pressures, and relative humidities are very similar. Instead, they pass below or above each other, causing rapid changes in the weather. They may confront each other in a sort of deadlock in which the weather does not change. In each case the moving

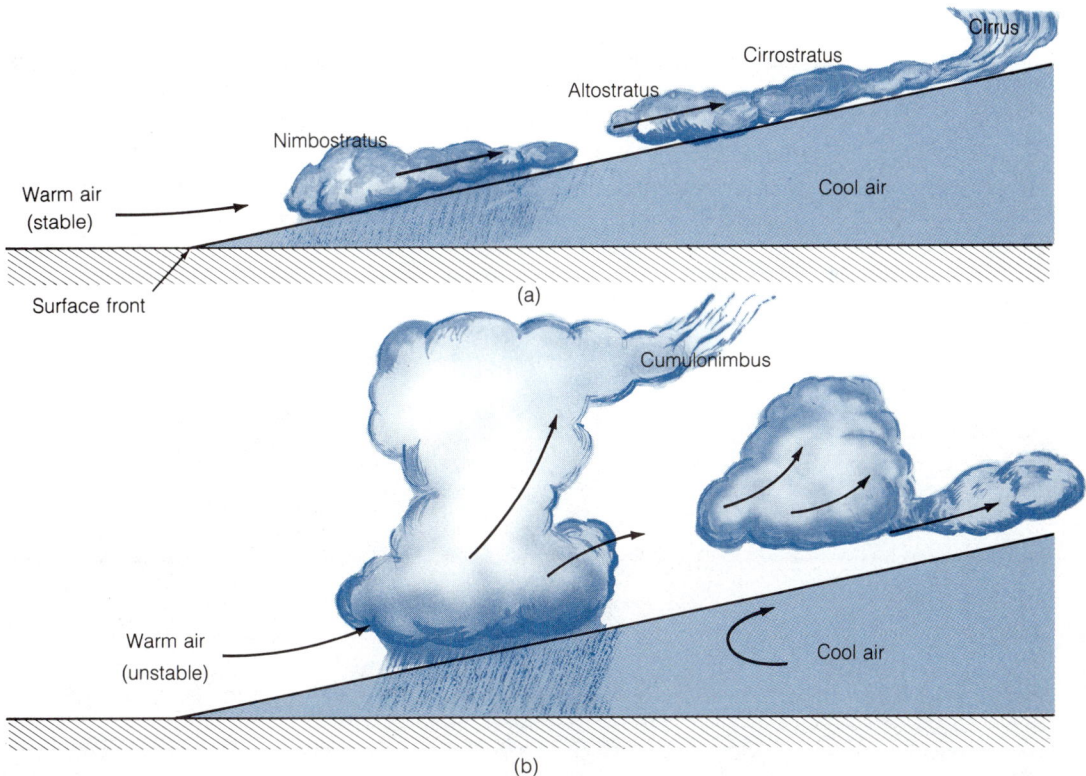

(a)

(b)

FIGURE 13–17
(a) Precipitation is moderate when a warm front with stable air slides over cooler air. (b) Precipitation is heavy when a warm front with unstable air overtakes cooler air.

From *Earth Science,* 3rd ed., by Edward J. Tarbuck and Frederick K. Lutgens. Copyright © 1982 by Bell & Howell Company. Used by permission of Charles E. Merrill Publishing Company.

boundary between the air masses is called a front, and the front is called a cold front, a warm front, or a stationary front according to the temperature of the air mass that is doing the displacing.

When a warm front moves forward it slides over the wedge of cold air that lies ahead of it, causing the cloud formations and precipitation shown in figure 13–17. Warm air usually has high humidity, and as it is lifted over the cold air it confronts, its temperature lowers. Low nimbostratus and stratus clouds form, and drizzle and rain develop. As the warm air progresses up the slope, the temperature decreases and moisture condenses as altostratus and cirrostratus clouds. As the air approaches the stratosphere, cirrus clouds may appear. If the air is unstable, a warm front may produce thunderstorms from the familiar "thunderheads" or cumulonimbus clouds.

When a cold front moves forward it slides under the warm air in its path. Warm air rises. This causes the warm air to rise and suddenly become cooler, forming the kinds of clouds indicated in figure 13–18. Usually a cold front moves faster than a warm front, and thus changes in the weather. If the warm air it displaces is stable, we can expect an overcast sky and rain. If the warm air is unstable, there will be scattered thunderstorms and showers. In some cases an almost continuous line of thunderstorms—called a squall line—may appear at the leading edge of the cold front.

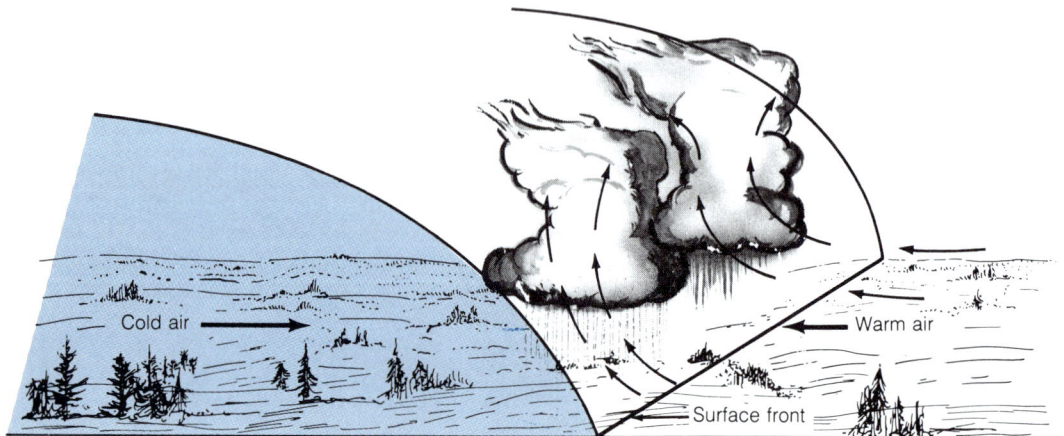

FIGURE 13–18
A fast-moving cold front meeting with unstable warm air usually produces thunderstorms.

From *Earth Science*, 3rd ed., by Edward J. Tarbuck and Frederick K. Lutgens. Copyright © 1982 by Bell & Howell Company. Used by permission of Charles E. Merrill Publishing Company.

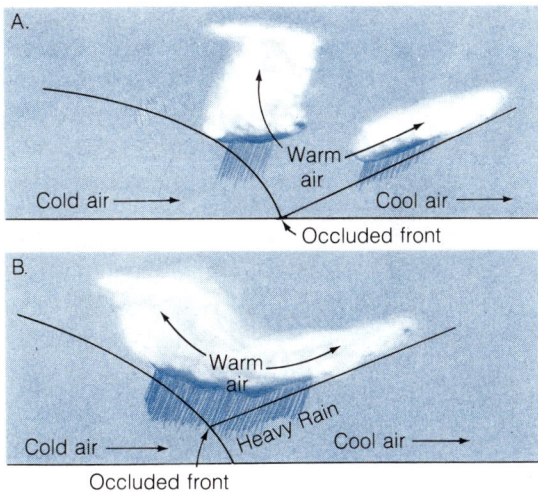

FIGURE 13-19
(a) As advancing cold air wedges the warm front upward, a new front emerges between the advancing cold air and the air over which the warm front is gliding. (b) Most of the precipitation comes from the warm air being forced upward, however the newly formed front can produce precipitation of its own.

From *Earth Science,* 3rd ed., by Edward J. Tarbuck and Frederick K. Lutgens. Copyright © 1982 by Bell & Howell Company. Used by permission of Charles E. Merrill Publishing Company.

Thirdly, there are *occluded fronts,* wherein an air mass is trapped between two colder air masses and lifted aloft, raining or snowing as it goes (as shown in figure 13-19) until it loses its identity.

Rain. Even when air has been **supersaturated** with moisture, it does not produce clouds unless there are many tiny **condensation nuclei** present to be the centers of the droplets of water that form clouds.

These nuclei may be salt particles from the sea or fine dust or smoke particles from forest fires or factories. They could be the products of volcanoes or water-attracting oxides left from lightning. Knowing that nuclei in the atmosphere are associated with rain, meteorologists have attempted to distribute these condensation nuclei so that the moisture in the air may collect around them and enlarge them into bigger elements. This causes them to fall as precipitation. In short, they have tried to be rainmakers.

In very turbulent air, the bigger elements collide and collect the smaller ones. In colder air, the cloud elements evaporate and become condensed on nearby ice crystals. Only when a cloud element grows to a raindrop size, that is, at least 1/125th of an inch in diameter, can it fall out of the cloud. But even if the cloud element (raindrop) falls out of the cloud it may not reach the ground. Many times a torrent may spill out of clouds high above a desert only to evaporate entirely before it reaches the ground. Raindrops that reach our earth in a fine spray or drizzle have fallen from low clouds with little time to collide with other droplets while falling. Raindrops that come as a vigorous downpour are coming from

SUPERSATURATED AIR An unstable condition in which air contains excess water vapor that would normally be precipitated at that temperature and pressure

CONDENSATION NUCLEI Airborne particles that cause moisture to form droplets around the particles, resulting in precipitation

deep clouds where the colliding of cloud elements and the capturing of little ones by bigger ones has been going on very actively.

Snow.   Snow is enjoyable to some people. For snow to form a cloud it must be a few degrees above or below zero Celsius (or 32° F). The cloud droplets then freeze together into crystals. Because these crystals carry a thin film of unfrozen water, they become snowflakes when they collide. In extremely cold weather, these crystals are dry and fall as granular snow. Rain that starts in warm air and falls through cold air turns into ice pellets, or sleet. Frozen raindrops from high clouds moving through a thunderstorm and hurled along with violent updrafts may pick up layers of snow and ice. Finally, they fall to earth as hailstones.

Tornadoes and Hurricanes.   Because they are particularly destructive, meteorologists are especially concerned with the violent winds of tornadoes and hurricanes.

TORNADO A destructive, whirling wind that moves in a narrow path over the earth and may involve internal wind speeds of 250 miles per hour

We call the most violent storms we know **tornadoes**. They are relatively small—about one-fifth of a mile in diameter and they usually travel less than 20 miles (32 km) before they dissipate. Tornadoes consist of rotating, funnel-shaped swirls of atmosphere that may involve winds of more than 250 miles per hour. The center of a tornado is a column of air that has a very low pressure, which means that structures and debris around it are propelled into it by their normal atmospheric pressure. Tornadoes, aside from the speed of winds within them, move at speeds of 25 to 40 per hour. Thus when the funnel-shaped spout of a tornado is sighted, residents of areas in the projected path of the tornado are immediately warned to take cover and have relatively little time to act.

HURRICANE Tropical wind and storm system that moves in a wide circle around a low pressure area; winds in a hurricane vary from 74–136 miles per hour

Hurricanes are not nearly so destructive as tornadoes, but cover much larger areas. During a hurricane rain falls heavily, and winds may exceed 100 miles per hour. Whereas a tornado devastates "only" about 3 square miles, a hurricane will affect an area 200 to 400 miles (323 to 645 km) in diameter, or 30,000 to 125,000 square miles (77,700 to 323,750 square km). The destructiveness of a hurricane is normally due to the great ocean waves and high tides it causes.

Perpetual Motion: The Hydrologic Cycle

VAPOR Visible particles of moisture floating in air, as fog, mist, or steam

EVAPORATION The process by which a liquid or solid becomes a vapor

Water in circulation on earth and in the atmosphere constitutes the hydrologic cycle. Water leaves the atmosphere as precipitation—largely rain and snow. Most of this precipitation falls on the oceans, and most of the water **vapor** that returns to the atmosphere does so through **evaporation** from the oceans. As shown in figure 13–20, roughly one-fourth as much precipitation falls on the land masses of the earth as falls on the oceans,

A tornado sweeping across the countryside may have winds swirling at up to 250 miles per hour in its funnel-shaped spout. Fortunately, tornadoes dissipate after 25 to 40 miles, but they can do a great deal of damage during the time they are active. (Joseph H. Golden, National Severe Storms Laboratory)

and water is returned from land to atmosphere by evaporation and **transpiration** (the process by which water vapor escapes from living plants into the atmosphere). The transfer of water from land to ocean is relatively small (5 percent) compared with direct transfers between the atmosphere and either the land or the oceans, and the ocean gives up more moisture than it receives. This is balanced by the land receiving more moisture than it gives. In effect, the ocean provides us with about one-third of our rainfall and other precipitation.

The water that moves between land, sea, and air is a tiny fraction of the water on earth. Nearly 95 percent is held in the lithosphere—in the animals, plants, and minerals in the earth's crust. Another 5 percent is held in the oceans.

The water in lakes, streams, groundwater, ice caps, and atmosphere combined is about 1 percent of 1 percent. Fresh water, upon which all human life depends for drinking water and all the rest of our

TRANSPIRATION The process whereby plants give off vapor

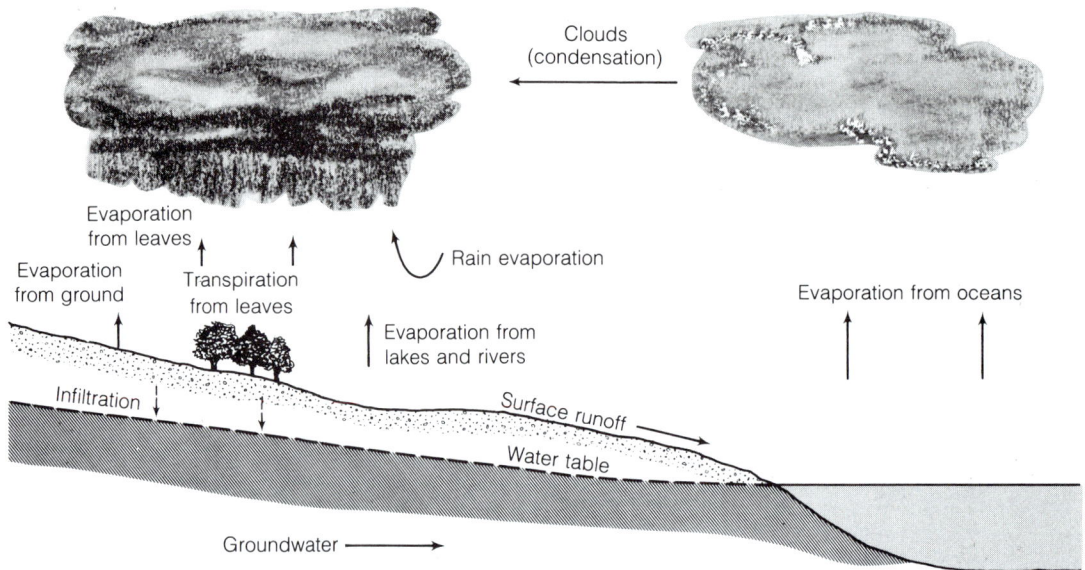

FIGURE 13–20

Some of the water falling on land seeps into the ground to replenish groundwater, and some runs off to rivers, lakes, and oceans. Evaporation from the surfaces of leaves, ground, and water returns water vapor to the atmosphere, where it condenses to fall as rain or snow.

From *Environmental Science*, 2nd ed., by P. Walton Purdom and Stanley H. Anderson. Copyright © 1983 by Bell and Howell Company. Used by permission of Charles E. Merrill Publishing Company.

water needs, is only one part in 100,000 of the water on earth; yet it is that one part that receives most of the fury of our industrial, agricultural, and urban pollution!

ASTRONOMY

Astronomy is the study of the celestial bodies in space: stars, our moon and sun, planets, comets, meteors, galaxies, constellations, and our earth itself, which is a planet of the sun. The universe is the total of all these objects in space. Astronomers study the movements of these objects, their distances from the earth and from one another, and their composition; and they try to determine the past and future of these celestial bodies. Although astronomers use telescopes and many other special instruments in their work, there is much of the universe that can be seen and understood with the unaided eye.

During the day the most obvious heavenly bodies are the sun, the center of our solar system, and often our moon, which *seems* to follow the sun around the earth each day, but falls further and further behind until the sun overtakes it, or laps it, about every 28 days. Just at sunset we can sometimes see the planet Venus—called the "evening star"—low in the west. At certain times of the year we may also see the planet Mercury low in the western sky just after sunset, and low in the eastern sky just before dawn. We are not able to see Mercury otherwise; it is too close to the sun.

During the day there are planets and stars other than the sun overhead, but the sun's light, scattered or diffused by the molecules in our air, fills our sky with daylight so that we cannot see the dim pinpoints of light from elsewhere in the solar system and the universe. The fact that short-waved blue light is scattered more than the rest of the sun's visible radiation is what makes the sky blue. If we had no atmosphere the sky would be black and we could see stars during the daytime. (But without the atmosphere to protect us from the sun's radiation, we would also burn up immediately.)

Observing the Planets

On a clear, moonless night we can see about 6,000 stars and 4 of our planets without binoculars or telescope: Venus, Jupiter, Mars, and Saturn. Venus is usually the brightest, then Mars, Jupiter, and Saturn. Mars appears reddish, but finding and identifying the planets depends on knowing where to look and what *phase* each one is in. These data depend on the day and time of the year and are given yearly in most almanacs, and daily and weekly in many newspapers. Recall that the planets, like our own moon, do not shine with their own light; we see them only because they are illuminated by the sun. Similarly, Venus and Mercury, because they are closer to the sun than we are, have phases. How well we can see them depends on how much of their sunlit surfaces we can see from wherever we happen to be in our respective orbits. We all orbit around the sun, each at a different rate and in a different orbit. Figure 13–21 shows the orbits of the planets, and figure 13–22 shows their relative sizes. With a small telescope, we can see Saturn's beautiful rings and two more planets, Uranus and Neptune. Our ninth and last planet, Pluto, is so distant and small that it can be observed only with a large telescope or long photographic exposures.

We can distinguish the planets from the stars because they are closer and, orbiting the sun, they seem to move back and forth across the heavens much more "rapidly" than the stars. Mars, for example, orbits the sun every 687 days, thus, seeming to move back and forth across our sky about every 22 months. The stars, of course are also moving—many at tremendous speeds relative to the speeds of the planets—but the stars are too far away for their movements to be noticed

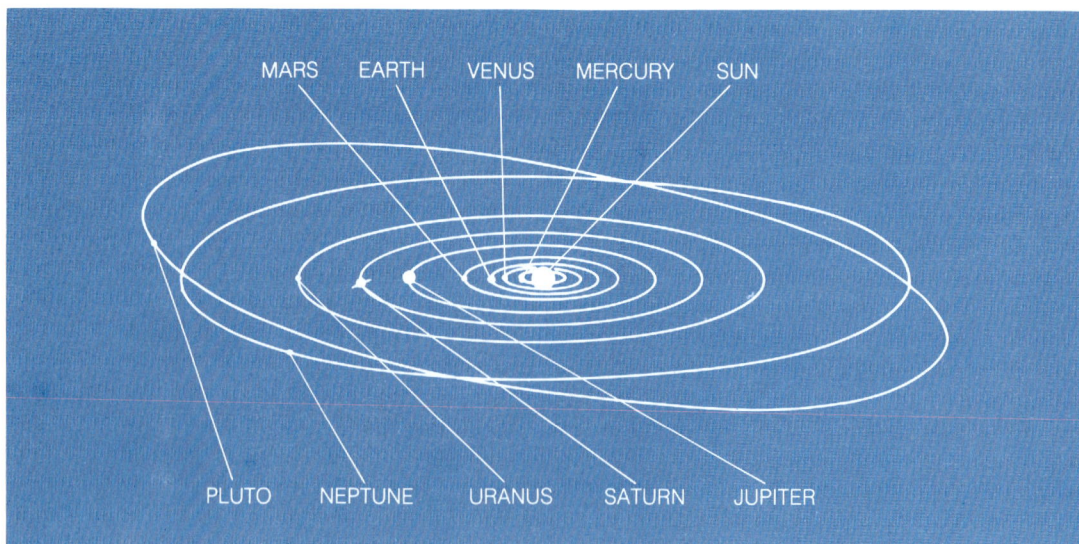

FIGURE 13–21
The nine planets of the solar system orbit the sun at varying speeds.

From *Earth Science*, 3rd ed., by Edward J. Tarbuck and Frederick K. Lutgens. Copyright © 1982 by Bell & Howell Company. Used by permission of Charles E. Merrill Publishing Company.

even in a single lifetime. The nearest star is 7,000 times farther from the earth than our farthest planet, Pluto. Because starlight is bent by the layers of air in our atmosphere they appear to twinkle, while the light from planets is nearer, brighter, and steadier.

Although we cannot see them without using a telescope, 5 of the other planets in our solar system have moons or (more properly) satellites. Mercury and Venus, the two planets closer to the sun than the earth, have no satellites; but Jupiter has 12; Saturn, in addition to its rings, has 10, including a double moon; Uranus has 5; and Neptune has 2. There may be even more that have yet to be detected. Besides Saturn, Jupiter and Uranus also have rings.

Other Objects in the Solar System

In addition to the sun, the planets, and their 32 satellites (including our own moon), our solar system includes thousands of asteroids, dozens of comets and millions of meteors, all revolving around the sun.

ASTEROID One of thousands of small planets that orbit the sun between Mars and Jupiter

Asteroids.   **Asteroids** are very small planets—the largest being less than 500 miles in diameter. (For comparison, our moon has a diameter

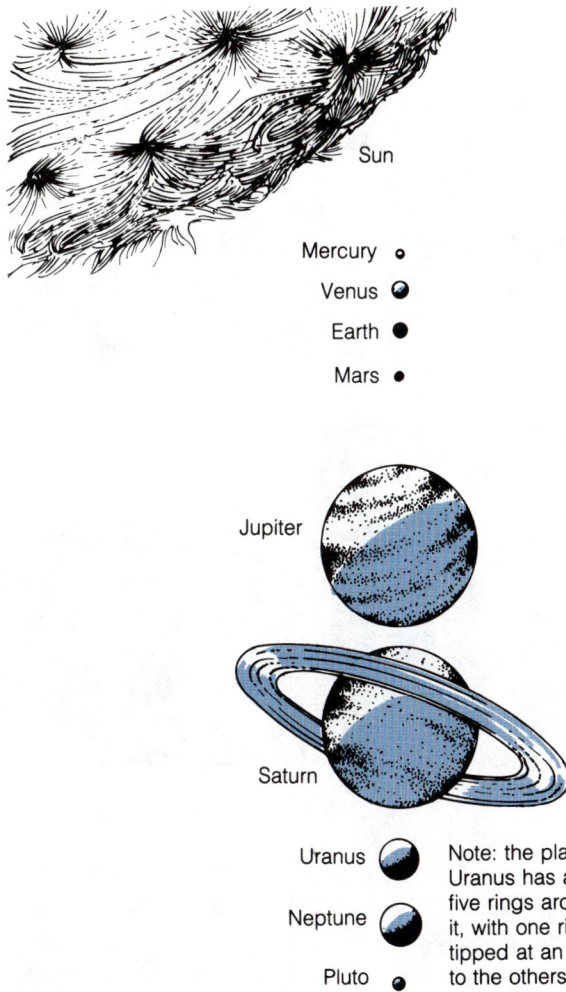

FIGURE 13–22
Relative sizes of the planets.

From L.D. Leet, S. Judson, and M.E. Kauffman, *Physical Geology*, 5th ed., © 1978, p. 400. Reprinted by permission of Prentice-Hall, Inc., Englewood Cliffs, N. J.

Sun

Mercury o

Venus ◉

Earth ●

Mars •

Jupiter

Saturn

Uranus ◐

Note: the planet Uranus has at least five rings around it, with one ring tipped at an angle to the others.

Neptune ◐

Pluto •

of nearly 2,200 miles.) Just like the planets, thousands of asteroids circle the sun with orbits between those of Mars and Jupiter. Asteroids cannot be seen with the unaided eye.

Comets.   Most of the scores of **comets** that have been observed cannot be seen without telescopes, but a few—1 or 2 every 10 to 15 years—are easily seen. Comets are balls of ice and dust that orbit around the sun, but usually have elongated, elliptical orbits. Although comets go far out into space, they also swoop near the sun and are observable from earth. As a comet approaches the sun it may develop a tail that can be seen over 15 to 20 percent of our sky. Even the bright-

COMET A ball of dust and ice, part of whose orbit brings it close enough to the sun to be detected

In the spring of 1986, Halley's comet is expected to return on its regular schedule—once every 76 years. The earliest record of Halley's comet was 240 B.C. (Courtesy of Hale Observatories)

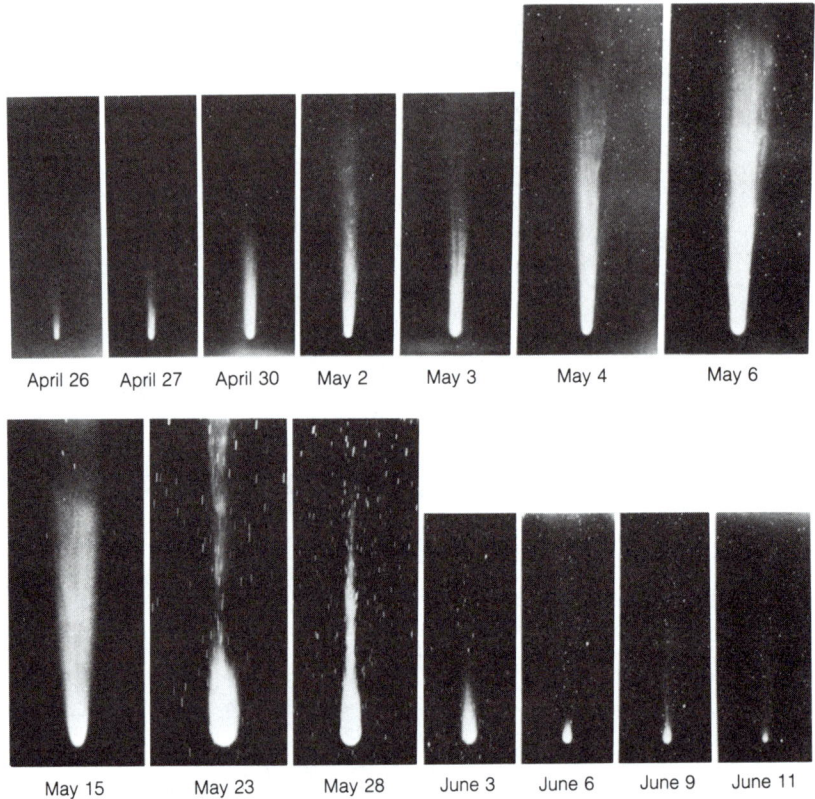

April 26   April 27   April 30   May 2   May 3   May 4   May 6

May 15   May 23   May 28   June 3   June 6   June 9   June 11

est comets are visible for only a few days or weeks. We will be visited next by Halley's comet in the spring of 1986. Halley's comet is a very bright comet that returns every 76 years and was first recorded in 240 B.C. The comet, Kohoutec, about which there was much excitement in 1975, will return at about the same time as Halley's.

METEOR A chunk of material from space that usually burns up as it passes earthward through the atmosphere

Meteors.   **Meteors,** also called "falling stars" and "shooting stars," are far more common than comets. On a clear night several may be seen every hour without the aid of binoculars or telescope; during some summer months one may be seen every few minutes. A meteor is a chunk of mineral or metal that burns up as it enters and passes through the earth's atmosphere on its way to the earth. Thus they are the closest of the "heavenly bodies." Far from being 25 trillion miles from us—the distance to Proxima Centauri, the star that is closest to the sun—"shooting stars" become visible about 60 miles above the earth! And although *almost* all of the millions of shooting stars that enter our atmosphere each year do in fact burn up before they reach the earth, a few of the larger ones occasionally make it to the earth's surface. The Hoba West meteorite that landed in South West Africa (now Nambia) weighs more than 50 tons. (Meteors are called *meteorites* if they reach the earth; they are called *meteroids* if they are in space.) Another meteorite on display in New York City's Hayden Planetarium weighs 34 tons. Even larger meteorites have been held to be responsible, millions of years ago, for large craters in Canada, the United States, and the Soviet Union.

Constellations

For purposes of location, the stars are divided into 88 regions in the sky. The group of stars in each of these regions is called a **constellation**, and each star is in only one constellation. A number of these constellations were defined by astronomers in ancient Greece, Rome, and Egypt, and named after their heroes, heroines, animals, and legends: Cassiopeia, the mythological Queen of Ethiopia; her husband Cepheus; Pegasus, the winged horse; and Orion, the mighty hunter. The signs of the Zodiac are also represented as constellations in the heavens: Aries, the ram; Taurus, the bull; Gemini, the twins; Cancer, the crab; Leo, the lion, and on through the other seven signs. These are shown in figure 13–23.

CONSTELLATION Any of 88 arbitrary configurations of stars into which the celestial sphere is subdivided

The North Star.   As we watch the stars on any given night, the heavens appear to move from east to west, like the sun. This is because our earth is rotating on its axis from west to east. Only the North Star—also called the Pole Star—seems not to move, and that is because it is almost directly above the North Pole. Navigators have known and taken advan-

FIGURE 13–23
The constellations of the Zodiac.

From *Exploring the Universe*, 2nd ed., by W.M. Protheroe, E.R. Capriotti and G.H. Newsom. Copyright © 1981 by Bell and Howell Company. Used by permission of Charles E. Merrill Publishing Company.

tage of this fact for thousands of years. The North Star, in the constellation Polaris, is the end of the 3-star handle of the Little Dipper.

Stars

Stars, almost by definition, are glowing objects in the heavens that seem to stay fixed in their relationships with one another. They are, like our sun, balls of glowing gas. They vary in brightness, location, color, temperature, mass, and size.

Astronomers divide stars into classes of brightness, or *magnitude,* from bright to dim.

The color and temperature of a star varies with its **mass,** which is the quantity of matter in it. The hottest stars have the most mass and appear blue, and the cooler stars have less mass and appear red. The hotter stars also tend to be brighter.

Just as a pound of feathers occupies more space than a pound of lead, even though they both weigh the same, the size of a star is not the same as its mass. Rather, the size of a star seems to be related to different stages in its lifetime.

MASS The quantity of matter in a body. Measured in grams or kilograms, the amount of force required to change the speed or direction of a given object

Lifetime Stages of Stars.   Stars are being born all the time. Hydrogen and helium gas make up nearly 99 percent of the weight of the universe (76 and 23 percent, respectively); all the remaining elements amount to only a little over 1 percent. Our star, the sun, is mostly hydrogen and helium. A star is born as a ball of gas—mostly hydrogen—that contracts because of the gravity of its own mass. As it contracts its core becomes hotter and denser. At temperatures in the millions of degrees, the hydrogen atoms collide with enough force to begin a *nuclear reaction* in which hydrogen atoms fuse to form helium. During this *nuclear fusion* reaction, an enormous amount of energy is released, and a star is born. Astronomers call such a new star a *main-sequence star.* Our sun, which is still mostly hydrogen and helium, is a main-sequence star. Main-sequence stars are medium-sized, as stars go.

After a star uses up the hydrogen in its core it becomes bigger and brighter and is called a *red-giant;* if it is about the size of our sun, it will eventually lose its outer layers and become smaller, at which time it is called a *white dwarf.*

But a red giant star that has more than 3 times the mass of our sun will, instead, continue to expand. At this stage it is called a supergiant—until it explodes. An exploding supergiant star is called a *supernova.* If less than 3 times of the mass of the sun is left after the explosion, the residue of tightly-packed neutrons is called a *neutron star,* or a *pulsar,* because it emits regular pulses of radio waves. If more than 3 times the mass of the sun is left after a supernova explodes, the remnant becomes invisible and is called a **black hole**: a spot in the universe with a gravitational force so great that not even light can escape from it.

Galaxies

In addition to the planets, their satellites, and the stars, the universe contains **galaxies.** Our solar system, which is about 10 billion miles (16 billion kilometers) across, is a relatively tiny point in our galaxy, which we call the Milky Way.

The Milky Way. We have all seen the Milky Way—that band of countless points of light that appears to us as an irregular milky stripe through our heaven on a clear night. The Milky Way appears as a stripe through the heavens because it is more or less flat and disc-shaped, like a coin or phonograph record. We observe it some 30,000 **light years** away from its center. We are not only *not* the center of the universe, we are 30,000 times 5.88 trillion miles (the distance light travels in one year) from just the center of our *galaxy*. Our galaxy, in turn, is just one of a group of 32 galaxies that are called the Local Group. Our Local Group is part of a larger group of galaxies called the *Virgo Cluster.*

Quasars

Perhaps even more disturbing to our sense of earthly importance and relevance in the universe than clusters of galaxies, are quasi-stellar radio sources, or **quasars,** for short. Quasars seem to be the reverse of black holes. Whereas black holes seem not to allow any energy out, quasars are extraordinarily intense sources of energy. Most quasars are "only" a few times larger than our solar system, yet they emit more energy than an entire galaxy of over 150 billion stars. Some astronomers believe quasars are the most distant objects in the universe, about 10 billion light years away, yet their incredible brilliance allows us to see them clearly.

The Universe

What about the universe as a whole? Although astronomers disagree on the details of what happened after the universe was born, most agree that it began with an explosion between 10 and 20 billion years ago. This is the **big bang theory** of the universe. According to this theory the universe has been expanding since it began. As it expands radiation is converted into matter. Although most of the initial radiation has by now changed into matter, some remains. We detect faint radio waves today that seem to be coming from all over the universe, and call these signals *primordial background radiation.* Some astronomers theorize that the universe will continue to expand forever; others think it may approach

BLACK HOLE Theoretically, the remnant of a collapsed star having so great a gravitational field that it will not permit light to escape

GALAXY A family of stars grouped in space

LIGHT YEAR The distance light travels in one year: 5.88 trillion miles (9.46 trillion km)

QUASAR Acronym for *quasi-stellar radio source,* an unusually intense source of radiation energy in the heavens

BIG BANG THEORY A theory that the universe originated with the explosion of a single mass of material 10 to 12 billion years ago, and that the universe is still expanding from that explosion

Courtesy of Burndy Library

Albert Einstein (1879–1955)

"The most incomprehensible thing about the world is that it is comprehensible." Embedded in this statement is a brief glimpse into the genius of Albert Einstein, a person who had the ability to take the usual and recast it into a new discovery. It was his lifelong desire to fit all the forces of nature into a unified concept. He always held firm to the belief that order and not chaos was the rule for he said, "God doesn't play dice with the universe."

Einstein grew up near Munich, Germany. Teachers saw him as a bit dull witted for he was never happy to conform to the routine drill and memorization of schooling. He asked questions that did not have simple answers. When he attempted to enter the Polytechnic Institute in Zurich, he was turned down because of "deficiencies" in modern language, zoology, and botany. After teaching himself for a year he was admitted. Once in school, he did not impress his professors, for he rarely attended their lectures and preferred to do his own studies. When he graduated in 1900, he was asked to teach physics at his school.

Einstein did not care to go through the monotonous routine of lectures, so he left teaching and began working in a Swiss patent office. Here he was recognized for his ability to understand the theories on which inventions were based. By 1905, his own thinking was reflected in articles he wrote, including his most famous work on the theory of relativity.

By 1913, Einstein had become widely known for his concepts of time, space, and math. He moved to Berlin and continued his scientific research which was based more on mathematics than on laboratory experiments. In 1921, he was awarded the Nobel Prize in physics. Because he disagreed with the political climate in Germany and because he was Jewish, Einstein came to the United States in 1933. He accepted a position at the Institute for Advanced Study in Princeton, New Jersey. This gave him the chance to do what he liked best—no regular lectures but many opportunities to help individual students think through their problems.

When he died in 1955, Einstein was still searching for laws that would more completely explain the universe. Because of his unique genius in reflecting upon the mathematical representation of the universe, he has forced a re-examination of the world of scientific investigation.

some limit and then begin to contract. Evidence that the universe is in fact expanding is described in the next section.

Measuring Distances in Space

The enormous distances between objects in space are measured in three ways: by **parallax** measurement, by the luminosity or relative brightness of the objects, and by their **red shift.**

PARALLAX An apparent shift in an object's direction due to a shift in the observer's position

Parallax. When you look at an object with both eyes, your eyes tell you how big and how far away that object is because they have slightly different views of the object. This is called parallax. You cannot tell how big and how far away heavenly bodies are because your eyes are too close together relative to the large distances and tiny differences between the angles from which your eyes view such distant objects. But if your eyes were separated by thousands or millions of miles, you could determine the distances to the objects you could see. Astronomers use this same principle by observing the same object from points that are distant from one another on the earth, and also by observing the same object from points widely separated along the earth's orbit as it circles the sun. Using trigonometry, the distances and sizes of perhaps the nearest 10,000 stars have been calculated, particularly with respect to stars that are much more distant.

RED SHIFT Any displacement of spectral lines toward red, indicating that galaxies are receding from observers on earth

Luminosity. Astronomers also measure the distance to stars by measuring their brightness in two ways and comparing them. The actual brightness, or **luminosity**, of a star is measured and then compared with its apparent brightness as observed through a telescope on earth. The further away a star is, the dimmer it is.

LUMINOSITY a measure of the brightness of a star

Red Shift. Red shift refers to the phenomenon that the observed **spectral lines** from the light of a distant galaxy are skewed towards the "red" or lower frequency end of the spectrum. A red shift indicates that the galaxy is moving away from us. If the galaxy was approaching us, the spectral lines would be shifted towards the blue or higher frequency end of the spectrum. This is explained as a **Doppler effect** (described first by Christian J. Doppler (1803–1853)). It is similar to the experience you have probably had when passing a loud noise at high speed: as you approach and pass a train that is blowing its whistle or horn it sounds higher in pitch before you pass it and lower in pitch afterwards.

SPECTRAL LINES Lines that appear when light passes through a prism and becomes "ordered" into colors of different wave lengths or frequencies

DOPPLER EFFECT Apparent change in the frequency of a wave due to the relative motion of the source and the observer

Except for those nearest the earth, all galaxies have pronounced red shifts, which is a powerful indication that they are moving away from us in all directions. In 1929, the American astronomer Edwin P. Hubble

(1889–1953) proposed that this could be explained only if everything were moving away from everything else. This is the theory of the constantly expanding universe that most astronomers accept today. If, for example, the galaxies were moving around in patterns we would see blues and reds that change over time. If we were approaching another portion of the universe we would see blue shifts in that area. If we were all drifting along together, we would see few shifts at all. But instead we see red shifts everywhere, at least from all the distant galaxies.

Light Years.   It is extremely difficult to understand a **light year** in any real-life sense, even though we can easily calculate the orders of magnitude light years involve. No human being has ever experienced a light year: 5.88 trillion miles. If the moment you were born you were hustled aboard a jet aircraft traveling 600 miles an hour, and stayed aboard for the next 70 years of your life, you would travel 368 million miles (594 million km), about 6 hundred-thousandths of one light year. If you were put on board 2,000 years ago, you would just now have come about 10.5 billion miles (16.9 billion km). That is less than one-fifth of 1 percent of one light year.

Traveling at 186,000 miles (300,000 km) per second, the light from the sun travels its 93 million miles (150 million km) to us in 8 minutes and 20 seconds, and to the outer edge of our solar system is about 7.5 hours. Compare this with the 30,000 years it takes light to travel to the center of our galaxy, the Milky Way. And our galaxy, as noted, is just one of many.

Astronomers have devised a unit of length that is much bigger than a kilometer or a mile, but not so mind-boggling as a light year. It is the *Astronomical Unit,* or AU, and it is the distance from the earth to the sun, 150 million kilometers, or about 93 million miles. This unit is used in measuring distances within the solar system.

A Sense of Proportion

We began this chapter by talking about the earth and its processes, and concluded the chapter by looking at distant objects in space. The study of the earth and planetary sciences, perhaps more than any of the other sciences, provides a sense of proportion that is different than our daily perspectives and yet compelling for us all. As we think about our solar system being created some 4.5 billion years ago and our mountain ranges forming 500 million years ago, we necessarily gain a new sense of time. When we try to imagine the vastness of outer space, our new sense of time and space expands. Relative to this vastness, our daily problems and concerns might seem smaller and less significant.

Clark, David L. *Fossils, Paleontology, and Evolution.* Dubuque, IA: Wm. C. Brown Co., 1976.

Colbert, E. H. *Evolution of the Vertebrates.* New York: John Wiley & Sons, 1969.

Eicher, Don L. *Geologic Time.* Englewood Cliffs, NJ: Prentice-Hall, 1968.

Federal Aviation Agency, Flight Standards Service. *Private Pilots Handbook of Aeronautical Knowledge.* Washington, D.C.: U.S. Printing Office, 1965.

Findley, Rowe. "St. Helens: Mountain with a Death Wish." *National Geographic.* (January 1981): pp. 17–65.

Leet, Don L.; Judson, Sheldon; and Kaufmann, Marvin E. *Physical Geology.* 5th ed. Englewood Cliffs, NJ: Prentice-Hall, 1978.

Miller, Franklin Jr., *College Physics.* New York: Harcourt Brace Jovanovich, 1982.

Odum, Eugene P. *Fundamentals of Ecology.* Philadelphia: W. B. Saunders Co., 1971.

Rumney, George R. *The Geosystem.* Dubuque, IA: Wm. C. Brown Co., 1970.

Stearn, Colin W.; Carroll, Robert L.; and Clark, Thomas H. *Geological Evolution of North America.* New York: John Wiley & Sons, 1979.

Sumner, John S. *Geophysics, Geologic Structures, and Tectonics.* Dubuque, IA: Wm. C. Brown Co., 1969.

Tuttle, Sherwood D. *Landforms and Landscapes.* Dubuque, IA: Wm. C. Brown Co., 1970.

Wetherill, George W., and Drake, Charles L. "The Earth and Planetary Science." *Science* 209 (July 1980): 96–104.

Zukav, Gary. *The Dancing Wu Li Masters: An Overview of the New Physics.* New York: William Morrow & Co., 1980.

BIBLIOGRAPHY

14

Teaching the Earth Sciences

The purpose of chapter 14 is to provide examples of science lessons designed to teach earth science to students from kindergarten through eighth grades. Planning Charts 14–1, 14–2, and 14–3 illustrate how each science lesson will allow you to meet several objectives at the same time. As students learn about solids, liquids, gases, and the limits to the earth's energy and other resources, they also acquire new concepts, develop their learning processes, improve their psychomotor skills, broaden their attitudes and values, and gain experience making decisions.

KEY CONCEPTS AND IMPORTANT IDEAS

☐ earth materials: solids, liquids, and gases ☐ weathering and erosion ☐ limits to energy sources ☐ collecting and storing solar energy ☐ human interaction and the earth's spheres ☐ population growth ☐ finding resources ☐ consuming resources

CHART 14–1
Planning Chart for Earth Science: Kindergarten through Grade 2

| Activity | Objectives | | | | | | Teaching Methods | Related Curriculum |
	Conceptual Knowledge	Learning Processes	Psychomotor Skills	Attitudes and Values	Decision-Making Skills			
1. Earth Materials: Solids	Nonliving materials (solid objects) have identities based on their properties and physical state; solids can change in form.	Observe, compare.	Handle nontoxic substances.	Explore and ask questions about objects in the environment.	. . .		Guided discovery, field trip	Language arts
2. Earth Materials: Liquids	Nonliving materials (liquids) have identities based on their properties and physical state; liquids can change in form.	Observe, compare, predict.	Handle nontoxic substances. Combine materials in new ways.	Explore and ask questions about liquids in the environment.	. . .		Learning center, experiments, discussion	Art, cooking
3. Earth Materials: Gases	Nonliving materials (gases) have identities based on their properties and physical state; gases can change in form.	Observe, compare, classify.	Handle nontoxic substances.	Explore and ask questions about gases in the environment.	. . .		Guided discovery, "hands on" experiences	Language arts

CHART 14–2
Planning Chart for Earth Science: Grades 3 through 5

Activity	Objectives					Teaching Methods	Related Curriculum
	Conceptual Knowledge	*Learning Processes*	*Psychomotor Skills*	*Attitudes and Values*	*Decision-Making Skills*		
4. Weathering and Erosion	The earth's surface is changing.	Observe, compare, gather and organize information.	Handle a variety of nontoxic substances; construct simple projects and models with some guidance.	Explore and assume responsibility for the environment.	Describe consequences of various alternatives.	Field trip, projects, questioning method	Social sciences
5. Limits to Energy Sources	There are limits to the capacity of earth materials to supply energy.	Observe, compare, measure, gather and organize information.	Handle nontoxic substances; construct projects.	Explore the environment. Assume responsibility for use of natural resources.	Describe consequences of various alternatives.	Guided exploration, discussion	Social sciences
6. Collecting and Storing Solar Energy	Earth Materials have different capacities to absorb and store solar energy.	Measure, form hypotheses, predict, construct and interpret graphs.	Construct simple project; handle nontoxic substances.	Assume responsibility for wise use of resources.	Form alternatives which are based on information.	Experiments, laboratory reports, discussion, debate	Social sciences, mathematics

CHART 14-3
Planning Chart for Earth Science: Grades 6 through 8

Activity	Objectives					Teaching Methods	Related Curriculum
	Conceptual Knowledge	Learning Processes	Psychomotor Skills	Attitudes and Values	Decision-Making Skills		
7. Human Interaction and the Earth's Spheres	Environments have identities based on their properties; environments change because living and nonliving matter interact.	Gather and organize information, describe relations among variables.	. . .	Actively explore and ask questions; seek alternative points of view and multiple sources of evidence.	Describe situations that require decision. Describe immediate and long-term consequences of various alternatives.	Oral and written reports, note-taking	Social sciences
8. Population Growth	There are limits to the environment's capacity to change and still conserve its identity.	Graph, infer, predict, form hypotheses.	. . .	Combine ideas in new ways. Seek alternative points of view.	Describe situations that require decisions.	Problem solving, film analysis	Mathematics
9. Finding Resources	Environments change when people use the earth's resources; there are limits to the earth's resources and to environmental change.	Observe, compare, infer, predict, and graph.	Handle nontoxic substances; use tools that require fine adjustment and discrimination.	Combine objects and materials in new ways. Explore the environment. Demonstrate the need and means to use natural resources wisely.	Describe situations that require decisions. Describe immediate and long-term consequences of various alternative actions.	Guided discovery	Social sciences
10. Consuming Resources	There are limits to the earth's resources and to environmental change; environments can lose their identity when they exceed their capacity to change.	Graph, infer, predict, describe relations of variables.	. . .	Seek alternative points of view. Demonstrate the need and means to use natural resources wisely.	Clarify consequences of alternative actions.	Simulation, problem solving	Social sciences

ACTIVITY ES–1

Earth Materials: Solids
(Kindergarten–Grade 2)

ACTIVITY OVERVIEW

In this activity, solid earth materials are considered; liquids or gases are considered and defined in the next two activities. The adjectives *heavy, light, hard, soft, wet,* and *dry* are defined through interaction with solid earth materials in the environment. Students make comparisons of materials in distinctively dichotomous situations. *Estimated time:* two class periods.

SCIENCE BACKGROUND

Anything that occupies space and has weight is called **matter.** Rocks, water, and air are all examples of nonliving matter. Birds, flowers, and the food you eat all are examples of living matter. Nonliving matter can be observed in three forms or physical states: solid, liquid, and gas. Matter that is **solid** always has a definite shape and size. Familiar examples of solids are rocks; clay products such as dishes; metal products such as coins, cars, pots and pans; and chemically synthesized products such as orlon, nylon, and rayon. A solid can be hard or soft, heavy or light, and wet or dry.

SOCIAL IMPLICATIONS

Being able to differentiate between the three physical states of matter is a prerequisite to describing the environment and learning more about changes that have occurred in earth materials.

MAJOR CONCEPTS

Nonliving earth materials have identities based on their properties and their physical state. Any object can have two or more different properties.

OBJECTIVES

☐ Observe and compare objects.

☐ Handle nontoxic substances.

☐ Explore the environment and ask questions.

MATERIALS

A tray for each child (small cardboard trays may be obtained from meat markets or grocery stores), common solid objects from the classroom and home (try to find objects that vary in weight, hardness, and dryness); examples include: pieces of cloth that are wet and dry, rocks of various sizes and shapes, cubes of sugar, pieces of chalk, scissors, pieces of aluminum foil, pieces of synthetic sponge, marbles or ping pong balls, pieces of broken dish or flower pot, Styrofoam bits, and corks.

VOCABULARY

solid objects	wet, dry
heavy, light	same, different
hard, soft	

PROCEDURES

A. Prepare, introduce, and carry out the activity as follows.

1. Prepare a tray of assorted objects for each child or pair of children. Each tray should contain objects that can be compared as hard and soft, light and heavy, or wet and dry. Trays can vary.

2. Begin the lesson by introducing the words *hard,* and *soft, light,* and *heavy,* and *wet* and *dry* on the chalkboard. Ask the students to think of examples of things that illustrate these words. Then explain that they will be exploring some objects that fit these words.

3. Distribute the trays and encourage the children to explore the materials. Exploration should continue for about 5 to 10 minutes or until interest wanes.

4. Bring the children together and encourage them to describe the objects on their trays. Then ask them to identify objects that are hard or soft, light or heavy, and wet or dry.

5. In summary, repeat the new words as you point to them on the chalkboard. Then ask volunteer students to select objects from their trays and use the new words to describe the objects. Have several students write the words on the board that describe the object of their choice.

6. Finally, introduce the word *solid* to the students. Tell them that all of the objects they have explored have a recognizable shape and weight. Scientists refer to objects as solid when they have a specific shape and weight.

Allow children adequate time to observe the objects.

EXTENDING THE ACTIVITY

B. Take a field trip around the school and use the new words as the focus of the trip. Follow the field trip with descriptions of solid objects seen in the environment that are wet and dry, heavy and light, and hard and soft.

ACTIVITY ES–2

Earth Materials: Liquids
(Kindergarten–Grade 2)

ACTIVITY OVERVIEW

In this activity the properties of **liquids** are introduced. Students have the opportunity to examine and explore the properties of various liquids. *Estimated time:* approximately five class periods.

SCIENCE BACKGROUND

Earth materials or matter exist in three physical states: solids, liquids, and gases. Rocks, water, and air are examples of matter in these three physical states.

Liquids differ from solid objects in an important way: liquids assume the shape of the container into which they are poured, while the shape of solid objects is unchanged by the shape of their containers. Liquids and solid objects are alike, however, in that their volume—meaning their *weight* and the *amount* of space they occupy—can be measured. To illustrate how liquids and solids are alike and yet different, consider water in two of its forms: as a familiar liquid and in its frozen or solid state as ice. Both the liquid and solid forms of water can be measured in terms of their volume. However, the shape of the container only affects the shape of water in its liquid form. If ice cubes are placed in a glass, their shape is unchanged by the shape of the container. Other examples of liquids are milk, honey, gasoline, and oil.

SOCIAL IMPLICATIONS

Liquids form a substantial part of the child's world. Before children understand changes in the environment they should know liquids as one of the three states of matter and thus be able to describe many of the changes that occur to matter.

MAJOR CONCEPTS

Nonliving earth materials have identities based on their properties and their physical states. Liquids can change to solids.

OBJECTIVES

- ☐ Observe, compare, and predict.
- ☐ Handle nontoxic substances.
- ☐ Actively explore the environment.
- ☐ Combine materials in new ways.

MATERIALS

About 30 to 35 clear, disposable plastic tumblers; 2 clear plastic trays; 6 plastic teaspoons (for distributing liquids to students); and approximately 5 ounces of each of these 6 liquids: water, mineral oil, white vinegar, liquid starch, liquid detergent, and corn oil.

VOCABULARY

liquid, solid	wet, dry
flow	clear, cloudy
hard, soft	pour
thick, thin	frozen

PROCEDURES

A. Pour each liquid into a tumbler without allowing the students to see the original bottles from which the liquids came. The identities of the liquids should be a mystery, so keep them out of view until you introduce the activity.

Seat the students so that everyone can see the 6 liquids as you display them in their plastic tumblers. Walk among the students carrying the tumblers, two at a time, one in each hand. Allow students to dip one finger in each tumbler and try to guess what the "mystery" liquids are, using their senses of smell, sight, and touch. Continue introducing the other mystery liquids in the same way. Invite the students to describe the liquids in terms of their color, clarity, and **viscosity:** What is the color of each liquid? How clear or cloudy is each liquid? How thick or thin is each liquid? Write the vocabulary words on the chalkboard and repeat them for the students. Ask individual students to hold one tumbler and point to the words that best describe the liquid.

Remind students to use only their senses of smell, sight, and feel.

B. On the second day, arrange the 6 liquids, 6 teaspoons and 2 plastic trays at a learning center. Allow students to take turns at the learning center, mixing *a few drops* from the tumblers onto plastic trays. They should observe the properties of the liquids when mixing together only 2 liquids at a time. Each student should clean up when finished.

The purpose of this exploration is to help students discover how liquids change in color, viscosity, and clarity when 2 liquids are combined. A group discussion should follow the Learning Center activities. During the discussion, the words suggested in the vocabulary should be recorded as they are used by the children to describe the liquids. Review the vocabulary introduced earlier in this activity.

EXTENDING THE ACTIVITY

C. Pose the question: Can liquids ever become solids? If no one recalls that water can be changed from a liquid that we drink to a solid (ice) that we use to chill a drink or to skate on, suggest that the class conduct a series of experiments.

Pour a small amount of each of the 6 liquids into 6 tumblers. Ask the students to predict what will happen to each liquid if you place the tumblers in a freezer. Then

Can liquids ever become solids? Students fill their tumblers with liquids of different kinds and place them in the freezer to find out. (Donald Birdd)

put the tumblers in the freezer portion of the refrigerator in the faculty lunchroom.

On the following day, bring the tumblers in to be examined by the students. Allow the students to feel the contents of each tumbler. Encourage discussion of the characteristics of each tumbler's content. Refer again to the vocabulary words related to liquids and solids. Ask students to predict what will happen if the tumblers are left at room temperature in the classroom—rather than in the freezer—for the remainder of the day. Allow the frozen material to return to liquid.

Finally, ask students to compare the liquids they have just studied with the solid objects studied in Activity ES–1. Ask: How are the two groups alike or different? Guide their attention to the fact that many solid objects can become liquid if they are heated at high temperatures. For example, sugar cubes and many rocks can be changed to liquid under very high temperatures. Other solid objects such as metal scissors, pots, pans, plates, and flower pots were once liquids that were poured into molds (forms shaped like the containers) and baked or cooled until they became solid. This science activity can lead naturally to art or cooking activities where liquids are changed to solids.

ACTIVITY ES–3

Earth Materials: Gases
(Kindergarten–Grade 2)

ACTIVITY OVERVIEW

In this activity, the properties of **gases** are developed through a series of activities. Bubbles are made by blowing air into soapy water. Gases are often colorless, odorless, tasteless, and they assume the shape and volume of the receptacle in which they are contained. Gases are the most abstract of the physical states that we will present. The main focus of Activity ES–3 is to introduce air as a gas. *Estimated time:* four class periods.

SCIENCE BACKGROUND

Earlier you learned that earth materials—or matter—can be seen in three forms: solids, liquids, and gases. You also learned that these three forms or states of matter have weight and take up space. Yet, *gases* are different from solids and liquids in important ways: A gas has neither a definite shape nor a definite (or constant) size. Oxygen, hydrogen, and ammonia are examples of gases. If you put gas into a container, it will spread out to fill the container, and it will take the shape of the container. For example, when hydrogen is forced into a balloon, the hydrogen takes the shape of the balloon. If the balloon bursts, the hydrogen is dispersed into the air and no longer holds the shape of the balloon. Because hydrogen is colorless, it is impossible to see as it spreads through the air around us.

SOCIAL IMPLICATIONS

Gases surround us and are essential for life. The air we breathe is a combination of several gases including oxygen, nitrogen, argon, and trace amounts of hydrogen and other gases. Oxygen is a life-saving gas used by deep-sea divers when they work under water. Oxygen is also essential for pilots and passengers who travel in jets at altitudes where the atmosphere has too little oxygen to support life. And, of course, hospitals and emergency rescue vehicles must have tanks of oxygen because it is vital to the survival of patients suffering from heart attacks or lung disorders. Yet, the physical state of gas is difficult for children to conceptualize. Unlike solids and liquids, gases are often impossible to see, feel, taste, smell, or hear. Thus, Activity ES–3 is designed to provide students with some firsthand experience with gases.

MAJOR CONCEPT

Nonliving materials (gases) have identities based on their properties and their physical states.

OBJECTIVES

☐ Observe and compare.

☐ Handle nontoxic substances.

☐ Explore the environment and ask questions.

MATERIALS

For Procedure A:

A plastic straw for each student (plastic straws can be made into a bubble pipe by cutting one end of the straw and spreading the tabs.

A small container of soapy liquid for each child [The soapy liquid can be made by mixing the following ingredients with an egg beater: 1 cup of detergent, 1 cup of glycerine (from a drug store), 5 cups of clean water, and a pinch of sugar (to toughen the bubbles), and newspapers or appropriate covers for students' desks.

For Procedure B:

1 balloon for each child, a pinwheel for every 4 children, a plastic bag for each child. (Note: The bag should be a small sandwich-sized bag. It should *not* fit over a child's head.)

VOCABULARY

air	shape
gas	soap
blow	bubbles
odorless	straw
colorless	blowing

PROCEDURES

A. Distribute straws and containers of clear water. Invite students to blow into the water, and describe what happens. Provide soapy water and allow students to play freely and at length with the materials. Focus their attention on the specific characteristics of the soap bubbles. Possible questions include: How would you describe the bubbles? What shape is your bubble? What is inside of each soap bubble? What happens when a soap bubble breaks? What did you use to make bubbles? The activity should be concluded with a free discussion about the children's observations.

B. Provide each student with a plastic bag. Caution the students not to place plastic bags over their head, mouth, or nose.

Allow students time to feel the pressure of the air inside the plastic bag when they gently touch it.

What is inside the balloon? Students learn to recognize gases which are colorless and odorless even though it is impossible to "see" them. (Jack Hamilton)

Take the plastic bag and make a sweeping motion with the bag open. Close the bag and twist the top until it is filled with air. Introduce the concept of air as a gas. Acknowledge the use of the word *air* and tell students that air is a gas. Invite the students to capture their own bag of air. Then ask the students to describe the properties of the gas inside the bag. Words such as *colorless, odorless,* and *tasteless* should be introduced and written on the chalkboard. Encourage discussion of the color, shape, odor, and taste of gas in the students' bags and the gas (called air) that we breathe.

Next distribute a balloon to each student. Invite students to blow into their balloons until they are about one-half filled. Ask the students if the gas in the balloon is the same as the gas in the plastic bag. Since both gases are colorless and odorless, it is impossible to "see" differences; but the gas we breathe out is different from the air we breathe in. When we inhale, we breathe in a mixture of several gases including oxygen, hydrogen, and nitrogen. This gas is held in our lungs until the oxygen is taken out by the blood. When we exhale or breathe out, the gas that comes out of our lungs is different because it contains less oxygen. Ask students if the gas in the balloons is the same as the gas in the bubbles students blew the day before (yes). Ask if anyone can explain why both gases were the same. Finally, ask students to name other objects that can hold gas. Examples include the tires on their car or bicycle and blimps in the sky.

EXTENDING THE ACTIVITY

C. Take the students on a field trip around the school and have them describe earth materials that are solid, liquid, and gas.

ACTIVITY ES–4

Weather and Erosion
(Grades 3–5)

ACTIVITY OVERVIEW

In this activity students explore the immediate outdoor area around the school for examples of erosion by water. Later, in the classroom, students set up models of eroding systems and examine in detail some of the processes. *Estimated time:* two class periods.

SCIENCE BACKGROUND

Weathering is the physical or chemical breakdown of rocks. Erosion is the transportation and deposition of weathered rock. Weathering includes a complex set of chemical processes that occur as a result of air and water interacting with the elements in minerals. When rocks from beneath the earth's surface are exposed to air and water, the process of **chemical disintegration** begins. This in turn increases the physical disintegration of rocks. The latter is simply the physical breaking apart of a rock.

The erosion process results when energy in the form of water, ice, wind, or gravity move the weathered rock material from one place and deposit it in another place. The process of chemical disintegration can continue during the process of erosion, thus releasing many chemical elements. The sediment of weathered rock is ultimately deposited in lakes and ocean basins.

SOCIAL IMPLICATIONS

Almost anywhere you look outdoors examples of erosion can be found. Human use of the land has increased the natural processes of weathering and erosion. Farming; the construction of roads, bridges, and houses; overcutting of forests; and overgrazing by livestock all are examples of human uses of the land that increase the erosion process. Humans also have begun to reverse many of these detrimental results. For example, sterile land has had soil nutrients replenished; steep slopes have been reforested; and the rate of erosion that occurs naturally has been slowed down in some places by the construction of drainage ditches and retaining walls.

MAJOR CONCEPT

The earth's surface is changing.

OBJECTIVES

☐ Observe, compare, gather, and organize information.

☐ Handle a variety of nontoxic substances.
☐ Construct simple projects and models with some guidance.
☐ Actively explore the environment and ask questions.
☐ Assume responsibility for the wise use of natural resources in the environment.
☐ Describe consequences of various alternatives.

MATERIALS

A container for each group of 2 or 3 students, a milk carton or planter, several chunks of soil (about milk-carton size), some grass or weeds, 2 bottles with sprinkler attachments (or be prepared to simulate rain by sprinkling water with your fingertips).

VOCABULARY

material   erosion
weathering   movement

PROCEDURES

A. Take a field trip to the school yard and/or to the street. An ideal time to take this field trip would be after a rainstorm or after a street cleaner has caused water to run in the gutters. However, if neither of these events has occurred recently, ask the custodian to have the water running on the yard near the gutter.
  1. There are two ways to make the trip. The decision depends on you and your perception of students. The first is very open: simply tell the students they are going outside to make observations of the effects the recent rain (or running sprinkler). Take the stu-

dents outside and allow them to make observations for 15 to 20 minutes.

2. The second approach is more direct. Prior to the trip tell the students they are to focus on questions such as:

What is the largest material carried by the water?

What is the smallest?

What types of materials are being carried in the water?

From where were the materials carried?

Then take the field trip. During the trip use the milk containers to collect some material and water that is observed. When the class returns to the classroom, allow students some time to analyze the collected material. You may wish to provide larger containers so the students can pour out the contents of their jar. Encourage students to describe the types of material that were carried in the water. Ask questions such as:

Where do you think the material came from?

Where do you think the material goes?

What do you think determines the size of material carried by the water?

Why do some materials accumulate in certain places?

What determines where materials are deposited?

3. Introduce the terms weathering and erosion. Write them on the chalkboard. Summarize the process of weathering and erosion, using the schoolyard example. Illustrate the process on the chalkboard.

B. Invite students to create a project or model that illustrates erosion using miniature systems with milk cartons or plastic

Advise students to elevate one end of the dishpan or make a hill with the soil to better simulate erosion with the running water.

containers and a chunk of soil from a vacant lot. Students can then use sprinkling bottles or their fingers dipped in water to "make it rain." They need not sprinkle much water on the soil before erosion will occur. If they observe their models closely, they will see some of the same processes that occurred outside. Relate the models to actual situations as much as possible.

EXTENDING THE ACTIVITY

C. Take a field trip to a nearby river or stream to observe erosion along the banks and edges. Then ask students to write reports on erosion by wind, ice, and water.

ACTIVITY ES–5

Limits to Energy Resources
(Grades 3–5)

ACTIVITY OVERVIEW

Through simple activities and observations, children realize that sources of **energy** are limited and that some sources can be used only once. *Estimated time:* two class periods.

SCIENCE BACKGROUND

Energy **resources** are often grouped as renewable, nonrenewable and synthesized or derived. **Renewable resources** are those that theoretically cannot be used up because they are replenished or recycled. The source is basically inexhaustible (the sun) or it can be renewed through natural or human intervention. A **nonrenewable resource** can be thought of as any resource that appears to have a finite and foreseeable limit. Nonrenewable resources can be depleted to the point that recovery and use are not economically feasable. Fossil fuels such as oil and coal are examples. In short, it is the quantity and quality of the resource that defines the limits of its use. By analogy, using a renewable resource is like living on the interest from your savings when you have no other income. Using nonrenewable resources is like living entirely on your savings rather than on interest or a salary. When the use of a resource exceeds its supply, then the resource is gone forever; there are no reserves.

Energy must be available in useful form, at reasonable costs, and without harmful effects on the environment. Energy scarcity, then, does not really mean we are running out of energy sources. It means that there may be a shortage of one kind of resource and, perhaps as a consequence, a rise in price (oil, for example). Or, a substantial amount of environmental damage may occur from efforts to extract a limited resource.

SOCIAL IMPLICATIONS

The oil embargoes make us aware of another fact: energy from oil or petroleum is needed for much more than cars. We are dependent on petroleum to run machines and grow food. If oil as a source of energy becomes more expensive, so do all of the other products for which oil is a source of energy.

The decade of the 1970s was a turning point for industrialized countries. During the period from 1973 to 1980, people became cognizant of limits to fossil energy resources. Hopefully we have taken the lesson of limits seriously and started the transition to conservative practices and the search for alternative sources of energy.

MAJOR CONCEPTS

There are limits to the earth's fossil fuels that supply our energy. Conservation is one method of avoiding the depletion of these resources. Changing energy sources is another method we can use to avoid the depletion of fossil fuels.

OBJECTIVES

☐ Observe, compare, measure, gather and organize information.

☐ Construct simple projects with guidance.

☐ Handle nontoxic substances.

☐ Explore the environment.

☐ Assume responsibility for the wise use of natural resources.

☐ Describe consequences of various alternatives.

MATERIALS

One set of the following items for each group of 2 or 3 students: small birthday candles, matches, Styrofoam cups, clock, thermometer, test tubes, and test tube holder.

VOCABULARY

energy	renewable
heat	nonrenewable
light	limits
resources	conservation

PROCEDURES

A. Before class, place a small hole in the bottom of each Styrofoam cup. Then place a candle in the bottom of each cup. Alternatively, you may wish to place a small amount of sand in the cup and place the candle in the sand.

1. Distribute the materials to each group of students. Tell the students they are going to use the candle to heat water in their test tube.

2. Place 5 cm of water at room temperature in each test tube. Tell the students they are to heat the water to 10°C. During the activity they should record (1) How long it takes to heat the water to 10°C and (2) How much (in centimeters) of their candle is used.

3. Allow them to proceed with the activity. Tell them that *under no circumstances* are they to heat the water to the boiling point (100°C).

4. Discuss their results when they have completed the activity. During the discussion ask them: *How long would it have taken to heat the water to 20°C? to 30°C?* How long would your candle last?

5. Compare the candle heating the water in the test tube to a much larger example in nature—the sun keeping the oceans, lakes, and rivers at a temperature that can sustain plant and animal life.

B. The second part of this activity is a challenge and exploration.

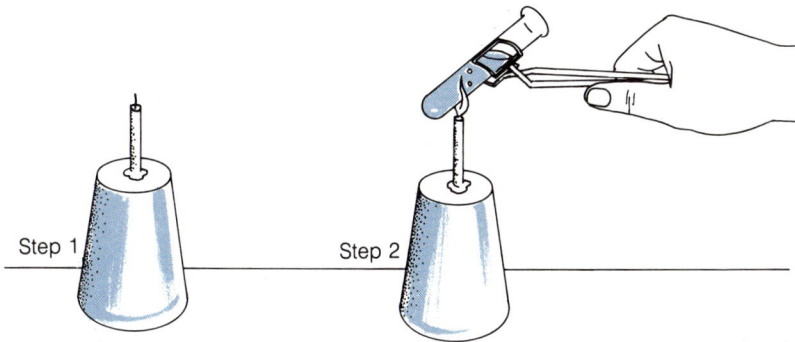

Step 1   Step 2

Depending on your students, you may allow them to conduct the experiment or you may wish to do this activity as a class demonstration so you can supervise closely.

1. Challenge the students to heat their water to 30°C without using all of their candle. Tell them to record the results they obtain and what they do.
2. Have the students share their results with the rest of the class. During the discussion, point out those ideas that represent conservation of this and other energy sources. Point out ideas that allow for combinations of the conservation and use of renewable resources.

EXTENDING THE ACTIVITY

C. Extend the discussion to the limits and conservation of natural resources such as fossil fuels

1. Present the students with a list of renewable and nonrenewable resources. See the list of resources provided in this activity.
2. Have students describe how the resources are limited. Ask students how we can avoid the depletion of natural resources and still have enough energy for survival.

List of Resources

coal	natural gas
garbage	oceans
petroleum	wildlife
sun	soil
oil shale	metals (such as
rivers	aluminum,
forests	iron, and
lakes	zinc)
wind	

ACTIVITY ES–6

Collecting and Storing Solar Energy
(Grades 3–5)

ACTIVITY OVERVIEW

In this activity the students collect solar energy using different earth materials. They

then determine which materials are best for storing solar energy. The activity ends with an exploration in which the students are challenged to set up a system that will store energy efficiently. *Estimated time:* three class periods.

SCIENCE BACKGROUND

Solar energy is abundant. This fact is one of the major advantages of solar energy as a resource. Solar energy falls on the upper atmosphere at a rate of about 1.3 kilowatts per square meter (A watt is a metric unit of power. A kilowatt is 1,000 watts.). On the average, about 13 percent of this energy arrives on the earth's surface. Here is another way of thinking of this resource: the energy equivalent to 10 barrels of oil falls on each acre of the United States each day, or over 22 billion barrels of oil per day falls on the contiguous 48 states. However, the amount available varies greatly with latitude, season, and weather. Technological advances may be able to overcome some problems and convert large enough stores of solar energy for our use.

Solar energy can include more than the direct rays of the sun. The term also may include the energy that results from the interaction between solar radiation and different parts of the earth's spheres.

There are, however, problems with the use of solar energy. The collection and concentration of the resource is difficult. Certain geographic locations are better suited for the use of solar energy than other areas. Presently the cost of technology for the use of solar energy is high; but, in terms of the lifetime cost of solar energy for home use, these costs are reduced. Direct solar energy is usually collected and concentrated for use in space heating and water heating. Through the use of photovoltaic cells it can also be converted to electricity. Many people who oppose nuclear energy, because of its hazards and potential irreversible damage to the environment, want to see solar technology developed because of its safety and economy.

SOCIAL IMPLICATIONS

Estimates vary as to the total potential use of solar energy in our society. Some think that 20 to 25 percent of U.S. energy needs could be served by the year 2000 and perhaps 50 percent by the year 2020—if we have an all out program of research and development.

MAJOR CONCEPT

Earth materials have different capacities to absorb and store energy.

OBJECTIVES

☐ Measure, form hypotheses, predict, construct, and interpret graphs.

☐ Handle nontoxic materials and construct simple projects.

☐ Assume responsibility for the wise use of resources.

☐ Explore the environment and combine materials in new ways.

☐ Form alternatives that are based on information.

MATERIALS

About 24 Styrofoam cups, about 24 thermometers, 4 cardboard boxes (save the boxes from ditto paper), plastic wrap, a variety of earth materials: sand, water, soil, leaf litter, small pebbles, or gravel; graph paper; and clocks or wristwatches.

VOCABULARY

collection   solar
control   **variables**
storage

PROCEDURES

A. This activity is to be completed outdoors on a sunny day. Divide the class into four teams. Each group should have 6 cups, 6 thermometers, and 1 box.

   1. Each team fills each cup with a different earth material. Have the students place sand, pebbles, water, cut paper, and nothing, respectively, in five containers. They are to select an earth material for the sixth container. Place the cups in the box and place thermometers in the cups. (Note: place a thermometer in the empty cup and cover the cup with plastic wrap or a similar material.)

   2. Have the students record the temperature after cups have been outdoors for 20 minutes. Then have students bring the cups and boxes indoors and wait an additional 20 minutes before recording the changes in temperature indoors.

3. Have the students graph their results. The graph should be set up as shown below.

4. On the next day, discuss the results. Ask the following questions:
Which material increased in temperature fastest? Slowest?
Which material held the heat energy longest? Shortest?
What material could best be used for storing solar energy?
How could your results be applied to heating the school? Classroom? Homes?

B. The next part of the activity is a challenge. Each student group is to design a system that they think is most efficient for the collection and storage of solar energy.

   1. Inform the students that they will have one week to design the system. They may use earth materials similar to those from Procedure A, but their sys-

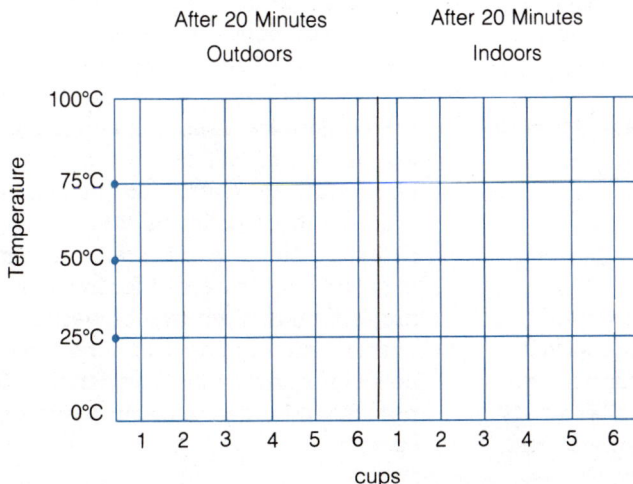

After 20 Minutes Outdoors — After 20 Minutes Indoors. Temperature axis: 100°C, 75°C, 50°C, 25°C, 0°C. Cups axis: 1 2 3 4 5 6 1 2 3 4 5 6.

Students can use the information from their graphs to easily answer the questions about different types of earth material and solar energy.

tems can vary in whatever ways they choose. Encourage creativity and problem solving.

2. Demonstrations of the student solar systems can be completed by taking each system outdoors and placing it in the sun for 20 minutes or whatever interval of time seems best suited to the individual projects to show their greatest effort.

3. Have the students describe how they constructed their collection and storage systems. Ask the class to decide which systems were most effective.

EXTENDING THE ACTIVITY

C. Have a debate on the merits of using various forms of energy, perhaps solar, nuclear, and fossil fuel.

D. Take a trip to a home that uses solar energy.

E. Have the students construct a small house (or modify an old doll house) so that it is heated by solar energy. Test the "solar doll house" to see if it works. The students should devise methods to test the solar efficiency.

ACTIVITY ES–7

Human Interaction and the Earth's Spheres
(Grades 6–8)

ACTIVITY OVERVIEW

In this activity the students study a selected topic related to the broad topics: population, resources, and pollution. Individual students then are the resident "experts" as the class continues the study of human interaction with the earth spheres in various future activites. *Estimated time:* two weeks to prepare the reports; two class periods to introduce their topics.

SCIENCE BACKGROUND

According to the Global 2000 Report to the President (1980), at current and projected rates, the world population is estimated at 6.35 billion people by the year 2000. In 1975 the number was 4 billion. It was one billion in the year 1850. And the world population in 1650 was about one-half million. These statistics give some idea of the accelerating rate at which the human population is growing. The increasing population brings increasing demands on the various earth spheres for resources and disposal of wastes. To meet the needs of a world population of 6.35 billion, food supplies and different energy resources will have to be increased and developed. Air, water, and soil need to be protected from pollution. While water is abundant on earth, it is scarce in many locations where it is needed for irrigation, food production, and removal of wastes. The interaction among population, resources, and pollution is important to understand for it is primarily a human interaction with earth's spheres.

SOCIAL IMPLICATIONS

Is the prospect hopeful for the world population in the year 2000 A.D.? It depends on whether we have the capacity to change those things we must, to develop new technologies where appropriate and, in general, to start these changes now. Humans have the capacity to imagine creatively. It is important for students to acquire fundamental information about the problems and potential solutions being considered today, and to see examples of human creativity at work on solving problems. The future will, in part, depend on the decisions our students make.

MAJOR CONCEPTS

The lithosphere, atmosphere, hydrosphere, and ecosphere can change and still conserve their identities. There are limits to the capacities of the earth's spheres to change. Human interaction with the earth's spheres has resulted in significant changes.

OBJECTIVES

- ☐ Gather and organize information.
- ☐ Describe relations among variables.
- ☐ Actively explore the environment and ask questions.
- ☐ Seek alternative points of view and multiple sources of evidence.
- ☐ Describe situations that require decisions; describe immediate and long-term consequences of various alternatives.

MATERIALS

Pencil, paper.

VOCABULARY

population	interaction
resources	**lithosphere**
pollution	**hydrosphere**
atmosphere	

PROCEDURES

A. Students are to select and prepare a written and oral report on a topic related to one of three broad themes: populations and population growth, the world's resources, and pollution. Possible topics include the following:

☐ *Population*
**Zero population
  growth**
Limiting population growth
Different views of population problem
Density of populations

☐ *Food*
Green revolution
Use of the sea
Malnutrition
Famine and disease
World hunger
Geography of hunger
"New" foods

☐ *Energy*
Fossil fuels
Nuclear
Solar
Wind
Geothermal
Hydroelectric
Tidal energy

Ocean currents
**Biomass
Synthetic fuels**
☐ *Minerals*
Exploration
Mining
Renewable and nonrenewable
Resources *vs.* reserves

☐ *Pollution*
Types of pollution
Air pollution
Waste disposal
Environmental pollution
Technology
Balance in nature
Environmental impacts of growth

☐ *Land*
Wilderness areas
Parks
Urban development
Surface mining

☐ *Land (cont.)*
**Conservation ☐ *Water*
and preserva-** Desalinization of
tion sea water
Soil pollution Pollution
Forests
Wildlife

1. A brief 5 to 10 minute presentation of each topic will be given by students 2 weeks after the assignment.
2. The activity and the student's role extends beyond the oral presentation. The student becomes the "resident expert" in his or her topic as future activities are completed

on the related themes of population, resources, and pollution. The role of the "resident expert" is to provide some of the basic information during future discussions of the activities.

EXTENDING THE ACTIVITY

B. Invite guest speakers to visit the class and discuss some of the many topics. Alternatively, invite students to interview people in the community concerning their attitudes about population growth, resource use, and pollution.

ACTIVITY ES–8

Population Growth
(Grades 6–8)

ACTIVITY OVERVIEW

A riddle on exponential growth rates starts the activity. The students then plot graphs for exponential growth rates and discover that slow early growth followed by rapid later growth presents problems. *Estimated time:* three class periods.

SCIENCE BACKGROUND

There are approximately 200,000 new persons on earth each day, 1.4 million each week, and about 72 million per year. The actual numbers of new persons on earth vary. According to Miller (1982), the net birth rate is about 235 babies per minute, or 338,000 per day. This number should be balanced by the net death rate of about 92 persons per minute or 133,000 per day. All of these new people are to be housed, fed, clothed, and

cared for by the resources presently available: resources are *not* increasing at a commensurate, geometric rate. Those resources that can be increased, such as food, do so at an arithmetic rate. Over one-third of the earth's population is already in need of food, housing, clothing, and other essentials.

The population growth rate is an example of exponential growth. If the growth rate were plotted on a graph, the form of the graph would be a *J-shaped curve.* This is the curve that results when any population grows by doubling: 2, 4, 8, 16, 32, 64, 128, and so on. Growth seems slow at first and then it increases very rapidly as the bend in the *J* is reached. (See the graph in the procedure section.) The population growth rate seems to be leveling off. In order to sustain those living on earth this leveling trend will have to continue.

SOCIAL IMPLICATIONS

One of the most important problems facing humanity is population growth. Population growth is connected to pollution of the environment and depletion of natural resources. We must begin to ask some very hard questions about the carrying capacity of the planet earth. The causes and consequences of population growth influence our lives everyday. It is especially important to understand the rate and dynamics of population growth. This activity introduces some of the basic problems associated with the concept of exponential growth.

MAJOR CONCEPT

There are limits to the environment's capacity to change and still conserve its identity.

OBJECTIVES

☐ Construct and interpret graphs, predict, and form hypotheses.

☐ Combine ideas in new ways; seek alternative points of view.

☐ Describe situations that require decisions.

MATERIALS

Film: *World Population* (ordered from Southern Illinois University, Carbondale, Illinois), graph paper, pencils.

VOCABULARY

population	**exponential**
growth	**linear**

PROCEDURES

A. This activity has one, major goal—to lead the student to an understanding of the exponential growth of a population. Later lessons can deal with birth and death rates and problems of overpopulation.

1. Start with what is now a famous French riddle. Just before the end of class give the students the riddle.

   > There is a lily pond that has a single leaf.
   > Each day the number of leaves doubles.
   > On the second day there are two leaves.
   > On the third day there are four leaves.
   > On the fourth day there are eight leaves. . . .
   > On the thirtieth day the pond is full.
   > When was the pond half full?

   Leave the students with this riddle as homework (the answer is the twenty-ninth day).

B. On the next day start the class with the film *World Population*. This is a four-minute silent film that illustrates world population growth extraordinarily well. Do not say anything about the film; simply start the projector. The students will be amazed at the rate of population growth depicted in the last seconds of the film. They will then have a first conceptualization of the problems of exponential growth. Discussing the film and the riddle should provide an excellent introduction to population growth rates.

C. Ask the students to graph an exponential growth rate. You can use a simulation of two garden snails that have two offspring per month each. And these, in turn, have two offspring a month and so on. Tell the students to construct a table describing the growth of the snail population for a school year of nine months.

Simulated Growth of Snails for a School Year

$2 \times 2 = 4$	1st month (September)
$4 \times 2 = 8$	2nd month (October)
$8 \times 2 = 16$	3rd month (November)
$16 \times 2 = 32$	4th month (December)
$32 \times 2 = 64$	5th month (January)
$64 \times 2 = 128$	6th month (February)
$128 \times 2 = 256$	7th month (March)
$256 \times 2 = 512$	8th month (April)
$512 \times 2 = 1,024$	9th month (May)

D. Have the students graph the results. Graphing will clearly demonstrate the slow, then sudden population increase depicted by the J-shaped curve.

E. Discuss the results of the students' activity and apply their findings to world population growth.

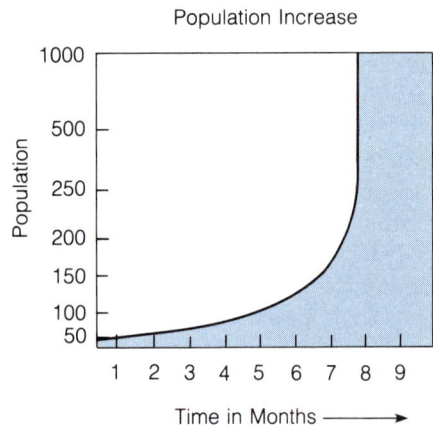

Population Increase

The curved line vividly shows the exponential growth rate.

EXTENDING THE ACTIVITY

F. Find the present rate of world population growth. Determine the rate of population growth for your city or area. Contrast the geometric growth rate of the population with the arithmetic rate of food production. Read about Thomas Malthus and his theory about world populations.

ACTIVITY ES–9

Finding Resources

(Grades 6–8)

ACTIVITY OVERVIEW

A simple activity of finding pennies hidden around the room is used to simulate the difference between reserves and resources. As the activity continues, the students experience the important factors in finding new resources; technology, economics, and environmental effects are developed. The concepts of this activity relate to many resources, including energy. *Estimated time:* two class periods.

SCIENCE BACKGROUND

As population and affluence increase, there is also an increased demand for resources—those earth materials needed by people, populations, and ecosystems. With resource depletion there is continued exploration and extraction of resources by those who provide resources. But as resources are depleted, new reserves are not as easy to find. Exploration takes longer, is more difficult, and more sophisticated technologies are needed. Likewise, extraction takes longer, is more challenging, and more complex equipment is needed. The combination of these factors increases the potential for environmental damage or improvement.

As the availability of many resources has decreased, the amount of **recycling,** reuse, conservation, and substitution has increased. The second half of the twentieth century has witnessed a marked increase in efforts to conserve the earth's resources.

SOCIAL IMPLICATIONS

The use of the earth's resources is related to economics, and therefore affects people's attitudes and values toward the use of resources. To illustrate this point, small, gas-efficient cars are in greater demand, reflecting changes in public attitudes and values about the conservation of our finite sources of fossil fuel. Similarly, the increased use of wood burning stoves and fireplaces to heat homes has been brought about in part by soaring costs of gas and electricity for home heating. Thus, wood as a source of heating, once rejected for its smoke-producing character, now has had wider acceptance because of its economy. However, although forests are a renewable resource they are being used faster than they are being replenished.

MAJOR CONCEPTS

Environments change when people use the earth's resources. There are limits to the earth's resources and to environmental change.

OBJECTIVES

☐ Observe, compare, measure, predict, and graph.

☐ Use tools that require fine adjustment and discrimination; handle nontoxic substances.

☐ Explore the environment; combine objects and materials in new ways; demonstrate the need and means to use natural resources wisely.

☐ Describe situations that require decisions; describe immediate and long-term consequences of alternative decisions.

MATERIALS

Tweezers, 100 pennies, toothpicks, rulers, small mirrors, and other items commonly available in the classroom.

VOCABULARY

economics **reserves**
environmental impact resource
exploration technology

PROCEDURES

A. Hide 100 pennies around the room. About 25 should be in locations that are visible and easy to reach. The remaining 75 should be placed out of view and in places more difficult to reach (under chairs, under paper and boxes, on ledges, between tiles, in erasers, and so on).

1. Divide the students into groups of four. Tell them that the activity simu-

lates the exploration and extraction of earth materials (resources) that are needed by individuals, populations, or ecosystems. There are pennies located around the room that represent these resources.

2. Have the students go on an *exploration*. Give each group 5 minutes to look around the room to determine how many pennies are in the room. The rules of the exploration are: (1) they cannot touch, turn over, dig up, or otherwise change anything in the room; and (2) they cannot collect any pennies; all pennies must remain where they are. Give the groups 5 minutes to look for the pennies.

3. Call the class together and have the different groups report on the number of pennies they observed. Place the numbers on the chalkboard along with their definitions. *Reserves* are the amount of an earth material in known locations. *Resources* are the total amount of an earth material.

Reserves and Resources

Groups	1 2 3 4 5 6 7
Reserves	
Estimated Amounts of Resource	
Actual Amounts of Resource	

4. Ask the groups to estimate the resources (the actual number of pennies) that are in the classroom based on the number they actually saw. Record their estimates on the chalkboard

using a format similar to the one shown.

5. In the next portion of the activity (about 15 minutes) the students will actually "mine" the resources. Tell the students that they are to keep track of the number of pennies they find and obtain each minute. Do not indicate how many are available or where they are located. But do indicate that the students must move things and use tweezers, toothpicks, rulers, mirrors, or other classroom items to look for and extract the pennies. That is, students cannot simply turn a book over and look for a penny; they cannot lift up a desk and remove a penny from under a leg. Instead of using only their hands or other parts of their bodies to "extract" resources, they must find *tools* in the classroom to do the job.

6. Have students complete a graph of the number of pennies found each minute over the 15 minutes. In general, their graphs will look like the one shown. With time there will be a steady decrease in the number of pennies found.

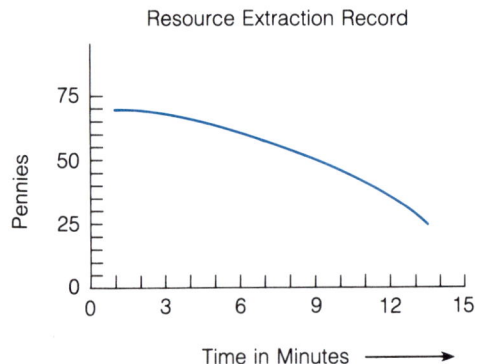

Resource Extraction Record

Graphing the results clearly shows students the diminishing amount of resources that can be extracted from the earth over a period of time.

7. Discuss the activity. Start by having the students explain what happened as their extraction of resources continued. Determine if they found all of the resource. *How close were they in their estimates of the resource? What problems did they experience with time? How is this like the actual extraction of resources?* Be sure to point out their increased use of "technology" (tweezers, rulers, mirrors, and so on), the possible environmental effects (turned over books, papers, chairs) and the economics (it took more time, energy, and technology to obtain fewer pennies as the period progressed). Introduce examples of natural resources, which are described in chapter 13, and that illustrate the concept.

B. Provide the students with a topic such as fossil fuel use. Have them describe the situation in terms of the known available resource, reserves, technology, economics, and environmental effects. Other topics might include air, water, trees, whales, and minerals including chromium, iron, gold, or silver.

EXTENDING THE ACTIVITY

C. Have a speaker from an oil company come to class and tell about the exploration for and extraction of petroleum.

ACTIVITY ES–10

Consuming Resources
(Grades 6–8)

ACTIVITY OVERVIEW

By keeping track of the amount of a natural resource they use, students are led to a basic understanding of **supply and demand.** The concepts of **scarcity** and **depletion** are also introduced. The activity ends with a discussion that relates these ideas to the use of natural resources. *Estimated time:* two class periods.

SCIENCE BACKGROUND

Many of the problems associated with human use of natural resources are related to the fundamental concepts of supply and demand. When scarcities of an earth material such as fossil fuels exist, some try to increase the supply while others try to decrease the demand. In the case of the former position, the proponents argue that scarcity increases prices, which decreases demand and subsidizes further exploration for the resource; this leads to a greater supply. On the other hand, those who think that supplies are finite, argue that no amount of technology or price increase can produce materials that do not exist. So, they argue, we must reduce the demand for the material and maintain the present and any future supply through reuse, recycling, and conservation.

SOCIAL IMPLICATIONS

The usual view of supply and demand as it relates to consuming resources is simple—

prices fall when supply is great and prices rise when supply is limited. But this is not the case; the relationship is affected by complex social interactions. Several factors that influence prices are not directly a result of the availability of the resource: (1) government and industry exercise some control over the supply of resources; (2) the actual price of a product has many hidden costs that have little to do with the resource; and (3) the energy used to recover a resource can limit its use. In the activity that follows you will find another example of the relationship between science (in this case, the study of the earth's resources) and society, including society's attempts to regulate the use of natural resources through the application of economic principles.

MAJOR CONCEPTS

Environments change when people use the earth's resources. There are limits to the earth's resources and to environmental change. Environments can lose their identity when they exceed their capacity to change.

OBJECTIVES

☐ Graph, infer, predict, describe relations among variables.

☐ Seek alternative points of view; demonstrate the need and means to use natural resources wisely.

☐ Clarify consequences of alternative actions.

MATERIALS

A large, clear container; 1,500 raisins; graph paper; and pencils.

VOCABULARY

resources
scarcity
supply and demand

PROCEDURES

A. To begin the activity explain that each student will be buying **shares** in an imaginary supply of oil that has been discovered recently. The jar of raisins represents the oil supply. There will be several rounds of share-buying, and the company selling shares will place limits on the number of shares that can be purchased at any one time of bidding.

1. Distribute the graph paper and help the students set up a graph such as the one shown. Use an overhead projector or the chalkboard to assist them. Or, create the graph on a ditto-master and run the graph paper through the ditto machine.

2. Begin the first round of bidding by asking the students if they want one or two shares per person. Have them raise their hands to vote for one or two. Have a student take from the jar the number of shares (raisins) for which the majority voted. For example, in a class of 30 students, if the majority vote was 2 shares per student, the volunteer should remove 60 shares (raisins) from the supply of 1,500 shares. (It will take too much time to actually distribute the shares during class.) Next, graph the results. (First round = 1,500 − 60).

3. Proceed to the next rounds of bidding. Repeat the procedures using the following options:
   2 or 4 shares in the 2nd round
   4 or 8 shares in the 3rd round

Supply and Demand

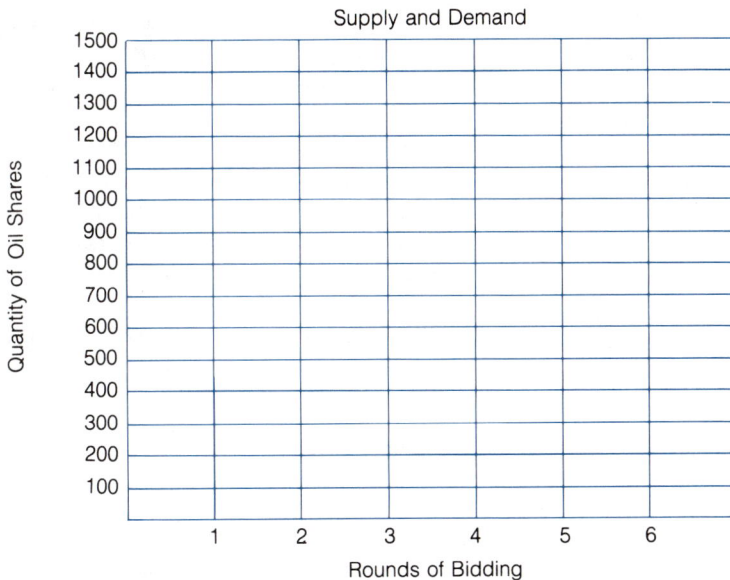

Quantity of Oil Shares

Rounds of Bidding

Guide students to draw conclusions about the supply and demand based on the data on their charts.

8 or 16 shares in the 4th round
16 or 32 shares in the 5th round
32 or 64 shares in the 6th round

4. Some students will realize that they cannot continue taking the largest number. Ask students to use their graphs to project how many shares (raisins) will be needed to meet the *demand* of each round.

Round 2 for 30 students = 60 or 120 shares

Round 3 for 30 students = 120 or 240 shares

Round 4 for 30 students = 240 or 480 shares

Round 5 for 30 students = 480 or 960 shares

Round 6 for 30 students = 960 or 1,920 shares

5. When the students realize that the supply decreases markedly as the demand increases, introduce the concept of *scarcity*. There is a scarcity when the demand for an earth material is approaching the supply of the material.

6. Stop the activity and discuss the following questions: How would you define supply? What is demand? What does it mean when we say a resource is being depleted? What could be done to reduce the scarcity of oil needs? Are the shares renewable or nonrenewable? What would happen if there were no more raisins (oil shares or other natural resources) in the world? How could you use the shares you saved? Would they be worth more, less, the same as when we had a large supply? Why? What could be done to make the supply of oil shares last longer? As the students answer these questions, you can extend their understanding of the concepts of supply and demand by referring to other scarce resources that are also familiar to them.

EXTENDING THE ACTIVITY

B. Give the students the following information: (1) the number of students in the school; (2) the estimated pieces of paper used a day; (3) the estimated increase in paper use over the next nine months will be 2 percent per month; and (4) the amount of paper for the next nine months.

You can use different numbers in the problem depending on the level of students. Have the students plot out the increased demand for paper. Then have them explain what will happen if the supply of paper is *fixed* for each month or if the supply can be *increased* or *decreased* each month.

C. Invite the social science teacher to come and discuss economics. Be sure to inform him or her of the activity you have introduced so the economics discussion will relate concepts in the students' experience.

D. Challenge the students to find examples of supply and demand and scarcity in their community.

BIBLIOGRAPHY

Teachers' Resources

Amidei, Rosemary. *Environment: The Human Impact.* Washington, D.C.: National Science Teachers Association, 1977.

Barney, G., ed. *The Global 2000 Report to the President: Entering the Twenty-First Century.* Washington, D.C.: U.S. Government Printing Office, 1980.

Brown, Lester R. *Building a Sustainable Society.* New York: W W Norton & Co., 1981.

Charles, Cheryl. *Project Learning Tree.* Washington, D.C.: The American Forest Institute, 1975.

Council on Environmental Quality. *Global Future: Time to Act.* Washington, D.C.: U.S. Government Printing Office, 1981.

Fowler, Kathryn M. *Hunger: The World Food Crisis.* Washington, D.C.: National Science Teachers Association, 1977.

Fowler, Kathryn M. *Population: The Human Dilemma.* Washington, D.C.: National Science Teachers Association, 1977.

Friends of the Earth. *Progress As If Survival Mattered.* San Francisco: Friends of the Earth, 1977.

Garigliano, Leonard, and Knope, Beth J. *Environmental Education in the Elementary School.* Washington, D.C.: National Science Teachers Association, 1977.

Matthews, William H. *Helping Children Learn Earth-Space* Science. Washington, D.C.: National Science Teachers Association, 1971.

Mervine, Kathryn, and Cawley, Rebecca. *Energy-Environment Materials Guide.* Washington, D.C.: National Science Teachers Association, 1975.

Miller, G. Tyler, Jr. *Living in the Environment.* 3rd ed. Belmont, CA: Wasdsworth Publishing Co., 1982.

Salk, Jonas, and Salk, Jonathan. *World Population and Human Values.* New York: Harper & Row, 1981.

Children's Resources

Blumbery, Rhoda. *The First Travel Guide to the Moon: What to Pack, How to Go, and What to See When You Get There*. New York: Four Winds (Scholastic),1980.

Coal: The Rock that Burns (film). Lawrence: Centron Education Films, 1976.

The Daily News (Longview, WA) and *The Journal American* (Bellevue, WA). *Volcano: The Eruption of Mount St. Helens*. Seattle: Madrona Press, 1980.

Discovering Fossils (filmstrips). Chicago: Coronet, 1977.

Exploring Space: The Moon and How It Affects Us (film). Chicago: Coronet.

Fields, Alice. *The Sun*. New York: Franklin Watts, 1980.

McFall, Christie. *Underwater Continent*. New York: Dodd, Mead & Co., 1975.

Navarra, John Gabriel. *Earthquake!* Garden City: Doubleday Publishing Co., 1980.

Oceans: A Key to the Future (filmstrips). United Learning, 1977.

Schultz, Charles M. *Charlie Brown's Second Super Book of Questions and Answers: About the Earth and Space . . . From Plants to Planets!* New York: Random House, 1977.

Simon, Seymour. *Projects with Air: A Science at Work Book*. New York: Franklin Watts, 1975.

15

Physical Science

The purposes of this chapter are to present some basic topics and issues in physics and to provide you with a more knowledgeable basis for teaching physical science. These are necessarily brief discussions. The reader is also encouraged to read current, introductory texts in physics and the physical sciences and to use the bibliography at the end of this chapter. Information in this field changes rapidly; keeping up-to-date is therefore an important task for the science teacher.

KEY CONCEPTS AND IMPORTANT IDEAS

☐ physical science ☐ matter ☐ energy ☐ atoms ☐ elements ☐ atomic number and weight ☐ periodic table ☐ molecules ☐ solids, liquids, and gases ☐ mass and weight ☐ density ☐ motion ☐ inertia ☐ measurement of energy ☐ heat ☐ light energy ☐ chemical energy ☐ electrical energy ☐ nuclear energy ☐ radioactivity

The **physical sciences** deal with **matter** and **energy**. These two things are really interchangeable, according to Einstein's most famous equation, $E=mc^2$. However, scientists have found it appropriate and convenient to handle matter and energy separately and so this chapter is organized in that way, too.

Matter is defined as anything that occupies space. This seems to be a pretty big subject, but modern scientific discoveries allow us to survey material and to convey basic concepts about matter. First we will discuss the composition of matter using the concepts of atoms and molecules. Second, the states of matter; that is, **solids, liquids,** and **gases,** are discussed along with the conditions required for transitions between those states. Third, the dimensional properties of matter, such as mass and density are described.

Up until now matter has been described as if it were fairly static. In order to describe matter in motion, we must study energy. Energy is defined as the capacity to do **work,** and work cannot be accomplished in the scientific sense of the word without motion. In part of this section, the various forms of energy, including nuclear energy, are discussed.

The physical sciences are concerned with accurately describing the natural world and discovering the **laws** that govern its events. These laws are universal in the sense that the events that take place under given conditions in laboratories here on earth are assumed to be equally true in the farthest reaches of the universe. These laws are considered permanent in that once they are verified, it can be assumed that these conditions have been true since the origin of the universe — and will

PHYSICAL SCIENCE The study of matter and energy

MATTER Anything that occupies space

ENERGY Capacity for doing work

SOLID Something that offers resistance to pressure and is not easily changed in shape

LIQUID A substance that takes the shape of its container

GAS A substance that is not a solid and is not confined to the shape of its container

WORK Force exerted over a distance

LAW A description of a physical event that is mathematically testable

When physicists use the term *work* they mean a combination of force exerted over distance. Is work being done in this picture of a man pushing his car? (Young/Hoffines)

continue to be true in the future. It is inherent in scientific methods that laws are continuously being refined. In other words, physical laws are the best descriptions that scientists have at any given time to explain as many phenomena as are known at that time.

Another important point to keep in mind while reading this chapter is that scientists often define words in such a way that their meanings are very restricted and even very different from their everyday, nonscientific meanings. The reason for this can be made obvious—by narrowly defining terms, we gain more specificity and more assurance that we are all talking about the same thing. For example, let's take the word "work." For most people, even scientists, the first thing that comes to mind is their job or maybe mowing the lawn or washing the floor. But when the word is used scientifically, the meaning narrows considerably. Here work means a combination of force exerted over a distance. A feather carried one inch has work done to it. A 50-pound suitcase held motionless 2 feet off the ground does not, scientifically speaking. This is because, although there is a great deal of effort required to hold the suitcase up, there is no motion involved. In this chapter the scientific definitions of words are supplied; otherwise you can assume that the dictionary definition applies.

MATTER—WHAT IS IT?

What is matter? Its basic definition, as given in the introduction is anything that occupies space. Unfortunately, this is not very satisfying to anyone who really wants to know what matter is: What is it composed of? How does it occupy space? What forms does it take? How can it be described quantitatively? For centuries, scientists and philosophers have answered these questions. But it is only in this century that theories have been proposed that provide accurate descriptions of current events along with the ability to predict future ones.

The Atom—The Theme

ATOM The smallest particle of an element that has the properties of that element and that can undergo a chemical reaction

Matter is composed of many kinds of elementary particles. Under some conditions the individual particles exist independently. On earth, almost all matter is made up of these particles organized into **atoms.** Their diameter is less than 1/100 millionth of an inch. It would take more than one million of them put edge to edge to be equal to the thickness of one page in this book.

Although particles come in a bewildering array, over 200 at last count, there are three basic sorts— positively charged, negatively charged, and uncharged.

The concept of charge is sometimes difficult to understand at first. The physicists Maxwell and Faraday found it most fruitful to think of charge as creating a disturbance or condition such that when there is another charge present, it feels a **force;** it is caused to move. The direction of the movement determines if the two charges are of the same **polarity.** If they are of different charge, they will attract each other, and if they are the same, they will repel each other.

One must be careful not to make the implication that a positive charge has something more than a negative charge. It was simply decided that the charge of the **electron** is negative and the charge of a **proton** is positive. In addition, the amount of charge carried by an electron is exactly equal to the charge carried by a proton.

The most important particles that make up atoms are the electrons, the protons, and the **neutrons,** which are uncharged.

At the center of the atom is the **nucleus,** which is made up of protons and neutrons. Electrons are found surrounding the nucleus. (see figure 15–1). The distances between the nucleus and its electrons are *relatively* many times greater than the distance between the sun and the earth. In order to get some idea of the relative sizes and distances in an atom, imagine a grain of salt suspended from the dome of St. Peter's basilica. Floating around that grain of salt are a few flecks of dust. The grain of salt represents the nucleus and the dust particles the electrons. This gives some idea of the relative sizes of the nucleus and the electrons, and the tremendous amount of empty space contained by an atom.

The atom was believed to be a solid indivisible particle from about 400 B.C. until the beginning of the twentieth century! In 1911, Ernest Rutherford, (1871–1937), a New Zealand scientist, demonstrated that the atom's positive charge was concentrated at its center with electrons

FORCE The cause (or push and pull) of motion

POLARITY The property of having two poles, positive and negative

ELECTRON A negatively charged particle that surrounds the nucleus of an atom

PROTON A positively charged particle in the nucleus of an atom

NEUTRON The uncharged particles in the nucleus of an atom

NUCLEUS The central part of an atom which is positively charged

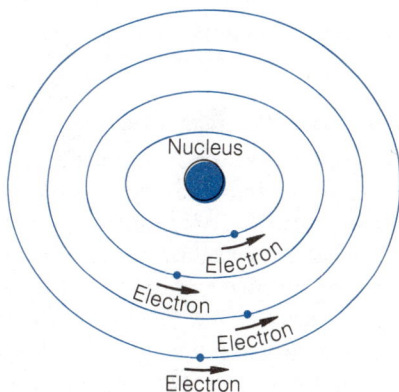

FIGURE 15–1
This rendition of Bohr's early twentieth century view of an atom shows precise, predictable orbits.

swarming around the nucleus. Two years later in 1913, Niels Bohr (1885–1962), a Danish physicist, modified Rutherford's model with the idea that all electrons move within distinct spherical shells around the nucleus. But how is this possible? The protons of the nucleus ought to repel each other and make the formation of the nucleus impossible. On the other hand, the electrons which remain some distance from the nucleus should be attracted to the protons in the nucleus and the whole structure should collapse upon itself. It is true that all these contradictory forces do exist, but there are overriding considerations. For example, the attractions between the electrons and protons are counteracted by the incredible speeds at which electrons travel—up to 600 miles per second.

Today Bohr's view of discrete electrons with precise, predictable orbits is gone. There is a growing belief that matter is best described in terms of probabilities that are much like the probability we hear on weather forecasts. The area of physics concerned with describing the probability that particles will be in a given place and have a given amount of energy is known as quantum mechanics.

A number of vital questions about the atom are still unexplained, particularly with respect to the organization of the atomic nucleus. According to one older theory, the nucleus might be a system organized like an atom itself, with nuclear particles revolving in orbits or shells just like Bohr's model for the motion of electrons around the nucleus. Other useful theories pictured the nucleus as a drop of liquid with nuclear glue holding the protons and neutrons together.

NUCLEI The plural of nucleus

Scientists continue to bombard atomic **nuclei** with particles that have enormous velocity in an attempt to separate from one another the constituent protons, neutrons, and many other strange and unidentified particles. These particles are like clowns tumbling out of a car in the circus ring. Newly discovered particles that have been identified have names right out of science fiction: quarks (from a word coined by James Joyce in "Finnegan's Wake"), hadrons, baryons, mesons, tau particles, neutrinos, leptons, gluons, deuterons, and basons.

Elements—Variations on the Theme

ELEMENT Any substance that cannot be separated into different substances without nuclear disintegration

Let's take the concept of an atom one step further and show how electrons, protons, and neutrons combine into specific kinds of atoms that are recognizable as **elements.** The only rule we need for doing so is that the atom has to be electrically neutral—meaning that the number of electrons has to equal the number of protons.

HYDROGEN Discovered in 1766, it is the third most abundant and lightest element; it is almost never found free on earth, but the sun and stars are almost pure hydrogen

Imagine that we have a box with three compartments filled with electrons, protons, and neutrons. If we select one electron and one proton and combine them, the element, **hydrogen** is produced. If we now add a neutron, the atom remains electrically neutral, and the element *deuterium,* a form of hydrogen is produced. We can add yet another neutron and the element *tritium,* another form of hydrogen is produced. All three forms of hydrogen occur naturally but the form with just one proton is far and away the most abundant.

Let's add another proton; the nucleus now has two positive charges and only one negative charge so one more electron must be added. In this way the element **helium** is formed. This compound rarely reacts chemically and is so extremely stable that it cannot accept another electron.

Continuing in the same vein, if we add a proton, an electron, and a neutron or two, **lithium** is formed. Lithium is an extremely reactive compound.

This process could be continued to produce the 88 naturally occurring elements. If it were continued even further, the 18 elements that have been produced artificially will also be generated. Amazingly, there is not a single gap in the series nor is there any known case in which two elements have the same number of protons.

Let's go back for just a moment to the hydrogens we generated. Why is tritium, which has 2 neutrons, considered a form of hydrogen when helium, which also has 2 neutrons, is considered an entirely different element? The answer is that tritium behaves chemically in the same way as hydrogen, whereas helium hardly ever reacts at all. The most important factor in determining the way each element reacts chemically is the number of protons in its nucleus. The number of electrons plays a real but secondary role. The number of neutrons has no significant bearing on chemical reactivity.

Another way of describing an element is to give its **atomic number,** which is the number of protons in its nucleus. This is very useful because it allows chemists to predict the behavior of an element.

All elements, whether natural or artificial, can be grouped together in families. The hallmark of each family is the way it behaves when introduced to elements of the other families. When the elements are arranged in order of increasing numbers of protons, a regular pattern can be established with respect to the various chemical families.

The first person to arrange this pattern was the Russian scientist, Dmitri Ivanovich Mendeleev (1834–1907) in the year 1864. He collected all the evidence that he could find about every known element in nature, because he was interested in finding a regularity or pattern among the elements (other than an alphabetical listing). Some data were already available. He knew that halogens, such as **chlorine** and **bromine,** had common characteristics. He also knew that certain metal elements were quickly oxidized when exposed to air. A common example of this is when iron rusts in air. He was aware of the noble metals, **copper, silver,** and **gold,** because they were durable and corrosion resistant.

HELIUM A very rare inert gas that is mainly found in natural gas fields; it has the lowest boiling point of any known element—minus 268.9°C

LITHIUM The lightest of all solid elements, it forms a black substance when exposed to air; used in ceramics, alloys, H-bombs, and to treat psychological disorders

ATOMIC NUMBER The number of protons or positive charges in an atom

CHLORINE Weak, corrosive substance that can be used as a bleach, a disinfectant, and a nerve gas

BROMINE A caustic liquid used as a disinfectant, in nerve gas, and in gasoline anti-knock compounds

COPPER A reddish-brown metal that is malleable and easily combined with other metals (e.g., gold, silver, tin, zinc)

SILVER A good conductor of heat and electricity used in photographic film

GOLD The most malleable metal, its use is primarily in coins, jewelry, and dental units

ATOMIC WEIGHT The weight of an atom of an element compared with the weight of an atom of an element arbitrarily selected as standard

FLORINE The most reactive of nonmetals, florine corrodes platinum and is highly flammable

SODIUM The sixth most abundant element on earth, soidum is used in compounds such as salt, baking soda, borax, and lye

GALLIUM A metal that melts in your hand and is the only metal that expands when it freezes

CALCIUM Preserved in the body—about two pounds in each adult—mostly in teeth and bones; regulates heartbeat and muscle contraction

TITANIUM Element that is used as paint pigment and in supersonic aircraft

SCANDIUM Element that is almost as light as aluminum and has a much higher melting point; no use for this metal is known yet

GERMANIUM A metal in the carbon family that is used in transistors

URANIUM The heaviest atom among the natural elements, uranium has a half life of 45 billion years

PERIODIC TABLE An arrangement of the chemical elements according to their atomic numbers

It is said that he created a system on his laboratory wall as a way of organizing his information. Each element was assigned a card that contained the **atomic weight** of each element and the other characteristics of the element and its compounds. He arranged and rearranged the cards on the wall until he eventually found that the cards could be placed in rows and columns so that the chemical and physical similarities of the groups of elements were reflected by the arrangement.

He put hydrogen in the first row in a special place because of its unique characteristics. The next seven known elements (helium was unknown at the time), from lithium to **florine,** he put in a sequence reflecting their increasing atomic weights.

In the second row, he arranged another seven—**sodium** to chlorine. In these two rows there seemed to be a regularity or periodicity of chemical behavior because in the first column there were two alkaline metals, lithium and sodium. In the seventh column were two halogens. It was the genius of this scientist that he could see the relations among the elements without knowing why they occurred; the idea of protons, electrons, and neutrons had not yet come to the fore.

As he organized his cards, he discovered places where there should have been elements to fit specific combinations of characteristics. While there were many scientists who were trying to accomplish the same task, the unique thing about Mendeleev's chart was that he left those places blank on his chart. He believed that one could predict the characteristics of undiscovered elements, and he was absolutely correct. In 1875 one of his empty spaces was filled by **gallium.** Another blank space between **calcium** and **titanium** was filled by an element discovered in 1879 in Sweden called **scandium.** In 1866 **germanium,** a material that is basic to today's transistors was discovered.

Within a few years of Mendeleev's death in 1907 laboratories were producing new elements to fit his chart. His chart began with number 1 (hydrogen) and ended with 92 (uranium). It is now believed that no more natural elements will ever be discovered between hydrogen and **uranium.**

Today the **periodic table** is found hanging on the wall of every chemistry classroom and in every chemistry text. Figure 15–2 presents such a table. The table is now organized into 107 boxes, each representing an element. (Any element with an atomic number greater than 92 is synthetic.) The top line gives the atomic weight of the element, the middle one the accepted abbreviation for the element, and the bottom line gives the atomic number.

Molecules—More Variations on the Theme

In our discussion up until now, we have examined elementary particles and put them together to form atoms. It is also possible to join atoms

Periodic Table

(Based on Carbon 12 = 12.0000)

Metals | Nonmetals | Noble Gases | Transition Elements

	IA	IIA	IIIB	IVB	VB	VIB	VIIB	VIIIB	VIIIB	VIIIB	IB	IIB	IIIA	IVA	VA	VIA	VIIA	VIIIA
1	1 **H** Hydrogen 1 0079																	2 **He** Helium 4 00260
2	3 **Li** Lithium 6 941	4 **Be** Beryllium 9 01218											5 **B** Boron 10 81	6 **C** Carbon 12 011	7 **N** Nitrogen 14.0067	8 **O** Oxygen 15.9994	9 **F** Fluorine 18 998403	10 **Ne** Neon 20 179
3	11 **Na** Sodium 22 9898	12 **Mg** Magnesium 24 305											13 **Al** Aluminum 26 98154	14 **Si** Silicon 28 0855	15 **P** Phosphorus 30 97376	16 **S** Sulfur 32 06	17 **Cl** Chlorine 35 453	18 **Ar** Argon 39 948
4	19 **K** Potassium 39 0983	20 **Ca** Calcium 40 08	21 **Sc** Scandium 44 9559	22 **Ti** Titanium 47 90	23 **V** Vanadium 50 9415	24 **Cr** Chromium 51 996	25 **Mn** Manganese 54 9380	26 **Fe** Iron 55 847	27 **Co** Cobalt 58 9332	28 **Ni** Nickel 58 71	29 **Cu** Copper 63 546	30 **Zn** Zinc 65 38	31 **Ga** Gallium 69 737	32 **Ge** Germanium 72 59	33 **As** Arsenic 74 9216	34 **Se** Selenium 78 96	35 **Br** Bromine 79 904	36 **Kr** Krypton 83 80
5	37 **Rb** Rubidium 85 4678	38 **Sr** Strontium 87 62	39 **Y** Yttrium 88 9059	40 **Zr** Zirconium 91 22	41 **Nb** Niobium 92 9064	42 **Mo** Molybdenum 95 94	43 **Tc** Technetium 96 9062*	44 **Ru** Ruthenium 101 07	45 **Rh** Rhodium 102 9055	46 **Pd** Palladium 106 4	47 **Ag** Silver 107 868	48 **Cd** Cadmium 112 41	49 **In** Indium 114 82	50 **Sn** Tin 118 69	51 **Sb** Antimony 121 75	52 **Te** Tellurium 127 60	53 **I** Iodine 126 9045	54 **Xe** Xenon 131 30
6	55 **Cs** Cesium 132 9054	56 **Ba** Barium 137 33	71 **Lu** Lutetium 174 967	72 **Hf** Hafnium 178 49	73 **Ta** Tantalum 180 9479	74 **W** Tungsten 183 85	75 **Re** Rhenium 186 2	76 **Os** Osmium 190 2	77 **Ir** Iridium 192 22	78 **Pt** Platinum 195 09	79 **Au** Gold 196 9665	80 **Hg** Mercury 200 59	81 **Tl** Thallium 204 37	82 **Pb** Lead 207 2	83 **Bi** Bismuth 208 9808	84 **Po** Polonium 208 98244*	85 **At** Astatine 209 98704*	86 **Rn** Radon 222*
7	87 **Fr** Francium 223 01976*	88 **Ra** Radium 226 0254	103 **Lr** Lawrencium 256 099*	104 257*	105 260*	106 263*	107 258*											

Rare Earth Elements

LANTHANIDE SERIES	57 **La** Lanthanum 138 9055	58 **Ce** Cerium 140 12	59 **Pr** Praseodymium 140 9077	60 **Nd** Neodymium 144 24	61 **Pm** Promethium 144 91279*	62 **Sm** Samarium 150 4	63 **Eu** Europium 151 96	64 **Gd** Gadolinium 157 25	65 **Tb** Terbium 158 9254	66 **Dy** Dysprosium 162 50	67 **Ho** Holmium 164 9304	68 **Er** Erbium 167 26	69 **Tm** Thulium 168 9342	70 **Yb** Ytterbium 173 04
ACTINIDE SERIES	89 **Ac** Actinium 227 02779*	90 **Th** Thorium 232 0381	91 **Pa** Protactinium 231 0359*	92 **U** Uranium 238 029	93 **Np** Neptunium 237 0482	94 **Pu** Plutonium 244 06424*	95 **Am** Americium 243 06139*	96 **Cm** Curium 247 07038*	97 **Bk** Berkelium 247 07032*	98 **Cf** Californium 251 07961*	99 **Es** Einsteinium 254 08805*	100 **Fm** Fermium 257 09515*	101 **Md** Mendelevium 258*	102 **No** Nobelium 255 093*

FIGURE 15–2

The Periodic Table reflects the same arrangement of all known elements on earth as was first conceived by Mendeleev in 1864.

From *Chemistry: A Modern Course* by Robert C. Smoot, Jack S. Price, and Richard G. Smith. Copyright © 1983 by Bell & Howell Company. Used by permission of Charles E. Merrill Publishing Company.

together into **molecules.** The atoms in the molecules are held together by bonds which are the result of electrical attractions that exist between positive and negative charges in the atoms. Figure 15–3 shows some examples of different sorts of molecules.

There are two major classes of molecules. The first type, those that have atoms that are identical, are called elements. **Oxygen** as it is found in the atmosphere is a molecule containing two oxygen atoms. Atmospheric nitrogen also contains 2 **nitrogen** atoms, and hydrogen also contains 2 hydrogen atoms. Ozone, which is produced in the atmosphere by electrical discharges and is found in smog, contains 3 oxygen atoms. The chemical formula for ozone is O_3: the O being the abbreviation for oxygen and the subscript 3 indicating that the molecule contains 3 atoms of oxygen.

The second type of molecule contains 2 or more different kinds of atoms and is called a compound. Water is a compound because it contains 2 hydrogen atoms and 1 oxygen atom, thus the familiar formula,

MOLECULE The smallest particle of a pure substance (element or substance) that can exist and still retain the physical and chemical properties of the substance

OXYGEN A tasteless, odorless gas that makes up 20 percent of the earth's atmosphere, it is needed to burn most carbon-containing compounds and is required for the complete metabolism of foods

NITROGEN A gas making up 78 percent of the earth's atmosphere and used in the manufacture of fertilizers

Dmitri Mendeleev (1834–1907)

In nearly every high school chemistry classroom, there is a copy of Mendeleev's contribution to science—the periodic table. Mendeleev was the youngest of seventeen children in a Russian family that lived in Siberia. When he was fifteen, he and his mother traveled to Moscow where she wanted him to develop his skills as a scientist. The college admission officer did not think he had these abilities and enrolled him in the Pedagogical Institute instead. Mendeleev developed his teaching abilities alongside his skills in chemistry. He be-

came so respected that in later years when he married his second wife before his divorce from his first wife was completed, the Czar is said to have commented, "Mendeleev has two wives yes—but I have only one Mendeleev."

In 1868 he published his major textbook in chemistry. As he was writing this book, he was confronted with what to him was an unacceptable lack of organization of the facts of inorganic chemistry. He acquired all he could find out about each then known element—and then searched to see if there was any order to this knowledge. On the wall of his lab, he developed his own pinups. On each card, he wrote all he knew about each element. Eventually he developed a system of rows and columns to reflect similarities between groups of elements. In his system he discovered these groups or "periodic happenings" without knowing why they occurred.

As he worked out his periodic happenings, he also reasoned that there must be blank spaces in the order to represent elements that had not yet been discovered. He left spaces in his periodic table to account for periodic happenings that could be imagined as missing from his pattern.

Mendeleev's logical reasoning made it possible to order more than sixty-five known elements into a scheme that permitted later scientists to search for and discover new elements whose properties corresponded to the missing elements in Mendeleev's periodic table.

Elements

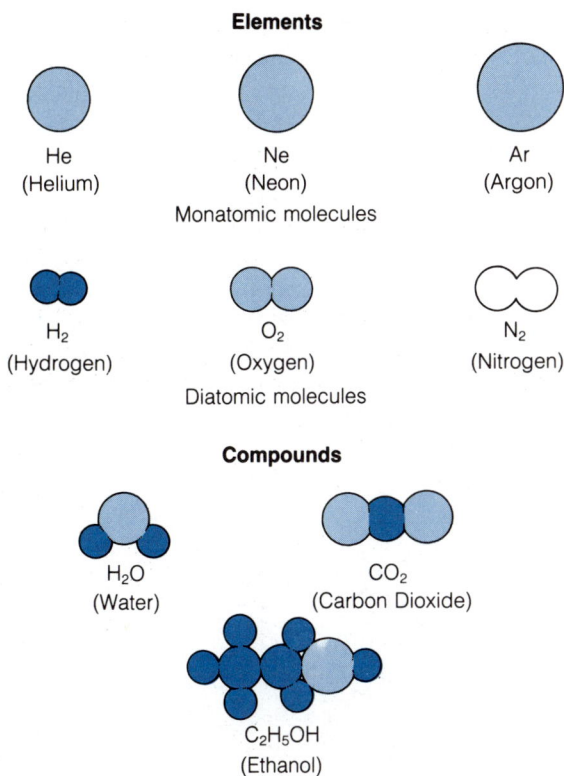

He
(Helium)

Ne
(Neon)

Ar
(Argon)

Monatomic molecules

H₂
(Hydrogen)

O₂
(Oxygen)

N₂
(Nitrogen)

Diatomic molecules

Compounds

H₂O
(Water)

CO₂
(Carbon Dioxide)

C₂H₅OH
(Ethanol)

FIGURE 15–3
Of the substances that surround you, only a few are composed of pure elements; slightly more than 100 elements occur throughout the earth. Compounds—those substances containing two or more elements—outnumber the elements by millions.

CARBON An element that is found in pure form as diamonds, charcoal, and graphite; it is also in an endless variety of compounds, and is an indispensable source of such varied products as nylon, gasoline, perfume, plastics, shoe polish, DDT, and TNT

MAGNESIUM Eighth most abundant element; found in sea water and in dolomite

NEON The best known of the inert gases; used in sign advertising and lights

MERCURY A liquid element used in thermometers, barometers, silver dental inlays, silent electrical switches, and blue-hued mercury vapor street lamps; sometimes called liquid silver

H_2O. The formula for carbon dioxide is CO_2, where C is the abbreviation for **carbon** and O, of course, is oxygen. However, not all molecules have just 2 or 3 atoms. For example, benzene, written C_6H_6, has 12 atoms. Chlorophyll is the component in plants that makes photosynthesis possible; it has the formula $C_{55}H_{72}MgN_4O_6$, which amounts to 55 carbon atoms, 72 hydrogens, 1 **magnesium,** 4 nitrogens, and 6 oxygens for a grand total of 138 atoms.

The most important point about all compounds is that their atomic composition is always the same regardless of their source. Water from the Mississippi River has the same atomic structure as rain water. The carbon dioxide that you breathe out has the same atomic composition as the carbon dioxide used by a plant during photosynthesis, and so on.

STATES OF MATTER

Under earth's atmospheric conditions, some compounds, such as chlorine and **neon,** are gases. Others, for example **mercury** and bromine,

LEAD A heavy, soft, bluish-gray solid; often used in alloys and piping

are liquid. Still others, such as **lead** and carbon, are solid. In the next few sections we will discuss the nature of these different states of matter and find out how the transition from one state of matter to another occurs. Figure 15–4 illustrates the differences in the three states of matter.

Gases

Gases are the most intangible form of matter. While they may be puzzling, they appear everywhere in our environment. Gases make up the air we breathe, the smells that come from cooking in the kitchen, and the oxygen in the tanks carried by scuba divers. Freon, used in air conditioners, is a gas; balloons and tires are filled with gases. Gases make a welding torch work and, as anesthetics, they bring special relief in the operating room.

FIGURE 15–4
In solids, molecules are tightly packed and vibrate about fixed points. The molecules in liquids are less closely packed and slide around each other. Gases contain molecules which are least densely packed and bounce off each other and the surface of containers holding gases.

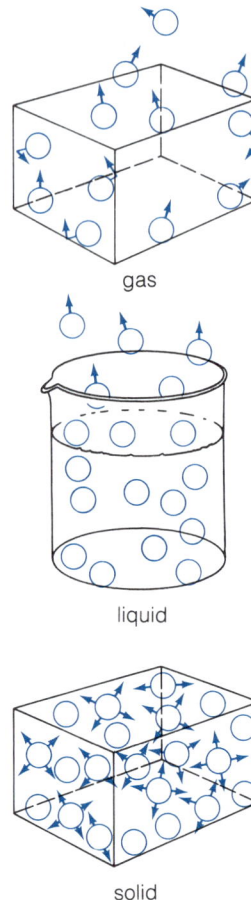

gas

liquid

solid

A gas consists of molecules that are so widely spaced that each molecule acts more or less independently of the others. Therefore it is easier to study the collective behavior of gases than it is to study other forms of matter.

Gases usually respond to the slightest movement, making the molecules knock against one another. This results in waves moving through the gas, which in turn causes the eardrum to vibrate; these vibrations are interpreted in the brain as *sound*. Such disturbances are not very difficult to set up; the mere scraping of a cricket's wing is sufficient to create a disturbance in the air that can be heard one-half mile away.

A second property of gases is that they readily mix and diffuse with each other. An open bottle of ammonia can be detected anywhere in the room. As the ammonia evaporates, its gas molecules diffuse and are wafted about by the air. Eventually, the atmosphere of the room becomes a mixture of air and ammonia.

A third property of gases is that they have no natural shapes of their own. They expand freely to fill any container or can be compressed into a smaller shape. The volume, **pressure,** and **temperature** of a gas in a container are all interdependent; that is, they affect one another. A balloon will expand as it goes away from earth because there is less air pressure outside the balloon. The same balloon will contract slightly as the temperature decreases at higher altitudes. Gases cool as they expand and then heat up when compressed.

PRESSURE A force applied per unit area on another object

TEMPERATURE A measure of the warmth or coldness of an object with reference to some standard (not to be confused with heat)

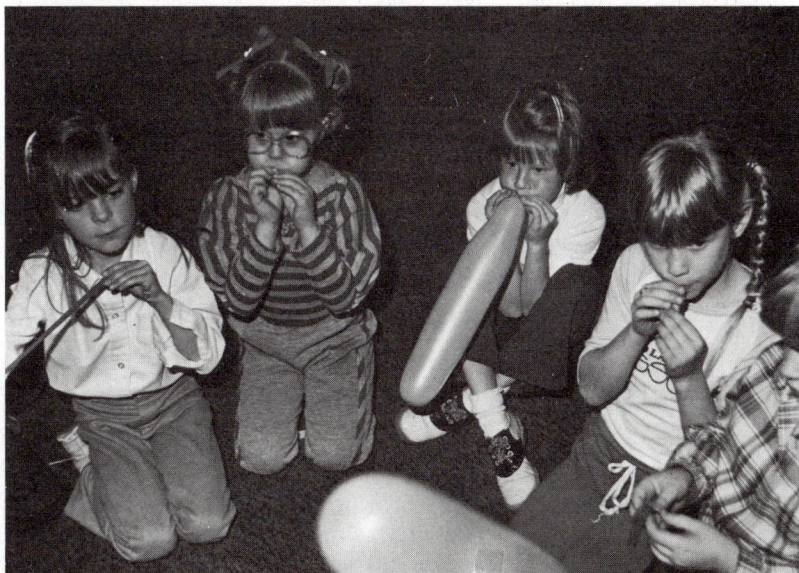

A property of gases is that they have no natural shapes of their own. They expand to fill any container or can be compressed into smaller containers. (Donald Birdd)

Gas molecules move at high speeds and undergo many collisions. Molecules in the air strike a window, for example, at the rate of 2 million, billion, billion per square inch every second. These molecules are relatively far apart, but because the average molecule moves at a speed of a 1,000 miles per hour, they collide with other molecules 5 billion times each second. These collisions are best described as elastic—that is, the molecules bounce off each other without losing much energy. It is this tremendous motion and collision that accounts for the characteristic behavior of gases. As a gas is compressed more and more, its molecules bump into the surface of the container more often; thus the pressure registered on the container wall increases.

The amount of molecular motion in a gas is measured by its temperature. Expanded gas *cools* because the molecules are further apart and consequently they collide less frequently; this results in lower molecular speeds. Conversely, when a gas is compressed, it *warms* because the molecules are pushed closer together. They collide more, and this results in higher speeds and a higher temperature.

Liquids

Liquid matter is both commonplace and puzzling. Liquids are a phase between gases and solids. Liquids seem to flow much like gases and they also seem to be about as shapeless.

On the other hand, liquids retain their volumes regardless of the shape of the container. Their molecules are usually more tightly packed; consequently, liquids are much less compressible than gases.

The most important liquid, water, covers nearly 75 percent of the earth's surface and makes up 50 to 80 percent of our body weights. Water and petroleum are the only two liquid compounds that occur freely in nature.

Of the 88 naturally occurring elements, only 2 are liquid when in pure form, mercury and bromine. The rarity of naturally occurring liquid elements is underscored when it is realized that neither mercury nor bromine ever occur freely on earth; they always are found together with other elements. Most compounds have a relatively narrow range of temperatures within which they remain a liquid, which is another reason why there are so few liquids in nature. Most liquids known today are manufactured, among them vegetable oil, alcohol, gasoline, and acetone, are examples.

In addition to basic liquids, there are three kinds of liquid mixtures. In a **solution,** the molecules of the **solute** separate from one another, causing the dissolved substance to look as though it has disappeared. A second type is a **suspension** in which the particles eventually settle out of the mixture. The third is a **colloid** in which the particles maintain their separate identities but remain in suspension.

SOLUTION Dispersing and dissolving one or more substances in another, usually a liquid

SOLUTE The component of a solution that is lesser in quantity

SUSPENSION Dispersing but not dissolving one or more substances in another, usually a liquid; the suspended particles eventually settle out

COLLOID Dispersed, nondissolved particles that remain suspended in another medium

Household ammonia is a solution of a gas (the ammonia) in a liquid (water). Calomine lotion is an example of a suspension. The white of an egg is a colloid. The albumin, which is the major protein in the egg white, remains suspended as small particles.

Liquids are best described by the nature and behavior of their molecules. The molecules of a gas are spaced quite far apart and their interaction with other molecules is almost negligible. In a solid, the molecules are very closely packed, take part in a regular structure, and exist in close cooperation. In liquids, molecules are in constant motion, but at the same time, they are more closely packed than in a gas. The effect of this is similar to having a plate or tray filled with small ball bearings or marbles rolling and slipping around one another.

Sometimes electrons pile up on one side of a molecule, creating a temporary bulge, or a temporary negative center of charge. This charge or bulge is matched by an equally positive charge on the other side of the molecule—a result of electrons that are no longer there. That molecule is then said to be polarized; it has a positive and negative charge on opposite sides.

Since molecules in a liquid are in close contact, this polarity can spread to neighboring molecules. When two molecules come near each other, their sides of opposite charges are momentarily attracted to one another and thus the molecules are bound together. Another force at work in forming liquids is the hydrogen bond discussed earlier in this chapter.

When the temperature of any liquid is high enough, the natural forces that bind the molecules to one another is less than the force of their collision; consequently, the liquid becomes a gas. This can happen in two ways: During *evaporation,* only the surface liquid turns into a gas. By *boiling* the liquid, the higher temperatures cause all parts of the liquid to turn into a gas. If the temperature becomes low enough, the molecules are drawn into a more rigid alignment. They stop sliding about each other and form a frozen liquid or a solid.

The attraction of the molecules for each other and their subsequent clustering together are critical characteristics of liquids. In this packed state, the molecules seem to resist any attempt to reduce their volume—even when pressure is applied. Rather than compress, liquids transmit pressure evenly in all directions. It is for this reason that when a car's brake is applied, the hydraulic system transmits the same pressure throughout the braking mechanism.

Solids

It would seem reasonable that solids would be the easiest state of matter to study since unlike gases, solids can be seen and touched, and unlike liquids they are firm, compact, and cohesive.

History is full of stories of how people have attempted to describe the solids of their environment. Yet it was not until about fifty years ago that the structure of solids was described fully. Until then, it was thought that the content of a solid determined its properties. In other words, what made a diamond hard, leather tough, and iron magnetic was a function of the kind of material it contained.

Characteristics of solids are now believed to be determined more by their structure and less by their content. The ways in which their atoms are bonded together is the significant key.

An entirely new advance in understanding matter came with the development of **crystallography,** or the study of substances through the use of **X rays.** A new group of scientists emerged called solid state physicists. They concentrated their efforts on the behavior of solid matter rather than on the behavior of other forms of matter. Gradually their research has helped us to develop a picture of solid matter as a *lattice work* of atoms. In this lattice, each atom is unable to move except for a small amount of vibration. What ties each atom to its neighbors are the bonding forces discussed earlier—those electrical attractions that exist between positive and negative charges in the atoms.

Solid state physicists have been in the forefront of fascinating advances. Because of their discoveries of the crystal structure of solid matter and of metallic bonding, there has developed an entirely new field of electronics. The development of solid-state **transistors** and low-temper-

CRYSTALLOGRAPHY The study of the form, structure, and properties of solid matter

X RAYS Electromagnetic radiation of very short wavelength produced by the bombardment of a substance by a stream of electrons moving at a great velocity

TRANSISTORS A tiny electronic device that controls current without the use of a vacuum

Today's computer industry was made possible when discoveries by solid state physicists in the areas of crystal structure and metallic bonding led to the development of solid-state transistors and low-temperature semiconductors. (Photo courtesy of Hewlett-Packard Co.)

ature **semiconductors** has been crucial in the areas of communications and computer technology.

   As our technology enables us to examine matter at finer and finer levels, it is apparent that our conventional classifications of matter as solid, liquid, and gas are merely phases in which all molecules can exist. What is important is identifying the basic molecular nature of a compound, and even more important, understanding its atomic structure. As we continue to think about matter in its three states—solids, liquids, and gases—our common experience leads us to seek ways to describe matter in familiar quantitative terms like volume (length × width × height) and weight. Additional units of measurement are needed by physicists to describe quantitative differences in matter as it changes from one phase to another and quantitative differences among various elements and compounds. Next we will examine several quantitative measures of matter.

SUPERCONDUCTORS Elements that offer little resistance to a flow of electrons

MASS AND WEIGHT

A fundamental unit used to describe all matter is its **mass.** This quantitative measure is often confused with **weight** but, as we will find shortly, mass is quite different from weight (see figure 15–5). Mass is a measure of matter contained in an object; in the metric system, mass is measured in grams or kilograms.

   It is easiest to visualize mass in relation to commonplace experiences. For example, imagine 1000 cubic centimeters (written 1000 Cm^3) of water which is a volume of about 4 inches wide, 4 inches long, and 4 inches high, at 4° Celsius (written 4°C) or about the temperature of

MASS A measure of matter contained in an object. Measured in grams or kilograms, the amount of force required to change the speed or direction of a given object

WEIGHT A measure of the force of gravity on an object

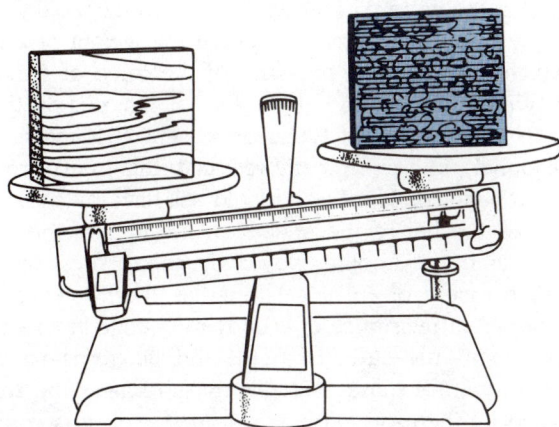

FIGURE 15–5
Though the volumes of the wood and sponge are equal, the balance shows the masses are not the same. Wood has more mass per unit volume.

The kilogram mass originated as a basic standard when the metric system was instituted by France in 1795. The unit of mass was a cubic decimeter of distilled water, at 4°C (the temperature of its greatest density). From this precise measurement of mass, a bronze kilogram was made and then later replaced by a platinum cylinder whose diameter equaled its length. (U.S. Dept. of Commerce, National Bureau of Standards)

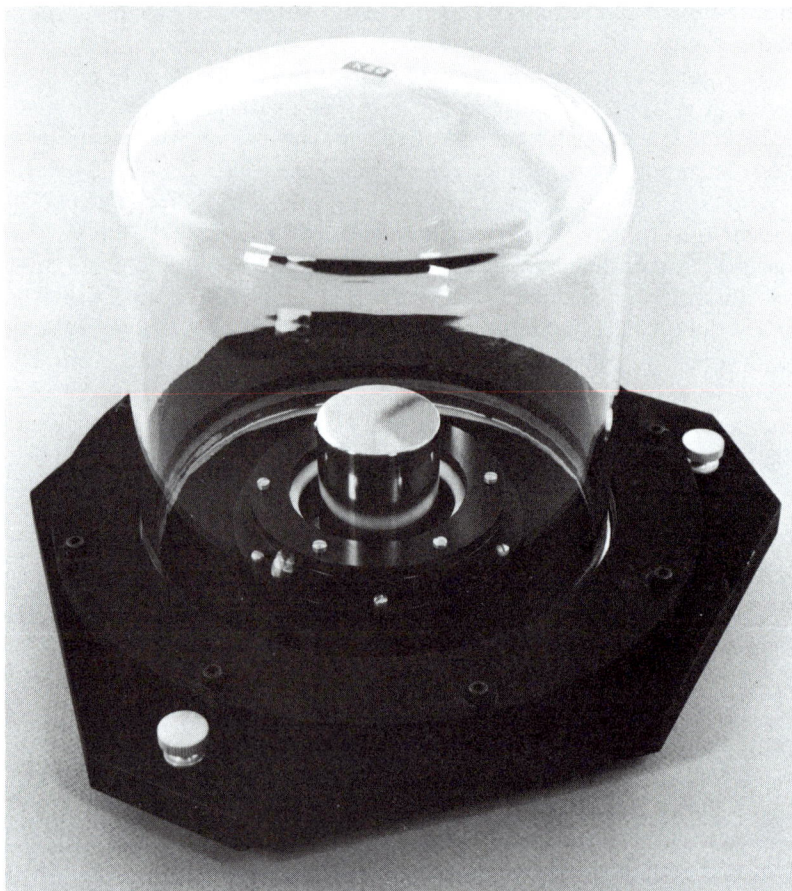

very cold water at the bottom of a lake. This volume of cold water would weigh 1000 grams (or 1 kilogram) or about the weight of 2 pounds of butter. Its weight is actually a measure of the force of gravity on the 1000 Cm^3 of water, that is, the earth's gravitational pull on the water is 1000 grams. If this volume of water were sent into space where the earth's gravitational pull were greatly reduced, this 1000 Cm^3 of water might appear to be weightless but it would still have mass. The amount of force required to change the speed or direction of this amount of water would be its mass. Another way of describing mass then is to say that mass is a measure of an object's inertia. Shortly, we will see that inertia is the apparent resistance of a body to change in its state of motion. That resistance—measured in grams and kilograms—is a measure of the 1000 Cm^3 water's mass. Because mass remains the same wherever it is measured (here on earth or out in space), it is considered to be an inherent quality of matter.

Mass is considered to be an inherent quality of matter because it remains the same wherever it is measured—here on earth or out in space. (NASA)

DENSITY

Another fundamental unit that describes all matter is **density.** This term is used to express the amount of mass in a given volume, or more simply, density is a measure of how tightly packed the mass of a substance is. Density is usually described in terms of so many grams per cubic centimeter, or so many kilograms per cubic meter. An easy way to visualize density is to picture once again the 1000 Cm^3 of water described earlier. The density of water is 1000 grams per 1000 cubic centimeters, or 1 gram per cubic centimeter. Now, for comparison, picture this same volume (1000 Cm^3 or approximately 4 inches × 4 inches × 4 inches) of mercury instead of water. At 20°C, mercury has a density of 13.6 grams per cubic centimeter; or in other words, mercury is about 13 times more dense than water.

An interesting way to understand density is to think of some common substances and try to guess their density. Ask yourself, would this substance float on top of water (evidence that its density is less than water) or sink in water (evidence that its density is greater than water)? Using the list on the following page, predict whether each substance will sink or float.

DENSITY The mass of a body per unit volume

Substance/ Matter	Density: grams/cubic centimeters
Cork	0.24 gm/1 Cm^3
Gasoline	0.69 gm/1 Cm^3
Ice	0.91 gm/1 Cm^3
Water	1.0 gm/1 Cm^3
Tin	7.3 gm/1 Cm^3
Mercury	13.6 gm/1 Cm^3
Gold	19.3 gm/1 Cm^3

We have considered several fundamental units of measurement that describe matter—volume, weight, mass, and density. Next we turn to the dynamics of matter: motion and energy.

MOTION

There are three useful descriptions of motion. The first is speed, which is defined as the distance traveled per unit of time. It is quite easy to measure speed. Simply drive your car on a straight and level road. Note the **odometer** reading and start a stopwatch at the same time. Keep the car's motion constant. After a minute or so, note the odometer reading again and stop the stopwatch. Divide the distance traveled by the time elapsed. The result is the **speed** of your car under those conditions. If the stopwatch is calibrated in seconds and the odometer in miles, the speed of the car would be expressed as the number of miles per second, or with a bit of mathematical juggling, it could be expressed as miles or kilometers per hour. Your speedometer does this calculation for you automatically.

ODOMETER An instrument that measures the distance traveled by an object

SPEED Distance traveled per unit of time

The second way to describe motion is **velocity.** In everyday language, speed and velocity are used interchangeably, but scientists define them differently. Velocity incorporates the idea of direction. For example, the speed of a car may be 30 miles per hour, but its velocity is 30 miles per hour north. The speed of an elevator may be 3 feet per second, but its velocity would be 3 feet per second up or down.

VELOCITY Speed of an object in a particular direction

The third way to describe motion is **acceleration.** We are all familiar with this concept; we use it to indicate speeding up. In this case, too, scientists have narrowed the definition to mean the change in the velocity of an object. Since velocity includes both speed and direction, if either speed or direction changes in any way, that object is accelerating.

ACCELERATION The rate of change in the velocity of an object

The acceleration of an object that does not change direction is determined as follows. The initial speed is first subtracted from the final

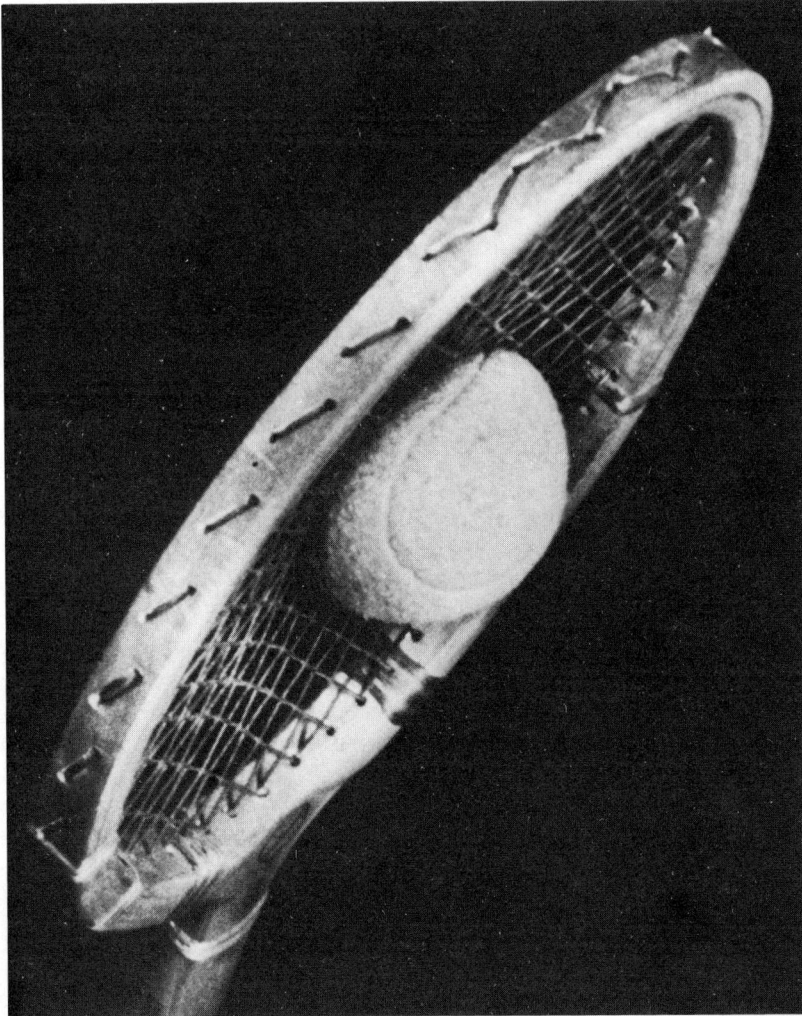

In this photograph a tennis ball has just been stopped by a racket and is about to change direction. It illustrates three useful descriptions of motion: the *speed* of the ball before it hit the racket (the distance traveled per unit of time), the *velocity* of the ball (its speed and direction), and *acceleration* (when either the speed or direction of the ball change). (Harold E. Edgerton, Massachusetts Institute of Technology)

speed. This number is then divided by the time over which the change in velocity took place, or:

$$\text{Acceleration} = \frac{\text{Final velocity (less) initial velocity}}{\text{time elapsed}}$$

A car that moves north from a stop light to a velocity of 30 miles per hour in 1 minute has an acceleration of

$$\frac{30 \text{ miles per hour} - 0 \text{ miles per hour}}{1 \text{ minute}}$$

or

$$\frac{0.5 \text{ miles per minute} - 0 \text{ miles per minute}}{1 \text{ minute}} = 0.5 \text{ miles per minute}$$

Inertia

INERTIA The tendency for a body at rest to remain at rest and for a body in motion to remain in motion

FRICTION The sum total of all the forces that oppose the motion of one body past another when the two are in contact

A property of matter that affects its motion is **inertia;** this is described as the tendency for a stationary body to remain at rest and a body moving in a straight line at constant speed to remain in such motion unless acted on by an outside force. This is known as Newton's first law of motion.

We all know that if an object is stationary, we have to give it a push to get it moving; additional force has to be applied in order to make it stop again. One common stopping force is **friction.** It occurs because surfaces are never perfectly smooth.

Consider a common example of inertia. A passenger on a train falls backward when the train speeds up, falls forward when the train slows down, and moves to the right when the train turns left. Why is this? When both the train and the passenger are at rest, they both tend to remain that way. However, when the conductor starts the train, the engine forces the train to move, but the passenger, who still has his own independent inertia, tends to remain at rest. Friction between his shoes and the floor keeps the passenger from sliding backward; that is, it keeps him in his original location, so he appears to be falling backward. The same reasoning applies when the passenger seems to fall forward as the train slows down. If the train stopped quickly enough, the friction from the passenger's shoes on the train floor might not be sufficient to keep him from sliding forward through the front of the car.

Earlier, we said that gravity is a force that exists between any two objects having mass. The magnitude of the force exerted on any given object is directly proportional to its mass and inversely proportional to its distance from the other object. For example, the earth exerts a force on all objects. The force on each object is proportional to its mass; the magnitude of that force is what we call *weight.* As an object moves away from earth, the force exerted on it becomes smaller and smaller. At some point the force of the earth pulling on the object becomes miniscule. However, the object still has mass, even though you cannot measure its weight. In order to measure its mass, one would have to determine how much force is required to accelerate it.

We have already defined force as a push or pull that causes motion. We can now refine that definition to read: force is a push or pull that causes an object to accelerate. Newton recognized the interrelationships between force, mass, and acceleration when he stated the second law of motion: The acceleration of an object is directly proportional to the net force acting upon it and inversely proportional to the mass of the object.

Thus acceleration can be described in the equation:

$$\text{acceleration} = \frac{\text{force}}{\text{mass}}$$

or

$$\text{force} = \text{mass} \times \text{acceleration}$$

Force

From the foregoing discussion we can see that force can now be defined as any influence that can make a body accelerate. Just as acceleration has both magnitude and direction, so does force. There are two kinds of forces: forces that only act when two bodies are actually touching, or *contact forces,* and forces that act at a distance. We are more familiar with forces that act at a distance than most of us realize. These forces are **gravitational forces, magnetic forces,** and **electric forces.** We will describe these forces at length later on in this chapter.

In order for us to examine the forces acting on an object we must isolate it from its surroundings. Suppose you have an apple in the palm of your hand. Draw an imaginary box around the apple. What forces are acting across that box? Are you pushing up on the apple? If you didn't, it would fall to the floor. You are exerting an upward force on the apple to counteract the downward force exerted by gravity. Since these forces are equal in magnitude, but opposite in direction, the apple does not move. In other words, no net force is exerted on the apple. *Forces in nature always occur in pairs.* This is Newton's third law of motion. Stated in more familiar terms it reads: For every action there is an equal and opposite reaction. Let's use our apple for an example of this law. If you take your hand away and the apple drops toward the earth, we know that the earth is exerting a gravitational force on the apple and making it move. However, the apple is also exerting a force on the earth. You are not aware of this force because the earth is so massive that it does not move appreciably as a result of the force exerted by the apple.

Space rockets make use of this principle as well. Rockets burn fuels that produce a stream of gases that are propelled at high speed out the back of the rocket. Because every action produces an equal and opposite reaction, the motion of the gases rushing backward causes the rocket to

GRAVITATIONAL FORCE The force that every particle exerts on other particles; the strength of this force depends on a ratio of the mass of the two objects and the distance between them

MAGNETIC FORCE The force with which a magnetic source attracts or repels other magnetic substances

ELECTRIC FORCE The force that results from the attraction of protons (positive charges) and electrons (negative charges) for each other

The roller coaster illustrates gravitational force (the attraction that every particle has for other particles). The strength of the force between the roller coaster and the earth is a ratio of the mass of the two objects and the distance between them. (Strix Pix)

move forward. The foward motion of the rocket is *not* caused by the gases pushing against anything, since there is nothing to push against out in space!

ENERGY

KINETIC ENERGY The energy in a moving object

POTENTIAL ENERGY The energy that results from position

Energy comes in many forms although we do not always recognize them as energy sources. A moving car, a lump of sugar, a log, and a piece of uranium all have energy. In this section, we discuss how energy can be recognized when it is released. Energy can be stored in the form of a moving object; this is known as **kinetic energy.** The energy an ob-

Many athletic games illustrate Newton's third law: that forces in nature always occur in pairs. Where a golf ball is hit with a club, both the ball and the club are accelerated. During the swing, the club exerts a force on the ball while the ball also exerts a force on that club. (Harold E. Edgerton, Massachusetts Institute of Technology)

ject has because of its position or state is called **potential energy.** Energy can be in the form of **heat** and **light,** or it can be stored in molecular bonds and called **chemical energy.** Energy can also be found in the flow of electrons; **electrical energy.** On the atomic scale, energy is stored in the interactions of nuclear particles; we call this nuclear energy.

Matter has always been an easier notion to understand than is energy because it can usually be seen, smelled, and touched. It is one thing to see a brick falling toward you and another to feel the pain when it hits. It is much more difficult to imagine that there is some quality in that moving brick that will vanish as soon as it reaches you.

Physicists have defined energy as the ability to do work. We have already discussed work earlier in this chapter. However, let's review this concept briefly here. Work is the application of force over a distance. The formula for work is:

$$\text{work} = \text{force} \times \text{distance}$$

HEAT ENERGY The energy contained in an object that is a result of the movement of its constituent molecules

LIGHT ENERGY The energy of radiation which moves at the speed of light—186,000 miles per second

CHEMICAL ENERGY The energy contained in the bonds between atoms in a molecule

ELECTRICAL ENERGY The energy that results from the attraction of electrons and protons

Just as we must isolate an object to determine what forces are acting on it, we must be careful to specify which object is having work done to it and what object is doing the work. Let's use the brick as an example. The brick, which weighs 3 pounds was raised to the top of a 20 foot-high roof. How much work was done on the brick? What is the force on the brick? Gravity exerts 3 pounds of force on it. When it is lifted, an equal amount of force must be used. The distance the force is exerted over, in this case, is 20 feet up. Therefore:

$$Work = force \times distance$$
$$Work = 3 \text{ pounds} \times 20 \text{ feet}$$
$$Work = 60 \text{ foot-pounds}$$

We have done 60 foot-pounds of work on the brick. One of the most basic laws of physics is that *energy is never created or destroyed, it only changes form.* Where did the energy come from to do this work? The person or machine that carried the brick up the 20 floors lost 60 foot·pounds of energy in the process of doing this job. The other important question is what happens to the energy in the brick? It can also be used to do work—at least 60 foot-pounds of it, if the brick were dropped.

Measurement of Energy

While the various kinds of energy can be classified according to their sources, the measurement of energy is of greater interest. This is because we need to be able to measure energy in order to describe it.

Energy can be changed from one form to another. For example, the chemical energy stored in wood is converted to heat energy when it is burned. The amount of heat in steam can be converted into electrical energy in a turbine. When energy is transformed from one kind to another no energy is lost, only changed in form. Scientists refer to this finding as the first law of **thermodynamics.**

THERMODYNAMICS A study of the relationship between heat and mechanical energy

HEAT A quality which, when absorbed, causes a change in the temperature of the object or a change in its physical state

Heat

Heat was the first form of energy to be observed and measured systematically. Early instruments lacked a clear, useful scale for heat measurement. Usually the high and low points of these scales were based on natural phenomena that seemed to take place at the same time. In Italy, they chose the cold of winter and the heat of summer as their high and low points. Winter cold meant the temperature of snow and ice. For heat, the temperature of a cow or deer was used. Another early measurement scheme used the temperature of air that would freeze water as its low point and the melting point of butter for the higher point.

One scientist, Gabriel Gr. Fahrenheit (1686–1736), picked for the lowest temperature of the scale a frozen mixture of water and salt. For the highest temperature, he used the human body temperature. On this scale, water freezes at 32° and boils at 212°. The Fahrenheit temperature scale is still used today, but is is being replaced gradually by a modification of the scale suggested some years later by Anders Celsius (1701–1744), a Swedish scientist. He suggested using zero as the boiling point of water and 100 as the freezing point. This scale was soon turned upside down and today we use the Celsius scale of 0° for freezing water and 100° for the boiling point of water (see figure 15–6).

People frequently use the words "hotter temperature" or "warmer temperature" when they mean higher, or a greater number of degrees. To avoid this kind of contradiction, remember that temperature is measured in degrees and therefore, any reference to temperature comparisons must indicate a higher or greater number of degrees compared with a lower or lesser number of degrees.

What is the relationship between heat and temperature? If a hot stone is immersed in a bowl of water, we can measure how much heat is transferred from the stone to the water. If we assume that no water evaporated when the stone was immersed, the difference between the original and final temperatures of the water gives a measure of the amount of heat absorbed by the water. The unit used to describe this measurement is the **calorie.** Suppose that the original temperature of

CALORIE The amount of heat necessary to raise the temperature of one gram of water one degree Celsius

FIGURE 15–6
Same familiar reference points along the Celsius temperature scale are shown.

From *Physics: Principles and Problems* by James T. Murphy and Robert C. Smoot. Copyright © 1982 by Bell & Howell Company. Used by permission of Charles E. Merrill Publishing Company.

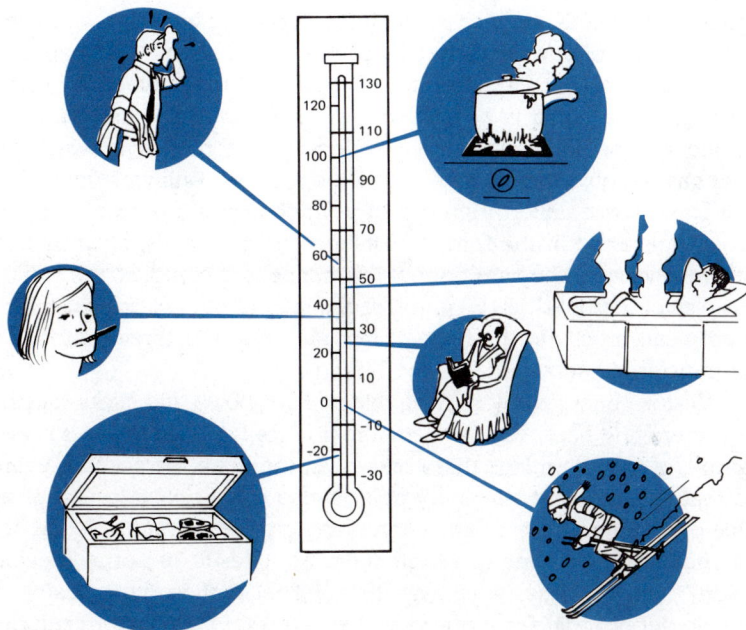

the water was observed to be 15°C; after the stone was immersed and the water was stirred again, its temperature was found to be 26°C. Thus each cubic centimeter of water (or gram since water has a density of one gram per cubic centimeter) showed an increase of temperature of 11°, and each cubic centimeter required 11 calories of heat to show this rise in temperature. If there were 250 cubic centimeters of water in the bowl, then 250 × 11 or 2,750 calories were transferred from the stone to the water.

Incidentally, these calories are not the same as the calories we are concerned about in dieting; one thousand of these "water calories" make one diet-calorie. As an example, if a bucket of water (about 700 milligrams) was raised from 35° to 50°C, an increase of 15°C, we can calculate that the water absorbed 700 (milligrams) × 15 calories, or 10,500 calories. This would be 10.5 diet-type calories. Diet calories are called **kilocalories** or large calories (C).

An operational definition of *heat* is that it is a quality that, when absorbed, causes either a temperature increase or a change in physical state (solid, liquid, gas). The heat required to raise the temperature of a substance is called **specific heat.** Once a substance reaches the temperature necessary for a change in state to occur, additional heat must be added or lost for the change to actually take place. This additional amount of heat is known as **latent heat**.

Heat is transferred in three ways: conduction, convection, and radiation. Heat may be transferred by **conduction,** that is by direct contact. In general, materials that are good electrical conductors, such as metals, are good heat conductors. Materials such as cloth, wood, and paper are poor conductors. Heat may also be transferred by **convection.** In this case a gas or a liquid is heated so that only part of it warms up. The warmed gas or liquid decreases in density and moves upward. The cooler gas or liquid then is heated and the process continues until all the air or liquid has a uniform temperature. **Radiation** is a process whereby heat is transferred in the form of electromagnetic waves, such as light. While the exact mechanism of how this happens is complex, we can see examples of it around us. Cast iron radiators and pot-bellied stoves emit radiant heat, mostly in the form of infrared rays. The three methods of heat transfer are shown in figure 15–7.

Heat is generated by moving things. By rubbing two sticks together or by exercising hard, we can generate a lot of heat. However, for each example of moving things that make heat, there are many more examples that illustrate that stationary objects also contain heat: a cup of hot coffee or an automobile engine after a long trip are examples. Is the heat in a moving object simply stored there as a result of some previous motion? If this is true, what was the original motion that caused the coffee to get hot? This is a question that puzzled scientists for centuries.

KILOCALORIE The amount of heat needed to raise the temperature of one kilogram of water one degree Celsius

SPECIFIC HEAT The amount of heat required to raise a mass of a substance one degree compared with a similar mass of water

LATENT HEAT The amount of heat required to change the physical state of a substance, e.g., the heat needed to change a substance from a solid to a liquid once it has reached its melting point

CONDUCTION The transmission of energy from particle to particle

CONVECTION The transfer of heat via currents of air or liquid

RADIATION The process by which energy is sent out from atoms and molecules because of their internal changes

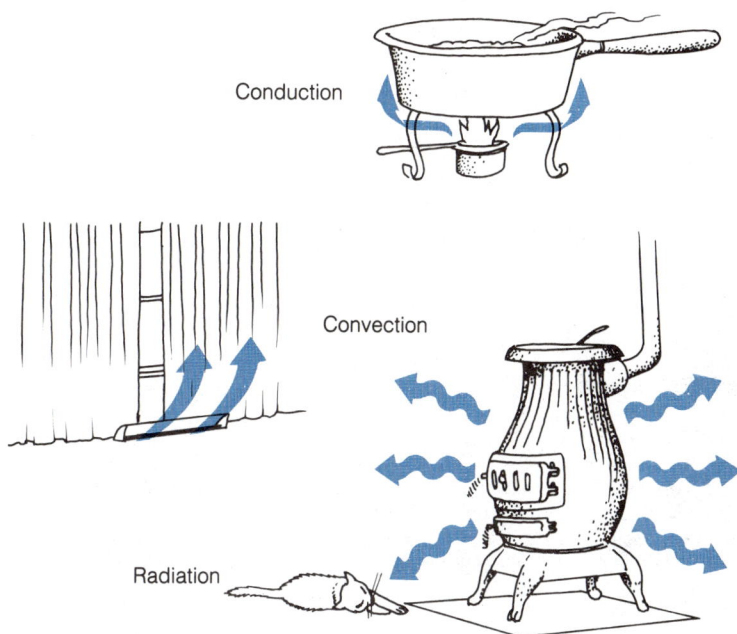

FIGURE 15–7
Direct contact is made when heat energy is transferred by conduction. Convection is the transfer of heat energy by air or liquid currents. Radiation is the transfer of heat energy by electromagnetic waves.

An object that is not visibly moving still contains a great deal of motion. Indeed everything we see around us contains a whole collection of atoms and molecules that are moving around at fantastic speeds. Heat is an indicator of the speed of that internal motion. In cooler objects, the molecular movements are slower than in warmer objects. Have you noticed that your automobile tires require less air in the summer than in the winter? In warmer weather the air in your tires becomes hotter and expands. Thus you need less air to keep your tires at the recommended air pressure while they are hot or in hot temperatures.

As scientists studied heat they found that it must flow from higher to lower temperatures in order to do work. This is shown as the second law of thermodynamics. The essence of this law is that heat will not flow spontaneously from a colder to a warmer place. A heat pump, shown in figure 15–8, illustrates this principle.

Light or Radiant Energy

Light has long been identified as a source of energy. A simple magnifying glass can be used to concentrate light into a beam so hot that it will cause a piece of paper to burn.

The exact nature of light has been studied by scientists for many years. It is now believed that light behaves as both a wave, radiation, and/or as little "packages" of matter called **photons.** While these two

PHOTON Basic unit of light energy

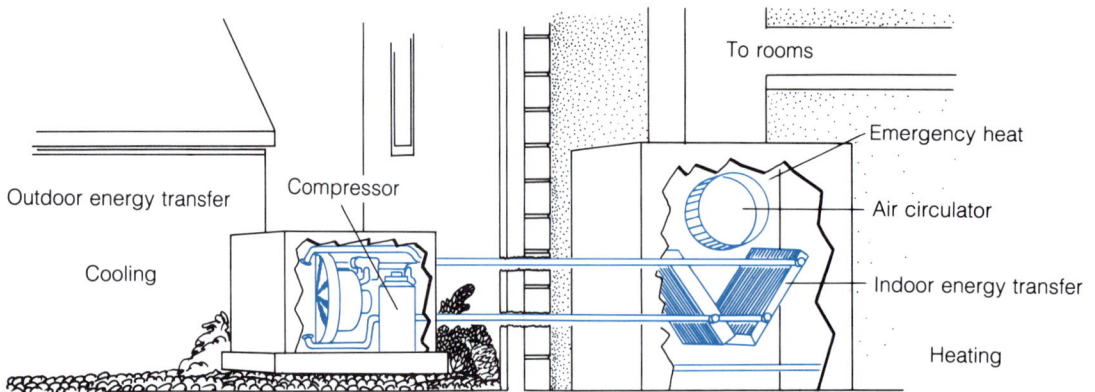

FIGURE 15–8

A heat pump runs in either direction depending on whether it is used in heating or cooling. In cooling, heat is extracted from the air in the house and pumped outside. In heating, heat is extracted from the outside air and pumped inside.

From *Physics: Principles and Problems* by James T. Murphy and Robert C. Smoot. Copyright © 1982 by Bell & Howell Company. Used by permission of Charles E. Merrill Publishing Company.

models may sound contradictory, neither model alone explains all of the properties of light energy. Thus scientists use both models depending on the kinds of experiments they are doing.

If we examine the wave nature of light, we will be able to understand how light releases its energy. A wave is a regular oscillating pattern, as shown in figure 15–9. The highest point of a wave is called the crest and the lowest, the trough. The distance from crest to crest and trough to trough is called the **wavelength.** The wavelength of any electromagnetic radiation, which light is, determines the energy of that radiation. If you stood in front of a wave-making machine, you would have more trouble staying upright if the peaks of the waves came very often than if they came infrequently.

Ordinary daylight (or white light) is composed of light of many colors, as can be shown by passing light through a prism. The continuous **spectrum of light** is an arrangement of ordinary white light into colors with different wave frequencies. Red, at one end of the spectrum, has the lowest frequency (the fewest number of wave repetitions). Red is followed by orange, yellow, green, blue, indigo, and violet, which has the greatest frequency of wave repetitions. The wavelength of each color of light is also different. Red light has the longest wavelength of the colors in the visible spectrum and violet light has the shortest. Thus red light is the least energetic and violet the most energetic.

Bordering the red end of the spectrum is **infrared** radiation; it has a longer wavelength and is even less energetic than red light. This radia-

WAVELENGTH The distance between two neighboring crests of a wave

SPECTRUM OF LIGHT A series of colored bands arranged in continuous order of their wavelengths

INFRARED RADIATION Radiation that has low energy and borders the red end of the spectrum

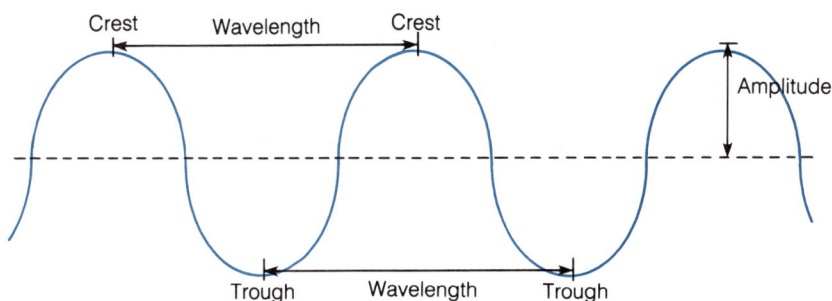

FIGURE 15–9
The total distance from one crest to another or from one trough to another is wavelength. Amplitude is the distance a wave rises or falls from its usual rest position. Waves can vary in length and amplitude.

tion cannot be seen but can be felt as heat. Infrared film responds to infrared radiation in the same way ordinary film responds to visible light; however, this film allows pictures to be taken in complete darkness. Living things show up extremely well because they give off a continuous infrared radiation.

Ultraviolet light borders the violet end of the spectrum. As you may have guessed by now, ultraviolet light has a very short wavelength and is very energetic. Part of the sun's radiation contains ultraviolet light. It is this component that is responsible for suntans—and sunburns. Ultraviolet light is so energetic that it is used to kill bacteria and viruses. Many hospitals and laboratories use ultraviolet lamps to sterilize whole rooms and pieces of equipment that would be difficult to keep germ free in any other way.

ULTRAVIOLET RADIA-TION Radiation that is very energetic and borders the violet end of the spectrum

Along with the kind of light just described, we also have radiant energy. Scientists have discovered that all the known elements (shown in figure 15–2) can be heated until they are incandescent or glowing with heat. In that state they emit a light that varies in wavelength depending on which element is being heated. These wavelengths can be measured (by length, i.e., by color) to determine which elements are present in a given sample of material. This identification process is possible because the wavelengths being measured reflect the energies of the electrons in an element's atoms. (You will remember that each element's atomic weight is unique.)

Chemical Energy

Another form of energy, chemical energy, is found everywhere. If you were to fly over most any part of the United States, you would see highways filled with moving cars. These cars are powered by explosions of gasoline mixed with air. The engines of these cars are using the chemical energy contained in gasoline. Plants and animals also use chemical energy. The energy that helps them to move and enables them to grow and to reproduce comes from the chemical energy of food. Of all forms of energy, none is more vital or useful to living things than chemical energy. It usefulness comes from its versatility.

Chemical energy is released when the complex bonds that cement molecules together are broken. Every bond between atoms and atoms and between molecules and molecules is a potential source of chemical energy.

Of all things that exist in nature, our most important sources of chemical energy are those substances that contain carbon. Important fuel sources, such as coal, oil, gas, wood, and alcohol, all contain carbon. The basic foods such as sugars, starches, proteins, and fats also contain carbon. These foods and fuels are the storage depots of chemical energy that originated from the radiant energy of the sun.

ION Electrically charged atoms resulting from loss or gain of electrons during chemical reactions or exchanges in radiation

Plants use radiant energy to split a water molecule into a hydrogen atom and an **ion** containing one oxygen and one hydrogen atom. The hydrogen ion is bonded to a carbon dioxide molecule to make a new molecule that has carbon, oxygen, and hydrogen atoms; this is a carbohydrate molecule. The process is known as photosynthesis.

The first law of thermodynamics says that when energy is transferred from one kind to another, no energy is lost, but only changed in form. This means that the energy required to break up a molecule must be equal to the energy obtained when the molecule is put back together. When carbohydrates are eaten, the body burns them through the process of cellular respiration and obtains the energy from breaking the chemical bonds, especially carbon-hydrogen bonds.

Electrical Energy

To understand electrical energy, we return to the structure of atoms. All atoms have a positively charged nucleus surrounded by negatively charged electrons. Usually, the negative charges add up to a total charge that equals the positive charge of the nucleus. Thus the atom itself is essentially neutral. However, some materials, especially fur and glass, have electrons that are loosely held and can be easily rubbed off and transferred to other objects. Because of this tendency of some materials to "give up" electrons, while other materials "grab" them, you can generate electrical energy and feel it as a jolt. You can also see electrical energy as a spark if you brush your hair vigorously in a dark room. Electrical energy is generated when you walk across a carpet and then touch a doorknob. The electrons from the carpet accumulate in your body; as soon as you touch a conductor (the metal doorknob), a current of electrons goes between the body and the doorknob and results in that familiar shock.

In some respects, electric current behaves like a mountain stream; both flow in only one direction. Just as water goes downhill, electricity

All atoms in the universe contain positive and negative charges. When charges accumulate in one place, and are then discharged, we can observe static electricity. Lightning is an example of static electricity. (NOAA)

will also seek a "downhill" path. Movement starts at a place with a greater number of free-floating electrons and ends in a place with fewer free-floating electrons. Electric current is nothing more than electrons moving to even out an imbalance. A measure of this imbalance between two unequal concentrations of electrons is called **voltage.** In the example of the mountain stream, voltage would be a measure of the height of the mountain. A small, one and one-half volt battery would represent a small incline while a 220-volt current (as needed by an electric dryer) would be like a steep waterfall.

VOLTAGE A measure of electrical potential that will cause a current to flow; measured in volts

Strix Pix

Nuclear Energy: Promise or Peril?

Nuclear energy is one of the important social issues of the present and coming decades. In the 1950s nuclear energy was her-
alded as the answer to many of our energy
problems. It was a source of energy that
when developed would be "too cheap to
monitor." Today, the production of nuclear

energy has become a controversial issue
because of concerns about danger to life.

Nuclear fission results when a slow or
fast moving neutron splits the nucleus of a
heavy atom, Uranium-235 for example, into
two lighter fragments. The process of splitting the nucleus results in a release of neutrons and tremendous amounts of energy.
Radioactive wastes from the fission process must be stored safely until the radioactivity is no longer harmful. The time estimated for this decay process varies from
thousands to a few million years. Because
geologic processes and climates change
over long periods of time just as do erosion
and groundwater activity, many people, including some scientists, are concerned
that radioactive wastes cannot be safely
stored.

There are other issues related to the use
of nuclear energy: adequate supplies, accidents, theft of nuclear fuel, and costs. Even
with all these issues the advocates of nuclear energy think the trade-off is worth it.
In return for inexhaustible energy, which
our society needs to maintain present levels of affluence, this and future generations
must commit themselves to the continuing
vigilance over the process and products of
nuclear energy.

Early projections of nuclear energy included 1,800 power plants that would pro-

AMPERAGE A measure of
the amount of charge flowing past a given point

CONDUCTOR A material
that offers little resistance
to the flow of electrons

Voltage is a measure of the difference in concentrations of electrons, or how high the hill is. **Amperage** is a measure of electric current—how much charge is going past a given point.

Another factor that influences the flow of electrical current is the
substance through which it passes. Some substances are good **conductors** because their electrons are loosely held. Glass is a poor conductor;

vide as much as 21 percent of the world's energy by the year 2000. In 1973 the U.S. Atomic Energy Commission (now disbanded) predicted there would be 1,500 nuclear power plants producing half of the nation's electricity by the year 2000. By 1980 there were 72 nuclear power plants in the United States producing 12 percent of the nation's electricity. While some 85 plants were under construction and 19 plants were on order in 1980, the future of nuclear energy seemed troubled.

Many of society's worst fears about nuclear energy were made vivid in April 1979 with the now historical incident at Three-Mile Island. A combination of faulty equipment and human errors resulted in a major accident at this nuclear plant. Until this accident, advocates of nuclear energy could claim that the probability of an accident was extremely small. Three-Mile Island made all of the complications and issues of an accident headlines for months. In a short period the public was informed about the problems of nuclear power.

In May 1983 the United States Supreme Court added to the problems of future development in the nuclear energy industry. In a unanimous decision the court held that California and other states may ban future plants until the federal government establishes permanent sites for the disposal of radioactive wastes. The immediate impact of this ruling is actually small. It applies only to nuclear plants planned for the future, and there are none. Plants under construction will not be affected. It should be noted that Congress has a mandate to establish a permanent waste site by 1987. Of course, both sides of the nuclear energy debate claim the ruling as a victory.

It seems that the promise of clean, future energy with cheap and safe nuclear power is being questioned. The construction costs and inability to resolve problems such as waste disposal have posed serious problems for the future of nuclear energy.

Questions for Discussion

1. What is your projection for the future of nuclear energy? What assumptions are the basis of your projection?

2. Which issues related to nuclear energy are most important to you? Society? Science?

3. Do you think it is fair to make a bargain for future generations—inexhaustible energy vs. commitment to thousands or millions of years of constant vigilance? How would you justify your response?

4. What sources of energy would you propose for the future of our society?

---

silver and copper are better conductors; and liquid helium is a superconductor.

Scientists have discovered that a flow of electric current creates a **magnetic field** around a wire and that a changing magnetic field will create an electrical current. A magnetic field is polarized, that is, there are two poles called North and South. These names are a result of the

MAGNETIC FIELD The space occupied by magnetic lines of force

discovery that the earth itself is a huge magnet, whose poles are oriented toward the north and south. Thus every magnet orients itself with respect to the earth's polarity. The movement of these poles passed over a good conductor will generate an electric current.

Nuclear Energy

Nuclear energy is a significant but controversial contributor to the world's supply of energy today. There are two sources of nuclear energy: either by the fission or fusion of atoms.

Scientists have learned that they can produce energy by subjecting certain atoms to a dividing process called **fission.** For example, when uranium is bombarded with neutrons, some of its atoms split into smaller fragments. In this process some mass disappears and is converted into energy. One neutron is known to be enough to split an atom, but even that neutron is not lost. Among the debris from the split, there exist two or three additional neutrons. These can split two or three other atoms—and thus release more energy—and make available four to nine more neutrons. It is easy to see why this process is called a *chain reaction.*

FISSION The nucleus of an atom is split into fragments; during this process, part of the mass of the atomic nuclei is converted into energy

The efficiency of chain reactions, however, can vary. Many neutrons simply fail to split another atom. Each time they miss their target, the chain reaction is weakened. Scientists have found that slower moving neutrons are more effective in causing fission than neutrons moving at higher speeds.

Nuclei of graphite do not absorb neutrons. Rather they bounce them off in the same way that rubber balls bounce off concrete walls. This reduces the energy of neutrons. Thus nuclear power plants use many graphite rods to surround the uranium used in the reaction. These rods were designed so that they can be pushed into or pulled out of the nuclear pile, depending on whether the reaction needs to be slowed down or speeded up.

FUSION The nucleus of one atom is joined or fused to another; during this process, atomic nuclei are combined in such a way that a mass is converted to energy

A second source of nuclear energy is created by a process which involves the **fusion** of nuclei. When nuclei are joined together through fusion, even more energy is released than through fission. This results in the release of even more energy.

Millions of degrees of temperature are required before hydrogen nuclei can fuse. Such temperatures are thought to exist naturally only in the interior of stars such as the sun. Ordinary machines cannot generate or tolerate temperatures hot enough to produce a fusion reaction here on earth. However, such temperatures are momentarily found in atomic fission explosions. For example, a hydrogen bomb is triggered by one or more small atomic bombs—called thermonuclear devices—that are built into it.

Presently, scientists have not found a way to control fusion reactions. If they do find a way, a limitless raw material will be available because hydrogen is the most abundant of all known elements in the universe. Even though the solution is not apparent to us right now, we have learned from history that the problem probably will be solved, and by a means that will seem most unconventional.

The energy released during fission or fusion comes about because the mass of the products is always slightly less than the mass of the materials before fission or fusion. This difference in nuclear mass appears in the form of nuclear energy. Einstein's famous formula $E = mc^2$ relates the energy released to the change in mass.

As we mentioned above, only a tiny fraction of the energy available in an atom is tapped in either kind of nuclear reaction: fission or fusion. If physicists were able to convert entire particles into energy in controlled settings, they could increase the energy released manyfold. Some believe that such discoveries would provide the energy society needs today. However, the production of nuclear energy, whether through fission or fusion, results in the accumulation of radioactive by-products or wastes. No safe method of disposing of **radioactive** wastes has been found. Since excessive exposure to radioactivity is known to be harmful to humans, the production of nuclear energy has become a very controversial issue.

RADIOACTIVE An atom that gives off energy in the form of a particle or ray

BIBLIOGRAPHY

Breuer, H. *Physics for Life Science Students*. Englewood Cliffs, N.J.: Prentice-Hall, 1975.

Constant, F. Woodbridge. *Principles of Physics*. London: Addison-Wesley, 1967.

Dickerson, R. E.; Gray, H. B.; and Haight, G. P. *Chemical Principles*. 3rd ed. Menlo Park, CA: Benjamin/Cummings, 1979.

Faughn, J. S., and Kuhn, K. F. *Physics for People Who Think They Don't Like Physics*. Philadelphia: W. B. Saunders, 1976.

Halliday, D. and Resnick, R. *Fundamentals of Physics*. New York: John Wiley & Sons, 1981.

Turk, J., and Turk, A. *Physical Sciences with Environmental and Other Practical Applications*. Philadelphia: W. B. Saunders, 1977.

Zukav, G. *The Dancing Wu Li Masters. An Overview of the New Physics*. New York: William Morrow, 1979.

16

Teaching the Physical Sciences

The purpose of chapter 16 is to provide examples of science lessons designed to teach physical science to students from kindergarten through eighth grade. Planning Charts 16–1, 16–2, and 16–3 illustrate how each science lesson will allow you to meet several objectives at one time.

KEY CONCEPTS AND IMPORTANT IDEAS

☐ estimating volume: how much is inside?
☐ exploring movement ☐ describing the weather ☐ learning to make predictions
☐ experimenting with forces ☐ constructing electrical circuits ☐ solving mysteries
☐ testing hypotheses with dissolving substances

CHART 16–1

Planning Chart for Physical Science: Kindergarten through Grade 2

Activity	Objectives						Teaching Methods	Related Curriculum
	Conceptual Knowledge	Learning Processes	Psychomotor Skills	Attitudes and Values	Decision-Making Skills			
1. Estimating Volume: How Much Is Inside?	Things (living and nonliving matter) have identities based on characteristics; volume is one way to describe how much is inside an object.	Observe, compare, measure.	Use and care for simple tools/equipment that require no adjustment.	Actively explore environment and ask questions.	Select alternatives and accept consequences.		Experiment, learning center	Mathematics
2. Exploring Movement	Movement is the change in location of matter; when objects interact, they change.	Observe, compare, classify.	Use and care for simple tools/equipment that require no adjustment.	Listen to alternative points of view; explore the environment and ask questions.	. . .		Games, discussion	Physical education
3. Describing the Weather	The environment changes because weather factors interact with the land and seas.	Observe, compare, measure.	Use and care for simple tools/equipment that require no adjustment.	Accept direction in caring for health and welfare of self and others in class.	Become aware of information needed to make decisions.		Hands-on activity	Mathematics, crafts

NOTE: Activities ES–1, ES–2, and ES–3 in chapter 14 can also be used to reinforce the concepts of the nature of solids, liquids, and gases presented in chapter 15.

CHART 16–2

Planning Chart for Physical Science: Grades 3 through 5

Activity	Objectives					Teaching Methods	Related Curriculum
	Conceptual Knowledge	Learning Processes	Psychomotor Skills	Attitudes and Values	Decision-Making Skills		
4. Learning to Make Predictions	All matter changes because it interacts with other matter in the environment and events occur in predictable patterns.	Observe, compare, measure, gather and organize information, infer, and predict.	Construct simple projects/models without guidance; construct materials in/for experiment.	Attempt to compare alternative points of view with evidence.	Rank alternatives. Select one, and accept the consequences.	Demonstration, experiments, lab reports, discussion	Mathematics, reading
5. Experimenting with Forces	All matter changes because it interacts with other matter in the environment (living and non-living matter); to start motion, pairs of forces must be unbalanced.	Observe, compare, measure, gather and organize information.	Use and care for tools/equipment that require small adjustments or coordination of senses.	Combine objects and materials in new ways; actively explore the environment and ask questions.	Describe consequences of various alternatives.	Learning center, lab reports	Mathematics, reading

CHART 16–3

Planning Chart for Physical Science: Grades 6 through 8

Activity	Objectives					Teaching Methods	Related Curriculum
	Conceptual Knowledge	Learning Processes	Psychomotor Skills	Attitudes and Values	Decision-Making Skills		
6. Constructing Electrical Circuits	All environments change when living and nonliving matter interact; electrical circuits are completed when the parts of the system interact.	Observe, compare, gather and organize information, infer, predict, and form hypotheses.	Construct complex projects/models with some guidance; use equipment that requires adjustment and discrimination.	Combine objects and materials in new ways; assume responsibility for health and welfare of others.	Describe immediate and long term consequences of various alternatives to self, others, and the environment.	Experiment, demonstration, discussion, learning center, lab reports	Reading, social studies
7. Solving Mysteries	Observations of repeated events are the bases for predictions.	Observe, compare, infer, and form hypotheses.	Construct complex projects/models with some guidance.	Combine objects and materials in new ways.	Rank alternatives and select one on the basis of most positive effect for those concerned.	Demonstration, discussion, peer learning	Reading
8. Testing Hypotheses with Dissolving Substances	Rules that explain relationships of cause and effect in events are hypotheses.	Observe, compare, measure, gather and organize information, infer, predict, and form hypotheses.	Construct complex projects/models with some guidance.	Seek alternative points of view and multiple sources of evidence.	Rank alternatives and select one on the basis of most positive effect.	Experiment, demonstration, discussion, learning center	Reading, social science

ACTIVITY PS–1

Estimating Volume: How Much Is Inside?
(Kindergarten–Grade 2)

ACTIVITY OVERVIEW

In this activity, students explore volume as they compare amounts of water, juice, sand, and beans. Ordering objects by **volume** is a challenge for the young mind. Being able to make direct comparisons is essential. How much is inside a container cannot be determined accurately by a mere visual comparison of the height or width of a container. *Estimated time:* five days.

SCIENCE BACKGROUND

As scientists explore the natural world and discover new objects—whether these are new stars, flowers, birds, enzymes, or viruses—they describe newly discovered objects in ordinary terms of size, color, shape, and texture. Similarly, young children explore the natural world and learn to name common objects by recognizing their color, shape, texture, and size. But among these characteristics, size presents a special challenge because size, or how big something is, can mean how tall an object is, its height; how fat something is, its width; or how much space it occupies, its volume. (Even more difficult for young students are the concepts of **mass** and **density** of familiar objects in motion or in water. However, the concepts of mass and density are not introduced in this activity.)

Thus the identity of an object can be described in terms of its *volume* which involves three dimensions: length, width, and height.

The comparison of volume of various things can be made successfully by estimating "how much is inside" of a container.

SOCIAL IMPLICATIONS

The emphasis on this topic is the **relative volume** of objects. Many adults have never had the opportunity to develop these skills. Yet, understanding volume is a key aid in making decisions in the supermarket. Commercial packagers often attempt to convince us that their product is the best buy based on the design of the container rather than the volume of its contents.

It should be clear to us that our perception of volume as adults may be quite different from that of our students; Piaget's studies clearly identify this contrast. As part of the student's social development, a valued outcome of being able to describe "how much is inside things" is to be able to see the world around us with new insight.

MAJOR CONCEPTS

Things (living and nonliving matter) have identities based on their characteristics. Volume is one way of describing how much is inside an object.

OBJECTIVES

☐ Observe, compare, and measure.

☐ Use and care for simple tools and equipment that require no adjustment.

☐ Actively explore the environment and ask questions.

☐ Select alternatives and accept consequences.

MATERIALS

Large (2 qt.) container of water, juice (two large 46 oz. cans), large package of dried beans, several clean containers or cups of various sizes (examples might be an orange juice can, mayonnaise jar, a cup, a small soup can, etc). In this activity it is helpful to have several containers which hold the same number of ounces but which differ in shape.

VOCABULARY

inside
volume

PROCEDURES

A. Show your students several containers of different sizes and a container of water. This activity can be fun if you pose the question: Which container would you want to use if you had to drink all of the water in 15 seconds?
1. Pour water into each container chosen by one of the students.
2. Ask several other students to count the number of seconds it takes each person to drink their water, and keep track of the amounts of time on the chalkboard.
3. Finally, display the several containers and ask your class to arrange them in descending order from the one that holds the most drink to the one that holds the least drink.

B. The next activity can best be done at an independent Learning Center where stu-

dents can extend their explorations. Place an assortment of four to six containers of various sizes on a table along with a tub of dried beans and a sign that reads: Which holds more?
1. Direct the group to explain what they think this question asks them to do: "Which holds more?" (Compare the amount of beans that will fit into each can.)
2. Then establish how many days the center will be available for students to experiment. Be sure to establish the rules by which a student may to go to the center. For example, if you want only two or three children at a time, make two or three yarn signs that they can hang around their necks—"Scientists at Work." The only authorized students at the center, then, are those wearing signs.)
3. When students have completed their ordering of containers, arrange a sharing time in which they describe their answers to the question: "Which one holds more?"
4. When there is disagreement, have students demonstrate the basis of their conclusion. By the end of the session the group should have the containers in an agreed-upon order from largest to smallest.
   If you have included some containers which have the same volume (hold the same number of ounces) but are different in shape, children will have an opportunity to gain important experience which leads to the development of the concept of conservation—that is, that a given volume is constant regardless of changes in the shape.

C. Rearrange the five containers from largest to smallest in amount of water or beans each will hold.

1. Now pose the problem: "You know which holds more. Now tell how much more each one has than the other." Note: You might need to suggest that it works best to start from the smallest container and work up to the largest one. They can record their results on a chart. After filling the smallest container, pour its contents into the next largest; note how much space is left. Repeat the action by pouring the next-to-smallest container into the next-largest one, and so on.

2. When completed, have students describe the differences between the amounts that various cans hold.

EXTENDING THE ACTIVITY

D. Explore volume with new substances. It is a useful way to let students reinforce their intuitive understanding of volume. Using sand, juice, or other kinds of materials are ways to make these comparisons more meaningful.

ACTIVITY PS–2

EXPLORING MOVEMENT
(Kindergarten–Grade 2)

ACTIVITY OVERVIEW

The description of events that includes moving objects represents a keen challenge to the young child. In this activity, students are provided with a variety of experiences that involve movement and ways to describe these events. *Estimated time:* four days.

SCIENCE BACKGROUND

Science is, in part, the description of events involving living and nonliving matter. Such descriptions often involve being able to tell if matter is moving or not moving. Motion is the change in location of matter.

The world is filled with motion—matter that is moving. Trees move with the breeze; clouds are moved by wind; people move as they walk; and in the distance, we can hear a car driving down the street or a plane flying overhead. Our eyes move as we watch events near us. We move in our chair as we shift our weight or begin to write. Physical science, as a field, studies movement.

SOCIAL IMPLICATIONS

By using objects in the immediate environment, students acquire a more precise way of communicating about motion. As they attempt to understand the world around them, students expand their basic foundation for classification, one that is based on moving and nonmoving objects.

Language is an essential tool for describing motion. As a second communication tool, *arrows* are a step toward being able to use **vectors** to describe a change in motion. At this point, the head of the arrow as a communicator of direction for change is all that is used. Later in the student's learning experience, the stem of the arrow will become a significant description of distance or speed.

MAJOR CONCEPT

Movement is the change in location of matter.

OBJECTIVES

☐ Observe, compare, and classify.

☐ Use and care for simple tools.

☐ Listen to alternative points of view; explore the environment; and ask questions.

MATERIALS

Electric fan, balls.

VOCABULARY

up	right
down	left
forward	fast
backward	slow

PROCEDURES

A. To provide your students with many opportunities to practice using directional words that describe movement, play such games as "Simon Says" and "May I," in which students are directed to "Take one step forward," and "Hop up and down," and so forth.

1. Say to your students, "Stand behind your chair. Move to the right. . . to the left. . .hop up and down."

2. Say to your students, "Raise your right hand above your head. Now bend and touch your left toe. Raise your left hand above your head. Bend and touch your right toe."

3. Have the students walk in each other's tracks—that is, following a person's steps, placing one's left foot on the spot where the first child stepped with his or her left foot and placing one's right foot on the spot where the first child stepped with his or her right foot.

4. Point down. Point up. Have the children place an object *down* in boxes and *up* on a shelf.

5. Sing motion songs such as "Looby-Loo"; "Heads, Shoulders, Knees, and Toes"; and "Marching."

B. To involve your students in identifying moving objects, begin using arrows.

1. Construct large arrows for the students from red and blue sheets of tagboard. First ask a student to choose any arrow. Have him or her point to the left or to the right as you call the direction.

Have students trace and cut out their own arrows if you wish.

2. Other directions to be used are: "Take three steps back and then touch your right ear with your blue arrow" "Point

the red arrows to the left" "Point the blue arrows up."

3. Continue this activity with fan streamers.
4. Tie streamers to the front of the fan and turn the fan on. Say, "Look at the streamer on the fan. Draw a picture of what's happening."
5. Now, turn the fan off. Say, "Draw a picture of what's happening now."
6. Continue with other events such as wind-up cars and balls. Use the wind-up toys to help children distinguish the difference between moving and non-moving.
7. Let each student do this activity with a wind-up toy and a partner. Have them describe what happens when the toy is turned loose to go by itself; that is, if it goes forward, right, left, backward, and so forth.
8. Use a rubber ball. Bounce the ball up and down. Then have each child control a rubber ball by rolling it in the directions that you call: "Roll it to the left," "Roll the ball to the right," "Roll it fast or slow."

EXTENDING THE ACTIVITY

C. You may wish to provide more practice with the vocabulary: *right, left, fast,* and *slow.*

1. Have the student stand and perform the following movements according to your directions. Say: "Turn left toward the window" "Turn right toward the door" "Look up toward the ceiling" "Look down toward the floor" "Walk fast to your seat" "Raise your hands above your head slowly."
2. Let the students pretend they are cars—or let them play with toy cars—turning left, turning right, going forward, going backward, moving quickly, and moving slowly.

You will find many opportunities in the classroom for students to explore and describe movement. (Donald Birdd)

3. Ask students to name animals that move rapidly and those that move slowly. Invite students to pantomime animals in motion and encourage others to guess the identities of the animals.

ACTIVITY PS–3

Describing the Weather
(Kindergarten–Grade 2)

ACTIVITY OVERVIEW

Weather is a firsthand, experienced event that requires many observations to adequately communicate what has happened. In this activity, students make weather observations and use charts to describe the weather. *Estimated time*: five days to introduce the concept, but continues throughout the year.

SCIENCE BACKGROUND

Weather is an event that surrounds us daily. We see weather discussed on television; we talk about the weather with people we meet. But what is weather? Weather consists of several factors: air temperature, amount and direction of wind, cloud cover, and moisture. Describing each of these factors leads us to wonder *why* they occur. Most weather reports chart descriptions and patterns involving these factors. For example, temperature readings tend to be lower on days that are cloudy. Moisture occurs on days that are cloudy; wind force changes before storms occur. Thus, **meteorology,** or the study of weather, involves the description of weather, the search for patterns in weather changes, and theoretical explanations as to why the weather changes. Note that *temperature* may be reported in the Fahrenheit (F°) or the Celsius (C°) scale. Many people think that temperature and **heat** are the same. Temperature is a measure of heat—but it is not heat; heat is energy.

You will find additional information about weather in chapter 13, pages 474–80.

SOCIAL IMPLICATIONS

Weather changes because specific factors interact with each other and the environment. Daily weather observations help students learn to compare, measure, and organize information. Observations also require that students learn to use simple thermometers as tools.

From a social perspective, students need to understand the relationship of weather to their own lives and how to use weather information for personal health. Individual decision making can be highlighted as you help students to recognize the relationship between the weather and what they wear, how frequently they get "colds," or how to adjust the thermostat in a room or house.

MAJOR CONCEPT

The environment changes because factors such as temperature, moisture, and wind interact with each other, the land, and the sea.

OBJECTIVES

- ☐ Observe, measure, and compare.
- ☐ Use and care for simple tools.
- ☐ Accept direction for the health and welfare of others.
- ☐ Become aware of information needed to make decisions.

MATERIALS

Thermometer.

VOCABULARY

temperature   warmer than
thermometer   cooler than
precipitation

PROCEDURES

A. Introduce students to **thermometers** (or review with them their function) by arranging three containers of water that have contrasting temperatures (for example, container A: 40°C; container B: 20°C; container C: 10°C). Ask students to put their fingers into two containers at a time, and describe how they feel. "Is A warmer than B?" "Is B cooler than A?"

   1. When there is clear agreement ask students to order the containers of water from warmest to coolest. Then pose the question: "How much warmer is the water in container A than in container B?"

   2. Here is when their sense of touch needs an extension—a measuring tool, the thermometer. Let them make measurements of each containers' liquid temperature to verify the order they assigned collectively.

B. In the group, pose the question, "What would we need to know to remember the weather today?" Note the aspects of weather the students list or suggest, for example, temperature, wind, precipitation (or rain), and clouds. With your students, try to make a weather picture on the chart. To help in making weather charts, you may wish to introduce a set of common symbols (or language) for everyone to use:

Cloud Cover

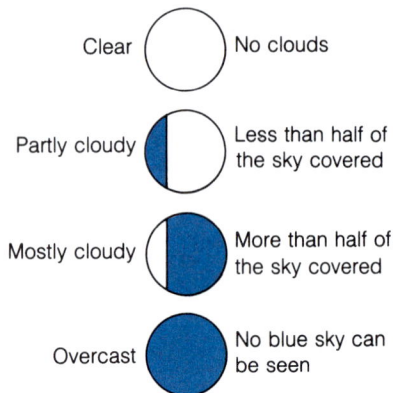

Wind

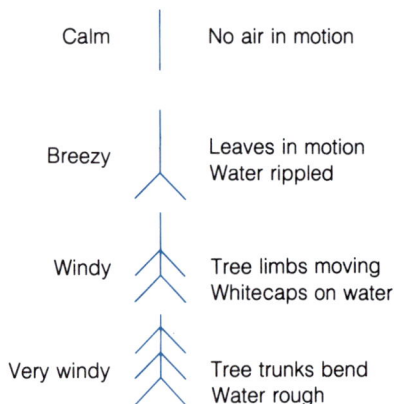

Precipitation

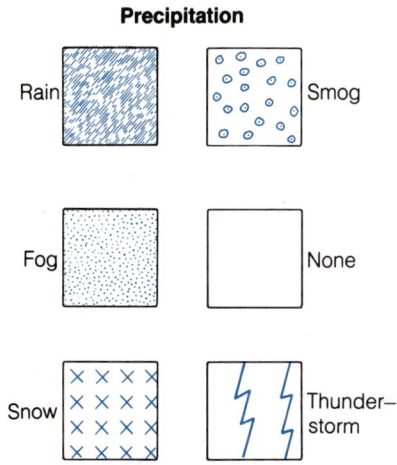

Rain

Smog

Fog

None

Snow

Thunder–storm

A sample weather chart is shown in figure 16–1.

If this activity is presented as a class project, it will be helpful to appoint a *weather team* to fill in the class chart each day. Students might also have their own personal charts. If the maturity of your students so indicates, you may wish to introduce and record each of the four aspects of weather separately for a week before doing a combined chart.

When you have a series of three to four days of weather charts completed, focus students' attention on reading the charts. "What day last week was the warmest?" "What day last week was cooler than today?" "Which days were cloudier than today?" "Today is windier than what days last week?"

Day Date	Monday	Tuesday	Wednesday	Thursday	Friday
Temperature					
Clouds	◯	◯	◯	◯	◯
Wind	│	│	│	│	│
Precipitation	☐	☐	☐	☐	☐

FIGURE 16–1
Sample Weather Chart.

EXTENDING THE LESSON

C. Keep a weekly temperature chart.

1. To make a chart, use a piece of cardboard approximately 48 × 52 cm, 6 pieces of red ribbon, and 6 pieces of white ribbon (each piece of ribbon should be about 1 cm wide and about 52 cm long). Sew, tape, or staple pairs of red and white ribbons together to make 6 pieces of ribbon, each red at one end and white at the other. Put the ribbons through slits cut in the cardboard, and tie them loosely on the back. The ribbons then can be pulled up and down to indicate changes in temperature throughout the week.

2. Practice interpreting weather charts with students. Study these daily weather reports and ask children to answer the listed weather questions.

Which day had the coldest temperature?

On which day would you see the branches of the trees moving the most?

On which day would you wear your raincoat?

On which day did the person marking this chart make a mistake?

Why did you answer the last question as you did?

Which day was warmer than Day 4?

Which day was cloudier than Day 4?

3. Practice drawing weather pictures with students. Draw a picture of yourself (dressed for the weather) swimming, building a snowman, or riding your bike in the rain. Then draw in the weather chart to match your picture.

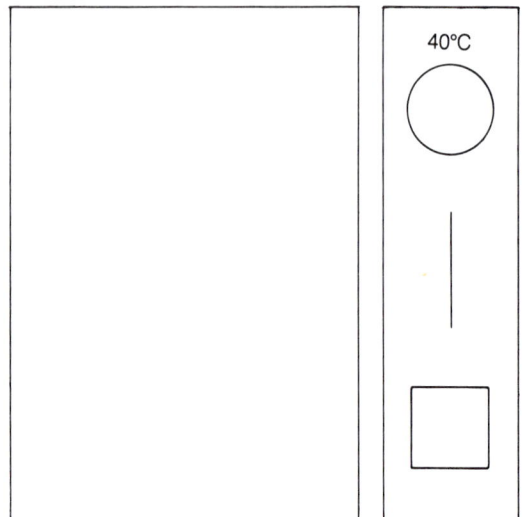

Day 1, 5°C   Day 2, 10°C   Day 3, 30°C

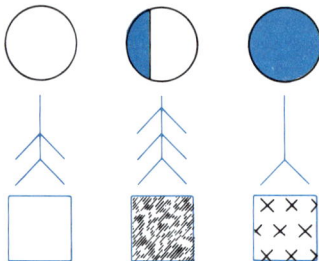

Day 4, 25°C   Day 5, 15°C

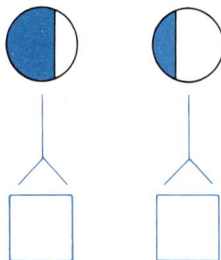

40°C

ACTIVITY PS-4

Learning to Make Predictions
(Grades 3–5)

ACTIVITY OVERVIEW

In this series of activities, students use pendulums and tug-of-war to search for some of nature's patterns. *Estimated time*: five days.

SCIENCE BACKGROUND

In the natural world, scientists observe that many kinds of events occur in predictable patterns. When the sun is bright, temperatures usually rise. Dark clouds usually bring rain. Lightning is often followed by thunder. Rivers usually flow in only one direction. Of course, there may be exceptions to some otherwise predictable patterns in nature. Rivers usually flow in only one direction—downhill. In cases where rivers flow into an ocean, rivers are subject to the effects of the ocean's tides. Then, twice in every 24 hours, the incoming tide forces the river to retreat as the ocean invades the river.

One of the most predictable patterns in nature can be observed easily and under a variety of circumstances. The pattern or "rule" states that when two or more objects interact, both are changed. Whether the objects are a river and an ocean, a plant and an animal, or any two other objects that interact in nature, both are changed. You can think of many examples. When you take a shower, you change when you interact with the water; your skin and hair become wet, and your body temperature rises or falls depending upon the temperatures of the water and your

body. The water that runs off your body is also changed; it now carries small amounts of soap, body oils, and salt. This rule applies to natural forces such as gravitational force and electricity. When a natural force interacts with an object in nature, both the object and the force are changed.

In their search for new knowledge about the natural world, scientists look for patterns of events, and pay special attention to repetition. Events that happen again and again arouse scientists' curiosities and lead them to search for the causes of such patterns.

SOCIAL IMPLICATIONS

A recurring theme in this book and in discussions among scientists is the interaction between science and society. As the amount of scientific knowledge increases and as society learns more about the predictability of events in the natural world, both science and society are changed. Likewise, as the world population increases and has a greater impact on the natural world, increasingly both the population and the environment will change. Thus, we must attempt to predict what changes will occur if we wish to improve the quality of life for all people, and to prevent unfavorable or lethal changes in the world.

In this series of activities, your students will construct some simple science projects involving motion and force. The openness of these activities, and the opportunity for students to construct their own rules, fosters the development of objectivity in your students. Their role is not to "swallow a conclusion" but rather to test or search for evidence to support observed patterns. Their decision-making development is thus enhanced by their awareness that a "rule" may not fit all the facts. One consequence of this awareness is the need to change the rule. On the other hand, students will discover that often more than one rule may fit the same facts.

MAJOR CONCEPTS

All matter changes because it interacts with other matter in the environment. Some events occur in predictable patterns.

OBJECTIVES

☐ Observe, compare, measure, gather and organize information, infer, and predict.

☐ Construct simple projects or models without close supervision.

☐ Compare alternative points of view with evidence; rank alternatives, select one, and accept the consequences.

☐ Combine materials in new ways.

MATERIALS

Task 1:   Either 10 to 15 yardsticks (often free at paint stores or lumberyards) or 1 wooden tongue depressor per student, ball of string, scissors, modeling clay, and washers (4 per student).

Task 2:   Strips (3' long) of masking tape for one-third of the class. Collection of balls and

scrap paper. (A balance is not necessary but will be useful for those who wish to compare the weights of the balls.)

VOCABULARY

prediction
alternative explanation

PROCEDURES

A. Rules for Swinging Objects
   1. Arrange a place for students to explore the movement of pendulums. Out of this activity they will generate rules or **predictions.** Place each yardstick between two desks. Books help to hold them in place. From these yardsticks, students will suspend their pendulums.

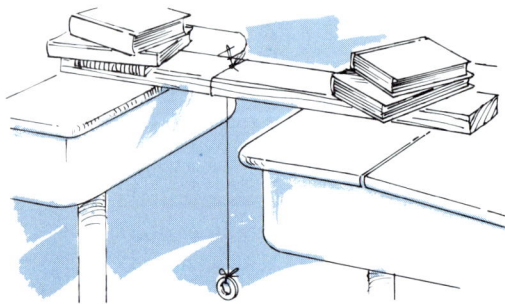

Or you can use tongue depressors (split at one end with a razor blade) into which you can slide the string. These depressors can be taped to the tops of desks or tables.

Rules for Swinging Objects

Materials you will need: string, scissors, and washers (4 per student).

First: Make two pendulums with washers. Tie one end of a piece of string through a washer. Tie the other end to a fixed object.

Your teacher will show you where you can swing the pendulums.

Second: Can you make two pendulums swing together? Draw a picture or tell how you did it.

What can you do to change the way the pendulum swings? Try it and then write down your rules for changing pendulums.

Third: Find two ways to make the pendulum swing faster. Tell about them here by writing down your two rules.

_____

_____

_____

Fourth: Change the weight on the pendulum by adding more washers. Try your predictions to see if your new pendulum will follow the same rules.

2. After your students have completed the directions on Task Card #1, schedule a sharing time. In this discussion time with the students, be sure that their rules are stated so others can follow them. In fact, it might be a good idea to have them write their rules on colored construction paper. These could be posted and others could try to see if the "rule" works. If it does, they could sign their initials; if not, they could write a new one that will work. (Sample rules: (1) the longer the string, the bigger the swing;

(2) The bigger the washer, the faster the swing.)

B. Rules for Sliding Objects

1. In this activity your students will have fun working together on a task that requires cooperation. It will work best if they work in teams of three—two to tug and one to act as referee and recorder.

2. When students have completed the activities on Task Card #2 and have their "rules" written, ask them to share their rules in group discussion and then test each other's rules.

Rules for Sliding Objects

Materials you will need: a strip of tape, 3 feet long, stuck to the floor.

First: If you and a friend hold hands and pull the same on each other, which one of you will slide first? Make a prediction.

_____

_____

What is the reason for your prediction?

_____

_____

_____

Can you change things so that you can make the other person slide first? Write your "First Slide Rule" here.

_____

_____

_____

Second: Place a strip of tape on the floor. Have a tug-of-war with a partner. Try each of these tug-of-wars:

One Partner with	The Other Partner with	Predict Who Will Slide First	Actual Outcome
1. Rubber shoes	Rubber shoes	_____	_____
2. Leather-bottom shoes	Leather-bottom shoes	_____	_____
3. Rubber shoes	Stocking feet	_____	_____
4. Leather shoes	Stocking feet	_____	_____
5. Stocking feet	Stocking feet	_____	_____
6. Stocking feet	Bare feet	_____	_____
7. Bare feet	Rubber shoes	_____	_____
8. Bare feet	Leather shoes	_____	_____
9. Bare feet	Bare feet	_____	_____

Third: What is your "rule" about what is the hardest to slide?

_____

_____

_____

_____

TASK CARD #2

What is your "rule" about what is the easiest to slide?

_____

_____

_____

_____

Can you write a "Second Slide Rule" that accounts for all of the tug-of-wars you have just completed?

_____

_____

_____

_____

Compare your "Slide Rules" with those of other students in the class. See if the class can agree upon a set of rules that allow anyone to predict the outcome of a tug-of-war.

C. Rules for Falling Objects

1. It may surprise your students to find that objects all fall at much the same rate, if they are about the same shape. As they record the results of dropping various pairs of objects, you may need to help them decide if objects fall at about the same time or very different times. In some cases, it is better to listen for the sound of objects that hit the floor rather than look at the objects.

2. After students have completed Task #3, share results. In the discussion, you should take the opportunity to have other students try out the prediction ("rule") to see how well it works.

Rules for Falling Objects

Materials you will need: scrap paper and a collection of balls that differ in size and weight.

People usually throw balls but, in this experiment, you will be finding out what happens when they drop.

First: Arrange the balls from the heaviest to lightest. If you dropped the heaviest ball and the lightest ball at the same time, which would hit the floor first?

            Try it.

            Drop each of these pairs and record what you see.

	Baseball	Ping Pong Ball	Golf Ball
Baseball			
Ping Pong Ball			
Golf Ball			

Now write your "rule" for falling objects.

_____

_____

_____

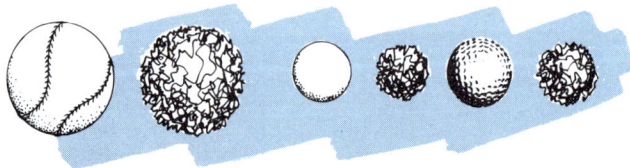

Second: Make a series of paper balls. Each paper ball should be tightly packed and made to be equal in size to one of the following: a baseball, a golf ball, and a ping pong ball.

            If you were to drop the baseball and a paper ball of equal size to the baseball, which would hit the floor first? Try it and record the result in the box below.

            Repeat, comparing the other pairs of balls and record your results.

TASK CARD #3

	Paper Baseball	Paper Ping Pong Ball	Paper Golf Ball
Regular Baseball			
Regular Ping Pong Ball			
Regular Golf Ball			

What "rule" can you write that best describes what happens when you drop any 2 balls?

_____

_____

_____

ACTIVITY PS–5

Experimenting with Forces
(Grades 3–5)

ACTIVITY OVERVIEW

In this set of activities, students experiment with a variety of forces and use arrows to indicate both the relative size of forces and their direction. *Estimated time*: ten days.

SCIENCE BACKGROUND

A **force** is a push or a pull.* How can a push or pull be measured? Probably the simplest way to measure a force is with a spring. If you pull on a spring like the one used for a

spring balance, it will extend. The amount of the extension is a measure of the force. Some springs, such as coil springs on an automobile, are made so they can be compressed. If you push on an automobile spring it gets shorter (compresses); pull on it and it gets longer (extends). This kind of spring could be used to measure either push or pull.

Forces have another characteristic—direction. A push or pull is always exerted in some direction. You pull *up* on the handle of a suitcase to lift it. You push forward on a shopping cart and you can describe this direction of push geographically, such as north or east. Contrast this with the way temperature is de-

*This section from "Science—A Process Approach," *Commentary for Teachers*, pp. 253–54. AAAS Pub. 68–7, 1968. Copyright © 1968 by the American Association for the Advancement of Science. Used with permission.

scribed. A thermometer may read 25°C, but it is not 25° *up* or 25° *north,* but just 25°C. When describing forces you must specify two things: *how much* and *in what direction.*

Forces Which Act on a Body. Perhaps the best way to figure out what forces are acting on a body is to enclose it with an imaginary line or surface which goes all the way around it. Then say to yourself, *"What forces (pushes or pulls) are exerted on the body inside the imaginary enclosure by objects outside the enclosure?"* There is nothing mysterious about forces and you do not have to try to invent ways for forces to be exerted on one body by another. The types of forces can be named: *contact forces* (when two bodies are actually touching), and *forces which act at a distance* (when two bodies are not actually touching).

The forces acting at a distance are gravitational forces (earth-pull), magnetic forces, and electric forces. The earth pulls on an apple and the apple falls to the ground; this is force acting at a distance because nothing touched the apple until it hit the ground. Another kind of force acting at a distance is the magnetic force that makes a nail jump from a table up to a magnet. Nothing touches the nail until it hits the magnet. You probably have had the experience of combing your hair and then finding that you could pick up a small piece of paper with the comb—an example of electrical force.

As an illustration of some of the points mentioned, suppose that you hold a heavy coil spring in the palm of your hand. What forces are acting on the spring? In your imagination draw a line around the spring. What forces are acting across that line? Are you pushing up on the spring? Can you find yourself holding it up? What would happen if you did not push up on the spring? It would fall to

the floor. What other forces act on the spring? The only other force that seems reasonable is the earth-pull, or weight of the spring. This is in the *downward* direction. You now have found the forces which are acting on the spring: the pull of the earth on the spring *down,* and the push you exert *upward.* If these forces are the same in amount, but opposite in direction, the spring will not move up or down.

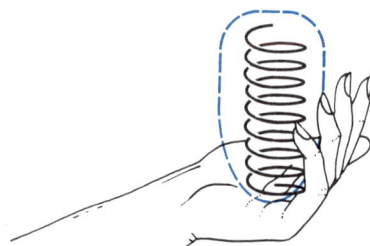

Suppose someone pushes down on the top of the spring as you hold it in your hand. Now what are the forces involved? There is a push *down* on the top. There is still the earth-pull down. You still push up on the bottom of the spring, but your push is greater now than before. Can you feel it? You have to push up harder on the spring now, or else your hand and the spring will move downward.

You can probably think of a number of examples of inanimate objects exerting forces on you or on other objects. Lean against a wall. You can feel the wall push against you to support you. If you do this with your eyes closed it is easy to imagine someone just behind the wall pushing to keep you up. Fasten a string to the knob of a door that is closed. Now pull on the string. With your eyes closed, you don't know whether someone is pulling on the other end of the string or whether it is the doorknob that is pulling the string.

In an example with a spring on a table, how could you find the forces exerted on the *table?* Draw your imaginary line around the table.

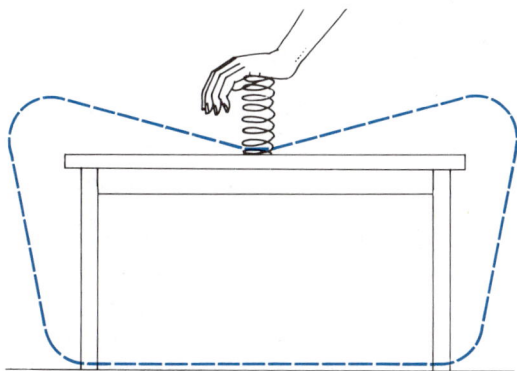

How can there be forces exerted on the table when it's easy to see that the table isn't doing anything? Let's see. As usual, there is the earth-pull *down* on the table. Also, the spring resting on the table pushes *down* on table. (Note that it is not the hand which pushes directly down on the table. The hand does not *touch* the table; it is the spring which touches the table.) Since the legs of the table touch the floor, the floor pushes *up* on them. You could imagine a circus strongman holding the table above his head; to support it he must push up on the table. Of course, he must be standing on something which supports him!

As a last example, suppose that you drop a golf ball from your hand. What forces act on the golf ball after the ball has left your hand, but before it hits the floor? If you visualize a surface enclosing the ball you will see that the only forces which can be acting on the ball are the earth-pull (downward) and a small force exerted upward on the ball by air molecules colliding with it. (We could elimi-

nate this by dropping the golf ball in a vacuum.) The downward force on the ball is much larger than the upward force on it, and the ball will fall rapidly in the downward direction.

To summarize, forces are pushes and pulls; they have direction as well as amount. The kinds of forces which exist are known and can be named. Sometimes forces are exerted by direct contact; other forces such as gravitational, electrical, and magnetic forces are exerted by one body on another without any visible contact. To analyze what forces act on a body, isolate that body by an imaginary enclosure and ask youself what forces are exerted by things outside the enclosure on the one inside it. For some object to be **at rest** (or moving without changing speed or direction), the forces on the body must all balance.

SOCIAL IMPLICATIONS

In this topic, the students' understanding of the forces is enhanced by their ability to describe more precisely the amount of any change using arrows as their tool. The concept that arrows represent forces and the relative balance between pairs of forces is the beginning of experiences with conservation. As such, the learning process is one of communicating information that has been gathered and organized. The equal arm balance is the only piece of equipment that is used.

The social development of the student is best described as creativity. In this topic, there is ample opportunity to encourage the student to try new ideas and to combine objects and materials in new ways in order to search for balanced pairs of forces. Some of this creativity will illustrate their development of added skills in converting what they now know into new questions or alternatives to explore.

MAJOR CONCEPTS

All matter changes because it interacts with other matter in the environment. To start motion, you must unbalance the pairs of forces.

OBJECTIVES

- ☐ Observe, compare, measure, gather, and organize information.
- ☐ Use and care for tools and equipment that require small adjustments.
- ☐ Actively explore the environment: ask questions and combine objects in new ways.
- ☐ Describe the consequences of various alternatives.

MATERIALS

Weight lifting equipment (borrowed), a pair of toy trucks (or roller skates), old magazines, an equal arm balance, rubber bands, and odds and ends from the classroom.

VOCABULARY

force   vector (optional)
arrow   balanced

PROCEDURES

A. To begin this activity called *Weight Lifters*, you might take the class to the gym or ask the physical education teacher for a demonstration of different weight-lifting and exercising units. After seeing various pieces of weight-lifting equipment, have the students name what forces or pushes and pulls are necessary to make the equipment move.

B. For this demonstration activity you will need: two toy trucks with a large heavy weight that you can attach to one side (roller skates may be used instead of trucks), and rubber bands to connect the trucks.

1. This activity is best done as a demonstration for small groups. Show the students the trucks at rest on the floor using the rubber bands to connect them. Have a student draw the trucks on a chalkboard and name the forces.

2. Now stretch the trucks apart and have the students list the forces. Also, have them mark the point where the trucks will meet if you let them go at the same time.

3. Repeat this, only add the heavy weight to one truck. Have the students mark where they they think the trucks will meet when you release them this time (closer to the truck with the weight). Have them list the forces for this experiment.

4. As they now talk about this, focus their thinking on how adding the weight to the one truck changed the results. Did it add more force? Or, did it take more force to get the truck started? What are the implications for auto accidents involving cars of differing weights?

C. Force Pictures is a simple task which requires students to cut out pictures from magazines that show things moving. Then they can name or draw arrows showing the forces on the moving object. Task Card #1 provides directions for this activity.

Force Pictures

Materials you will need: old magazines, paste, and scissors.

First: Find at least three pictures of things moving.

Second: Paste them on your force picture.

Third: Draw arrows on the object for each force and name them at the side of the picture.

TASK CARD #1

1. Make a large *Force Collage* bulletin board to display students' work. Any student may challenge the picture and forces if he or she can think of more forces for the picture.

D. For this learning activity, called *Balancing Forces*, you will need an equal arm balance and a collection of classroom objects: erasers, chalk, balls, paper clips, rulers, marbles, rubber bands, and so on.

1. Here is an opportunity for your students to explore the balance and to show different forces with arrows. Have them complete Task Card #2, in learning centers.

2. After they have completed Task Card #2 have them show their drawings

---

Balancing Forces

Materials you will need: an equal arm balance and a variety of small classroom objects (perhaps erasers, chalk, scissors, pencils, paper clips, etc.).

First: Find out how many paper clips it takes to balance each of the five objects.

Second: Draw force pictures of your results.

Object #1 was _____

_____

Object #2 was _____

_____

Object #3 was _____

_____

Object #4 was _____

_____

Object #5 was _____

_____

Third: Please share this with your teacher.

TASK CARD #2

and share what objects they found that had the most force on the balance. It is especially fun to see if they can figure out a way to order the objects based on the force pictures and then check their results with a balance.

An equal arm balance and a variety of small classroom objects help students understand how to balance forces. (Donald Birdd)

ACTIVITY PS–6

Constructing Electrical Circuits
(Grades 6–8)

ACTIVITY OVERVIEW

In this series of activities students will be constructing electrical circuits. At the same time they learn how to write and use **operational definitions.** *Estimated time*: five days.

SCIENCE BACKGROUND

It is not always possible to make direct observations of a phenomenon through your senses of seeing, hearing, smelling, and feeling. With the aid of tools, gathering of evidence may be extended. From this collected evidence, one or more inferences may be drawn. These inferences contain statements describing observations to be made and expected outcomes of these tests.

This is well illustrated in these activities involving electricity. Electricity will flow only through a conducting closed circuit. With this principle in mind, you can infer the connection pattern of wires hidden beneath a board, by connecting the exposed ends of pairs of wires to a dry cell (battery) and light bulb. When you test a pair of wire ends it may or may not result in the bulb lighting. If the bulb lights, you can infer a connection. If the bulb does not light, you infer that there is no connection. By testing pairs of wire ends you may (1) infer the connection patterns; and

(2) predict (from inferred patterns) the outcome of further testings.

Knowing that electricity flows in closed paths or circuits can be enhanced by thinking about how the kinds of materials in those paths influence the flow. It helps to classify materials as conductors and nonconductors of electricity.

The classifying can be done first by testing the conductivity of substances placed in the circuit with a 1.5-volt battery and a light bulb. A further test is repeated using a 6.0-volt battery and a light bulb. In the latter case some of the substances you originally classified as nonconductors will be found to be conductors. At still higher voltages, other substances might be found to be conductors. It is never expected that you or your students will work with electric potentials greater than 6.0 volts.

For your purpose of working with students, it is sufficient here to talk about "more or fewer dry cells" or "more or less electricity." As a useful meaning for the term "volt" these activities focus on operational definitions. An operational definition includes a statement of what to do (what operation to perform) and what to observe. A complete circuit is defined as an arrangement of a dry cell, a bulb, and wires (the *operation* is connecting these materials in an appropiate way) in which the bulb lights (the *observation* is that the bulb glows). Compare this definition with a nonoperational one: a complete circuit is an arrangement in which an electric current flows. It is not possible to observe an electric current flowing.

Think also of an operational definition of a conductor. A conductor is something that can be used to complete an electrical circuit. The object is put into the circuit (this is the operation) and the bulb is observed. If the bulb lights the object is a conductor; if not, it is a nonconductor. Contrast this with a nonoperational definition: a conductor is something through which an electric current can flow.

SOCIAL IMPLICATIONS

A meaningful strategy to help students understand phenomena in their environment is to help them to both describe events and to operationally define their causes.* An operational definition is a clear description of what to look for and how to know you have found it. Words are the currency of much human communication. It is through verbal exchange, either written or oral, that much of our teaching and learning occurs. Competency in forming operational definitions and using terms precisely is of critical importance in investigations and experiments.

A measure of the cultural level of a society lies in the sophistication of its vocabulary; the extent of the application of that vocabulary contributes to the continuation and advancement of that society. In fact, as the complexity of societal operations increases, so does the use of specialized words and terms. In our own society, we find that specialized groups develop their own jargon. The meanings of existing words may be changed by such groups to express entirely different ideas. It is in this context of communication and change, then, that forming operational definitions takes on its significance as a skill. In *forming operational definitions*, the pupils will be expected to define terms in the context of their own experience.

A definition that limits the number of things to be considered and, at the same time,

---

*Part of this section from "Science—A Process Approach," *Commentary for Teachers*, pp. 166–67 and 169–70. AAAS Pub. 68–7, 1968. Copyright © 1968 by the American Association for the Advancement of Science. Used with permission.

specifies the essential experimental evidence to be gathered is more useful than a definition that encompasses all the conceivable variations that might be encountered. More than one operational definition may pertain to a particular situation. You should encourage children to propose and use alternative definitions.

"You say I was speeding?" said Sam Smith.

"Yes, you exceeded the speed limit of 15 miles per hour in that school zone. I will have to give you a ticket! You were driving at an unsafe speed."

"But the children are all in school and no one was along the street. How could my speed have been unsafe?"

"The law is the law. Here is your ticket."

This conversation may suggest how not to argue with a police officer, but it does illustrate how we may get involved in everyday situations with *operational definitions of terms*. The officer is correct. The law does define "speeding in a school zone" as: *Driving a car in excess of 15 miles per hour in a school zone which is marked by appropriate signs.* The law tells the officer what to observe and what to do. All he or she has to do is observe a car inside a marked school zone and measure the speed of the car; if the speed exceeds 15 mph, the car is speeding and the driver gets a ticket. The definition tells the officer exactly what to observe.

Mr. Smith used his own operational definition in his futile debate with the police officer. Sam's definition of speeding is: *In a marked school zone, a car is speeding only if the speed exceeds 15 mph when children are near the street.* It is an operational definition because it tells you what to do (check the speedometer) and what to observe (presence of children, pedestrians). Mr. Smith's plight illustrates several important ideas to keep in

mind when teaching your students how to form operational definitions. Some of these are listed here and you may think of others.

MAJOR CONCEPTS

All environments change when living and nonliving matter interact. Electrical circuits are completed when the parts of the system interact.

OBJECTIVES

- ☐ Observe, compare, gather and organize information; infer, predict, and form hypotheses.
- ☐ Construct complex projects/models with some guidance; use equipment that requires fine adjustments and discrimination.
- ☐ Combine objects and materials in new ways.
- ☐ Assume responsibility for contributing to the health and welfare of others.
- ☐ Describe consequences of various alternatives.

MATERIALS

For each pair of students: 1 flashlight battery and 1 bulb; wire (2 pieces, each 10″ long); 1 battery holder; and 1 switch.

Miscellaneous materials, e.g., nail, cork, screw, string, paper clip, chalk, sugar, salt, aluminum foil, plastic spoon, penny, cloth, scissors, book, paper, and rubber band.

VOCABULARY

circuit: closed, open, series, parallel
operational definition

PROCEDURES

A. Light the Bulb
1. Start this task by giving each pair of students in the group a plastic bag that contains a flashlight battery, bulb, and 2 pieces of wire, each about 10″ long. Have each pair of students complete Task Card #1.
2. Now collect their diagrams and circulate them to others. It will help if they also record on their task sheets the results of their testing of others' pictures.
3. After they have completed their drawings of four ways to light the bulb, let the students share their results. You may want to award special honors for to the most creative strategy.
4. This will be an opportunity for your students to note first that a working plan is one in which the bulb lights, and, second, that there may be more than one way to do the job.

B. Switches
1. Give each group of three for four students a flashlight dry cell in a holder, a flashlight bulb in a holder, a switch, and 3 10-centimeter (about 4 inches) lengths of insulated wire. Ask the groups to connect all of these things together in such a way that pressing or releasing the switch can turn the light on or off.

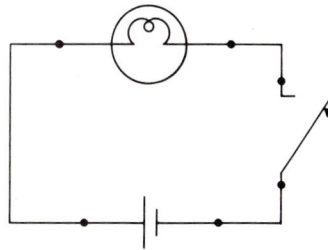

2. When they have completed this task, have them review Task Card #2. You

Firsthand experience with batteries, wires, bulbs, bells, and other familiar objects accelerates students' understanding of electricity. (Donald Birdd)

Light the Bulb

Materials needed: plastic bag and its contents (battery, bulb, and 2 pieces of wire).

In your bag you will find four objects.

Show me that you can light the bulb.

Now draw a picture of what you did to make the bulb light.

Keep a record of your results when you tried other plans.

Whose plan did you try?	Did the plan work?
1.	
3.	
4.	
5.	
6.	

How did you decide if a plan worked? _____

_____

_____

Show at least four different ways you have now found to light the bulb.

1.	2.
3.	4.

Switch

As a group, use the materials your teacher supplies to construct a circuit.

If you have never seen this circuit, or one like it, which of the three definitions would most help you identify the switch? Circle your choice.

1. A switch is part of an electric circuit.
2. A switch is something that turns a bulb on and off.
3. A switch is something in an electric circuit that turns on a bulb when it is pressed.

Why is the definition the most useful? _____

_____

Does it specify what to do and what to observe? _____

A definition that specifies what to do and what to observe is an *operational definition*.

Circuits

Use these operational definitions to identify the circuits in the room. Put the letter or letters for one or more circuits beside the operational definition of the circuit.

_____ 1. *Series circuit:* Any number of lamps, a battery (cell), a switch, and wires connected so that all the lamps glow when the switch is closed. If any one lamp is unscrewed, all the other lamps go out.

_____ 2. *Parallel circuit:* Any number of lamps, a cell, a switch, and wires connected so that all the lamps glow when the switch is closed. If any one lamp is unscrewed, the other lamps all still glow.

_____ 3. *Closed circuit:* Any number of lamps, cells, a switch, and wires arranged so that at least one of the lamps glows when the switch is closed.

_____ 4. *Open circuit:* Any number of lamps, cells, a switch, and wires arranged so that none of the lamps glow when the switch is closed.

_____ 5. *Series-parallel circuit:* Three lamps, a cell, a switch, and wires arranged so that all the lamps glow when the switch is closed. Two lamps are arranged so that if either is unscrewed, the other will go out. There is one lamp in the circuit that can be unscrewed without causing the others to go out.

may find it helpful to have them compare their definition of a switch with the definition in a dictionary.

C. Circuits
  1. Now set up these five demonstrations around the room. Using Task Card #3, have the students decide which definition best fits each demonstration. Remember, each demonstration should have a label, but not the name of the circuit.

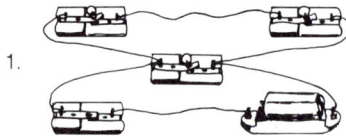

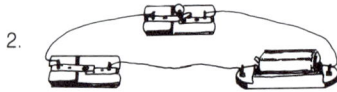

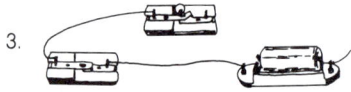

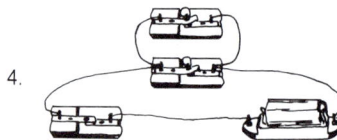

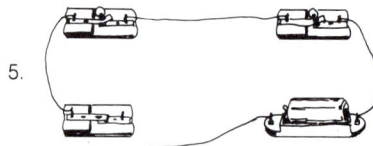

2. Do you agree that the students' choices should be:
  1. Series Circuit   E
  2. Parallel Circuit   D
  3. Closed Circuit   A, B, D, E
  4. Open Circuit   C
  5. Series-parallel Circuit   A

D. Conducting through Energy Paths
  1. In this activity, your students will choose the most useful definition from

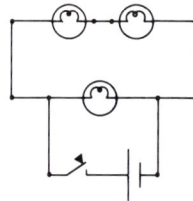

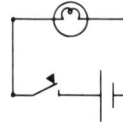

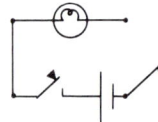

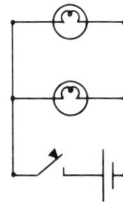

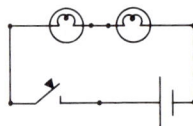

Lamp in Mounting

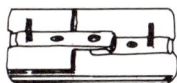

Lamp

Switch

Switch

Cell

Cell

alternative definitions in identifying conductors and nonconductors.

2. They will need to recall that an operational definition specifies what must be done and what must be observed in order to identify or construct what is being defined.

3. In order to be able to interpret the drawings that appear in Task Card #4, the students must be able to name the parts represented by the symbols. On the chalkboard sketch these symbols and the names, as shown above.

4. Give each group of three or four students the equipment necessary to construct the circuit and the additional exploratory "junk" materials:

_____nail	_____paper
_____cork	____plastic spoon
_____screw	_____penny
_____string	_____cloth
_____paper clip	_____scissors
_____chalk	_____book
_____sugar	_____your finger
_____salt	____rubber band
__aluminum foil	

5. As they complete their tests, have them record their results on a class chart:

	Group				
Item	A	B	C	D	E
Nail	con	con	con		

6. Now have them select which of these statements are the best operational definitions.

a. Conductor

_____ (1). A conductor is a wire.

_____ (2). A conductor is anything that makes a lamp light in an electric circuit.

_____ (3). Connect a lamp, 3 wires, and a cell as shown in the diagram. A conductor is any object that will cause the lamp to light when it is connected across points A and B.

b. Nonconductor

_____ (1). Connect a lamp, 3

TASK CARD #4

Circuits Again

Here are operational definitions of two kinds of circuits.

Series circuit: Two or more lamps, one or more dry cells, a switch and wires connected so that all the lamps glow when the switch is closed. If any one lamp is unscrewed all the other lamps go out.

Parallel circuit: Two or more lamps, one or more dry cells, a switch and wires connected so that all the lamps glow when the switch is closed. If any one lamp is unscrewed all the other lamps still glow.

Construct the circuit diagrammed below and then answer the questions listed below.

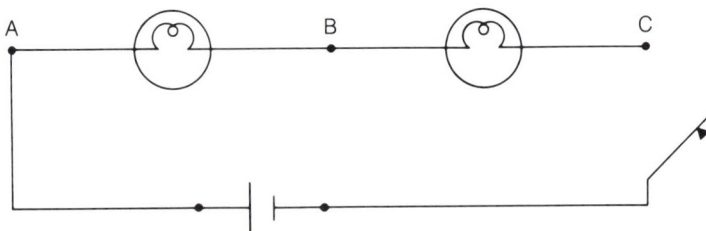

1. Using the operational definitions, what type of circuit is this?
_____

2. a. What happens when you connect a wire from A to B? _____
   b. What happens when you connect a wire from B to C? _____
   c. What happens when you connect a wire from A to C? _____
   d. What happens when you insert another lamp so that the circuit is still the same type? _____
   e. What happens when you put another dry cell in the circuit connected as follows? _____

   f. What happens when you reverse one of the dry cells so they are connected as follows? _____

wires and a cell, as shown in the diagram. A nonconductor is any object that will not cause the lamp to light when it is connected across points A and B.

_____ (2). A nonconductor is a piece of chalk.

_____ (3). A nonconductor is something that will not cause a lamp to light when it is connected in a closed circuit.

E. More on Circuits

After your students have completed the activity described in Task Card #4, have them compare and share results such as these:

1. A series circuit.
   a. Lamp 1 goes out and lamp 2 glows more brightly.
   b. Lamp 2 goes out and lamp 1 glows more brightly.
   c. Both lamps go out.
   d. Three lamps glow less brightly than two lamps.
   e. The lamps glow more brightly.
   f. The lamps do not glow.
2. A parallel circuit.
   a. Both lamps go out.
   b. Three lamps glow less brightly.

c. The lamps glow more brightly.
d. The lamps do not glow.
e. The lamps glow with the same brightness as they will with one dry cell.

3. You will find it useful to have them also discuss who needs to know about circuits or operational definitions.

EXTENDING THE ACTIVITY

F. Short Circuits

1. Give each group of students a piece of Christmas tree tinsel, an index card, and 2 clips arranged as shown below. Also, give them a bulb and holder, a battery and holder, a switch, and 4 ten-centimeter pieces of covered wire.
2. Which items change the bulb?
3. What is the meaning of "short circuit"?

G. Liquid Circuits

1. This is a demonstration for you to do with the group. Using a grapefruit or weak acid solution, connect wires to 2 strips of material (for example, aluminum or zinc) and a meter to register the flow of current. As you make the comparison between the two kinds of strips specifically focus the discussion on what observations support their inferences.

ACTIVITY PS–7

Solving Mysteries
(Grades 6–8)

ACTIVITY OVERVIEW

Through the use of hidden circuits in puzzle boards, students will have opportunities to collect evidence and to draw pictures—or conclusions—or inferences—as to what are unseen wiring patterns. Extending this strategy into their daily lives will be a significant task. *Estimated time:* five days.

SCIENCE BACKGROUND

Nothing is more fundamental to clear and logical thought than the ability to distinguish between an observation and an inference.* Many pitfalls in logical thinking can be avoided if this distinction is made and used.

An observation is a personal experience that is obtained through one of the senses. An inference is an explanation of an observation and results from thinking about the observation. Our thought processes are frequently conditioned by past experiences, and may take place in a fraction of a second. For example, you may be in your living room and see a bright flash outside the window. Almost immediately after the flash

you hear a loud crashing noise. In less than a second you may begin to state your inference, "Lightning struck something not very far away." This inference is an explanation of two observations—light and sound. It is based on past experience with lightning and thunder including the knowledge that the time interval between the flash and the sound is a measure of how far away the lightning struck. Do you suppose that your immediate reaction might have been to make a different inference about the flash of light and the loud noise if it were the fourth of July?

In many cases it is possible to make more than one inference to explain an observation or set of observations. It is important to remember this. Think how many times you have "jumped to a conclusion," that is, accepted the first inference you thought of, to explain an observation, only to find out later that your inference was not supported by further observations.

For example, suppose that it is recess. You look out the window and see David and Neal playing catch with a softball. You go back to your desk, and in a few moments there is a crash of broken glass as a ball goes through the window and rolls across the floor. You jump up, see David and Neal running away, and call through the broken window, "David and Neal, come in here this instant." In half a minute they appear in your room and you

*This section from "Science—A Process Approach," *Commentary for Teachers*, pp. 145-146. AAAS Pub. 68-7, 1968. Copyright © 1968 by the American Association for the Advancement of Science. Used with permission.

begin to lecture them about running away from the scene of an accident. Both boys protest that they did not break the window, and David produces the softball that he has been holding behind his back as evidence that the window was not broken by the ball that they had been playing with. Only then do you retrieve the ball that came through the window from under a desk and find that it is a golf ball. And there is a golf course just across the street from the school. What new inference would you make based on this further information?

The "hot line" between Washington and Moscow was constructed to decrease the chances that either the United States or Russia would jump to a conclusion (inference) about some international incident and act too hastily on the basis of that inference.

Magicians carefully cultivate the art of inducing the audience to jump to a conclusion with one hasty inference to explain an observation. The audience sees the magician reach into a tall silk hat and "pull out" a rabbit, a balloon, and several other large objects. They are expected to infer that those objects were in the hat before the magician "pulled them out."

Scientists cultivate the ability to make at least one, and frequently more than one, carefully thought out inference to explain an observation or set of observations.

Central to all scientific investigations are the activities stated in the objectives. Scientists make observations, construct several inferences about each set of observations, and decide what new observation would help support the inferences. They then make these new observations to see whether each of the inferences is an acceptable explanation.

SOCIAL IMPLICATIONS

In their work, scientists are always looking for patterns. They continually look for regularities in the interactions of events. In general, scientists believe that once something is understood well enough, the pattern of interactions will be quite simple and relatively easy to describe. During the search, many of the proposed patterns to explain a phenomenon may appear to be very complicated, since they have to account for what is known—which may be a very small part of the total picture. However, just because there are missing pieces, the job of finding the experience that explains the pattern is very exciting.

In order to facilitate the study of the relationship among variables, diagrammatic charts and models are sometimes constructed to aid in visualizing the interactions. For this purpose, graphs are very handy tools. They are easy to construct and they also make it much easier for the experimenter to picture the relationship among variables.

MAJOR CONCEPTS

In science, observations of repeated events are the bases for predictions.

OBJECTIVES

- ☐ Observe, compare, infer, and form hypotheses.
- ☐ Construct complex projects with some guidance.
- ☐ Combine materials in new ways; explore and ask questions.
- ☐ Describe the consequences of various alternatives.

MATERIALS

Demonstration circuit board:

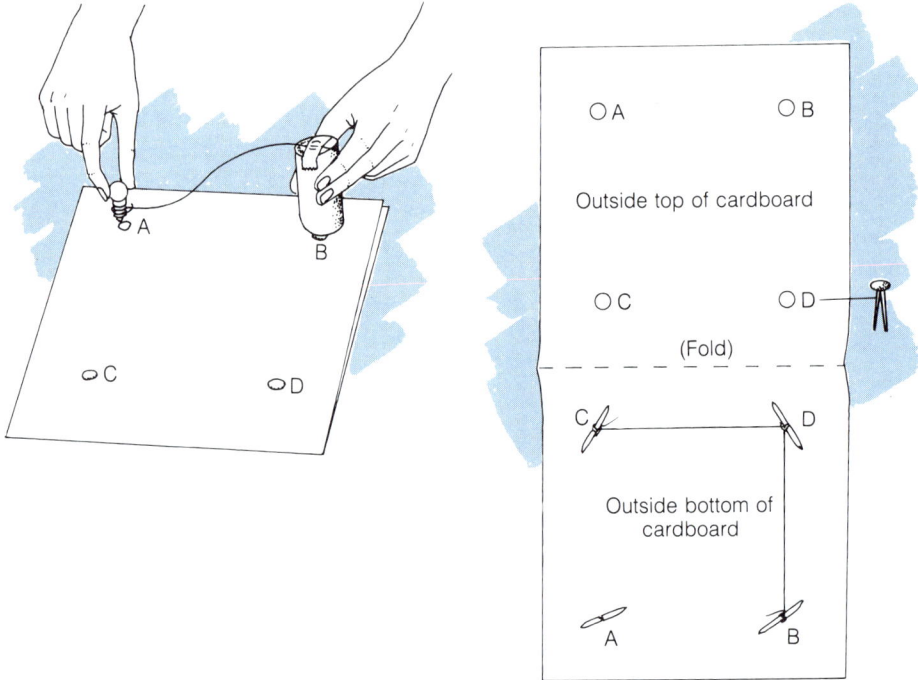

Several batteries and bulbs and a collection of 9 circuit boards wired as indicated by the following diagrams:

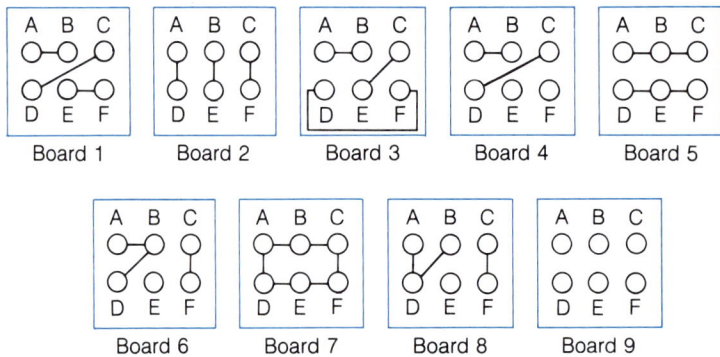

| Board 1 | Board 2 | Board 3 | Board 4 | Board 5 |

| Board 6 | Board 7 | Board 8 | Board 9 |

VOCABULARY

inferences
circuit board

PROCEDURES

A. Inference Boards
1. Show the students a demonstration circuit board that has four exposed paper-fastener heads on the front. Let them see the back of the board. Show them the wire connectors between some of the fastener heads. Ask, "How could you tell which wires connect between the points if you could not see them?" The students may say that there is no way to determine this connection pattern without looking at the back of the board. However, several of them should suggest that they use the bulb (lamp) and the battery (dry cells) to see if they can make a closed circuit by connecting any of the points. If you wish, let some students try this with various pairs of connection points on the demonstration board.
2. Ask the students to name possible pairs of points that could be tested. List these pairs on the chalkboard as they are named. Arrange the pairs in four columns, which may look like these:

    A–B   B–A   C–A   D–A
    A–C   B–C   C–B   D–B
    A–D   B–D   C–D   D–C

3. During the listing, the students may note that half the pairs are duplicated because A–B and B–A are equivalent. If not, ask, "Are there some pairs the same?" (Yes.) "What are they?" (A–B and B–A; A–C and C–A; A–D and D–A; B–C and C–B; B–D and D–B; C–D and D–C.) As the students identify

each equivalent pair, cross out one of them. To avoid confusion and to save time, have the students agree on a systematic procedure for testing and recording the observations.

4. Using the demonstration board, test each pair of points remaining on the list. As the class decides whether the circuit is closed or open in each case, record the observations by writing "closed" or "open" beside the label of each pair. Or put an "X" in the appropriate box if your chart is set up like this:

Pair	A–B	A–C	A–D	B–C	B–D	C–D
Closed	X		X		X	
Open		X		X		X

5. Based on their observations, have them draw a picture or table of what they think the back of the board looks like. An example of that picture may look like the following:

Inferred Connections

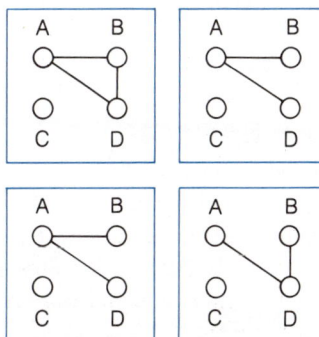

B. Alternative Pathways
1. Now show the students a set of eight inference boards. As they complete

their listing of each one, have them make a data chart and draw a picture of how they think the board is wired. Have the pictures posted on a bulletin board in the room. When most students have finished the task, have them examine the variety of pictures and deferred conclusions they have drawn. (More than one could be correct.)

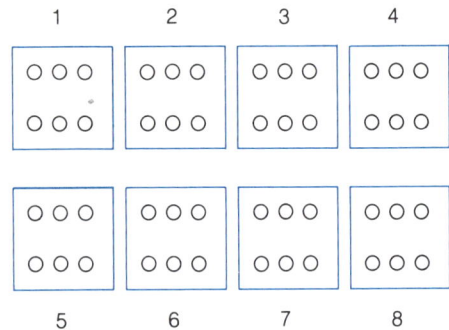

ACTIVITY PS–8

Testing Hypotheses with Dissolving Substances
(Grades 6–8)

ACTIVITY OVERVIEW

In these activities, students generate a rule to explain what causes material (sugar or Alka Seltzer) to dissolve. Students then make an opposite rule (null hypothesis) and search for new data or observations to see which of their two rules is supported by the experiment. This search is systematic, however, and requires that they name their variables and operationally define them. *Estimated time:* five days.

SCIENCE BACKGROUND

We measure the progress of science partly by the extent to which we can describe our environment in general rather than by specific statements. For example, the statement that blue stars are hotter than red stars is more general than the observation that Polaris is a blue star and Arcturus is a red star. Or the statement that inherited characteristics are determined by genes transmitted from the father and mother is more general than the

statement that the offspring of a horse and a donkey is a mule.

In *Science—A Process Approach,* a **hypothesis** is defined as a generalized statement that includes all objects or events of the same class.* It may be an explanation, in which case it is a **generalization** of an inference. Or it may be simply a generalization about observations. For example, we can observe that chalk is a poor conductor of electricity because it cannot be used to complete an electrical circuit. We can also observe that chalk is a poor conductor of heat because we can hold it in our hand when we put the other end in a flame. A related hypothesis that is a generalization about observations on nonconductors of heat and electricity might be that *all* poor conductors of electricity are poor conductors of heat. It is a hypothesis because it is a general statement that includes

---

*This section from "Science—A Process Approach," *Commentary for Teachers,* pp. 159-60. AAAS Pub. 68-7, 1968. Copyright © 1968 by the American Association for the Advancement of Science. Used with permission.

all substances in the class of poor electrical conductors.

If you invert a glass jar over a burning candle the candle will continue to burn for a short time and then go out. One of several inferences that you might make to explain this observation is that the candle went out because all the oxygen in the air in the jar was used up. A related hypothesis that is a generalized explanation might be that burning candles covered with jars go out when all of the oxygen in the jar is used up.

An important activity of a research scientist is devising tests of hypotheses. Testing a hypothesis that is a generalization consists simply of making more and more observations on whatever relationship is covered by the hypothesis. For example, the hypothesis that all poor conductors of electricity are also poor conductors of heat can be tested only by trying many poor conductors. If any substance is found that is a poor conductor of heat and a good conductor of electricity or *vice versa*, then the hypothesis is not supported and must be revised.

Testing a hypothesis that is an explanation involves conducting a test that will provide data that support or do not support the hypothesis. In the case of the hypothesis about burning candles under glass jars, it is only necessary to test the air in the jar after the candle has gone out to see whether or not oxygen is present. This experiment has been done, and oxygen is indeed present, so the hypothesis is not supported.

When data are collected that do not support a hypothesis, the hypothesis must be either modified or rejected. In the case of the poor conductors of heat and electricity, if only a few exceptions are found, the hypothesis might be revised to say: Most poor conduc-

tors of heat are also poor conductors of electricity.

The hypothesis about burning candles might be modified to state: Candles under inverted jars go out when the amount of oxygen in the air is reduced to 10 percent. Or it might be replaced by a new hypothesis such as: Candles under inverted jars go out when the amount of carbon dioxide in the air increases to 3 percent. Both of these hypotheses could be tested by further investigations.

In *Science—A Process Approach* a clear distinction is made between inference and hypothesis.* Inferences apply to single observations or sets of observations made of a single object or event. Hypotheses are generalized inferences. Scientists rarely bother to decide whether they have made a hypothesis or an inference. Whichever it may be, they are more interested in making new observations to find out if their ideas are refuted or whether they are perhaps on the right track. The excitement for a scientist is in thinking of a reasonable explanation and testing it by making further observations. It is important for you to provide opportunities for your pupils to learn from their own experiences that there can be fun in doing this kind of activity and that they can indeed do it.

Science is characterized by the use of certain modes of inquiry.† Basic assumptions are made, terms are defined, phenomena are observed, hypotheses are suggested, evidence is collected, data are interpreted, and information is related to existing conceptual structures of science. In most cases scientific

---

*From *Science—A Process Approach: Commentary for Teachers* (Third Experimental Edition, 1968), pp. 159–60).

†The remainder of this section from *University of California, SCIS Elementary Science Sourcebook* (Chicago: Rand McNally & Co., 1974). Now supplied by Delta Education, Nashua, New Hampshire.

work serves to refine or readjust segments of the conceptual structure. Occasionally, major scientific revolutions, such as the Copernican revolution which placed the sun rather than the earth near the center of the solar system, reorient entire fields of science.

To illustrate a mode of inquiry in the sciences, let us consider a fairly simple example and how it might be investigated.

A ball is released and it drops to the ground. Other objects such as pieces of wood, metal, and a feather are also released; they all fall to the ground. We may formulate a hypothesis: "When released, all objects fall to the ground." But, suppose a balloon filled with helium is released; instead of falling to the ground, it rises. That's the end of the hypothesis. It could, however, be modified to state: "All objects fall to the ground when released in a vacuum." (A helium-filled balloon will fall to the ground in a vacuum.) But, even this hypothesis is sensible only near the earth or another large celestial object. In space, far from the earth, the statement "falling to the ground" is meaningless. The objects might very well go into orbit around some celestial body such as the earth or the sun. In order to cover more classes of events, the original hypothesis has to be extended and refined. This process of formulating, testing, revising, and extending hypotheses is an important mode of operation in science. Eventually, a hypothesis that has stood up under repeated testing and covers a wide range of phenomena may be considered a law.

Certain basic assumptions are usually made in science. One assumption is that there is a real physical universe, that there is matter, and that matter is involved in natural phenomena. In the example, it is assumed that the ball, wood, metal, feather, balloon, and the earth actually exist. A second assumption

is that natural phenomena are reproducible, that is, if the same set of conditions are set up, the same phenomena will occur. These assumptions are implicitly made by the members of the scientific community, and are a part of what might be called the "scientific point of view."

Hypotheses are suggested explanations for observed phenomena. They are formulated out of other similar observations and experiences. A hypothesis is used to help direct further investigations. If further observations are consistent with the hypothesis, the hypothesis is confirmed and tentatively accepted. If the hypothesis turns out to be inconsistent with further observations, it must be revised and subjected to additional testing. For example, the hypothesis that "when released, all objects fall to the ground" was found not to hold for a balloon filled with helium. It had to be revised in order to include what happens when a balloon filled with a light gas is released.

In science, *the ultimate test of an idea is an empirical one.* Is the idea consistent with direct observations? Does it work when it is tried? Based on previous experiences theory, it is possible to predict with some certainty what will happen when certain operations are carried out. For example, we can predict that, when a ball is released, it will fall to the ground. The ultimate test of this prediction is the empirical one of what actually happens when a ball is released. Sometimes, of course, the empirical evidence does not uphold the prediction. Then, the theoretical framework on which the prediction was based has to be reconstructed in light of the empirical evidence.

The empirical test of ideas is of central importance in science. Even in very theoretical discussions an attempt is usually made to suggest empirical tests for ideas. For example, as a check on one of his ideas associated with relativity, Albert Einstein suggested observations of starlight that could only be made during a total solar eclipse. The observations supported his ideas and were used to substantiate his theory.

Theoretical predictions can be checked in empirical tests. Again, ideas associated with the theory of relativity were used to predict that objects, accelerated to velocities approaching the velocity of light, would gain in mass. These predictions were made even though there were no ways of achieving such velocities at that time. Later, with the invention of the cyclotron, it became possible to accelerate particles to velocities approaching the velocity of light. The prediction was substantiated—the particles did gain in mass.

SOCIAL IMPLICATIONS

Whether scientists discuss an event or just think about it, it is important for them to distinguish among actual observations, inferences (explanations of what has been observed), and general statements (hypotheses), which include all occurrences of the event, even the ones they did not personally observe.

Based on their observations and inferences, scientists may generalize to unobserved situations by saying that when variable A is added to variable B, event C will occur. They predict that whenever the same variables are combined, the same event will result. A general statement of this type is called a hypothesis. A hypothesis can be thought of as an inference that has been made general to cover all cases, not only those actually observed.

Once a hypothesis has been stated, it must be tested. An experiment is set up, variables are controlled, and data collected. If the data pattern is consistent with the pattern predicted by the hypothesis, the hypothesis is supported.

A special word of caution is needed at this point. If the data pattern is consistent with the hypothesis, many people will say that the hypothesis is *proved*. However, hypotheses can *only* be proved if *all* possible cases have been tested. Each individual set of observations will *support* the hypothesis but not prove it. On the other hand, the first set of data that is contrary to the hypothesis results in the hypothesis being *disproved*. A disproved hypothesis must be either dropped or restated to account for the new findings.

In summary then, a hypothesis is a statement that describes what will happen when variables interact. It is a general statement in the sense that it attempts to cover all cases, —both observed and unobserved. A hypothesis may be disproved when one set of observations is contrary to what the hypothesis describes; however, a hypothesis cannot be proved until *all* possible cases have been tested. Remember that the next set of observations may be the one that disproves the hypothesis! Following are some activities dealing with generalizing.

MAJOR CONCEPT

Rules that explain relationships of cause and effect in events are hypotheses.

OBJECTIVES

☐ Observe, compare, measure, gather and organize information; infer, predict, and form hypotheses.

☐ Construct complex projects or models with some guidance.

☐ Seek alternative points of view and multiple sources of evidence.

☐ Rank alternatives and select one on the basis of the most positive effect.

MATERIALS

Sugar, salt, Alka Seltzer and Polident tablets, stop watch or clock with second hand, suitable containers, hot and cold water, and thermometers.

VOCABULARY

inference
hypothesis

PROCEDURES

A. With your students watching, place on a desk 1 beaker of very warm water (as hot as you can get it) and 1 beaker of cool water. Drop a Polident tablet in each of the beakers and watch the results.

   1. Ask your students to describe how these two events were similar. (Both tablets dissolved.)

   2. How were they different? (Rate of change was different.)

   3. What variables did you work with?
      Responding Variable: rate of dissolving time
      Manipulated Variable: temperature of water
      Constant Variables: amount of water, kind of tablet, size of container, time of day

   4. Why did one tablet dissolve faster? Have your students write a sentence which tells about dissolving tablets in water of different temperatures, for example, Polident tablets dissolve more quickly in warmer water.

B. Show your students sets of other materials: sugar, salt, and Alka Seltzer (Fizzies and Kool Aid are further possibilities).

   1. Ask: "Does your rule about dissolving Polident in water also apply to seltzer tablets?
      How can we test this rule?
      How about other substances? Would our rule be the same for other substances?"

   2. In groups of two or three have them collect their data using Task Card #1. Note that as their rule becomes more general to apply to all dissolving (rather than just the Polident tablet) it changes from an inference (specific explanation) to a hypothesis (generalization).

   3. Their hypotheses could be:
      A. Hot water dissolves substances faster.
      B. Cold water dissolves substances faster. Or a more formal statement can be done in a "if-then" form: *If* you increase the temperature of water, *then* substances will dissolve faster.

   4. Making class charts of their results is a good way to help students compare their findings.

C. Now have your students do Task Card #2 as a focus for finding new data or evidence to support their hypotheses.

   1. You may want to have your students do this with each of the four or five substances.

   2. A total class chart is a useful stimulus in a sharing time as you ask, "What do we now know that we did not know before we started this investigation on dissolving time?"

Rules and Evidence

First: If you were going to dissolve one of the substances in water, would you want to use hot or cold water? _____

Make a hypothesis telling which would dissolve the substance faster, the hot or the cold water.

_____

_____

_____

Second: Make a second hypothesis which would be opposite your first one.

_____

_____

_____

Third: What materials will you need?

Test your idea three times and write your data here.

Trial	Temperature	Time
1	20°	
	40°	
2	20°	
	40°	
3	20°	
	40°	

Fourth: Make a graph of your results.

Which trials support hypothesis A? _____

Which trials support hypothesis B? _____

TASK CARD #1

EXTENDING THE ACTIVITY

D. On the chalkboard, present the information found in the following table from a test of the time it takes a Saccharin tablet to dissolve in water at different temperatures.

Temperature of water	Dissolving time
20°	29 seconds
45°	15 seconds

Testing Hypotheses About Reaction Time

Alka Seltzer Dissolving Time

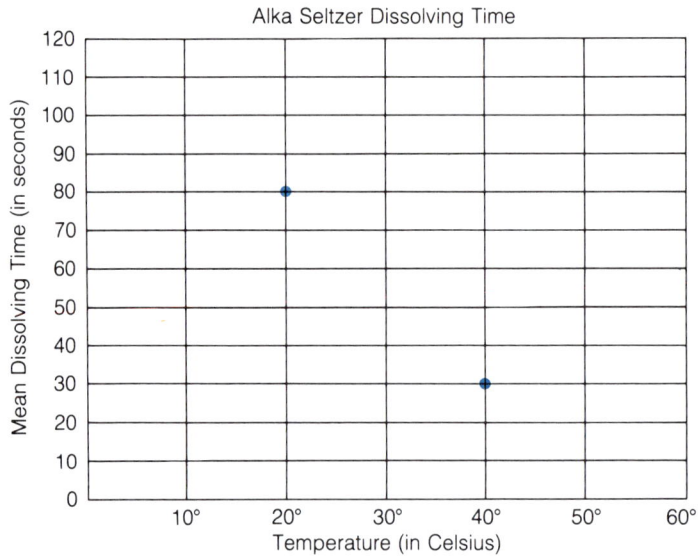

First: Study the graph above. In a test of the time it takes to dissolve an Alka Seltzer in water that is 20° and 40°, what was the mean or average time for

20°? _____

40°? _____

Second: Predict what you think the time would be if the temperature was 30°C.

_____

Did you predict by connecting the dots? If you didn't, connect the dots to help you make another prediction.

Third: Get a cup, a thermometer, 3 seltzer tablets, and water that is 30°C from your teacher. Make three trials to test your hypothesis. Find the mean or average time and put it on the graph above.

Temperature	Time for Tablet to Dissolve			
	Trial 1	Trial 2	Trial 3	Mean
30°				

TASK CARD #2

Fourth: Did your test support your hypothesis A or B? _____

Fifth: On the graph, did the three temperatures make a straight line? _____

    Why do you suppose this happened? _____

_____

Sixth: From this set of information, predict what the time would be at 10° and for 50°.

    10° _____

    50° _____

Seventh: Make a test of your hypothesis. Use three trials and record your data here. Find the mean time at each temperature.

	Time for Tablet to Dissolve			
Temperature	Trial 1	Trial 2	Trial 3	Mean
10°				
50°				

Eighth: Which hypothesis does your information support? _____

Weighing the empty glass

Pour measured amount of water into glass

Checking the water temperature

Asking, "Do my results support my hypothesis?"

Dropping in the tablet and timing

Writing or recording the results

Ask students to make a rule or hypothesis about how the temperature of the water changes the dissolving time. (The Saccharin tablet dissolves more quickly in water of higher temperatures.)

E. The pictures opposite show activities that students might use to test their hypotheses. Ask students to place a check in the corner of each box that shows an activity they would use in their experiments (2, 3, 4, 5, and 6).

F. A student tested this hypothesis: Salt dissolves faster in warmer water. A graph of the results is shown below. Ask students to circle the line segments that support the hypothesis. For example, the line between points C and D is a line segment.

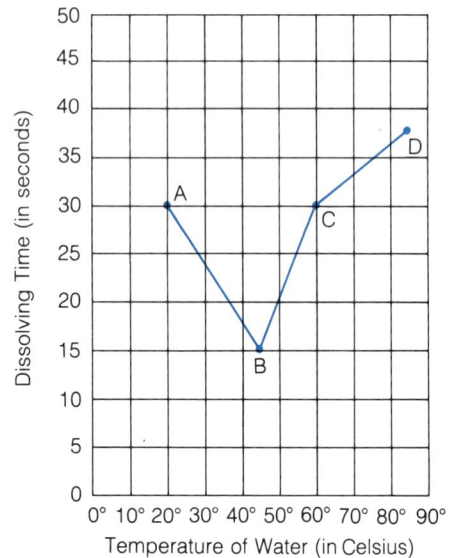

Teachers' Resources

Science—A Process Approach: Commentary for Teachers. Third Experimental Edition. Washington, D.C.: American Association for the Advancement of Science, 1968.

Science Curriculum Improvement Study. *Elementary Science Sourcebook.* Chicago: Rand McNally & Co., 1974.

Children's Resources

Allison, Linda. *The Reasons for Seasons: The Great Cosmic Megagalactic Trip Without Moving from Your Chair.* Boston: Little, Brown & Co., 1975.

Attraction of Gravity (film). Santa Monica, CA: BFA Educational Media, 1975.

Asimov, Isaac. *Building Blocks of the Universe.* Rev. ed. New York: Abelard, 1974.

Branley, Franklyn M. C. *Color from Rainbow to Lasers.* New York: Thomas Y. Crowell & Co., 1978.

Burns, Marilyn. *This Book is About Time.* Boston: Little, Brown & Co., 1978.

Caney, Stephen. *Inventions and Contraptions.* New York: Workman Publishing Co., 1980.

Color: A First Film (film). Santa Monica, CA: BFA Educational Media, 1979.

Cosmic Zoom (film). New York: National Film Board of Canada, 1970.

Epstein, Sam, and Epstein, Beryl. *Look in the Mirror.* New York: Holiday House, 1973.

Goran, Morris. *Experimental Astronautics.* New York: Bobbs-Merrill Co., 1967.

History of Measurement (filmstrips). Burbank, CA: Walt Disney Educational Media, 1976.

Mullin, Virginia L. *Chemistry Experiments for Children.* New York: Dover Publications, 1968.

Reuben, Gabriel. *Electricity Experiments for Children.* New York: Dover Publications, 1960.

Van Deman, Barry, and McDonald, Robert, eds. *Nuts and Bolts: A Matter of Fact Guide to Science Fair Projects.* Harwood Heights: Science Man Press, 1980.

Webster, David. *How to Do a Science Project.* New York: Franklin Watts, 1974.

Elementary and Junior High Science Resources: A Guide

This is a list of science resources for elementary and junior high school teachers; it is limited to in-print trade books, games, films, and filmstrips. The listing includes many resources that lead to other sources of information as well.

Materials were selected for their intrinsic qualities; some preference was given to inexpensive materials and to materials emphasizing activities and/or the interconnectedness between science and society.

The titles suggested here can be augmented by many excellent out-of-print titles available in any good children's library, film library, or film rental collection. For the sake of brevity, the addresses of publishers or audiovisual producers have been excluded but they are listed in *Books in Print* or the *Audio Visual Marketplace*. The addresses of associations and periodical sources, however, are provided.

TEACHER RESOURCES

Theory into Practice

The books in this section tend to address theoretical issues in science; those in the following sections are more practical in nature.

Barman, Charles R. *Science and Societal Issues: A Guide for Science Teachers.* Cedar Falls: Price Laboratory School, 1979. 145pp. Available from Price Laboratory School, University of Northern Iowa, Cedar Falls, IA 50613.

Helps students to confront controversial issues and make personal decisions; helps teachers to think through problem areas. Mostly for the secondary school teacher, but many issue-oriented areas such as energy choices, the metric system, automobile safety restraints, and so on, are relevant for the elementary teacher as well.

Good, Ronald G. *How Children Learn Science: Conceptual Development and Implications for Teaching.* New York: Macmillan, 1977. 337 pp.

Presents selected concepts in science and mathematics and ideas concerning mental structures related to these concepts. Part I reviews Jean Piaget's work. Part II considers research studies on cognitive development. Part III describes Piaget's stages of cognitive development. Part IV examines intellectual requirements of concepts in elementary science texts and programs in the first, third, and fifth grades. Part V describes five roles for science teachers, with appropriate conditions. Part VI provides concrete ideas for helping youngsters learn science through investigation. Appendices include a self-test and background materials.

Hofman, Helenmarie, and **Ricker, Kenneth S.** *Sourcebook: Science Education for the Physically Handicapped.* Washington, D.C.: National Science Teachers Association, 1979. 284 pp.

Includes chapters on law and the handicapped; teacher training; resources; careers; and visual, auditory, and orthopedic handicaps. The work makes a position statement and recommendations.

Lawson, Anton E. *1980 AETS Yearbook: The Psychology of Teaching for Thinking and Creativity.* Association for the Education of Teachers in Science, 1979. 319 pp.

This seventh yearbook of the Association contains papers that involve the teaching of thinking and creativity in the context of science education.

Olson, Robert W. *The Art of Creative Thinking.* New York: Harper & Row, 1980. 265 pp.

Based on research and experiments in ways to stimulate creativity, the book gives techniques and exercises to develop creative ability.

Romey, William D. *Teaching the Gifted and Talented in the Science Classroom.* Washington, D.C.: National Education Association (NEA), 1980. 64 pp.

A helpful perspective on giftedness as it relates to science teaching and the role of the scientist. Includes motivational devices for stimulating excellence and intriguing science activities. Includes a 54-item bibliography. Part of NEA's series on *Teaching the Gifted and Talented in the Content Areas.*

Thelen, Judith. *Improving Reading in Science.* Newark: International Reading Association (IRA), 1976. 55 pp. Available from IRA, Newark, DE 19711.

Little booklet intended to improve science reading skills. The five-step plan suggests ways to help students read to learn, with twenty exercises that illustrate how to carry out each step.

Todd, Vivian Edmiston, and **Armstrong, Terry.** *Beginning Everyday Science Early.* San Jose, CA: Lansford Publishing Co., 1980. 36-page booklet.

Booklet outlines the values and methods of teaching science to young children and suggests how to choose and organize a science curriculum. Transparencies are recommended for workshops.

Activities, Provisions, Sources, and Practices

Bartholomew, Rolland B., and **Cawley, Frank E.** *Science Laboratory Techniques: A Handbook for Teachers and Students.* Menlo Park: Addison-Wesley Publishing Co., 1980. 311 pp.

Covers major demonstration and laboratory techniques for all grade levels in general biology (sterilizing and making leaf prints), earth sciences, chemistry, and physics. Helpful appendices.

Blackwelder, Sheila. *Science for All Seasons: Science Experiments for Young Children.* New York: Prentice-Hall, 1980. 272 pp. $11.95, cloth.

Activities for grades K–6 that answer questions and teach problem solving techniques and that answer children's specific questions.

Gross, Phyllis, and **Railton E.** *Teaching Science in an Outdoor Environment.* Berkeley: University of California Press, 1972. 175 pp.

Includes many suggestions, starting points, and activities for learning science in an outdoor environment. Starts with the school environment and moves out to vacant lots and private or public lands. Includes interesting, creative, and self-actualizing activities appropriate for independent study, class projects, and field trips.

Humphrey, James H. *Teaching Elementary Science Through Motor Learning.* Springfield, IL: Charles C Thomas, 1975.

Based on the premise that children will learn better when academic learning occurs through physical activity. Provides specific game-like motor activities for studying the earth and the universe, conditions of life, chemical and physical changes, and so on. Includes, for each area, the applications to be made and the concepts to be learned.

Keyser, Tamara, and **Souviney, Randall.** *Measurement and the Child's Environment.* Glenview, IL: Goodyear Books, 1983.

A wide variety of activities and games which make measurement fun because it is related to children and their world. Contains worksheets that can be reproduced as well as ideas for bulletin boards.

McGiffin, Heather, and **Brownley, Nancie,** eds. *Animals in Education: A Resource for Biology Teachers.* Washington, D.C.: Humane Education, 1979. Available from National Association for the Advancement of Humane Education, 2100 L Street NW, Washington, D.C. 20037.

Includes sixteen articles addressing the controversy on the use of live animals in elementary and high school projects; also includes guidelines for the use of live animals in biology classes and science fairs and a discussion of state legislation on this topic.

Mikkelson, Mary Domb. *Science Fair Handbook: A Guide for Teachers and Students.* San Diego: County of Education, 1979. Available from Greater San Diego Science and Engineering Fair, c/o Copley Newspapers, P.O. Box 1530, La Jolla, CA 92038.

A very useful and practical composite.

Orlans, F. Barbara. *Animal Care from Protozoa to Small Mammals.* Reading, MA: Addison-Wesley Publishing Co., 1977. 374 pp.

A practical compendium for teachers on how to find, order, culture, feed, care for, handle, and teach about living animals. About half the book is devoted to vertebrates, half to invertebrates, excluding birds, wild mammals, domesticated carnivores, monkeys, and ungulates. Emphasizes the proper handling of laboratory animals and gives practical directions on how to pick up frogs, how to make cages, and so on. Includes good photos of children handling animals.

Wolton, R. W., with **Richtmyer, J.** *Science Equipment in the Elementary School* (rev. ed.). Boulder, CO: Mountain View Center for Environmental Education, 1975. 51 pp. Available from Mountain View Center for Environmental Education, University of Colorado, Boulder, CO 80302.

A topical, illustrated guide to inexpensive equipment and materials for elementary school science. Helpful comments.

Finding Out: Sources for More Information

Newman, Michele M., and **McRae, Madelyn A.** (compilers and editors). *Films in the Sciences: Reviews and Recommendations.* Washington, D.C.: American Association for the Advancement of Science, 1980. 184 pp.

Summarizes, reviews, and evaluates nearly 1,000 films in all areas of science, with reviews arranged by the Dewey Decimal Number. Includes indexes by subject, title, and film distributor.

Woodbury, Marda. *Selecting Materials for Instruction: Subject Areas and Implementation.* Litteton, CO: Libraries Unlimited, 1980. 335 pp.

Provides extensive criteria and resources for the different subject fields, with separate chapters on science, mathematics, environmental education, affective education, health education, narcotics, nutrition, and sex and family life education, among others. Each chapter suggests organizations and resource bibliographies as well as standards and procedures to simplify the process of selection.

This is one of a series of three volumes that covers thoroughly the process of selection. *Selecting Materials for Instruction: Media and the Curriculum* covers requirements and sources for different media (such as toys, games, and films) and has a thorough chapter on free materials that considers the problems in selecting free science materials. *Selecting Materials for Instruction: Issues and Policies* provides an overview of policies and criteria, with chapters on teacher, parent, and student roles in selection, and on individualization, gifted children, and special education.

Note: Ask your librarian about the *Books in Print* series: *El-Hi Textbooks in Print, Children's Books in Print, Subject Guide to Children's Books in Print,* and so on. These annual guides, available in most libraries, will help you keep up with in-print and soon-to-be-published books.

Information on Games

Many of the activity books also include activities that could be considered games or simulations.

Davis, Arnold R., and **Miller, Donald C.** *Science Games.* Belmont, CA: Fearon-Pitman (Pitman Learning, Inc.), 1974. 30 pp. Grades 1–6.

Describes fifty-two inexpensive games for elementary science classes which can be made by students or teachers and require a minimum of equipment. For each game, the book provides objectives, number of players, materials required, if any, and procedures.

Horn, Robert E., and **Cleaves, A.** (eds). *The Guide to Simulations/Games for Education and Training* (4th ed.). Beverly Hills: Sage Publications, 1980. vol 1: *Academic Simulations/Games;* vol. 2: *Business Games.*

The basic guide to simulations and games with more than 1,400 separate games in the current edition. Half of the entries in this edition are new; games are listed for all levels (with information on source, number and age level of players, etc.). Separate chapters include games on the future, ecology, and science with introductory essays on each subject area.

Hounshell, Paul B., and **Trollinger, Ira R.** *Games for the Science Classroom: An Annotated Bibliography.* Washington, D.C.: National Science Teachers Association, 1977. 24 pp.

Includes 1977 information on how to choose, order, or prepare more than 100 science games and summaries of their rules. Commentary includes purpose of each game and its usability in the classroom.

ORGANIZATIONS AND INFORMATION SOURCES

The organizations listed here are only some of the many that could be helpful to elementary and junior high school science teachers. Many more are incorporated and described in Woodbury's series on *Selecting Materials for Instruction,* the citation for which is listed earlier. Her *Subject Areas and Implementation* section is especially valuable for areas related to science, environmental education, health education, sex education, nutrition education, and drug education.

American Association for the Advancement of Science (AAAS), 1776 Massachusetts Avenue NW, Washington, D.C. 20036.

Its publications include many that review elementary (and other) science materials.

Animal Welfare Institute, P.O. Box 3650, Washington, D.C. 20002.

Like the National Association for the Advancement of Humane Education, this group produces some valuable items for teachers. Their *First Aid and Care of Small Animals* (46 pp.) is free and useful. Also worthwhile is their *Human Biology Projects.*

Center for Science in the Public Interest (CSPI), 1757 S Street NW, Washington, D.C. 20009.

A public service organization whose well-researched and attractive publications on food and nutrition balance out corporate offerings. Their monthly magazine, *Nutrition Action,* includes news, activities, and reviews. Their *Creative Food Experiences for Children* by Mary Goodwind (1974) is well designed for teaching with activities, games, recipes, and facts. The posters, *Nutrition Scoreboard and Chemical Cuisine,* are attractive and accurate.

ERIC Clearinghouse for Science, Mathematics, and Environmental Education (ERIC/SMEAL), Ohio State University, 1200 Chambers Road, Third Floor, Columbus, OH 43210.

Like other ERIC Clearinghouses, this center processes documents and articles for inclusion in *Resources in Education (RIE)* and *Current Index to Journals in Education (CIJE).* Additionally, it has an extensive publication program whose materials are listed in a free catalog. Compiles many summaries of research and practice. The ERIC Clearinghouse for Social Studies/Social Science Education (855 Broadway, Boulder, CO 80302) also has some materials in the areas where social studies link with science (for example, environmental education).

Food and Nutrition Information and Educational Materials Center, National Agricultural Library, U.S. Department of Agriculture, 10301 Baltimore Blvd., Beltsville, MD 20705.

Disseminates information on food and nutrition to those teaching nutrition to children. Information includes books, pamphlets, films, tapes, manuals, programmed materials, and so on. It also compiles bibliographies of loan materials and distributes these free to nutrition educators (as long as the supply lasts). It also offers search services and provides referral to other information sources.

National Nutrition Education Clearinghouse (NNECH), 2140 Shattuck Avenue, Berkeley, CA 94704.

A clearinghouse of the Society of Nutrition Education with a substantial library of more than 8,000 items and an ongoing publication program; it compiles bibliographies and evaluates teaching materials. Their *Journal of Nutrition Education* reviews books, current research, and curriculum materials. Other publications include guides to audiovidsual materials, elementary teacher materials, sources, and so on.

National Science Teachers Association (NSTA), 1742 Connecticut Avenue NW, Washington, D.C. 20009.

An organization of science teachers, founded in 1944, whose membership provides publications such as *Science and Children,* the *Science Teacher,* the *NSTA News-Bulletin,* and a discount on many supplementary publications. Publications include a how-to-do-it series, a Careers series, *Elementary Science Packets* that lists resources and includes *Environment and Energy.* Other publications worth acquiring are *Hunger: The World Food Crisis* and *Population: The Human Dilemma.* Write for their catalog and membership information.

Orienteering Services, 308 West Fillmore, Colorado Springs, CO 80907.

A useful group for outdoor science teachers who want to develop map skills along with other programs. Offerings include a very helpful workbook, *Your Way with Map and Compass* (student's edition and teacher's edition). A filmstrip program, *Adventures with Map and Compass,* is available. Write for their catalog.

PERIODICALS

This section includes periodicals for both teachers and students, and emphasizes those that suggest activities and/or review books, games, and media materials.

Other review sources are librarians' publications, for example, *Booklist* (which reviews audiovisual materials also), *Library Journal, School Library Journal,* and others. The Educational Film Library Association's *Sightlines* and *Landers Film Reviews* are the best review sources for educational films.

Science education articles in other publications can be followed through two periodical indexes: *Education Index* is frequently found in large public libraries; *Current Index to Journals in Education (CIJE)* is available at many schools of education and district-level libraries. Computer searches of *CIJE* are available through sources that search ERIC.

Appraisal: Children's Science Books. 3 issues per year. From Appraisal, Children's Science Book Review Committee, 13 Appian Way, Cambridge, MA 02138.

An interesting review journal that combines excellent evaluative annotations and ratings by children's librarians and science specialists. Includes rating and grade level for each book and covers 70 to 80 books per issue. Reviewers are identified at the back of each issue.

The Computing Teacher. Currently 7 issues per year. From Computing Center, Eastern Oregon State College, La Grande, OR 97850.

A journal for persons interested in the instructional use of computers which covers the impact of computers on curriculum, teaching computing, and using computers. Ongoing departments include film reviews, book reviews, software reviews, and general news and information for teachers of the elementary grades.

Journal of Environmental Education. Quarterly. From Heldref Publications, 4000 Albemarle Street NW, Suite 500, Washington, D.C. 20016.

An award-winning, multidisciplinary resource devoted to research and practice in environmental education and communication. Includes project reports, critical essays, and research articles.

Journal of Nutrition Education. Quarterly. From Society of Nutrition Education, 2140 Shattuck Avenue, Suite 1110, Berkeley, CA 94704.

Includes substantial, thoughtful reviews of books, games, reports, charts, audiovisual and classroom materials, and professional readings—about 70 items per issue.

National Geographic World. Monthly. From National Geographic Society, Department 00481, 17th and M Streets NW, Washington, D.C. 20036.

A children's magazine with articles that are useful for studying our natural and human-created environments. Features such topics as "Easter Ornaments," "Rooftop Ideas," and "Fun Pack."

Instructor. 9 issues per year. From Instructor Publications, Instructor Park, Dansville, NY 14437.

A journal for elementary school teachers that reviews about 500 books, nonprint materials, and professional books each year. Its reviews of science materials are in the October, December, January, and March issues; its reviews of films and filmstrips appear in October, December, February, and April. It occasionally issues bibliographies on science topics, usually in the October issue.

Odyssey. Monthly. From P.O. Box 92788, Milwaukee, WI 53202.

An astronomy and outer-space magazine highly popular with kids, and, therefore guaranteed to be read. Includes ongoing articles on astronomy and space travel, with columns like Kepler's Corner, NASA News; also distributes inexpensive full-color *Space Cadet Posters*. Back issues constitute a veritable library of space science.

Ranger Rick's Nature Magazine. 10 issues per year. From National Wildlife Federation (NWF), 1412–16th Street NW, Washington, D.C. 20036. Grades 4 and up.

Colorful nature magazine of about 48 pages with scientifically accurate articles and activities. **Activity Guides** to accompany this publication are about 8 pages each. **Your Big Backyard** (also issued by NWF) is for preschool and primary children (10 issues per year). Each issue contains about 20 pages, with a separate page of suggestions for parents and teachers, and many built-in exercises. It is available from 8925 Leesburg Pike, Vienna, VA 22180.

Science Books and Films. 5 issues per year. From American Association for the Advancement of Science, 1515 Massachusetts Avenue NW, Washington, D.C. 20005.

An easy-to-use review journal, arranged by the Dewey Decimal System, that reviews about 250 trade and textbooks and 50 to 60 films and filmstrips in each issue. Appraisals (from highly recommended to not recommended) and grade levels can be seen at a glance. Items reviewed are appropriate for elementary, junior and senior high, college students, professionals, and general audiences. The bulk of the material may be for older audiences. Reviews have been compiled into **Films in the Sciences.** It is valuable for its comments on content and comparisons with similar items.

Science and Children. 8 issues per year. From National Science Teachers Association, 1742 Connecticut Avenue NW, Washington, D.C. 20009.

The national professional journal for elementary teachers that contains good articles and intelligent, substantial reviews of 20 to 25 books per issue. The March issue includes an annual annotated list of about 100 recommended science books prepared by a joint committee of the NSTA and the Children's Book Council. The list is also available free of charge from the Children's Book Council, 67 Irving Place, New York, NY 10013 with first-class postage for 2 ounces. This publication also reviews curriculum materials in the Curriculum Reviews section, while books and audiovisual aids are listed separately in a Resources Review section.

Science 84 (85,...) 10 issues per year. From the American Association for the Advancement of Science, 1515 Massachusetts Avenue NW, Washington, D.C. 20005.

An outstanding periodical designed to bring science to citizens. Leading science writers make science come alive and understandable, tackling topics such as the state of the art in nuclear reactors, solar energy, and outer-space probes. Excellent color photography. (Some parental or teacher guidance is needed for students sensitive to liquor advertisements.)

Scientific American. Monthly. From 415 Madison Avenue, New York, NY 10017. The December issue of this journal includes an annual conspectus on books on science and technology for younger readers.

Appendix B

Using Computers in the Classroom

Some teachers have a natural interest in machines and technology that will lead them to utilize microcomputers or "micro's" for instruction in schools. Most beginning teachers are drawn to the micro through a different process, however. They explore avenues that seem promising as a means for improving instruction. Early data on the use of computers for instruction is impressive. Peter Rizza reports that in twelve computer-managed instruction (CMI) PLATO projects across the country, students are achieving one year of growth in reading by spending from twelve to thirty hours at the terminal. And indeed teachers all over the country are using whatever resources they can to purchase microcomputers for their classrooms. In 1983, there were more than 120,000 micros in our schools at a time when school resources were meager. That number grows each month and year. Many people involved in the educational use of computers believe that the chief limiting factor at the moment is not lack of money to purchase hardware (the machines themselves) but a scarcity of good software (the programs that one puts in the machines). When software is designed to teach (rather than to sort, for example), we call it courseware.

Our chief concern in this discussion is to provide background information about educational computing so that you can begin to explore the topic on your own. To assist you, we provide sources of information about training, and list information sources such as computer magazines and courseware catalogs.

BACKGROUND

There are three sizes of computer hardware: mainframe computers, minicomputers, and microcomputers. Mainframe or "on line" computers are large, require expensive telephone connections to classroom terminals, and are capable of storing and processing large amounts of information, including student test information. Schools probably don't need one of these unless they are going to use whole libraries of courses stored in a central system. District offices usually have a mainframe to handle payroll, accounting, and other tasks for which large amounts of information need to be stored.

The minicomputer "stands alone." That is, minicomputers are not connected to a central system but offer intermediate amounts of memory or storage and processing capacity. Schools can perform many tasks in addition to instruction with a "mini," including attendance and grade reporting.

The microcomputer or "micro" is the one everyone is talking about and it is the focus of this chapter. It is small, looks like a television set connected to a typewriter, and is inexpensive. It has a small but impressive memory and is now selling at the rate of about 300,000 units a year. About 500,000 micros are expected to be in use in U.S. schools by 1985.

Few of us appreciate how rapidly the computer industry has developed. The history of computers is a story focusing on making computers more widely available and easier to use. During the late 1930s and 1940s, great progress in these two areas was made in the United States and Europe. The first computer was developed at the University of Pennsylvania during World War II to determine missile trajectories and predict the weather. The invention of the transistor in 1948 made modern computers practical because of their dependability, small size, and low power requirements. In the late 1950s the integrated circuit was developed and soon after thousands of components could be placed into a tiny chip. In 1970 the first microprocessor chip made possible small, personal computers. In the 1950s programming languages such as FORTRAN and COBOL were developed; in the mid 1970s Kemeny and Kurtz at Dartmouth created BASIC, a computer language that teachers frequently learn and teach their students in schools.

The idea of computer literacy for the general student population emerged in the 1960s. The Conference Board of the Mathematical Sciences recommended universal computer literacy in its 1972 report; they suggested teaching computer literacy in the junior high school.

Some say the development of computers is evolutionary, not revolutionary. However, in 1977 a series of events began that led to a revolutionary reduction in cost of microcomputers and a consequent increase of their use in schools. This increase can be attributed not only to micros' low cost, but also their utility and reliability. They are available when you want them, unlike the large mainframe computers where one "signs on" and waits his or her turn to run a program. They are also less complex and therefore less likely to break down than are the big machines. This becomes important when your class is excited about an activity and you have just ten minutes to finish it before lunch. The microcomputer has become a fine tool for teachers and administrators.

Not too many years ago, the calculator that we take for granted today could have passed for a computer. Calculators work with numbers: they add, subtract, multiply, and divide. Computers work with numbers and alphabetic data such as names and words. A computer can be programmed to repeat a function using new or updated data every time. It can examine a list of numbers and find a particular item or it can alphabetize a list of names and addresses. It can logically evaluate information, following a program, and can then act on its findings.

A computer is to the human mind what a power saw is to the hand—a machine capable of extending the mind's speed and effectiveness. It can free a teacher from repetitious tasks such as grading tests and reporting the results. It can instantly print out scores so the teacher can see student problems and take corrective action. In a matter of seconds it can report which students are in the bottom quartile so that patterns can be discerned and instruction targeted to specific groups of students. If the teacher wishes, the computer can then be programmed to offer skills reviews for students. These utilize a game format that also increases motivation. Skills reviews can do this both for individuals and for groups; they reduce monotony and help to interest students.

Computers can be tutors for students, as in most computer assisted instruction. Some computer assisted instruction (CAI) has a tutee mode as well, in which the student directs the action by programming the computer. Use of the word processor is another example of the tutee mode through which writing can be typed, edited, stored, and corrected for spelling errors. Students can also have tutee experiences in geometry, art, and music. It is now unclear whether the dominant student use of micros will be in the tutor or tutee mode in the years ahead.

If you think about how hand calculators are changing the teaching of math, imagine how hand-held microcomputers will impact science, math, and other curricular areas. This day is upon us as the first pamphlet-sized micro is now coming on the market for about $100.

In addition to assisting in learning and problem solving, computers and information science are now a subject of study and an important new discipline. The use of computers has affected the rate of discovery of information in all areas of knowledge. Virtually every area of science has harnessed this newest tool. Instruction in computer literacy is becoming a requirement for the training of new scientists on nearly every campus.

Computer programmers continue to be in great demand in all fields, of course, with 35 percent more jobs available this year than last year and computing power growing at 40 percent each year. Preparation for these opportunities begins at the elementary school level. Seymour Papert at Massachusetts Institute of Technology has created a special programming language for children called LOGO. When LOGO is turned on, a pointed arrow called a *turtle* is displayed and can be moved about the screen to draw objects of interest to the child. This allows the child to control activities in the tutee mode. This capability is then used to involve the child in actively solving a series of problems. It is fun to imagine the many ways that a language like LOGO can be used to improve elementary, middle, and junior high school instruction.

The possibility that school districts can begin to develop their own courseware is appealing but in the beginning, courseware is likely to be what is called Level 1—mainly drill and practice exercises. To make material more interesting to students, greater expertise and expense are required in courseware development. The potential cost of courseware development should be considered when computer hardware is being analyzed for purchase: in the long run, courseware development is more expensive. A team of four professional courseware developers may take one month to develop one hour of courseware. Thus, a typical one-year school course might take five or six months to develop.

The advantage to schools in developing their own courseware is the acquisition of programs that exactly meet their requirements. The disadvantages are the cost and the difficulty in finding staff who would enjoy this solitary, non-teacher-like activity.

WHY GET INVOLVED NOW?

There are many reasons for becoming involved with computers now. Computers help students to achieve more in a shorter time. They reduce costs in areas such as bilingual education and mathematics and make changes that will have a lasting impact, even if teachers try to resist computers. They will influence curriculum as it is being developed.

High quality reading programs like PLATO already help students learn more in a shorter time. Other quality courseware will become available because major publishers are now starting to develop them. Once there are 20,000 micros of a given make and type, then it is worthwhile for major publishers to target materials for that machine. There are several such units that are eligible as publishing targets at this time, and new machines are continually entering the field.

Computers reduce costs in areas such as bilingual education and mathematics because these are instructional areas in which there is an extensive need for drill and practice. This is an ideal place to utilize micros with their never-ceasing patience and reliability.

Microcomputers will make changes that have a lasting impact. This may be especially appealing to teachers who get used to a program and then see it discontinued before complete data on its effectiveness is gathered and studied. Because of the significant cost of computer technologies and the difficulties involved in introducing them to an organization, micros will tend to become rather permanent once they are established. Computers are creating a massive change that is coming even if teachers resist it. All over the country, professional parents and students are the main forces pushing for the introduction of microcomputers. Parents see how important the technology is in job advancement and students like the machines because they are interesting to use.

The materials used to teach, such as textbooks, are great levers for influence in schools. At this time there are great opportunities for teachers to influence publishers by learning to program what they want or by serving as content specialists and helping someone else to create the programs they want. These programs may become more important than textbooks in the years ahead.

Of these arguments for involvement, the most compelling one is the staying power of technological change. The rapid change of curricula before its usefulness has been thoroughly tested is one of the most destructive characteristics of American schools. Computer courseware should, by its nature, give us the opportunity to study its effectiveness at length prior to any move to discontinue it. High quality courseware with films, video tapes, slides, games, and simulations will allow the stability that is crucial for sound curriculum evaluation.

How to Get Started

At the school level there is a pattern of microcomputer use that is quite typical. The first micro in an accessible space will elicit mild teacher interest for six months or so. Then students may start to teach their teachers how to use the micro. By about the eighth month, teacher interest increases and three to five micros may be purchased if funds can be found to do so. Interest continues to increase and the next event is the purchase of ten to fifteen machines. Finally, forty to fifty machines are purchased and a technology center is in operation. There are a few of these centers now and there will be many more soon. This pattern of slow, incremental change in the acquisition of computers has been observed in schools throughout the United States.

Although most school people are now indicating that one should locate appropriate courseware before selecting a microcomputer, we still hear of examples to the contrary. As we learn more about how micros are used successfully in schools, we find there is yet one more major step that should precede hardware selection. This is the location of a *support system,* usually found in a nearby school. Most likely this will be a teacher who teaches in your instructional area and who is now using a micro. We see the sequence of steps in selecting a microcomputer as follows:

1. Locate a microcomputer at another school, at a friend's home, at a computer store where you are comfortable, or at some other accessible location.

2. Locate a supportive person to help you get familiar with the micro. You should have them show you how to "sign on," to put on several software programs that interest you, and to run these programs. Perhaps you might write a simple addition program so that you can begin to understand what a program is.

3. Purchase an informative paperback on the topic such as *Computers For Everybody* by Jerry Willis and Merle Miller or another that you like at a computer store or a university bookstore.

4. Study the steps they suggest and make a list of questions that come to mind as you read the book.

5. After you have read the introductory paperback, visit a computer store that sells several makes of micros and learn about their differences. Ask the questions from your list. Then ask the salesperson to demonstrate the software for word processing. Once you feel you can explain this to your principal, ask for a demonstration of the number processing program VISICALC. These two programs will be very useful to your principal and his or her secretary and may help you later to gain their support in your effort to purchase a micro.

6. Become familiar with locations where you can continue to improve your computer literacy and get training in the use of instructional software. Ask for self-instructional materials to use when you get access to a micro on which to practice your skills.

7. Begin a systematic search for courseware you could use in your classroom. Your best bet will be teachers in other schools. Ask the math and science coordinator in your county office of education if he or she knows of anyone with a list of computer-using teachers. Some states have organizations of computer-using teachers and they can give you teacher's names. You can also check district office personnel in districts that have a number of computers. Visit these teachers and check out their courseware. Work hard to develop a relationship in which you can share things you have in return for informal instruction in selecting courseware from your mentor, or support teacher(s).

8. Search for courseware through other channels such as computer sales organizations and institutions offering microcomputer instruction. Some colleges and universities have such courses and others are just developing them. If no courses are offered, locate interested faculty and talk to them about developing such courses. Some county offices of education are developing centers where teachers can go to preview available courseware and other offices should be encouraged to do so.

9. Learn about the hardware that is (1) used in your "support" school, and (2) that has courseware you would like to use. These are key factors in your choice of hardware since you need to be able to use hardware in another school before you purchase it, and you need to know that the courseware that runs on their hardware will work in your situation.

10. Identify major uses of the hardware. For example, will you really use it, in part, as a word processor?

11. Specify minimum requirements and preferred features, such as a fast touchkey board for a word processor.

12. Identify secondary uses and desired features such as a graphics capability.

13. Decide about how much you wish to spend—now and later. The cost range will be from $300 to $4,000 or thereabouts.

14. Survey the field by talking to knowledgeable people and by visiting computer stores. Get a quote on the price of equipment you want from one store and then take your quote to another store to see if they can beat the first price.

15. Survey sources, including a local or regional computer store, a mail order supplier, and the manufacturer. The mail order supplier may have the lowest price, since few manufacturers sell directly to the customer. The best service may come with the highest price at the computer store.

16. Buy or lease the computer. This decision will be influenced by the amount of funds available as well as tax write-off considerations. If you can get a knowledgeable person to advise you, you may even purchase a used computer through the classified adds.

Sources of Training

As you look for good training sources, your local community college is one good source. (Call the head of the mathematics department if there is no computer department and ask for suggestions.) Four-year colleges and universities also may be a rich training source. One graduate program in educational technology we know of has ten courses available emphasizing the microcomputer.

In some areas, adult education at your local high school can be a good source too. Several computer stores conduct limited types of training.

You may be surprised at faculty receptivity if you make training suggestions. A four-level computer literacy curriculum for teachers might include the following components:

☐ Level 1 Games your students play
☐ Level 2 Curriculum software now available in your teaching area
☐ Level 3 Creating a simple program in your teaching specialty

☐ Level 4 Management systems software to fit your needs

Professional organizations and county offices of education are additional sources of information, especially staff concerned with mathematics curricula. Training materials will come with your computer and they tend to be well written. The widely-sold machines also have available many materials written by independent authors. In the school world, the Minnesota Educational Computing Consortium (MECC) has a variety of useful materials for sale. Their address is 2520 Broadway Drive, St. Paul, Minnesota, 55113, and you can expect to get useful responses from them on most any topic related to the use of micros in schools.

The following list provides information about a variety of microcomputer magazines and other helpful resources.

Computer Magazines and Resources

80 Microcomputing
P.O. Box 981
Farmingdale, NY 11737

Apple Orchard
P.O. Box 1493
Beaverton, OR 97075

Apple Sauce
20013 Princetown Avenue
Carson, CA 90746

Classroom Computer News
P.O. Box 266
Cambridge, MA 02138

Computer Town USA
P.O. Box E
Menlo Park, CA 94025

The Computing Teacher
Dept. of Computer and Information Science
University of Oregon
Eugene, OR 97403

Courseware Magazine
School of Business and Administrative Sciences
California State University
Fresno, CA 93740

Creative Computing
P.O. Box 789–M
Morristown, NJ 07960

Educational Computer
P.O. Box 535
Cupertino, CA 95015

Electronic Education
Suite 220
1311 Executive Center Drive
Tallahassee, FL 32301

Hands-on
8 Eliot Street
Cambridge, MA 02138

Info World
375 Cochituate Road
Box 880
Framingham, MA 01701

Interface Age
16704 Marquardt Avenue
Cerritos, CA 90701

Kilobaud Microcomputing
80 Pine Street
Petersborough, NH 03458

Mean Brief
256 North Washington Street
Falls Church, VA

Nibble
P.O. Box 325
Lincoln, MA 01773

On Computing
70 Main Street
Peterborough, NH 03458

On Line
24695 Santa Cruz Hwy
Los Gatos, CA 95030

Output
666 Fifth Avenue
New York, NY 10103

Personal Computing
1050 Commonwealth Avenue
Boston, MA 02215

Personal Computing
P.O. Box 13916
Philadelphia, PA 19101

Program Design, Inc.
11 Idar Court
Greenwich, CT 06830

Recreational Computing
1263 El Camino Real
Box E
Menlo Park, CA 94025

Softside
6 South Street
Milford, NH 03055

Softtalk
10432 Burbank Blvd
North Hollywood, CA 91601

Paperback Books

Albrecht, Robert L.; Finkel, Leroy; and **Brown, Jerald R.** Basic: A Self Teaching Guide (2nd ed). New York: John Wiley & Sons, 1978.

Dertouzos, Michael L., and **Moses, Joel** (eds.). *The Computer Age: A Twenty Year View.* Cambridge, MA: MIT Press, 1980.

Poole, Lon; McNiff, Martin; and **Cook, Steven.** *Apple II User's Guide.* Berkeley, CA: Osborne/McGraw-Hill, 1981.

Sadlier, James, and **Stanton, Jefferey.** *The Book of Apple Software: 1982.* Lawndale, CA: The Book Co., 1981.

Sippl, Charles, and **Sippl, Roger.** *Computer Dictionary.* Indianapolis, IN: Howard W. Sams & Co., 1980.

Taylor, Robert P. (ed.). *The Computer in the School: Tutor, Tool, Tutee.* New York: Teacher's College Press, 1980.

There are also dozens of books relating to microcomputers in the Creative Publications Catalogue, P.O. Box 10328, Palo Alto, California, 94303.

Learning about the many publications related to micros has been discussed, with one notable exception. You should know about the *Microcomputer Index* which offers bibliographic information on over 1,200 articles and reviews each quarter. It can be obtained from Microcomputer Information Services, 2464 El Camino Real, Box 247, Santa Clara, California, 95051.

Selection of Courseware

Becoming familiar with software of interest, including courseware in your curricular area, involves continual data gathering. Most people who use computers join *user clubs* to keep up-to-date on general software available. To help you get started and keep up-to-date, this section includes a comprehensive listing of available software catalogs.

Microcomputer Software Catalogue List

Education

Education Software Directory
Apple 11 Edition
Sterling Swift Publishing Co.,
P.O. Box 188,
Manchaca, TX 78652

School MicroWare
Dresden Associates
P.O. Box 246
Dresden, ME 04342

MARCK No 2
MARCK 280 Linden Avenue
Branford, CT 06405

Queue Catalog No. 4
Queue
5 Chapel Hill Drive
Fairfield, CT 06432

Selected Microcomputer Software
Opportunities for Learning, Inc.
8950 Lurline Avenue
Chatsworth, CA 91311

Scholastic Microcomputer Instructional Materials
Scholastic Inc.
904 Sylvan Avenue
Englewood Cliffs, NJ 07632

K–12 Micromedia
P.O. Box 17
Valley Cottage, NY 10989.

Other Fields

Radio Shack TRS–80
Applications Software Sourcebook
Radio Shack
Box 17400
Fort Worth, TX 76102

TRS–80 Software Source
COMPUTERMAT
P.O. Box 1664
Lake Havasu City, AZ 86403

80 Software Critique
RWC Microcomputing Services
P.O. Box 134
El Dorado, CA 95623

Atari Program Exchange
Atari Inc.
P.O. Box 427
155 Moffett Park Drive
Sunnyvale, CA 94086

SOFTWARE SELECTION CRITERIA

Learning how to select software is made easier for persons new to the computer field by applying the following eleven criteria.*

1. *Ease of Use*—Are the screen layouts of documentation, if necessary, sufficiently clear and

*James Sadlier and Jeffrey Stanton, *The Book of Apple Computer Software* (Lawndale, CA: The Book Co., 1981), p. 6.

well laid out to enable the new user to "run" the program with a minimum of difficulty?

2. *Documentation*—Does it answer all the questions? Is it clear? Is it sufficiently extensive?

3. *Reliability*—Does the program do what it is supposed to do?

4. *Price/Usefulness Ratio*—Is the buyer getting value for his or her money? For example, a fair program from one vendor priced at $7.95 might be a better value than only a slightly better program of the same type from another vendor priced at $24.95.

5. *Vendor Support*—Does the software company back its products? Are company representatives available to answer questions? Will they replace a program that is defective?

6. *Visual Appeal*—Does the program take advantage of the graphic capabilities of your computer?

7. *Error Handling*—Does the program "bomb" during execution? Are there proper "error trapping" routines?

8. *Creativity*—Is the program creative and imaginative?

9. *Challenge*—Does the program have sufficient challenge for the intended audience?

10. *Copyable*—Is it possible to make copies of the program, if needed?

11. *Availability*—Can the program be obtained when needed?

HELPFUL TECHNICAL INFORMATION

In this section we will list a variety of services, including *MicroSift News,* CONDUIT, CUE, and SOFTSWAP, and discuss the possible advantage of setting up a network of micros utilizing the CORVUS system. There is a growing list of organizations that provide teachers with information about the use of microcomputers; the ones mentioned are intended to be representative of a much larger group.

The Northwest Regional Educational Laboratory (NWREL), located at 300 SW Sixth Avenue in Portland, Oregon 97204, operates a Computer Technology Program that includes a clearinghouse to assist educators. The purpose of the clearinghouse has changed from physical dissemination of courseware to dissemination of information about courseware. This reflects the increasing amount of courseware available from commercial sources although many teachers still get the courseware they use from the "public domain." That is, courseware is copied from programs developed under federal funding or from some other source for which a commercial fee is not required. The NWREL Clearinghouse is also becoming involved in the evaluation of courseware. Of particular value to teachers and administrators is the clearinghouse publication *MicroSift News* which you can obtain from NWREL. Each issue is filled with useful information.

The Minnesota Educational Computing Consortium (MECC), 2520 Broadway Drive, St. Paul, Minnesota 55113, is another center of expertise that will help you. MECC has purchased thousands of micros over the years and has a fine catalogue of both training materials and instructional programs.

As you will learn when you subscribe to a computer magazine for educators, there are many state and regional organizations directed to teacher use of micros.

Computer Using Educators (CUE), 1776 Educational Park Drive, San Jose, California 95133, has 5,000 members across the country and holds four conferences annually. These are major events with dozens of workshops and as many as 1,200 teachers attending. It also publishes a newsletter that is extremely useful. CUE, together with the San Mateo County (California) Office of Education, sponsors SOFTSWAP, an exchange service that allows teachers to trade microprograms they have created for those of other teachers in the same curricular area. Since all the programs are in the public domain, they can also be purchased for a small fee per program for the Apple, TRS 80, PET, Atari, and COMPU-COLOR microprocessors. This is a fine service that allows teachers to help each other in obtaining much needed courseware. A catalogue can be ordered from SOFTSWAP, San Mateo County Office, 333 Main Street, Redwood City, California 94603.

As you consider your microcomputer purchases, it is well that you know about the CORVUS constellation possibility. If you are ready to purchase two or more micros, you may benefit from tying them together in order to have:

1. more memory for students' records,
2. sharing of courseware among micros,
3. sharing of printer and graphics among micros, and
4. control of student users.

Information about this equipment can be obtained from CORVUS Systems, 2029 O'Toole Avenue, San Jose, California 95131.

As you consider the exciting possibilities of utilizing microcomputers in your school, you may benefit from answering seven questions of *accountability* developed by John J. Beck, Jr.:

1. Will the addition of computers aid in mastering the school curriculum?
2. Does a computer-literate administration and faculty exist?

3. Is the District willing to budget funds to achieve computer literacy among its administrators and faculty if the need exists?
4. What are the relative characteristics of each system being considered?
5. What are the standard software package options of each system being considered?
6. What are the relative support services available for each system being considered?
7. Does any educationally-sound courseware exist for the curriculum you need to deliver?*

Each of us could ask these questions in a different manner and we encourage you to reword them to meet your needs and then go to work and find the answers.

Great potential lies ahead for the increased use of microcomputers in schools. Many districts have begun to purchase them and this will, in turn, encourage major publishers to develop high quality courseware. As good courseware becomes available, it will carry some of our educational activity into the home, business, and governmental settings. This is great news because school people have often believed that education should be conducted in a variety of settings. Powerful education occurs when there is a coordinated effort between the school, the home, and the work place. The microcomputer may be a key mechanism for advancing this vision.

---

*John J. Beck, Jr. "The Microcomputer Bandwagon; Is It Playing Your Tune?" *The Directive Teacher* 4 (1), 1982.

Allen, Judith. *Microsift: An Implementation Plan for Resources In Computer Education*. Portland, OR: Northwest Regional Education Laboratory, 1981.

Ballard, Richard. *The Educational Software Market*. Oak Park, IL: Talmis Publishers, 1981.

Beck, John J. Jr. "The Microcomputer Bandwagon: Is It Playing Your Tune?" *The Directive Teacher*. Winter/Spring, 1982.

BIBLIOGRAPHY

Branscomb, Lewis. "Electronics and Computers: An Overview. *Science.* 215 February, 1982.

Conference Board of the Mathematical Sciences Committee on Computer Education. *Recommendations Regarding Computers in High School Education.* April, 1972.

McWilliams, Peter A. *The Word Processing Book: A Short Course in Computer Literacy.* Los Angeles: Prelude Press, 1982.

Papert, Seymour. *Mindstorms.* New York: Basic Books, 1980.

Rizza, Peter. *Basic Skills Evaluation.* Minneapolis, MN: Control Data Corporation, 1980.

Sadlier, James, and Stanton, Jefferey. *The Book of Apple Computer Software: 1981.* Lawndale, CA: The Book Co., 1981.

Taylor, Robert, P. (ed.). *The Computer In the School: Tutor, Tool, Tutee.* New York: Teacher's College Press, 1980.

Glossary

Acceleration The rate of change in the velocity of an object.

Accountability The ability to give an explanation; explainability.

Acid A substance that dissolves in water and increases the hydrogen ion content of the solution.

Acid precipitation Rain, snow, or fog higher in acidity than normal rain.

Algae One-celled plants.

Alkali Any chemical that dissolves in water and decreases the hydrogen ion content of the solution.

Alternative explanation A hypothesis of equal credibility.

Alveoli Microscopic air sacs that are located at the end of the airways in the lungs.

Amino acid A simple nitrogen carboxyl-containing molecule which is the building block of protein.

Amniocentesis The process by which doctors extract and examine amniotic fluid to determine whether a baby will be affected by Down syndrome; it is also useful in determining whether a baby will be male or female.

Amniotic fluid The fluid surrounding the fetus that protects it from shocks.

Amperage A measure of the amount of charge flowing past a given point.

Angiosperm Plants that produce seeds encased in a covering or ovary.

Antibiotics Chemical substances that inhibit the growth of microorganisms produced by simple organisms.

Antigen Substance that provokes an immune response.

Aorta The largest artery in the body.

Aqueous humor The liquid in a space between the lens of the eye and the cornea.

Artery A vessel that carries blood away from the heart.

Associationism A theory that starts with supposedly irreducible mental elements and asserts that learning and the development of higher processes consist mainly in the combination of these elements; the point of view of those who define the variables or constructs of learning theory and experimentation in terms of stimulus and response, and the relationships of temporal contiguity between them.

Asteroid One of thousands of small planets that orbit the sun between Mars and Jupiter.

Asthenosphere The weak layer or shell of the earth below the lithosphere where adjustments in crustal vertical movement take place.

Astronomy The study of celestial bodies.

Astrophysicist One who studies the physical properties of heavenly bodies.

At rest Without motion.

Atmosphere A mixture of invisible gases, dust particles, and water vapor; generally seen as consisting of five layers: troposphere, stratosphere, mesosphere, thermosphere, and exosphere, according to how temperatures change with altitude.

Atom The smallest particle of an element that has the properties of that element and that can undergo a chemical reaction.

Atomic number The number of protons or positive charges in an atom.

Atomic weight The weight of an atom of an element compared with the weight of an atom of an element arbitrarily selected as standard.

Atrium The upper chambers of the heart.

Attitude An enduring, learned predisposition to behave in a consistent way toward a given class or objects; a persistent mental and/or neural state of readiness to react to a certain object or class of objects, not as they are, but as they are conceived to be.

Aurora borealis Luminous streamers, arches, and other patterns, consisting of glowing molecules and atoms in the mesosphere.

B cells Lymphocytes that can mature into antibody-secreting cells.

Bacteria Small, single-celled organisms that do not contain a nuclear membrane; may be harmful (causing disease) or beneficial.

Basalt A fine-grained igneous rock dominated by dark-colored minerals.

Behavioral or performance objective Behavior or action that is a developmental aim.

Big bang theory A theory that the universe originated with the explosion of a single mass of material 10 to 12 billion years ago, and that the universe is still expanding from that explosion.

Biomass A measure of the total amount of living material (plants, animals, etc.) in a given area.

Biome A large geographic area where the community of plants and animals is determined by climate and soil.

Biosphere The total of all the various ecosystems on the planet along with their interactions.

Black hole Theoretically, the remnant of a collapsed star having so great a gravitational field that it will not permit light to escape.

Blood The body's circulating fluid that contains many types of cells and carries food, oxygen, waste products, and hormones to and from the cells of the body.

Bones Hard tissue forming the skeleton of most vertebrates.

Botany The study of plants, their life cycles and processes, structure, and classification.

Bromine A caustic liquid used as a disinfectant, in nerve gas, and in gasoline anti-knock compounds.

Bryophytes Plants without true leaves, such as mosses and liverworts.

Burnout The state of exhaustion that is reached when one expends energy reserves; in teaching, a stage that may be brought on by working for an extended period of time under frustrating conditions that cannot be changed.

Calcium Preserved in the body—about two pounds in each adult—mostly in teeth and bones; regulates heartbeat and muscle contraction.

Calorie The amount of heat necessary to raise the temperature of one gram of water one degree Celsius.

Carbon An element that is found in pure form as diamonds, charcoal, and graphite; it is also in an endless variety of compounds, and is an indispensable source of such varied products as nylon, gasoline, perfume, plastics, shoe polish, DDT, and TNT.

Carbon dioxide A colorless, tasteless, and odorless gas that is released by living things when food is converted to energy.

Cardiopulmonary endurance The amount of stress one's heart and lung system can withstand.

Carnivore Flesh-eating animal.

Carrying capacity The ability of a landscape to support a given plant or animal species.

Cartilage Dense, fibrous, connective tissue that confers flexibility and strength on joints.

Catalyst A substance that speeds up a chemical change, but does not change itself.

Cellular respiration The process where hydrogen atoms, obtained from the breakdown of food, combine with oxygen to produce water and a form of energy that cells use to do all work.

Change Any alteration in a structure, a process, or an event; an observed difference in a given perception with the passage of time.

Chemical disintegration A process by which substances undergo change; e.g., when rocks are exposed to air and water, their basic elements begin to break down or change into other elements or compounds.

Chemical energy The energy contained in the bonds between atoms in a molecule.

Chlorine Weak, corrosive substance that can be used as a bleach, a disinfectant, and a nerve gas.

Chlorophyll The green substance in plants that enables them to use the energy of sunlight to convert carbon dioxide and water into carbohydrates.

Chloroplast The part of the plant cell that contains chlorophyll.

Chromatid The 2 daughter strands of a duplicated chromosome that are still joined at their centers.

Chromosome Thread-like structures containing genetic information.

Circulatory system The means by which animals move nutrients, chemical messages, and waste products through their bodies.

Climate The average or prevailing weather condition of an area, including air pressure, heat, wind, and moisture.

Climax The stage in succession that gives an ecosystem its final appearance.

Climax community A relatively stable community, especially of plants, that is achieved through successful adjustment to an environment; especially the final stage in ecological succession.

Clinical evaluation Determining the attainment of goals by studying the individual as a whole; relying upon the intuitive judgment of the clinician rather than upon measurement; the intuitive integration of measurement findings with direct observation.

Closed questions Questions that call for only one answer, not subject to further consideration.

Cloud Visible mass of particles of water or ice in the atmosphere.

Coded message The capability of cells to reproduce themselves or to produce new cells according to a predetermined or inherited formula. Also referred to as genetic code.

Coenzyme Substance that bonds with an enzyme, causing a chemical reaction to occur.

Colloid Dispersed, nondissolved particles that remain suspended in another medium.

Comet A ball of dust and ice, part of whose orbit brings it close enough to the sun to be detected.

Commensalism When a plant or animal lives in or on another but is not parasitic or injurious.

Community A group of plant and animal populations living and interacting in a given locality.

Competency-based programs Curricula which have been constructed or analyzed to insure students' acquisition of specific skills or knowledge; students are then tested on their competence following each unit of study.

Compound Substances that can be reduced into simpler, pure substances, each with a definite proportion by weight.

Condensation nuclei Airborne particles that cause moisture to form droplets around the particles, resulting in precipitation.

Conditioning The complex of organismic processes involved in the experimental procedure, or the procedure itself, wherein: (1) Two stimuli are presented in close temporal proximity. One of them has a reflex or previously acquired connection with a certain response, whereas the other does not. After many repetitions of the paired stimuli, the second stimulus acquires the potentiality of evoking a similar response. (2) A stimulus, having either evoked a response that brings into view or rewarding stimulus, or having evoked a response that prevents or removes a noxious or punishing stimulus, thereafter is more likely to evoke that response.

Conduction The transmission of energy from particle to particle.

Conductor A material that offers little resistance to the flow of electrons.

Conifers Plants that carry their seeds in cones, such as a pine or fir tree.

Conservation The process whereby the use of natural resources is controlled, restrained or reduced in order to preserve the known reserves. The idea that matter and energy are constant in the universe; neither is created or destroyed in nature but both may change in form (chap. 1).

Conservation of volume The idea that a given amount of a substance remains the same regardless of how that amount is distributed in space; liquids may be poured into containers of various sizes or solids may be broken into any number of pieces but the amount of the liquids or solids remains the same regardless of these transformations.

Constant variable A thing or quantity that does not change.

Constellation Any of 88 arbitrary configurations of stars into which the celestial sphere is subdivided.

Continental drift Slow, horizontal movement of the continents; involves plates as vehicles of transport in the mantle.

Contract performance A way of packaging or presenting the curriculum in which the student agrees to learn specific content and have his/her performance tested. Also referred to as a performance contract.

Convection The transfer of heat via currents of air or liquid.

Convergent evaluation An approach to the assessment of what has been learned; questions are asked for which there are specific "right answers."

Copper A reddish-brown metal that is malleable and easily combined with other metals (e.g., gold, silver, tin, zinc).

Core Innermost zone of earth, surrounded by the mantle.

Coriolis effect The tendency of any moving object, on or starting from the surface of earth, to continue in the direction in which the earth's rotation propels it.

Cornea A layer of tissue that forms the outer transparent cover of the eyeball to protect the iris and pupil.

Criterion referenced tests Tests which measure what students have learned and compare the results against some absolute scale; generally reported as a percentage of correct answers.

Crust The outer zone of the earth consists of the continental crust and the oceanic crust.

Crystallography The study of the form, structure, and properties of solid matter.

Cytoplasm All the material contained in the cell, except for the plasma membrane and the nucleus.

Deciduous Plants which shed their leaves annually.

Decision making Formulating a course of action with intent to execute it.

Density The mass of a body per unit volume.

Deoxyribonucleic acid (DNA) Part of the chromosome that contains coded information about heredity.

Depletion The gradual or sudden disappearance of substances like natural resources; the process of something being used up.

Detritivore An organism that consumes nonliving particulate or dissolved organic matter.

Detritus Nonliving particulate and dissolved organic matter in the ecosystem.

Development The process by which living things change in form.

Dicot Plants that produce embryos with two cotyledons (embryonic leaves).

Digestive system The means by which animals feed, absorb nutrients, and expel waste materials.

Directive management A style of management in which the leader controls the actions of others; in classrooms the teacher dominates and gives explicit instructions to students regarding learning tasks.

Discovery learning A theory of learning associated with Jerome Bruner, Wolfgang Köhler and other cognitive field theorists; based upon the idea that the essence of learning is through insight or the reorganization of thought patterns rather than through stimulus-response conditioning.

Divergent evaluation An approach to the assessment of what has been learned; questions are asked for which many "right answers" are acceptable.

DNA *See* Deoxyribonucleic acid.

Doppler effect Apparent change in the frequency of a wave due to the relative motion of the source and the observer.

Down syndrome A genetic defect that results primarily in mental retardation.

Earth's crust Outermost layer or shell of the earth.

Earthquake A sudden motion or trembling in the earth caused by the abrupt release of slowly accumulated strain.

Ecology The study of the relationships between living things and their environment.

Ecosystem The system of interaction between living organisms and the environment.

Egg Gamete supplied by the female.

Electric force The force that results from the attraction of protons (positive charges) and electrons (negative charges) for each other.

Electrical energy The energy that results from the attraction of electrons and protons.

Electromagnetic spectrum Includes the range of wave frequencies from radio (lowest), to infrared, the color spectrum we see, ultraviolet, x-radiation, gamma radiation, and cosmic ray waves (highest frequency).

Electron A negatively charged particle that surrounds the nucleus of an atom.

Element Unique combination of protons, neutrons, and electrons that cannot be broken down by ordinary chemical methods; one of the more than 100 "building blocks" of which all matter is composed.

Elementary science movement The style of teaching science in the late 1800s in which the teacher presented science as an organized body of knowledge and gave less emphasis to first-hand experience with nature.

Embryo A plant or animal still in an early stage of development.

Endocrine glands Glands that produce hormones that are secreted into the blood stream.

Energy Capacity for producing motion that can be derived from and converted back into mass.

Environment All the conditions and influences that surround the development of living things.

Environmental biology The study of life and all the surroundings of organisms, living and non-living.

Environmental impact The result of a force or action upon the naturally prevailing conditions (weather, soil, water, atmosphere, light) existing in a given area.

Enzyme A protein that acts as a catalyst.

Erosion The general process or group of processes whereby the materials of the earth's crust are loosened, dissolved, or worn away and simultaneously moved from one place to another.

Esophagus The passageway from the mouth to the stomach.

Eutrophication Rapid accumulation of phosphates in lakes and reservoirs leading to an increased ratio of plant growth to water volume.

Evaluation The process by which one takes into account what has been learned or accomplished and makes judgments about the outcomes.

Evaporation The process by which a liquid or solid becomes a vapor.

Exocrine gland A gland that has its secretions carried to body surfaces by ducts.

Exponential A rate of increase involving exponents; a mathematical term referring to multiples of a given number; shown as a superscript such as 4^3 ($4 \times 4 \times 4$).

Extinction No longer in existence.

Fallopian tubes The 2 tubes that serve as a connection between the ovaries and the uterus.

Fatty acids Basic subunits of fats.

Fault A rupture or crack in the earth's crust along which there has been differential stress.

Fauna The animals in a biome.

Feedback The process by which a control mechanism is regulated through the very effects it brings about.

Fertilization The combination of female and male sex cells which ultimately produces a new individual.

Fetus An unborn developing animal or human.

Fission   The nucleus of an atom is split into fragments; during this process, part of the mass of the atomic nuclei is converted into energy.

Flora   The plants in a biome.

Florine   The most reactive of nonmetals, florine corrodes platinum and is highly flammable.

Flow-limited resources   Sunlight and other resources that are naturally dispersed and available at limited rates.

Fluorocarbons   Compounds containing carbon and fluorine, used in aerosols, lubricants, and insulators.

Food chain   A series of organisms dependent on each other for food.

Food web/food pyramid   A network of organisms in a community among which nutrients are exchanged, usually including producers, consumers, and decomposers.

Force   The cause (or push and pull) of motion.

Formal evaluation   Measurement of the attainment of educational goals, including a study of the relative effectiveness of regulated conditions in furthering or hindering such attainment.

Formative evaluation   The process of assessing progress during the implementation stage of learning rather than at the end of a unit of study; generally used to improve the process of instruction.

Fossil   Any trace or imprint of a plant or animal that has been preserved in the earth's crust.

Fossil fuels   Organic remains compacted 50 million years ago, now burned for heat in the form of coal, petroleum, and natural gas.

Friction   The sum total of all the forces that oppose the motion of one body past another when the two are in contact.

Front   Boundary between air masses that differ in temperature, pressure, or moisture.

Fusion   The nucleus of one atom is joined or fused to another; during this process, atomic nuclei are combined in such a way that a mass is converted to energy.

Galaxy   A family of stars grouped in space.

Gallium   A metal that melts in your hand and is the only metal that expands when it freezes.

Gametes   Cells with one-half the normal amount of DNA that unite at fertilization to produce an organism with a full complement of DNA.

Gas   A form of matter in which the molecules move freely and expand indefinitely in the atmosphere.

Gene   (old) A unit of inheritance; (new) a segment of a chromosome that codes for a functional product.

Genera   Plural of genus.

Generalization   A general statement or conclusion.

Genetics   The study of heredity and variation in all living things.

Genus   A taxonomic group that includes one or more species; a subdivision of a family.

Geochemist   One who studies the chemical composition of the earth's crust and the chemical changes that occur there.

Geologic time scale   A history and relative calendar of the earth.

Geology   The study of the earth, its physical properties, and the processes that have changed the earth over time.

Geophysics   The study of the physical properties of the earth; the applied physical measurement of geological events.

Geothermal   Heat produced within the earth's interior.

Geothermal power   Energy generated from the hot, molten areas within the earth.

Germ cells   Cells that are destined to become gametes even though they contain a normal amount of DNA.

Germanium   A metal in the carbon family that is used in transistors.

Glacier ice   A unique form of ice made by compression and recrystallization of snow.

Gland   Specialized tissue that secretes specific hormones or proteins.

Glucose   The most common 6-carbon sugar.

Goal   The end result, immediate or remote, which an organism is seeking; the end result of an action specified or required in advance by someone directing behavior.

Gold   The most malleable metal, its use is primarily in coins, jewelry, and dental units.

Granite   Coarse-grained, igneous rock dominated by light-colored minerals.

Gravitational force The force that every particle exerts on other particles; the strength of this force depends on a ratio of the mass of the two objects and the distance between them.

Green revolution The expansion of crop yields brought about by the development of new subspecies of food plants and/or the improvement of agricultural technology.

Growth The process by which living things increase in size.

Gymnosperm Plants that produce seeds that are not enclosed in a seed case.

Habitat The place where a plant or animal lives, finds its food and shelter, and raises its young.

Heat A quality which, when absorbed, causes a change in the temperature of the object or a change in its physical state.

Heat energy The energy contained in an object that is a result of the movement of its constituent molecules.

Helium A very rare inert gas that is mainly found in natural gas fields; it has the lowest boiling point of any known element—minus 268.9°C.

Hemoglobin The major protein of the red cells, which carries oxygen from the lungs to the body tissues and carbon dioxide from the tissues to the lungs.

Herbaceous Having little or no woody tissue and persisting usually for a single growing season.

Herbivore A plant-eating animal.

Hierarchy of needs A theory proposed by Abraham Maslow in which one's behavior is influenced by the degree to which basic physiological and psychological needs for survival and growth are met.

Hormone A chemical secreted in one part of the body that exerts specific controls on other parts.

Host A plant or animal that provides food or shelter for a parasite.

Humanistic learning An approach to teaching which acknowledges the importance of the totality of the individual learning and which takes into account students' intellectual, emotional, physiological, and social needs when planning instruction.

Humus The organic part of the soil that results from the partial decay of leaves and other living matter.

Hurricane Tropical wind and storm system that moves in a wide circle around a low pressure area; winds in a hurricane vary from 74 to 136 miles per hour.

Hydroelectric power (hydropower) Generation of electricity by free fall of water, creating about 4 percent of the world's electricity.

Hydrogen Discovered in 1766, it is the third most abundant and lightest element; it is almost never found free on earth, but the sun and stars are almost pure hydrogen.

Hydrosphere The totality of the water on the surface of the earth, including oceans, lakes, rivers, and other bodies of water.

Hypothalamus The control center of vital functions of the body, located at the top of the brain stem.

Hypothesis An unproven idea or theory accepted as true and used as a basis for more reasoning and study until a better theory is found.

Identity A sameness of essential or generic character in different instances; the distinguishing character or personality of an individual; the condition of being the same with something described or asserted.

Igneous rock A rock or mineral that solidified from molten or partly-molten material.

Immunity The state of being able to resist disease.

Immunization A process by which antibodies are formed to protect individuals from specific serious diseases.

Impulses In human physiology, the neuro-transmitted message to act; generally means the tendency to respond or act without conscious thought.

Inertia The tendency for a body at rest to remain at rest and for a body in motion to remain in motion.

Informal evaluation Measurement of the attainment of educational goals in which a series of tasks show the level of ability. There are no norms, but the evaluator uses intuitive standards.

Infrared radiation Radiation that has low energy and borders the red end of the spectrum.

Instructional management The formulation of a more or less cohesive plan for teaching which includes basic principles, beliefs, or assumptions about education and teaching that serve as a foundation upon which future actions are based.

Interpersonal relations The bearing or influence of one person on another; the relation between two or more persons.

Invertebrate Animal with no backbone.

Ion Electrically charged atoms resulting from loss or gain of electrons during chemical reactions or exchanges in radiation.

Iris The round, colored portion of the eye that surrounds the pupil of the eye.

Isostasy Ideal condition of balance attained by earth materials of differing densities.

Isotope Alternative form of an element produced by variations in numbers of neutrons in the nucleus.

Kilocalorie The amount of heat needed to raise the temperature of one kilogram of water one degree Celsius.

Kinetic energy The energy in a moving object.

Landforms Features of the earth's surface attributable to natural causes.

Landscape The combination of features seen on the earth's surface.

Latent heat The amount of heat required to change the physical state of a substance, e.g., the heat needed to change a substance from a solid to a liquid once it has reached its melting point.

Law A description of a physical event that is mathematically testable.

Leaching The filtering of water through soil that causes weathering and furthers soil formation.

Lead A heavy, soft, bluish-gray solid; often used in alloys and piping.

Leaf The plant part in which food is produced.

Legume Plants in the pea family noted for their ability to store nitrates.

Lens The structure of the eye which is situated between the iris and the vitreous humor and which focuses rays of light (coming from the pupil) upon the retina.

Lesson plan A detailed formulation of the procedures and content for specific lessons.

Ligaments Connective tissue that links 2 bones at a joint.

Light energy The energy of radiation which moves at the speed of light—186,000 miles per second.

Light year The distance light travels in one year: 5.88 trillion miles (9.46 trillion km).

Limiting factor An ecological influence that limits or controls the abundance and/or distribution of a species.

Limits Something that binds, restrains, or confines; the point beyond which further practice seems to effect no improvement.

Linear A mathematical term referring to a rate of increase, involving terms of the first degree.

Liquid A form of matter in which the molecules move freely, creating a continuous substance that flows; in fluid form, matter assumes the shape of the container into which it is poured.

Lithification Process by which unconsolidated, rock-forming materials are converted into the consolidated, rock state.

Lithium The lightest of all solid elements, it forms a black substance when exposed to air; used in ceramics, alloys, H-bombs, and to treat psychological disorders.

Lithosphere The outer, solid portion of the earth, which includes the crust and the upper mantle, active component in plate tectonics.

Luminosity A measure of the brightness of a star.

Lymphocytes White blood cells involved in destroying disease-causing organisms.

Magnesium Eighth most abundant element; found in sea water and in dolomite.

Magnetic field A region in which invisible forces cause some objects (those having positive or negative charges) to move toward or away from other objects.

Magnetic force The force with which a magnetic source attracts or repels other magnetic substances.

Malnutrition A physical state which results when the quality of food eaten lacks essentials like protein, minerals, and vitamins. Undernutrition is the lack of quantity rather than quality of food eaten.

Manipulated variable A thing or quantity that can be managed or controlled.

Mantle Intermediate zone of the earth; surrounded by crust resting on the core.

Mass A measure of matter contained in an object. Measured in grams or kilograms, the amount of force required to change the speed or direction of a given object.

Matter The substance or material something is made of; the solids, liquids and gases that occupy space in the universe; anything that occupies space.

Meiosis Process whereby germ cells divide to form gametes.

Menstrual cycle A monthly cycle in the female during which the egg matures and is released to be made available; hormonal secretions affect the uterus during the cycle so that in the absence of fertilization, a brief period of bleeding occurs.

Mercury A liquid element used in thermometers, barometers, silver dental inlays, silent electrical switches, and blue-hued mercury vapor street lamps; sometimes called liquid silver.

Metabolism The sum of chemical reactions within the cell, including both breakdown and synthetic reactions.

Metamorphic rock Any rock derived from pre-existing rocks in response to marked changes in temperature, pressure, stress, and chemical environment.

Meteor A chunk of material from space that usually burns up as it passes earthward through the atmosphere.

Meteorite Stony or metallic body fallen to earth from outer space.

Meteorologist One who scientifically studies weather and climate.

Meteorology The study of the atmosphere and its phenomena, especially weather and weather predictions.

Microbiology The study of microscopic living things such as bacteria, protozoa, and viruses.

Microclimate Wind and moisture patterns of a designated environment.

Migrate The practice of some animals to follow a well-established route across one or more biomes—usually on an annual cycle.

Mineral Naturally occurring, inorganic element or compound having an orderly internal structure.

Mitochondria Sausage-shaped particles in the cytoplasm that generate, store, and supply energy for metabolism in the cell in the form of high energy bonds.

Mitosis A process of cell division in which 1 cell produces 2 cells that contain all the inheritance characteristics of the parent cells.

Molecules The smallest unit of a pure substance (an element or compound) that can exist and still retain the physical and chemical properties of the substance. Examples include oxygen (O_2) and water (H_2O).

Monocot Plants that produce embryos with one cotyledon.

Mucus A thick, slippery liquid produced by glands of the mucous membranes.

Muscles Tissue composed of contractile cells.

Mutation A change in genetic makeup that results in a new characteristic that can be inherited, either a physical change in chromosome relations or a biochemical change in the nucleotides that make up the gene.

Mutualism When organisms live together for mutual benefit.

Natural resources Energy and matter considered important for human use.

Nature study movement A way of teaching science in the late 1800s; teachers emphasized firsthand observation of nature rather than reading about nature.

Nectar A sweet liquid found in flowers and used by hummingbirds as food.

Neon The best known of the inert gases; used in sign advertising and lights.

Neuron A nerve cell.

Neutral The state when the concentration of hydrogen ions in solution equals that of pure water.

Neutron The uncharged particles in the nucleus of an atom.

Niche The ecological role of an organism in its

ecosystem and its relationship to the total environment.

Nitrogen A gas making up 78 percent of the earth's atmosphere and used in the manufacture of fertilizers.

Nonrenewable resource A natural resource that cannot reproduce itself; it is unlikely that we will totally exhaust the earth's nonrenewable resources because of the difficulty in locating and extracting them. Examples include water, minerals, fuels, and gases. Also called stock-limited resource.

Norm referenced tests Tests which measure what students have learned and compare results with test scores of other students or a "normal group"; generally thought of as grading on a curve.

Nuclear energy Energy from the nucleus of an atom, released when the nucleus is split or combined with other atomic nuclei.

Nuclear reaction Combination or splitting of atomic nuclei with release of energy.

Nuclei The plural of nucleus.

Nucleus Cellular organelle surrounded by a specialized membrane, controls many cell functions and contains the organism's hereditary material; found in all human cells except mature red cells and platelets. The central part of an atom which is positively charged.

Object teaching A way of teaching science in the mid-1800s; students learned about objects in nature (e.g., volcano, desert, octopus, etc.) even though the objects were remote.

Objective That at which one aims.

Obsidian A glassy equivalent to granite.

Oceanography The study of the oceans.

Odometer An instrument that measures the distance traveled by an object.

Omnivore A flesh- and plant-eating animal.

Open classroom Generally refers to a form of school architecture where individual classrooms are separated by movable walls, bookcases, or other furniture, creating a feeling of openness between classrooms.

Open questions Questions permitting or designed to permit spontaneous and unguided responses; defining the general topic inquired about but leaving to the option of the person replying both the form and substance of the reply. Also called open-ended questions.

Operant conditioning A theory of learning associated with B. F. Skinner; the idea that individuals learn to act or behave a certain way in order to receive reinforcement for their actions.

Operational definition Describes a property, object, or parts of an object in the context in which it is used.

Optic nerve The structure that transmits information from the eye (the image recorded on the retina) to the brain (the visual cortex).

Organ A body part composed of several tissues grouped together into a structural and functional unit.

Organelle A well-defined subcellular structure.

Organic Matter composing or derived from living things.

Organism An individual living thing.

Organization An arrangement of individuals, groups, matter in the universe, or information wherein the structure or relationships of the parts to the whole are apparent.

Ovaries The organ in the female that contains the gametes.

Overload principle The idea that strength of a muscle can be increased by being made to work harder than usual.

Overpopulation A condition that exists when the size of the population exceeds the available food, water, space, or other basic needs.

Oxbow lake A lake formed by the meandering or curving of a river which then gets cut off from the main flow.

Oxygen A colorless, tasteless, odorless gas that makes up 20 percent of the earth's atmosphere, it is needed to burn most carbon-containing compounds and is required for the complete metabolism of food; essential to all living things.

Ozone layer Thin layer of a rare form of oxygen in the stratosphere that is formed by solar ultraviolet radiation; decreases the amount of ultraviolet rays that reach the earth.

Paleontologist One who studies the fossil remains of animals and plants.

Paleontology The study of fossil remains of animals and plants.

Parallax An apparent shift in an object's direction due to a shift in the observer's position.

Parasite A plant or animal that lives on another without making any compensation to its host.

Participative management A style of classroom management in which the teacher shares decision making with the students.

Pathogen Disease-causing organism.

Periodic table An arrangement of the chemical elements according to their atomic numbers.

Peristalsis Rhythmic wave-like contractions and dilations of the walls of the digestive tract.

Permissive management The style of classroom or instructional management in which the teacher establishes goals and objectives for science, but leaves to the students decisions about how to reach the objectives and how to interpret the results.

Photochemical smog Air laden with compounds produced by the reaction of sunlight and hydrocarbons released from incomplete burning of fossil fuels.

Photon Basic unit of light energy.

Photosynthesis The process by which plants convert the sun's energy into chemical energy—or into food for the plant.

Physical science The study of matter and energy.

Placenta An organ made up of fetal and maternal components that aids in the exchange of materials between the fetus and the mother in mammals.

Planning chart A chart designed to help one make or carry out plans, achieve goals.

Planning checklist A complete list used to check off items and procedures necessary to carry out a plan.

Plasma Charged particles containing about equal numbers of positive ions and electrons; like a gas, but is a good conductor of electricity and is affected by a magnetic field, including the earth's magnetic field.

Plasma membrane The outer cell membrane, which is composed mainly of phosphate-containing fatty acids.

Plates Segments of the earth's crust, varying in thickness and including part of the upper mantle above the asthenosphere.

Plate tectonics Theory of worldwide dynamics of the many rigid plates of the earth's crust; a branch of geology dealing with the broad architecture of the outer part of the earth.

Polarity The property of having two poles, positive and negative.

Pollen The yellow powder-like male sex cells produced by plants.

Pollination The transfer of pollen from the male part of the plant to the female part.

Pollutant Harmful chemical or waste material discharged into the water, atmosphere, or land.

Polymer A regular arrangement of similar basic subunits produced by one or a few chemical reactions.

Population A group of individual organisms of the same species.

Potential energy The energy that results from position.

Precipitation Falling rain, snow, sleet, and other moisture.

Prediction To say that something will happen.

Preservation The process whereby a natural resource is protected, maintained or kept from harm.

Pressure A force applied per unit area on another object.

Protein Arrangement of the 20 most commonly occurring amino acids.

Proteins Chains of amino acids linked together in defined ways.

Proton A positively charged particle in the nucleus of an atom.

Protoplasm The complex of substances that make up the living cell.

Psychomotor activity Motor activity directly proceeding from mental activity.

Puberty The stage of development when it is first possible to reproduce sexually.

Pulse rate The number of heartbeats per minute as measured by placing the fingertips gently in key locations on the body such as the wrist or on either side of the throat.

Quasar Acronym for *quasi-stellar radio source,* an unusually intense source of radiation energy in the heavens.

Radiation The process by which energy is sent out from atoms and molecules because of their internal changes.

Radioactive An atom that gives off energy in the form of a particle or ray.

Radiometric dating The measurement of geologic time by the isotopic abundance and known disintegration rates of radioactive isotopes.

Range Distance from the highest to the lowest score or value in a distribution.

Receptor A specialized nerve ending that receives stimuli.

Recovery rate The amount of time it takes the heartbeat rate to return to normal after exercise.

Recycling To cause a substance to be used and then reconstituted in its original form.

Red shift Any displacement of spectral lines toward red, indicating that galaxies are receding from observers on earth.

Reef A chain of rocks, coral, or a ridge of sand near the surface of water.

Refractive errors A handicapping condition of vision involving the bending of light rays; a condition generally treated by optometrists.

Regeneration A process by which new individuals, life forms, or tissues replace those being used or dying.

Reinforcement The strengthening of something by adding to it, or that which strengthens when added; the natural occurrence or the experimental presentation of an unconditioned stimulus along with a conditioned stimulus or the strengthening of the conditioned response relationship.

Relative volume Capacity or amount of room or space compared to something else.

Renewable resource A natural resource that has the capacity to renew itself; it is possible to exhaust a renewable resource, as illustrated by the fact that we have completely eliminated many species of wildlife. Also called flow-limited resource.

Replication The exact duplication of an entire strand of DNA.

Reserves Amounts of materials assumed to be currently available for extraction at known locations; the amounts and locations of resources are less well-known.

Resources Various kinds of living and nonliving matter that are viewed as valuable because of their usefulness to us.

Responding variable A thing or quantity that can be used to give an answer.

Retina The layer of cells at the back of the eye which captures images brought into focus by the lens.

Ribonucleic acid (RNA) A biological material used to transfer information from DNA.

Ribosomes Spherical particles in the cytoplasm that are used for protein synthesis.

River delta An area, often triangular-shaped, where a river stream flows into a larger body of water and deposits sediment at its mouth.

Rock cycle A sequence of events involving the formation, alteration, destruction, and re-formation of rocks.

Rocks Aggregates of one or more minerals or a body of undifferentiated mineral matter.

Root The plant part, usually below the ground, that holds the plant in position, draws water and nutrients from the soil, and stores food.

Salinity The quality of being salty.

Scandium Element that is almost as light as aluminum and has a much higher melting point; no use for this metal is known yet.

Sedimentary rock A rock resulting from the consolidation of loose sediments that have accumulated in layers.

Segmentation The ability to move one area of the body independently of the others.

Seismic data Information gained from a seismograph or instrument for recording the earth's vibrations during earthquakes or volcanic eruptions.

Seismic sea wave Large ocean wave generated

by vibrations of an earthquake (sometimes mistakenly called a tidal wave).

Seismograph An instrument that detects, magnifies, and records vibrations of the earth; the resulting record is a seismograph.

Seismologist One who studies earthquakes and other earth vibrations.

Seismology Scientific study of earthquakes and other earth vibrations.

Self-evaluation inventory A complete listing of characteristics, traits, and aptitudes that enables one to make a judgment of oneself.

Sensory stimulus Physical or chemical changes that occur in the environment and are detected by the senses of touch, smell, vision, taste, or hearing.

Sexual reproduction Reproduction that depends on the union of 2 gamete cells to form a new individual.

Shaping behavior Conditioning of actions of an organism.

Shares Designated amounts of a commodity apportioned to individuals.

Silver A good conductor of heat and electricity used in photographic film.

Sodium The sixth most abundant element on earth, sodium is used in compounds such as salt, baking soda, borax, and lye.

Soil Material that forms on the earth's surface as a result of organic and inorganic processes.

Solar wind The continuous ejection of plasma from the sun's surface into and through interplanetary space.

Solid Something that offers resistance to pressure and is not easily changed in shape.

Solute The component of a solution that is lesser in quantity.

Solution Dispersing and dissolving one or more substances in another, usually a liquid.

Species A group of organisms that have certain characteristics in common and are able to interbreed; a subdivision of a genus.

Specific gravity Ratio between weight of a given volume of material and weight of an equal volume of water.

Specific heat The amount of heat required to raise a mass of a substance one degree compared with a similar mass of water.

Spectral lines Lines that appear when light passes through a prism and becomes "ordered" into colors of different wave lengths or frequencies.

Spectrum of light A series of colored bands arranged in continuous order of their wavelengths.

Speed Distance traveled per unit of time.

Sperm Gamete supplied by the male.

Spores Small reproductive bodies that are able to resist adverse circumstances and can generate a new adult individual.

Stem The plant part that connects the roots and the leaves.

Steroid A large complex molecule of biological origin composed of 4 attached rings of carbon atoms.

Stock-limited resources Minerals, fossil fuels, and other resources, such as water and air, that are present on earth in limited quantities.

Stress Any physical or mental strain, tension, or force that creates pressure or a sense of urgency for the individual.

Substrate A chemical acted upon by an enzyme.

Succession The changes in vegetation and animal life causing one population or community to be replaced by another.

Summative evaluation The process of assessing outcomes at the conclusion of a unit of instruction or project; generally used to assign grades.

Superconductors Elements that offer little resistance to a flow of electrons.

Supersaturated air An unstable condition in which air contains excess water vapor that would normally be precipitated at that temperature and pressure.

Supply and demand A process whereby individuals try to balance the abundance of a commodity with requests for the commodity.

Suspension Dispersing but not dissolving one or more substances in another, usually a liquid; the suspended particles eventually settle out.

Symbiosis When two different organisms must live together for survival.

Symmetry Matching form or arrangement in an organism.

Synapse The point of contact between neurons where nerve impulses are transmitted from one neuron to another.

Synthetic fuels Fuels produced in laboratories by chemical processes as opposed to fuels occurring naturally in the environment.

Systematics The science of classification; a system of classification; the classification of organisms with regard to their natural relationships.

T cells Lymphocytes that mature in the thymus gland.

Taxonomy The study of general principles of scientific classification, systematics; orderly classification of plants and animals according to their presumed natural relationships.

Teaching machines Any of various mechanical devices for presenting a program of instructional materials.

Teaching style The combination of several qualities or characteristics exhibited by a teacher that distinguishes her/him from colleagues; generally includes the teacher's attitude toward students, learning, and the purpose of education, and includes the predominant methods used by the teacher.

Team teach Teaching in which two or more teachers share the teaching responsibilities for a class or classes.

Temperature A measure of the warmth or coldness of an object with reference to some standard (not to be confused with heat).

Testes Pair of egg-shaped organs suspended from the male body, location where sperm cells are produced.

Thermodynamics A study of the relationship between heat and mechanical energy.

Thermometer An instrument for measuring temperature by degrees.

Tissue A group of similar cells that work together to perform a role in the structuring and functioning of the body.

Titanium Element that is used as paint pigment and in supersonic aircraft.

Topography The general shape and feature of land.

Tornado A destructive, whirling wind that moves in a narrow path over the earth and may involve internal wind speeds of 250 miles per hour.

Tracheophytes Vascular plants.

Transcription The copying of short stretches of DNA to produce RNA strands.

Transistors A tiny electronic device that controls current without the use of a vacuum.

Translation The process where RNA is used to direct protein synthesis.

Transpiration The process whereby plants give off vapor.

Turbine Engine or motor turned by the pressure of steam, water, air, etc., against the curved vanes of wheels fastened to a driveshaft.

Ultraviolet (UV) rays Radiation having wavelengths between visible light and x-rays; in excessive amounts UV rays disrupt terrestial life.

Ultraviolet radiation Radiation that is very energetic and borders the violet end of the spectrum.

Umbilical cord The stalk containing the blood vessels connecting the placenta and the fetus.

Undernutrition A condition where the individual lacks the quantity of food needed for sound health.

Uranium The heaviest atom among the natural elements, uranium has a half life of 45 billion years.

Uterus The organ in a female in which unborn young grow and develop.

Value The worth or excellence ascribed to an object or activity or a class thereof; an abstract concept, often merely implicit, that defines for an individual or for a social unit what ends or means to an end are desirable.

Vapor Visible particles of moisture floating in air, as fog, mist, or steam.

Vascular Having a system of ducts or tubes for carrying body fluids, such as sap.

Vein A vessel that brings blood to the heart.

Velocity Speed of an object in a particular direction.

Vena cava The largest vein in the body.

Ventricles The lower chambers of the heart.

Vertebrate Animal possessing a backbone.

Virus A protein-coated, minute organism containing either DNA or RNA that is a parasite dependent on nutrients in cells in order to live.

Viscosity The resistance of a liquid to flow, making it syruplike.

Vitamin An organic chemical that usually acts as a coenzyme and is an essential dietary requirement, i.e., the body cannot synthesize it.

Volcano Landform that is built by the accumulation of lava around a central vent or fissure.

Voltage A measure of electrical potential that will cause a current to flow; measured in volts.

Volume Capacity or amount of room or space.

Watershed Land area drained by a river or a river system.

Wavelength The distance between two neighboring crests of a wave.

Weather The general, day-to-day condition of the atmosphere in a particular place and time, involving temperature, moisture, clouds, wind, and so forth.

Weathering The process by which landforms and rock materials are decomposed.

Weight A measure of the force of gravity of an object.

Work Force exerted over a distance.

X rays Electromagnetic radiation of very short wavelength produced by the bombardment of a substance by a stream of electrons moving at a great velocity.

Zero population growth A condition where the number of new births equals the number of deaths in a population.

Zone of aeration Zone immediately below the earth's surface in which openings are filled with both air and water; area of intense weathering and soil formation.

Zone of saturation Underground region within which all openings fill with water.

Zoology The study of animals, their life cycles and processes, structure, and classification.

Index

RITA PETERSON is currently Director of Teacher Education at the University of California, Irvine and former Director of Research in Science Education at the National Science Foundation. She received her Ph.D. and M.S. in science education from the University of California, Berkeley, and her B.S. from California State University, Hayward. Dr. Peterson served as collaborator or author of several *Science Curriculum Improvement Study* publications, has been the botanical illustrator of several books on native plants of California, produced a film on *Piaget's Formal Thought* with Robert Karplus, and authored numerous articles. She has been active in NSTA (National Science Teachers Association), AAAS (American Association for the Advancement of Science), and NARST (National Association for Research in Science Teaching), has received the Outstanding Research Award of NARST, and been named a Fellow of the American Association for the Advancement of Science.

JANE BOWYER is currently Associate Professor and Department Head of Education at Mills College in Oakland, California. She received her Ph.D. and M.A. in science education from the University of California, Berkeley and her B.A. in education at Miami University in Oxford, Ohio. Dr. Bowyer has co-authored many publications for the *Science Curriculum Improvement Study* and published numerous articles on science education. She is an active member of NSTA, NARST, AAAS, and AERA (American Educational Research Association) and has received the NSTA Award for the Outstanding Research Paper that has Applications for Classroom Teaching, the Outstanding Research Paper Award of NARST, and a Fulbright Award.

About the Authors

DAVID BUTTS is currently Professor of Science Education, Chair of the Department of Science Education, and Associate to the Dean for Computing Technology at the University of Georgia. Dr. Butts received his Ph.D. and M.F. from the University of Illinois in science education and was awarded his B.S. in biology from Butler University. He has authored or co-authored both college textbooks and children's books, including *Teaching Science: A Self-Directed Learning Guide, Science and Children,* and *Chocolate, Vanilla, and Watermelon,* as well as numerous articles in science education. Dr. Butts is active in NSTA and NARST and served as Editor of the *Journal of Research in Science Teaching.* He has received the Robert Carleton Award, an NSTA award for excellence in teaching, research, and service to science teachers.

RODGER BYBEE is currently Associate Professor and Chair of the Department of Education at Carleton College in Northfield, Minnesota. He received his Ph.D. in science education and psychology from New York University, and his B.A. and M.A. from the University of Northern Colorado. Dr. Bybee has authored or co-authored several books in science education, including *Becoming a Secondary School Science Teacher, Piaget for Educators,* and *Violence, Values and Justice in the Schools,* and has published numerous articles in professional journals. He is an active member of NSTA, NARST, AAAS, and NABT (National Association of Biology Teachers). Dr. Bybee has previously taught at all levels in the public schools and received awards as Outstanding Science Educator of the Year and Leader of American Education.